Flash of the Cathode Rays

A History of J J Thomson's Electron

Also by Per F Dahl

Ludvig Colding and the Conservation of Energy Principle:
Experimental and Philosophical Contributions
(1972)

Superconductivity: its Historical Roots and Development
from Mercury to the Ceramic Oxides
(1992) American Institute of Physics

Related Titles Published by Institute of Physics Publishing

Twentieth Century Physics
L Brown and B Pippard (eds)

Front Page Physics:
A Century of Physics in the News
A J Meadows and M M Hancock-Beaulieu (eds)

The Young Einstein:
The Advent of Relativity
L Pyenson

Michael Faraday and the Royal Institution:
The Genius of Man and Place
J M Thomas

From the Watching of Shadows:
The Origins of Radiological Tomography
S Webb

A Century of X-Rays and Radioactivity in Medicine:
With Emphasis on Photographic Records of Early Years
R F Mould

The Origin of the Concept of Nuclear Forces
L M Brown and H Rechenberg

Flash of the Cathode Rays

A History of J J Thomson's Electron

Per F Dahl

Lawrence Berkeley National Laboratory,

CA, USA

CRC Press is an imprint of the
Taylor & Francis Group, an **informa** business

CRC Press
Taylor & Francis Group
6000 Broken Sound Parkway NW, Suite 300
Boca Raton, FL 33487-2742

First issued in paperback 2019

ISBN-13: 978-0-7503-0453-5 (hbk)
ISBN-13: 978-0-367-40109-2 (pbk)

British Library Cataloguing-in-Publication Data
A catalogue record for this book is available from the British Library.

Library of Congress Cataloging-in-Publication Data are available

Visit the Taylor & Francis Web site at
http://www.taylorandfrancis.com

and the CRC Press Web site at
http://www.crcpress.com

Contents

Preface

This book traces the history of the electron, from the earliest concepts of a quantum of electricity through the marshaling of firm evidence for the first subatomic particle in nature. It grew out of my earlier studies of events leading to superconductivity in 1911 that impressed on me the richness of glow discharge experimentation in the final decades of the 19th century. With several recent scholarly studies illuminating aspects of the burgeoning proliferation of radiations by 1900, I decided that an account sketching these multifarious developments, culminating in J J Thomson's cathode rays and Millikan's electric charges, was timely. As far as I am aware, the last full-length monograph specifically on the subject was published some 30 years ago. By happenstance, moreover, this year, 1997, marks the 100th anniversary of what is generally regarded as the year of the 'discovery' of the electron.

I have organized the narrative around various quasi-distinct epochs and themes. I start, after some preliminaries, with the prehistory of electrical science from Volta through Faraday and on to the German school of Plücker and company; they are followed by the Victorian amateur scientists, the doyen of which was the incomparable William Crookes. There are Hertz and Lenard, and the near-rivalry between Schuster and Thomson; Perrin in Paris and C T R Wilson in Scotland. The stage for the new era of experimentation is set by Röntgen, Becquerel, and their multitude of rays. The chronicle reaches its climax in the e/m experiments of 1897 and their aftermath nailing down e and m; we also have the parallel developments in Leiden and Amsterdam; next come β- and allied rays, mostly in the hands of Rutherford. Controversies embracing the evanescent N-rays and purported positive electrons enliven events; the canal ray studies of Thomson (but not of Aston) are not without their share of misconstructions. I conclude with the Millikan–Ehrenhaft dispute leading to the unambiguous isolation of the electron, and, by way of rounding out the chronicle, the crowning achievement in classical atomism in Rutherford's atom of 1911.

My primary sources include the wealth of archival material at Cambridge University: in particular, the notebooks of J J Thomson, original papers, and much of his correspondence—all held in Cambridge

University Library (CUL). (The same goes for the Rutherford archives at CUL, spanning his work at McGill, Manchester, and Cambridge, and encompassing notebooks, notes and calculations, manuscripts, and correspondence.) The key to the notebooks of Thomson is their handlist prepared by Paul Spitzer, a copy of which is on hand, curiously not at the CUL, but at Trinity College Library. Indispensable for accessing the Trinity College holdings of Thomson material is the *Catalogue of the Papers and Correspondence of Sir Joseph John Thomson* issued by the contemporary Scientific Archives Centre. Not to be overlooked is the museum collection of original apparatus, as well as the photographic archives, at the Cavendish Laboratory nowadays located in the rolling fields of west Cambridge. Equally important were the extensive holdings of the Niels Bohr Library, American Institute of Physics, and with them the recently issued *Guide to the Archival Collections*. Among the holdings of note are the venerable Archives for the History of Quantum Physics, portions of which I have also accessed periodically at Berkeley's Office of History of Science and Technology; however, much of that particular material falls outside the time frame of focus here.

Among secondary sources pertinent to the subject at hand, I have profited particularly from Bruce R Wheaton's work on Lenard and the photoelectric effect, Marjorie Malley's on β-rays, Helge Kragh's on the positive electron, Isobel Falconer's on positive (canal) rays, and Gerald Holton's on the Millikan–Ehrenhaft controversy. References to these and other recent studies are given in the notes and bibliography. For general guides to material such as this, the Berkeley *Literature on the History of Physics in the 20th Century*, and its companion *Inventory of Published Letters to and from Physicists, 1900–1950*, are a must. The best all-around introduction to J J Thomson and his time is probably still Raleigh's *The Life of Sir J J Thomson*; by contrast, the volumes on Rutherford are legion, constituting an industry in its own right. Anybody seriously interested in the Cavendish Laboratory and its personalities up to ca. 1910 will need to consult *A History of the Cavendish Laboratory 1871–1910*.

I am indebted to the following individuals and institutions for archival assistance: Alison Sproston and Jonathan Smith (Trinity College Library, Cambridge), Anthony J Perkins and Godfrey Waller (CUL), Simon Schaffer (Whipple Library, Department of History and Philosophy, Cambridge University), Shirley Fieldhouse and Hilary Coote (Cavendish Laboratory, Cambridge), Irena M McCabe (The Royal Institution, London), Library staff (University of Bergen), Joseph Anderson and his staff (Niels Bohr Library, American Institute of Physics, College Park, MD), Mary Batchelor (Founderen Library West, Southern Methodist University, Dallas, TX), C A Lamy (Engineering Library, University of Texas at Arlington, TX), Marietta Nelson (Library, National Institute of Science and Technology, Gaithersburg, MD), Eric Barbely (Energy Library, US Department of Energy, Germantown, MD),

Rita Labrie (Research Library, Lawrence Berkeley National Laboratory, Berkeley, CA), and Pat Kreitz and her staff (Library, Superconducting Super Collider Laboratory, Dallas, TX; since discontinued).

Similar thanks go to various individuals for permission to quote correspondence or to reproduce text or illustrations from published works; they are listed in the Acknowledgments section.

I thank my publisher, Jim Revill, and his editorial staff at Institute of Physics Publishing, including Sara Gwynn, my editor, for their splendid support and cooperation in all stages of manuscript processing and production of the volume. I also thank Robin Rees for assistance rendered by him and his staff at Institute of Physics Publishing's USA office.

Above all, I am indebted to Eleanor, my wife and collaborator in this as in other projects, for her lack of intimidation with software, for her meticulous attention to detail, for her patient and expert copy-editing and wordprocessing of the manuscript, and for her companionship and steadfast support during the unsettled life of recent years on the road between Berkeley, Dallas, and Washington, DC.

Per F Dahl

Acknowledgments

The author gratefully acknowledges permission to quote excerpts from the following works:

Fletcher H 1982 My work with Millikan on the oil-drop experiment *Physics Today* **June** 43–47. Copyright ©1982 American Institute of Physics. Reprinted with permission from American Institute of Physics, Paul C Fletcher and the Fletcher Family.

Transcript of oral history interview with Harvey Fletcher by Vern Knudsen and W James King, 15 May 1964; transcript of oral history interview with George Paget Thomson by John L Heilbron, 20 June 1967; transcript of oral history interview with Niels Bohr by Thomas Kuhn, 1 January 1962. Reprinted with permission of American Institute of Physics.

Letters to and from Thomson and Rutherford in the J J Thomson and E Rutherford archives, Cambridge University Library. Reprinted with permission from the Syndics of Cambridge University Library.

Letters from E Rutherford to B B Boltwood in Badash L (ed) 1969 *Rutherford and Boltwood, Letters on Radioactivity* (New Haven, CT: Yale University Press). Copyright © 1969, Yale University Press. Reprinted with permission from Yale University Press.

Holton G 1978 Subelectrons, presuppositions and the Millikan–Ehrenhaft dispute *Historical Studies in the Physical Sciences* **9** 161–224. Copyright © 1978 The Regents of the University of California. Reprinted with permission from Gerald Holton and from University of California Press Journals.

Millikan R A 1950 *The Autobiography of Robert A Millikan* (New York: Prentice-Hall) Copyright © 1950 Prentice-Hall, Inc., renewed 1978. Reprinted with permission from Simon & Schuster.

Seabrook W B 1941 *Doctor Wood, Modern Wizard of the Laboratory* (New York: Harcourt Brace). Copyright © 1941 Harcourt Brace & Company. Reprinted with permission from Harcourt Brace & Company.

Casimir H B G 1983 *Haphazard Reality: Half a Century of Science* (New York: Harper & Row). Copyright © 1983 H B G Casimir. Reprinted with permission from Harper Collins Publishing, Inc.

Eve A S 1939 *Rutherford: Being the Life and Letters of the Rt Hon Lord Rutherford, OM* (London: Cambridge University Press). Copyright © 1939 Cambridge University Press.

Letters from N Bohr to his mother and to M Nørlund in Rosenfeld L and Rüdinger E 1967 The Decisive Years 1911–1918 *Niels Bohr: His Life and Work as Seen by his Friends and Colleagues* (Amsterdam: North Holland) and (New York: John Wiley). Letters from N Bohr to H Bohr in Rosenfeld L, Nielsen J R and Rüdinger E (eds) 1972 *Niels Bohr: Collected Works* (Amsterdam: North Holland). Reprinted with permission from the Niels Bohr Archive, Copenhagen.

Birks J B (ed) 1962 *Rutherford at Manchester* (London: Heywood & Co) and 1963 (New York: W A Benjamin). Copyright © 1962 Victoria University of Manchester. Reprinted with permission from the University of Manchester.

Gamow G 1966 *Thirty Years that Shook Physics: The Story of Quantum Theory* (New York: Doubleday & Co). Copyright © 1977 the Estate of George Gamow. Reprinted with permission from Elfriede Gamow.

Chadwick J 1964 Some Personal Notes on the Search for the Neutron *Proceedings of the 10th International Congress of the History of Science (Ithaca, 1962)* (Paris: Hermann). Copyright © 1962 Hermann editeurs des Sciences et des Arts. Reprinted with permission from Hermann.

The author gratefully acknowledges permission to use the following illustrations:

Figures 1.1 and 3.2 by courtesy of AIP, Emilio Segrè Visual Archives and Argonne National Laboratory.

Figure 5.2 by courtesy of Ullstein Bilderdienst and AIP, Emilio Segrè Visual Archives.

Figures 6.1 and 8.1 by courtesy of Burndy Library and AIP, Emilo Segrè Visual Archives

Figure 7.3 by courtesy of the Master and Fellows of Trinity College, Cambridge.

Figures 9.4, 9.5, 9.6, 15.9, 17.5, 17.6 and 18.1 by courtesy of the University of Cambridge, Cavendish Laboratory, Madingley Road, Cambridge, UK.

Figure 15.3 reprinted by permission of the Syndics of Cambridge University Library.

List of illustrations

Chapter 1

J J THOMPSON AND HIS CONTEMPORARIES

1.1. Würzburg, 1895

Late in the evening of Friday, November 8 1895, Professor Wilhelm Conrad Röntgen, of the Physical Institute at the University of Würzburg in northwestern Bavaria, prepared an experiment on the electrical discharge in vacuo—an investigation only a few days old. The late hour was deliberate. Röntgen, aged 50 and Director of the Institute and Rector of the University, was a busy man with limited time for personal researches; they were best conducted at hours when colleagues and assistants were no longer around [1-1].

The year of Röntgen's experiment was a promising one in experimental physics. In Paris, Henri Becquerel prepared for his researches on uranium salts. Also in Paris, Marie Sklodowska of Poland, then studying physics at the Sorbonne, married Pierre Curie. Across the English Channel, Ernest Rutherford arrived in Cambridge to begin a research fellowship under J J Thomson. The year 1895, more generally, marked the midpoint in an intellectually stimulating decade second to none. This was the heyday in the cultural history of the 19th century. In literary circles, the decade was dominated by the giants Henry James, Anton Chekhov, Thomas Hardy, Rudyard Kipling, and Oscar Wilde: James experiencing his second productive period devoted to purely English subjects; Chekhov's period of full-length plays; Hardy turning from Victorian novels to poetry; Kipling's two *Jungle Books*; Wilde building his dramatic and witty reputation from *Lady Windermere's Fan* to *The Importance of Being Ernest*. In the musical arena, there was Giacomo Puccini with his *Manon Lescaut* and *La Boheme*, Giuseppe Verdi's last great masterpiece *Falstaff*, Nicolay Rimsky-Korsakov turning to opera while also rewriting most of his earlier work, and, finally, Jean Sibelius' first great symphonic poem *Finlandia* made its debut late in the decade. France, on the other hand, offered Henri Matisse, who came under the influence of Impressionism and Neo-Impressionism in this period, Claude Monet with his mature studies of light in landscape renderings, and Pierre Auguste Renoir in the flowering of his last great period.

The decade was as well the jumping-off point for climactic terrestrial explorations soon to come. Distant regions were charted by Robert Peary in his early Arctic voyages, by Mary Kingsley in her two remarkable journeys to West Africa, in the balloon expedition of Sweden's Salomon Andrée (lost in the Arctic reaches), and by Sven Hedin in his ceaseless travels in the Himalayas, The Gobi, and Tibet. Rudolph Diesel invented his engine at this time, and Guglielmo Marconi flashed his wireless telegraph. Sigmund Freud developed his psychoanalysis, and Ivan Pavlov performed his conditioned-reflex experiments.

These were no less challenging times in purely scientific circles, with leading personalities equally prominent. In the United Kingdom, Lord Kelvin (William Thomson) held sway as the omnipotent statesman of physical science (what with James Clerk Maxwell's premature demise some 16 years earlier); nearly on Kelvin's authoritative par were Lord Rayleigh, Sir William Ramsay, and their colorful colleague Sir William Crookes. French science was dominated by Louis Pasteur who passed away in our year 1895; another French scientist of broad scholarly repute was Jules-Henri Poincaré. The unquestioned spokesman of German science was Hermann Ludwig Ferdinand von Helmholtz, with his favorite pupil Heinrich Hertz now sharing the limelight. Helmholtz's heavyweight counterpart in the neighboring Netherlands was Hendrik Antoon Lorentz, whose work formed the transition between the classical physics epitomized by Maxwell's electrodynamics and the new physics soon to be kindled by two younger physicists, Max Planck and Albert Einstein (the latter still a student in 1895).

The brief period spanning 1895–1897, in particular, saw a rash of revolutionary discoveries in experimental physics—several out of the blue and some the swift and climactic culmination of a seemingly endless traditional line of research. We will have more to say about these developments presently; for now it suffices to note that, among other things, cathode ray studies were then a hot subject in experimental physics—indeed a revitalized subject of inquiry stretching back nearly 200 years. By Röntgen's time, two contending schools maintained conflicting interpretations as to the nature of the luminosity accompanying the passage of electricity through evacuated glass discharge tubes. The English school, championed by Sir William Crookes and with some French scientists joining in, maintained that the luminous rays were a stream of material, electrically charged particles. The Germans, most recently Heinrich Hertz and his student Philipp Eduard Anton von Lenard (though with the notable exception of Helmholtz, their mentor), held that they were some kind of electromagnetic wave propagation in the all-pervasive æther. Obviously, the matter was far from settled and Röntgen seized the opportunity of jumping into the fray. A physicist of the old school, he decided that the latest experiments of Hertz and Lenard needed checking—a custom he adhered to in starting any experimental investigation.

In particular, Lenard's intriguing experiment warranted repetition. Lenard, while still under Hertz's tutelage, had directed the rays emanating from the negatively charged cathode against a thin 'window' of aluminum foil in the other end of an otherwise standard vacuum tube. He found that not only did they penetrate the window (a fact already established by Hertz), but also passed into the external air and caused it to fluoresce over a distance of nearly an inch—something quite impossible for even the smallest atoms according to the prevailing dynamic theory of gases.

Röntgen was an experienced physicist and academician of considerable repute; thus he had left his mark on science and academia well before the experiments currently under way. Born in 1843 into a mercantile family in Lennep, a small town in the Rhineland, he spent most of his childhood in Holland. At the age of 20, he enrolled in mechanical engineering at the Eidgenössische Technische Hochschule (ETH) in Zürich, graduating in 1868. He obtained his doctorate the year after on *Studies of Gases* under August Kundt at the University of Zürich; subsequently he followed his mentor, first to the University of Würzburg and thence to the Kaiser-Wilhelm University of Strasbourg. At Strasbourg, Röntgen began his academic career proper as *Privatdozent*, the first rung on the academic ladder in German universities. (In the Franco–German War of 1870–1871 the Germans captured Strasbourg and annexed it, and with it the University, until after World War I.) His researches soon spanned many branches of experimental physics. Especially noteworthy was his demonstration in 1888 that magnetic effects are produced in a dielectric when it is moved in an electric field. The same year he accepted an appointment as professor of physics at Würzburg and director of its Physical Institute, succeeding Friedrich Kohlrausch, who had established the Institute during 1878–1879. The Institute occupied a commodious two-story building on the city's elegant Pleicher Ring avenue (now named Röntgen Ring avenue), with the upper story set aside as Röntgen's private residence. Here Röntgen held sway for over a decade.

Several conflicting accounts of the events on the eve of 1895 abound—none, however, by eyewitnesses, intimate associates, or assistants. Röntgen, when undertaking a fresh scientific inquiry, was in the habit of working quietly alone until a solution was well in hand. Thus it is that we owe the most reliable description of his experimental setup to one H J W Dam, a London-based American newspaper reporter, who was one of very few granted an interview with Röntgen shortly after his momentous discovery [1-2].

Like Lenard and Hertz before him, Röntgen carefully assembled two

Figure 1.1. *Wilhelm Conrad Röntgen.*

pieces of rather standard scientific apparatus in two of the smaller rooms on the ground floor of the Institute [1-3]. One was a Rühmkorff induction coil, a device for producing high-voltage electric currents and named after its developer, a German instrument maker living in Paris in the 1850s. It was fabricated by the Reiniger, Gebbery–Schall Company and equipped with a mercury interrupter, a refinement introduced by Armand Fizeau, better known for his measurements of the speed of light with a system of mirrors. Charged by a current of 20 A, it was capable of generating a spark 4–6 in long [1-4]. Two wires from the coil led to the other piece of apparatus, placed in an adjacent room: a bulbous vacuum tube of the Hittorf–Crookes type—quite likely one designed by Lenard. It was evacuated by a Raps pump, a slow process requiring several days. Aside from featuring Lenard's aluminum window, the tube was simply an improved version of the ubiquitous discharge tube first introduced by Heinrich Geissler of Bonn four decades earlier. Following Lenard's precaution, Röntgen covered the tube with close-fitting black cardboard. With the laboratory room completely darkened, he observed the fluorescence of the cathode rays passing through the aluminum foil onto a paper screen coated with barium platinocyanide, a screen routinely used to indicate the presence of ultraviolet light. Behold, the screen fluoresced equally well whether the treated side or the other side faced the discharge tube! Not only that, the fluorescence was visible when the screen was positioned *two meters* away from the apparatus—far beyond the range of cathode rays in air [1-5].

Röntgen's preliminary communication on the subject, submitted for publication to the secretary of the Physical–Medical Society of Würzburg on December 28 and marvelously complete despite its paucity of pages [1-6], is frustratingly unclear on exactly how many days and nights sufficed to establish the enthralling salient facts concerning these mysterious 'x-rays', as he termed them. We do know that between the initial discovery and the submission of his report, Röntgen virtually shut himself off from everyday affairs, eating and sleeping in the isolated laboratory (to the consternation of Frau Röntgen) while intently pursuing the new phenomenon at hand. He soon established that

> the most striking feature of this phenomenon is the fact that an active agent here passes through a black cardboard envelope, which is opaque to the visible and the ultraviolet rays of the sun or of the electric arc; an agent, too, which has the power of producing active fluorescence.... We soon discover that all bodies are transparent to this agent, though in very different degrees. I proceed to give a few examples: Paper is very transparent; behind a bound book of about one thousand pages I saw the fluorescent screen light up brightly, the printer's ink offering scarcely a noticeable hindrance. In the same way the fluorescence appeared behind a double pack of cards; a single card held between the apparatus and the screen being almost unnoticeable to the eye. A single sheet of tin foil is also scarcely perceptible; it is only after several layers have been placed over one another that their shadow is distinctly seen on the screen.... If the hand be held between the discharge tube and the screen, the darker shadow of the bones is seen within the slightly dark shadow image of the hand itself. [1-7]

In addition to their great penetrating power, the x-rays differed from cathode rays in one essential respect: they were not deflected by a magnetic field.

> I therefore reach the conclusion [continued Röntgen in his report] that the x-rays are not identical with the cathode rays, but that they are produced by the cathode rays at the glass wall of the discharge apparatus.... The justification for calling by the name 'rays' the agent which proceeds from the wall of the discharge apparatus I derive in part from the entirely regular formation of shadows, which are seen when more or less transparent bodies are brought between the apparatus and the fluorescent screen (or the photographic plate). [1-8]

Not the least startling, in fact, was their intriguing effect on the photographic plate.

> I possess, for instance, photographs of the ... shadow of the bones of the hand; the shadow of a covered wire wrapped on a wooden spool; of a set of weights enclosed in a box ... [1-9]

The eleventh-hour announcement, and particularly the sensational photographs, provoked an immediate and unparalleled public reaction in popular and scientific circles alike. Emil Warburg related the news of the discovery to the Berlin Physical Society on the January 4 1896. The revelation of the *Wiener Presse* on January 5 was followed by simultaneous accounts two days later in Frankfurt (*Frankfurter Zeitung*), London (the *Standard*), and New York (the *New York Times*). Scientific journals were not far behind. *Nature* carried a full-length translation of Röntgen's *Sitzungsberichte* report on January 23, as did America's *Science* on February 14. Röntgen himself, bowing to pressure from colleagues, addressed the subject only once, in a lecture before the Physical–Medical Society at Würzburg on January 23—an appropriate forum, in view of the self-evident medical importance of the new discovery. Amidst all the hoopla, congratulatory letters and telegrams poured in from colleagues at home and abroad, from Boltzmann, Warburg, Kohlrausch, Stokes, Poincaré, and even Lord Kelvin—all in response to preprints mailed out by Röntgen on January 1, each accompanied by x-ray photographs of the hand of his wife Bertha.

Röntgen was besieged with invitations to give lecture demonstrations virtually everywhere, but steadfastly declined, with one exception. He could not very well turn down an invitation to give a demonstration in Potsdam before the Kaiser (himself a staunch supporter of science, despite his not being very bright), after which he dined with His Majesty and received the Order of the Crown, 2nd class. (A still greater honor awaited him in 1901, with his receipt of the very first Nobel Prize. On that occasion he managed, as was his wont, to dodge a formal Nobel lecture, although he did give an eloquent speech of thanks at the banquet. He donated his prize money to the University of Würzburg.)

Not all colleagues were equally joyous over Röntgen's discovery. The conspicuous exception was Lenard, habitually a difficult man, who after all had stimulated the Würzburger experiments in the first place. He now sourly complained of lack of due recognition of his active assistance to Röntgen in the matter and, in years to come, of lack of due recognition of his own contributions presaging the discovery generally. One who definitely appreciated the discovery was J J Thomson of Cambridge University (who would himself soon provoke Lenard's scorn as well). While it was left for others to sort out the nature of the new-found radiation, Röntgen's discovery contributed indirectly to the solution, two years later, of what must be considered the most fundamental problem in physics at the turn of the century—a breakthrough harboring a new era in physical science. The solution, a will-o'-the-wisp of experimentalists for nearly two centuries, was provided by a mathematical physicist with demonstratively mixed experimental flair.

1.2. Prelude to the electron

J J Thomson's interest in cathode rays predated Röntgen's discovery by over a decade. He had, in fact, chosen the subject as his first experimental investigation on inheriting the Cavendish Professorship of Experimental Physics from Lord Rayleigh in 1884. (Though experimental physics at Cambridge stretched back to Newton's time, a physical laboratory, and with it an experimental professorship, was only 15 years old when Thomson assumed his position.) His first set of experiments on the electrical discharge in gases was published in 1886, though he was soon distracted by an unfruitful experiment on the electrodeless discharge and allied matters; however, Röntgen's announcement promptly rekindled his interest in cathode rays.

Just how Thomson, with excellent theoretical credentials but sufficiently lacking in experimental prowess for colleagues (including Rayleigh) to seriously doubt his qualifications for such a prestigious experimental chair, nevertheless succeeded very well in his new responsibilities, remains a classic dilemma in the annals of late 19th-century physics. Before coming to grips with it, however, a somewhat closer look at particular aspects of physics ca. 1895 is useful. Or, let us say, of *physical science*, since throughout the 19th century physics and chemistry remained intertwined to a greater degree than would be the case in the age of compartmentalization so characteristic of 20th-century science. Thus, only in 1900 was Germany's prestigious *Annalen der Physik und Chemie* reorganized simply as *Annalen der Physik*; France's corresponding *Annales de Chimie et Physique* resisted restructuring until 1914.

Subject areas dominating late-19th-century issues of *Annalen*, *Annales*, and Britain's rival *Philosophical Magazine* included the liquefaction of gases, gaseous, electrolytic, and metallic conduction of electricity, specific heats, electromagnetism with emphasis on electromagnetic waves, ionic and molecular dissociation, and several branches of the relatively young field of thermodynamics. The kinetic theory of matter and statistical mechanics were in a rather advanced state of development in 1895, thanks to Ludwig Boltzmann and Josiah Willard Gibbs, among others. Yet, these subfields were accorded scant attention by many physicists—an attitude possibly contributing to Boltzmann's suicide in 1906. Though the atomic 'hypothesis' was much in vogue based on chemical evidence and the kinetic theory alike, the reality of atoms was by no means universally accepted in these years. In London, the Chemical Society sponsored heated evening discussions on their pros and cons. Technically, the atomic versus anti-atomic arguments centered on the validity of William Prout's old (1815) hypothesis that elements are built from multiples of a single protyle unit which he identified with hydrogen, in light of evident deviations of atomic weights from integers. Many senior scientists viewed atoms as a useful heuristic concept at best; other anti-atomists argued against the atom precisely because it was a logically unintelligible basis for the teaching or demonstration of scientific knowledge [1-10]. James Clerk Maxwell, who constructed the mathematical basis for the kinetic theory of gases, suggests in his *Treatise* 'that we call [the] constant molecular charge, for convenience in description, *one molecule of electricity*' [1-11]. As late as the turn of the century Max Planck was 'not only indifferent, but to a certain extent even hostile to the atomic theory' [1-12], although by 1910 he proclaimed the reality of atoms with near-religious zeal [1-13]. Others were offended by the simplistic models or mechanical analogues of atoms and molecules then in circulation, especially those proffered by the British luminaries Lord Kelvin and Sir Joseph Larmor, and even by Maxwell. Particularly augmentative in his castigation of the British penchant for mechanical modeling was Pierre Duhem, professor of physics at the University of Bordeaux: 'We thought we were entering the tranquil and neatly ordered abode of reason, but we find ourselves in a factory'. We hasten to caution that these mechanical representations were hardly served up as so many realistic depictions of nature. Far from it, they were considered indispensable for providing pedagogical and intellectual insight—in essence research aids in their own right and quite in keeping with the Cambridge tradition of mathematically elaborating physical phenomena [1-14].

In addition, however, a new, if short-lived, alternative to atomism burst on the scene in our very year of focus, 1895. In a meeting that year of the Society of German Scientists and Physicians [1-15] in Lübeck, the prominent physical chemist Wilhelm Ostwald and his colleague Georg Helm voicefully argued for the reality of *energy* over matter—a view Ostwald steadfastly maintained until persuaded otherwise by irrefutable experimental evidence for atoms and molecules not long in coming. Another notable anti-atomist was the influential Viennese physicist Ernst Mach who was irked by the failure to explain spectra by atomic modeling, and who championed the empirical foundation of thermodynamics over the prevailing and rather crude atomic or molecular representations then making the rounds. In fairness, he too eventually was swayed by the accumulating evidence to the contrary.

A veritable rash of evidence for the particulate nature of matter suddenly swamped the Continental and British laboratories alike within two years of Röntgen's discovery of x-rays, as if to atone for the brief hiatus during which x-rays took center stage. First to be heard from was Jean Baptiste Perrin, a budding French physical chemist at the École Normale and a student of Marcel Brillouin, an advocate of Boltzmann's statistical mechanics and an outspoken adversary of Ostwald's and Mach's 'energetics' [1-16]. As early as 1895, Perrin reported an experiment demonstrating that the cathode rays are indeed negatively charged material particles, as argued by Crookes, and not waves in the æther as Hertz and Lenard would have it [1-17]. In his discharge tube Perrin mounted an insulated aluminum cylinder with an orifice facing the negative cathode—a so-called Faraday cylinder—connected to an electroscope and enclosed in a larger grounded metallic cylinder. Sure enough, when the tube was in operation, the Faraday cylinder was invariably charged with negative electricity.

> This group of results seems not to be easily reconciled with the theory which supposes that the cathode rays are an ultraviolet light. On the contrary they agree very well with the theory which considers them a material radiation. [1-18]

Perhaps so, though the Germans were not yet convinced. To be sure, that negative charges were collected could hardly be doubted, but whether they were indissolubly linked to the cathode rays, whatever the exact nature of the rays, had not been convincingly demonstrated in their opinion.

However, hard on the heels of the new French evidence supporting the particulate view of cathode rays, confirmatory evidence for the existence of discrete electrically charged particles came from a quite disparate line of investigation in nearby Holland. Whereas the cathode ray controversy involved *freely* moving electrical charges, ongoing spectroscopic studies at the Physical Laboratory of the University of Leiden dealt with the possibility of *bound* charged particles moving within atoms. Peter Zeeman was then a *Privatdozent* at Leiden, where he had studied under Heike Kamerlingh Onnes, director of the laboratory, and Hendrik Antoon Lorentz, Onnes' theoretical counterpoint at Leiden and a scientist of equal international esteem. Subsequently he became Lorentz's assistant. Zeeman had received his doctorate in 1893 for his work on the Kerr effect, or the effect of a magnetic field on polarized light. Somewhat later he was struck by an unsuccessful magneto-optical experiment in 1862 by none other than Michael Faraday, the last in his celebrated career at the Royal Institution of London: by his attempt to influence the yellow spectral lines of sodium vapor by a magnetic field. Surely, reflected Zeeman, the experiment warranted repetition with the superior apparatus now at hand in Leiden: better vacuum pumps, stronger magnets, and diffraction gratings with vastly greater resolving power than Faraday's prismatic spectroscopes. Behold, a slight broadening of the D lines was detected almost at once—something, in fact, anticipated by Faraday, but doubted by Joseph Larmor [1-19] who subsequently became Lucasian Professor at Cambridge and Secretary of the Royal Society and who already carried much weight.

Zeeman promptly communicated his results [1-20] to Lorentz, who was equally quick in providing an explanation. In 1892, Lorentz had published his first paper on the atomistic interpretation of Maxwell's electromagnetic equations in terms of currents of electrically charged 'particles' [1-21]. In 1895, he refers to the particles as 'ions' in his classic paper [1-22] which introduces the expression for what we nowadays term the *Lorentz force* or the electromagnetic force acting on an 'ion' of charge e and velocity v moving through electric and magnetic fields E and B, respectively:

$$\boldsymbol{K} = e(\boldsymbol{E} + v \times \boldsymbol{B}). \tag{1.1}$$

Lorentz's explanation was based on light emitted by 'ions' orbiting within the sodium atom; their motion was determined by the restraining harmonic forces within the atom in conjunction with the force exerted by the ambient magnetic field in accordance with the above expression. The resulting equation of motion yields a solution for the change in frequency $\Delta\nu$ of the emitted radiation in terms of the ratio of charge to mass (e/m) of the bound ions, *viz.*

$$\Delta\nu \sim (e/m)B. \tag{1.2}$$

Hence, from Zeeman's measured broadening [1-23], Lorentz was able to deduce a value for the ratio e/m for bound ions a full year before J J Thomson at Cambridge, as well as Walter Kaufmann in Berlin and Emil Wiechert at Königsberg, obtained values for the same ratio for freely moving cathode rays. The initial Zeeman–Lorentz analysis 'gives 10^7 as the order of magnitude of the ratio e/m when e is expressed in electromagnetic units' [1-24], in well-nigh astonishing agreement with the modern value of 1.76×10^7 emu g^{-1}. Inexplicably, though, neither Zeeman nor Lorentz expressed surprise or concern in print over this numerical value for the ratio, fully a thousand times larger than expected on the basis of contemporary atomic charge and mass values—e.g., assuming Proust's protylean hydrogen subunit. Kaufmann, who arrived at precisely the same value from his cathode ray studies, made a point of the seemingly embarrassing discrepancy, but had no explanation to offer. Wiechert, again with the same result in hand, did have an explanation in terms of an effective mass m 2000–4000 times smaller than the hydrogen atom, albeit unjustifiably so on the basis of an *ad hoc* assumption for the charge e. Across the Channel, Arthur Schuster of Owens College, Manchester, initially a spectroscopist like Zeeman and later a cathode ray experimentalist and protagonist of J J Thomson, narrowly missed the anomalous value and its concomitant implication by an unfortunate experimental judgement.

It fell on Thomson at Cambridge to the south, the awkward and inexperienced experimentalist, to perform the requisite measurements nailing down the value for e/m and draw the inspired conclusion for which he is justly credited. At this point, then, we must make our acquaintance with J J Thomson and the circumstances surrounding his experiments without further ado.

1.3. Thomson's formative years

Joseph John Thomson was born in Cheetham, a suburb of Manchester, in December 1856 [1-25]. He entered physics by a quirk of fate. It was intended that he should become an engineer which, in those days, was only possible by serving a rather costly apprenticeship to an established engineering firm. Unfortunately, the chosen firm, a leading producer of locomotives, had a considerable waiting list. Instead of having him mark time in school while waiting for an opening, his father, Joseph James Thomson, a bookseller and publisher, arranged for him to attend Owens College in Manchester. Here the boy enrolled at the age of 14, prepared to spend three years completing the standard engineering course. However, at the end of his second year his father died. His plan for becoming an engineer had to be scrapped for want of money for the apprenticeship premium. Instead, with the aid of some small scholarships he was able to complete the engineering course and received, in addition to the engineering certificate, a scholarship in engineering and a prize for an essay on some engineering subject.

Owens College, the embryonic Manchester University (later to become the Victoria University of Manchester) was located in a rather shabby section of town. It had, all the same, an excellent faculty. There was Thomas Barker, professor of mathematics and former Senior Wrangler and Fellow of Trinity College, Cambridge. Physics was taught by Balfour Stewart, the Scottish meteorologist and geophysicist, chemistry by Henry Enfield, and Osborne Reynolds, former seventh Wrangler, was the professor of engineering. The physical plant was another matter. The engineering department was housed in a former stable. Chemistry was better off, quartered in an adjoining house respectable enough, but the physics department was nothing more than a cramped room housing apparatus for lecture demonstrations. 'The cramped space was not without its advantages', recalled Thomson nevertheless. 'We were so closely packed that it was very easy for us to get to know each other' [1-26].

As a student of engineering, Thomson was mainly exposed to the colorful Osborne Reynolds—as original, independent, and frequently as obscure, as anybody could possibly be. (Reynolds is best remembered for his contributions on turbulent and laminar flow of liquids, and the dimensionless parameter that bears his name.) Young Thomson also came under the tutelage of Thomas Barker, quite an opposite type: careful, neat, and an excellent teacher from whom Thomson obviously inherited his early inclination for mathematics. At the end of his three-year stint, Barker advised him to remain at Owens for another year, studying mathematics and physics with the goal of attempting an entrance scholarship at Barker's alma mater, Trinity College. Concurring, Thomson did just that and had a try at Trinity in 1875, albeit with disastrous results. Undaunted, he tried again the next year. This time he passed, and, fortified with a modest yearly scholarship and other allowances, settled in at Cambridge where he would remain ensconced for his entire career.

The highest aspiration of any undergraduate student of mathematics at Cambridge was to successfully 'wrangle'—that is, place high in the nerve-racking Mathematical Tripos, a highly competitive examination covering virtually all branches of pure and applied mathematics then known. (It is named after the three-legged stool which was formerly used during ceremonies at degree time.) Preparing for it took 'three years and a term' [1-27], aided by the indispensable coaching of Edward John Routh, Fellow of Peterhouse (Saint Peter's College), the oldest of the Cambridge colleges. Routh was a born teacher of mathematics who served as private tutor for the old Tripos during its entire period in force, from 1855 to 1888 [1-28]. Thomson sat for the Tripos in the raw English month of January, 1880. It was an extremely stressful ordeal, lasting nine days in the unheated Senate House, mercifully interrupted by a 10-day stint during which the majority of the less than top-notch mathematical students were eliminated by the examiners. He emerged Second Wrangler, quite to his satisfaction, with Joseph Larmor placing Senior. As a result, Thomson was elected to a Fellowship with the opportunity of staying on at Trinity for the immediate future.

A drawback of the old Tripos was its exclusive concentration on problem solving in mathematical science at the expense of experimental or practical scientific subjects. Indeed, in those years Fellowships were awarded purely for successes in examinations, not for demonstrated promise in experimental research, say in physics. This situation was rectified in 1895, when a new regulation was instituted granting a degree to 'research students', at first an MA but soon a PhD in line with long-standing German university practice.

Thomson scarcely frequented the Cavendish Laboratory, if at all, as an undergraduate, and never met Maxwell. He did, however, receive a thorough grounding in mathematical physics, with emphasis on Maxwell's *Treatise on Electricity and Magnetism*, through the lectures of William D Niven (later the editor of Maxwell's collected papers and of the second edition of the *Treatise*). Another lecture series on physics which left a strong impression on him was that on light by Sir George Stokes, who then occupied the prestigious Lucasian Chair. Mostly, however, his undergraduate training was in pure mathematics, taught principally by, in addition to Routh, J W Glaisher, Second Wrangler in 1871 and at the peak of his academic career at this time [1-29].

The next step on the academic ladder at Cambridge, a Prize Fellowship, required a dissertation. Thomson chose as his topic the nature of energy, his thesis being 'of the view that all energy was kinetic' [1-30].

> The view which regards all potential energy as really kinetic has the advantage of keeping before us the idea that it is one of the objects of Physical Science to explain natural phenomena by means of the properties of matter in motion. [1-31]

Thomson developed his argument in terms of Hamiltonian and Lagrangian analyses of systems solely possessing kinetic energy, using modified versions of Lagrangian formalism first introduced by Routh. (The dissertation itself was never published, but portions of it are found in the *Transactions* of the Royal Society and in his book *Applications of Dynamics to Physics and Chemistry* [1-32].) Passing, in addition, the required Fellowship Examination, he was elected Fellow the same year he passed the Tripos, 1880.

The privileges of being a Fellow, whose tenure lasted seven years, were in those days limited until he became a Master of Arts. For one thing, while still a Bachelor of Arts, a Fellow was not permitted to dine at the High Table, but, like all BAs, dined at the Bachelor Table managed by the Senior Bachelor. The Senior Bachelor, a precarious distinction which Thomson held for two years, was responsible, with the Head Waiter, for ordering dinner for the next day. More ominously, he was also allotted a fixed sum to pay for each dinner eaten by a Bachelor. Should the kitchen bill exceed the allotment at the end of the term, as it often did, the difference was charged to the Senior Bachelor's personal account [1-33].

Thomson lost no time in immersing himself in several additional investigations in mathematical physics. One of the first of these, and his most significant one, drew on Maxwell's theory that magnetic forces could be produced, not only by electric currents, but also by the time-rate-of-change in an electrical force in a dielectric. Thomson concluded that an electrical charge moving in an electric field will experience an increase in mass as its speed increases, and that this excess 'electromagnetic' mass resides in the space immediately surrounding the particle and not in the charged particle itself [1-34]. The effect, he noted, is analogous to that of a sphere moving through water; the moving sphere sets the surrounding water in motion, making the sphere behave as if its mass were increased. Similarly, the electromagnetic mass is identifiable with the drag of the æther surrounding the charged particle. In this, not only did Thomson anticipate Einstein's energy–mass equivalence by 24 years, but in the same paper he derives an expression for the force exerted on the moving charge by a magnetic field, the so-called Lorentz force (equation (1.1)), albeit off by a factor of two from Lorentz's result in 1895. (The expression was also anticipated by Oliver Heavyside in 1889 [1-35].)

Another important mathematical work of Thomson in this period was his investigation of 'the action of two vortex rings on each other' [1-36]; this was the subject for the Adams Prize for 1882, a prize competition commemorating the prediction of Neptune by John Couch Adams [1-37]. The vortex representation of matter originated with Helmholtz in 1858, and it was further developed in the physical ruminations of Lord Kelvin. It pictured matter as that it

> ...might be made up vortex rings in a perfect fluid, a theory more fundamental and definite than any that had been advanced before. There was a spartan simplicity about it. The material of the universe was an incompressible perfect fluid and all the properties of matter were due to the motion of this fluid. The equations which determined this motion were known from the laws of hydrodynamics, so that if the theory were true the solution of the problem of the universe would be reduced to the solution of certain differential equations, and would be entirely a matter of developing mathematical methods powerful enough to deal with what would no doubt be very complex distributions of vortex motion in the fluid. [1-38]

This model, akin to smoke rings in air, appealed to Thomson who, following the practice of Maxwell, his great mentor [1-39] and his other senior contemporaries, had a penchant for visual representations of physical phenomena (while not necessarily supposing them to be physically real representations), rather than for the esoteric mathematics involved. Another appealing feature 'was the analogy between the properties of vortex filaments and those of the lines of electric force introduced by Faraday to represent the electric field' [1-40].

The vortex model of nature, which, among other things, is not predicated on forces between material particles (only on ideal hydrodynamic principles), was an increasingly popular hydrodynamical model among British physicists in these years; it was also in direct conflict with the contemporary molecular approach generally favored on the Continent, based on forces between electrical particles (a subject we defer for now). Curiously, it was Thomson who would provide the experimental *coup de grâce* for the British approach and vindicate the Continental alternative [1-41]. Even before he did, the British vortex school began a decline. Its chief architect, Kelvin, abandoned it by 1890, though Thomson held sway on the subject for a few additional years.

1.4. 'J J' and the Cavendish

Despite his preoccupation with mathematical physics, 'J J' (as he was known, even by his own son) Thomson also began a modest research program in the Cavendish Laboratory in 1880, immediately upon receiving his degree. That was the year Lord Rayleigh assumed the Cavendish Professorship as director of the laboratory, following Maxwell's untimely death the year before. As noted earlier, Thomson had occasionally visited the Cavendish as an undergraduate, but never met Maxwell personally; Maxwell was, in these years, almost totally occupied with editing the unpublished papers of Henry Cavendish and not engaged in regular research at the laboratory [1-42].

The Cavendish Laboratory had its genesis in a report by a Cambridge University Syndicate that met in the winter of 1868–1869 to consider a strengthened undergraduate curriculum in experimental physics at the University. In February 1869, it recommended a laboratory at Cambridge and with it a Professorship (as well as a Demonstratorship) in Experimental Physics. Experimental science was then in ascendancy in academic circles elsewhere, and the importance of science in education was increasingly recognized. University laboratories were under construction at Oxford, London, and Manchester, and in the planning stage in Berlin. Yet, as late as 1873, an editorial in *Nature* bemoaning that 'science is all but dead in England', went on to say that

> It is also known that science is perhaps deadest of all at our Universities. Let any one compare Cambridge, for instance, with any German university; nay, with even some provincial offshoots of the University in France. In the one case he will find a wealth of things that are not scientific, and not a laboratory to work in; in the other he will find science taking its proper place in the university teaching, and, in three cases out of four, men working in various properly appointed laboratories, which men are known by their works all over the world. [1-43]

Indeed, as noted earlier, experimental research was not in the mainstream of the singular mathematical tradition at Cambridge (though Isaac Newton was not a slouch when it came to experimentation). At best, the work of the experimentalists consisted in resolving rival theories, or in uncovering some minor auxiliary effect which might augment the theory. Even so, in 1868 the subjects of heat, light, electricity, and magnetism had been added as subjects for examination in the Mathematical Tripos (ergo, they were still basically regarded as topics in applied mathematics). How to ensure adequate instruction in these subjects was the official charge of the aforesaid syndicate, and the laboratory and its Professorship was the proffered solution.

Lord Kelvin (Sir William Thomson at the time) was approached to occupy the first Cavendish Professorship (as the experimental chair soon became known), but he declined to relinquish his private researches and professorship at Glasgow; so did his German counterpart, Helmholtz. Instead, the position went to Maxwell in 1871, another Scot then aged 39 and living on his estate Glenair in Galloway, writing and doing research. His inaugural lecture was not a good omen. Possibly due to some snafu on the part of the University administration, it was poorly announced and delivered in some obscure lecture room, not in the Senate House at the confluence of Trinity Street and King's Parade where distinguished lecturers normally spoke, and thus the audience was merely a bunch of undergraduate Tripos candidates.

> When, a few days later, it had been announced with proper formality that Professor Maxwell would begin his lectures on heat at a certain time and place, the *dii majores* of the University, thinking that this was his first public appearance, attended in full force out of compliment to the new Professor, and it was amusing to see great mathematicians and philosophers of the place such as Adams, Cayley, Stokes, seated in the front row while Maxwell, with a perceptible twinkle in his eye, explained to them the difference between the Fahrenheit and Centigrade scales of temperature. [1-44]

The original Cavendish building on Free School Lane was constructed under the supervision of Maxwell and formally opened in October of 1874. Alas, the University was in no position to foot the hefty bill, £8450 for the building alone. William Cavendish, Chancellor of the University, a mathematician as a young man and a steel tycoon later, seventh Duke of Devonshire and heir to the distinguished physicist Henry Cavendish, personally saved the day by donating funding for the building as well as stocking it with apparatus. The laboratory, which had initially been called the Devonshire Laboratory, was, appropriately, renamed the Cavendish Laboratory in honor of both the Duke and Henry Cavendish.

The role of the Laboratory, in Maxwell's opinion, was as a place for providing research experience and opportunities for undergraduates who had successfully weathered the Mathematical Tripos. Following a short period of training in handling scales, verniers, galvanometers, and magnetometers, the students were expected to undertake some modest research project, usually on a subject selected by Maxwell. Formal courses of instruction were not instituted under Maxwell, despite the importance of science in education having been the driving force behind the Laboratory in the first place. However, periodic lectures were given by the Demonstrator, William Garnett, and by Maxwell himself. Oddly, Maxwell's lectures (actually rather elementary) were poorly attended, recalls Thomson [1-45]. In the main, though, 'the work done by undergraduates was very unsystematic and they were to a great extent left to shift for themselves as to what they should do and how they should do it' [1-46].

After Maxwell's death from cancer in 1879, the Professorship was again offered to Kelvin, who stubbornly declined once more. This time the post was accepted by John William Strutt, the third Lord Rayleigh, on his proviso that the appointment be limited to five years. Ill health had prevented Rayleigh, who graduated in mathematics at Cambridge in 1865, from pursuing an academic life. Following a convalescent stint on a Nile houseboat (during which he wrote his masterpiece, *The Theory of Sound*), he spent most of his productive years in a private laboratory adjacent to the family mansion in Essex [1-47].

During his 5 year term at the Cavendish, Rayleigh tackled the manifold problems haunting a fledgling institution with great energy. Richard Tetley Glazebrook and William Napier Shaw, both Cambridge graduates who joined the Cavendish in 1876, were appointed Demonstrators. With their assistance, systematic instruction in both experimental and theoretical subjects was soon under way. Before long the number of students grew rapidly, so much that eventually two more Demonstrators were needed to cope with the load [1-48].

Thomson began his tenure under Rayleigh with several flawed experimental investigations [1-49]. The most important of these, undertaken at Rayleigh's suggestion, involved determining the ratio of the electrostatic to the electromagnetic units of electricity—a ratio which ought, on Maxwell's theory, to be equal to the velocity of light [1-50]. In the midst of this piece of laboratory work, he competed unsuccessfully for a chair in applied mathematics which had been established at Owens College just then. In the event, the chair went to Arthur Schuster instead. Schuster's career and research program were destined to be curiously intertwined with Thomson's. Like Thomson, he had studied at Owens (mainly physics, less mathematics) and held research positions there off and on before joining the Cavendish Laboratory as a researcher under Maxwell. At the Cavendish, he initially saw the completion of his work on a test of Ohm's law, pursuing, in effect, an experimental dud [1-51], and then on the absolute determination of the ohm under Rayleigh. Late in his stint at the Cavendish he resumed spectroscopic studies, his original research field. We will have much to say about Schuster and Thomson, perennial protagonists, in due course.

As to Thomson's latest investigation, Rayleigh himself had already designed some of the apparatus, intending personally to collaborate in the project. In the event, 'Thomson rather ran away with it' [1-52], and published hastily, being overly confident in his experimental prowess.

With such a shaky experimental debut, plain to all, it would seem that Thomson had reached the modest apogee in the experimental career of someone rightfully a theorist. Far from it, in fact. In the summer of 1884 Rayleigh, true to his word, submitted his resignation from the Cavendish Professorship. For a third time Kelvin was approached, with near-unanimous Cambridge opinion persisting in his favor, but he still could not be swayed. Schuster applied for the position, as did FitzGerald, Glazebrook, Larmor, and Reynolds. So did Thomson, 'without dreaming that [he] should be elected, and without serious consideration of the work and responsibility involved' [1-53]. To his dismay, in December he was informed of his appointment to the prestigious chair by the board of electors, among them Stokes, Kelvin, George Darwin, and William Niven. It is believed that Niven was the most influential in promoting Thomson's candidacy. It has also been suggested that mathematicians at Trinity College sought to deflect the Cavendish from its industrially oriented trend emphasizing electrical standards, a program decreed by Maxwell and Rayleigh alike [1-54].

What to do? Thomson, barely 28, had never given lectures in which experimental demonstrations figured prominently, nor lectured in experimental physics for that matter. Fortunately the Laboratory staff stayed on, with few exceptions; in particular, Glazebrook and Shaw. Glazebrook, heir presumptive to the Professorship, and Rayleigh's personal choice for the chair, accepted his defeat graciously.

> Dear Thomson,
> Forgive me if I have been wrong in not writing before to wish you happiness and success as Professor. The news of your election was too great a surprise to me to do so. I had looked on you as a mathematician, not an experimental physicist, and could not at first bring myself to regard you in that light. [1-55]

Equally cordial was Schuster's acknowledged defeat at about the same time.

> My dear Thomson,
> I send you my best and sincere congratulations on your appointment as Cavendish Professor.
> I have beaten you once, you have beaten me now in return, and so we are quits.
> I have no doubt the laboratory will flourish under your superintendence. [1-56]

In responding, Thomson (evidently not having heard from Glazebrook yet) vents his apprehension of the task at hand.

> Dear Schuster,
> Many thanks for your very kind congratulations. I can hardly realize my position yet as I never regarded my candidature as a serious one, and should have been very pleased if either you or Reynolds had been elected. I am glad to say that Shaw is staying on as demonstrator. I have not heard from Glazebrook yet. I hope you wont [sic] mind if I bother you with questions about laboratory work sometimes, as I feel I have an immense deal to learn. Wishing you all the compliments of the season, I remain,
> Your ever. [1-57]

Finally, there is a congratulatory letter from his former teacher at Owens College—himself, as noted, a candidate for the Professorship.

> My dear Thomson,
> I do not like to let the occasion pass without offering you my congratulations, which are none the less sincere that we could not both hold the chair. Your election is in itself a matter of great pleasure and pride to me as it must be to all those connected with the commencement of your brilliant career, and I have no doubt but every hope that you will amply justify the wisdom of the election. [1-58]

With Glazebrook and Shaw staying on for the duration, the courses established under Rayleigh continued smoothly without interruption. Even so, Thomson himself could not avoid teaching as well, with his very first lecture off to an inauspicious start.

> ... in my first lecture, though I got on quite well until about ten minutes before the time to stop, I then became very dizzy: I could not see my audience, and thought I was about to faint and had to dismiss the class. I recovered in a few minutes and have never suffered in this way again. I attribute it to having turned to look at the blackboard at my back too often ... [1-59]

On the whole, however, his tenure began well. An important responsibility was replacing the outgoing superintendent of the laboratory workshop; George Gordon, who had retained that crucially important post, now left to be private assistant to Rayleigh. A replacement, one D S Sinclair, was soon on board, to be followed by a long succession of superintendents and instrument makers in the years to come.

Even more important was choosing a new program of experimental research. In this, too, Thomson lost no time. Immediately upon being elected to the Professorship, he began a program on the transmission of electricity through gases at low pressure—the hot topic of experimental investigations just then. To adequately appreciate the background for this controversial subject, the *raison d'être* for this volume, it is necessary to regress roughly four centuries and review early notions on the nature of electricity.

Chapter 2

ELECTROMAGNETIC PHENOMENA UNRAVELED

2.1. The nature of electricity

The ancient Greeks were acquainted with the fact that amber, if rubbed with fur, will attract hair and other light bodies. For a very long time the attractive power of amber was regarded as an intrinsic property of that one substance, or possessed by at most one or two other substances [2-1]. Thus the Venerable Bede, the English monk and father of English literature, mentions in his *Ecclesiastical History of the English People* that jet (compressed coal), when warmed by rubbing, 'holds fast whatever is applied to it, like amber' [2-2]. William Gilbert, president of the Royal College of Physicians and court physician to Queen Elizabeth I (and briefly to her successor James I), took issue with Bede; in his famous work *De Magnete* he showed that the same effects are induced by friction in a rather large class of bodies, including glass, sulfur, wax, and precious stones [2-3]. Gilbert also coined the term *electric*, after *ηλεκτρον*, the Greek word for amber [2-4].

The salient facts concerning electric phenomena—principally activation by frictional stimulation and screening effects—convinced Gilbert that, in analogy with the prevailing doctrine of bodily 'humours', they are caused by some sort of material humour or fluid (Gilbert uses the term 'effluvia') that is liberated from electrifiable bodies under the influence of friction. As a result, every electrified body is surrounded by an atmosphere of highly attenuated effluvia. Gilbert's emanation theory attracted much attention; among his followers were the Jesuit physicist and mathematician Niccolo Cabeo and the celebrated Robert Boyle. Cabeo was also among the first to observe that electrified bodies *repel* as well as attract [2-5]—a decided complication. So was Francis Hauksbee, a student of Boyle and demonstrator of experiments at the Royal Society of London [2-6]. A feature of Gilbert's theory, devised to account for the lack of 'sensible diminution of the weight of the electrick body', as

Newton himself phrased it in his *Opticks* [2-7], was the inherent tendency for the emanation to reunite with the parent body. Ultimately, the corpuscular theory of light, as espoused in a tentative way by Newton and his followers, particularly Wilhelm Jacob 'sGravesande, came to be seen as a model of emissions too rare and subtle to cause perceptible loss of weight, and the doctrine of the return of effluvia was dropped [2-8].

'sGravesande viewed effluvia as permanently attached to electrified bodies, and attributed electric effects to vibrations excited in the atmosphere of effluvia by friction. The English translator of his influential Latin text [2-9], the chaplain Jean Theophilus Desaguliers, would himself play some role in these developments. One reason was that he was an acquaintance of Stephen Gray. In 1729, Gray, a 'gentleman pensioner' of the Charterhouse in London and member of the Royal Society [2-10], reported in a letter to Desaguliers and some other fellows of the Royal Society his discovery 'that the Electrick Vertue of a Glass Tube may be conveyed to any other Bodies so as to give them the same Property of attracting and repelling light Bodies as the Tube does, when excited by rubbing' [2-11]. In other words, he had conceived *electrical conduction*. The experiments were continued by Desaguliers after Gray's death; Desaguliers coined the term *conductors* for metals which were conspicuously effective in transmitting the 'Electrick Vertue' [2-12]. The fact of conduction spelled an end to the notion of effluvia permanently attached to bodies. Rather, it emphasized the fluidity of the emanations, whether particulate of merely some kind of state of agitation of the æther.

Another who took up Gray's experiments was Charles-François de Cisternay du Fay. Du Fay came from a prominent family, and his father, who had many high connections, secured for him a position as adjunct chemist at the Académie des Sciences, and later superintendent of the gardens to the King of France. Du Fay had no formal training in science; nevertheless he enthusiastically threw himself into the study of electricity when he learned of Gray's experiments. He concluded that

> ...there are two distinct electricities, very different from one another; one of which I call *vitreous electricity*, and the other *resinous electricity*. The first is that of glass, rock-crystal, precious stones, hair of animals, wool, and many other bodies; the second is that of amber, copal, gum-lack, silk, thread, paper, and a vast number of other substances. The characteristic of these two electricities is, that a body of the *vitreous electricity*, for example, repels all such as are of the same electricity; and on the contrary, attracts all those of the *resinous electricity*; so that the tube, made electrical, will repel glass, crystal, hair of animals, etc., when render'd electrick and will attract silk, thread, paper, etc., though render'd electrical likewise. Amber on the contrary will attract electrick glass, and other substances of the same class, and will repel gum-lac, copal, silk, thread, etc. [2-13]

In other words, like types of electricity repel, and unlike types attract one another. Du Fay's experiments included suspending a boy by silk cords, electrifying him by friction, and drawing sparks from him (a quite common diversionary practice among electrical experimenters and demonstrators—Gray, for one).

One who collaborated with du Fay and became a great popularizer of the new electrical phenomena was Jean-Antoine Nollet, later famous as the Abbe Nollet. Nollet was educated by the church, but suspended a clerical career in favor of the arts and sciences. In time he rose to high prominence, becoming preceptor in Natural Philosophy to the royal family and professor at the University of Paris; among his most noteworthy contributions was his six-volume treatise on physics, *Leçons de Physique Expérimentale* [2-14].

Nollet came to the attention of René-Antoine Ferchault de Réaumur and du Fay, and assisted them in a vast range of scientific investigations, including electricity. He even accompanied du Fay on an information-gathering tour of England and Holland, during which he became acquainted with, among others, Desaguliers and 'sGravesande. His ill-fated contribution to electricity stems from the spring of 1745, inspired by new electrical experiments by the German G M Bose, which made him a tireless popularizer of the subject as well as a theorist. His electrical theory interpreted du Fay's vitreous and resinous forms of electricity as two distinct currents of a fluid of electricity, an 'affluent' current and an 'effluent' current, subtle enough to flow in and out of all bodies via 'pores'.

Nollet's two-fluid theory proved short-lived, at least for the time being, due to experiments started at Leiden in which he, too, became involved. In 1745 the Leiden jar, the first electrical capacitor, was discovered by Pieter van Musschenbroek, professor of mathematics and physics at Leiden [2-15]. Musschenbroek arranged some spectacular demonstrations of the power of the device, in the course of which a friend named Cunaeus was subjected to a prodigious shock. Nollet learned of the Leiden experiments via a letter from Musschenbroek to Réaumur in 1746 [2-16] and lost no time in contriving even more spectacular demonstrations, such as depicted in figure 2.1. They culminated in one involving 700 monks joined in a circle to a Leiden jar [2-17]!

News of the Leiden experiments reached London equally quickly (1746); there, in the same year, they caught the attention of the versatile William Watson, physicist, physician, and botanist (who was eventually knighted). He, too, set about repeating the basic experiments of Musschenbroek. In so doing, he observed a pattern of electrical discharge—namely, that a participant, like Cunaeus, tended to feel the shock 'in both his arms and across his breast' [2-18]. It suggested that a *single* 'electrical æther' or fluid is transferred, but never created or destroyed, from one body to another only when one has a surplus and

Figure 2.1. *Nollet's electrical machine.*

the other a deficit of electrical æther [2-19].

Watson's one-fluid theory was propounded in greater depth and nearly simultaneously by the prodigious savant Benjamin Franklin of Philadelphia. His interest in these matters was stimulated from several quarters just then: by some 'imperfectly performed' electrical demonstrations in Boston in 1743 by a certain Scotsman, Dr Adam Spencer, by a subsequent gift of a glass tube from the manufacturer and naturalist Peter Collinson of London [2-20], and by an article on the latest electrical diversions in the *Gentleman's Magazine* of 1745, a London publication. Franklin, then clerk of the Massachusetts State Assembly, immediately undertook a series of experiments in collaboration with his associates Philip Syng, Thomas Hopkinson, and Ebenezer Kinnersley. Syng, Hopkinson, and Kinnersley were members of the 'Junto', a group of friends formed by Franklin who met on a regular basis for a drink or a meal. On July 11 1747 he wrote to Collinson, informing him of the following observations.

1. A person standing on wax, and rubbing a tube, and another person on wax drawing the fire; they will both of them, provided they do not stand so as to touch one another, appear to be electrified to a person standing on the floor; that is, he will perceive a spark on approaching each of them with his knuckle.
2. But if the persons on wax touch one another during the exciting of the tube, neither of them will appear to be electrified.
3. If they touch one another after exciting the tube and drawing the fire as aforesaid, there will be a stronger spark between them than was between either of them and the person on the floor.

4. After such a strong spark neither of them discover any electricity [2-21].

Franklin speaks of a *deficiency* of electricity as 'electrised negatively', and an *excess* as 'electrised positively'.

> To electrise *plus* or *minus*, no more needs to be known than this; that the parts of the tube or sphere that are rubbed, do in the instant of the friction attract the electrical fire, and therefore take it from the thing rubbing. The same parts immediately, as the friction upon them ceases, are disposed to give the fire, they have received, to any body that has less. Thus you may circulate it, as *Mr Watson* has shown; you may also accumulate or subtract it upon or from any body, as you connect that body with the rubber, or with the receiver, the communication with the common stock being cut off. [2-22]

Thus, charge is never created or destroyed; it is always conserved. Certain features of the discharge process led Franklin to tentatively identify positive electricity with the vitreous fluid of du Fay (with whose work he was unacquainted at the time)—an unfortunate association that would stick. We know today that electrons carry electricity of the resinous type, and it would make more sense for Franklin to have opted for the reverse of his sign convention that, however, remains in usage to this day.

Franklin's letters to Collinson were published in book format by Collinson [2-23], and some of his letters were also published in Watson's papers to the Royal Society. Nevertheless, his experiments attracted rather little notice in European circles—possibly testament to the novelty of his ideas—until a French translation was issued at the behest of the naturalist Georges Louis Leclerc de Buffon.

Franklin's aforesaid letters to Collinson hint at the notion of action at a distance, although he explicitly retains a physical picture of a single, subtle, and elastic fluid, somewhat in the vein of effluvia but composed of particles able to exert attractive or repulsive powers despite intervening screens (such as the glass wall of a Leiden jar). These vitreous particles of electrical matter repel each other, but two bodies with an excess and a deficiency, respectively, of vitreous charge attract each other. However, the mutual repulsion between resinously electrified bodies (of which Franklin only learned subsequently) was to him problematic. That difficulty was eventually resolved by Franz Ulrich Theodosius Æpinus, director of the astronomical observatory of the Imperial Academy in Saint Petersburg (although not himself strictly an astronomer). Æpinus, in fact, put the lingering concept of effluvia decisively to rest. Earlier experiments of his in Berlin in collaboration with the Swede Johan Carl Wilcke had shown that not only glass, but even air is impermeable to electricity. On their basis, he was led to deny the existence of effluvia surrounding charged bodies; the electric fluid is confined to the surface of an excited body, and electricity acts at a distance across the intervening medium, be it glass, air, or any other non-conducting substance [2-24]. The awkward repulsion between resinously charged bodies was explained by the proviso that, in addition to Franklin's repulsion between vitreously charged bodies and attraction between oppositely charged bodies, particles of ordinary matter, stripped of some of its normal complement of electricity, repel each other as well.

Æpinus and Wilcke were less successful in providing a quantitative description of the forces acting between charged bodies—that is, determining the precise law according to which the force between electrically charged bodies varies with the distance between them—although Æpinus in his *Tentamen* in effect uses it in a numerical application. That law was first correctly surmised by none other than Joseph Priestley, the discoverer of oxygen and an acquaintance of Franklin, on the basis of experiments first performed by Franklin and repeated by Priestly in 1766 on Franklin's urging. The essential result of these experiments, that bodies enclosed in an electrified metal cavity sense no electric force, suggested, by analogy with a similar result for bodies within a massive, gravitating shell and the known law of gravitational force [2-25], that 'the attraction of electricity is subject to the same law with that of gravitation, and is therefore according to the squares of the distances' [2-26].

Direct experimental verification of the inverse square law for the electrical force was first obtained, albeit with some reservations, by John Robinson of Edinburgh in 1769 who, unfortunately, did not publish his results for many years [2-27]. It was rediscovered four years later by Henry Cavendish, although his contribution proved to have no more influence on his contemporaries. The son of Lord Charles Cavendish, fellow of the Royal Society and administrator of the British Museum, Cavendish was educated at Cambridge but, like many aristocrats, finished without a degree. Subsequently he devoted himself to research in physics and chemistry, greatly assisted by a vast fortune inherited from his uncle, the Second Duke of Devonshire. (At his death, Cavendish left over a million pounds.) Being, however, an eccentric recluse, he shunned public (though not altogether scientific) contacts and disdained publication of much of his work. His interest in electrical researches was another inheritance, in this case from his father, who had assisted Watson in his experiments of 1747. His own experiments involved a metal sphere enclosed within another sphere, with or without electrical contact between them; they showed that electrical charge was invariably confined to the outer sphere—a result consistent with the inverse square law to within 1%. Alas, his proof of the law, and many of his other contributions to electrical science, were only brought to public attention in 1879, after his notebooks and unpublished manuscripts were edited and published by Maxwell on the urging of Lord Kelvin [2-28].

The first really convincing verification of Priestley's law was that of the French military engineer and physicist Charles Augustin Coulomb. Returning in poor health from nine years of military service in tropical Martinique, he settled in Paris in 1772, intending to exploit his engineering experience. However, a prize competition of the Paris Academy diverted him into physics by its requirement of an improved compass needle. Pursuing this led him to the researches for which he is best remembered, those based on his famous torsion balance. Essentially, it consisted of a very fine silver wire carrying a horizontally suspended, carefully balanced straw needle covered with wax; to the straw, rotating in a large-diameter glass cylinder with a graduated scale around its circumference, was attached a charged pith ball. Another charged pith ball was introduced at various distances from the rotatable ball, and the torsional twist of the straw was measured as a function of ball separation. Coulomb summarized his initial results in a memoir to the French Academy in 1785 [2-29], in a law which he expressed as follows:

> The repulsive force between two small spheres charged with the same sort of electricity is in the inverse ratio of the squares of the distances between the centers of the two spheres... [2-30]

He extended his law to the attraction of oppositely charged bodies in his second memoir of the same year [2-31]. Both memoirs were published in 1788, one year before the storming of the Bastille caused Coulomb to prudently withdraw from Paris.

Coulomb disagreed with the one-fluid theory of Franklin and Æpinus (and Cavendish), preferring to develop the two-fluid theory of du Fay which had already been extended by Robert Symmer [2-32]. As a result, two rival camps on the matter came into being, the two-fluid theory being generally championed in France, while the one-fluid theory held sway in Holland (on the authority of Van Marum) and in Italy (Volta) [2-33]. The dispute was only settled by mature cathode ray experiments over a century later—the culmination of a long tradition of experiments on the discharge of electricity in evacuated vessels, and experiments more specifically of relevance to our inquiry here. But before coming to grips with this singularly fruitful line of experimentation, it behoves us to backtrack briefly once again and trace progress on *magnetism* during the period covered thus far.

2.2. From magnetism to electromagnetism

The attracting powers of the lodestone, magnetized iron ore, have been known at least as long as that of amber. The Greeks called the ore '*λιθοσ Μαγνητισ*', or 'Magnesian stone', after the city in Asia Minor where it was mined. However, the traditional wisdom that its use in navigating at sea originated in ancient China, and that the technique was subsequently transmitted to the West via the Arabs, appears questionable [2-34]. For one thing, the Chinese were not noted navigators. Nevertheless, they are known to have been acquainted with the directive powers of the lodestone by the end of the 11th century; it also seems all but certain that by this time the English seafarers made use of the compass. The Icelandic sagas refer frequently to the *leidarstein*, though in Ari Fróthi Vilgerdarson's discussion of the discovery of Iceland (in 874 by Floki Vilgerdarson) it is unclear as to whether the lodestone was known to the Vikings prior to ca. 1225 [2-35]. The polarity of a lodestone magnet was discovered by Pierre de Maricourt (alias Peter Peregrinus) of Picardy. He observed that not only does every magnet have two poles, a north-seeking one and a south-seeking one [2-36], but also that a north-seeking pole of one magnet will attract the south-seeking pole of another magnet and that north-seeking poles or south-seeking poles repel each other [2-37].

Further progress in magnetism awaited William Gilbert and his opus *De Magnete* of 1600, which laid the foundation for both magnetism and electricity and a work we have already encountered. Gilbert recognized that the Earth is a great spherical magnet, explaining why a compass behaves as it does.

> In the heavens, astronomers give to each moving sphere two poles; thus do we find two natural poles of excelling importance even in our terrestrial globe, constant points related to the movement of its daily revolution, to wit, one pole pointing to Arctos (Ursa) and the north; the other looking toward the opposite part of the heavens. In like manner the loadstone has from nature its two poles, a northern and a southern; fixed, definite points in the stone, which are the primary termini of the movements and effects, and the limits and regulators of the several actions and properties. It is to be understood, however, that not from a mathematical point does the force of the stone emanate, but from the parts themselves; and all these parts in the whole—while they belong to the whole—the nearer they are to the poles of the stone the stronger virtues do they acquire and pour out on other bodies. These poles look toward the poles of the earth, and move toward them, and are subject to them. [2-38]

The foundation for a quantitative description of the law of attraction between magnets was laid earlier than for the corresponding law for electric charges. Newton, in the *Principia* [2-39], notes that 'the power of magnetism ... in receding from the magnet, decreases not on the duplicate, but almost in the triplicate proportion of the distance, as nearly as I could judge from some rude observations'. The law was pinned down by the geologist and astronomer John Michell of Queens

College, Cambridge in 1750 in his *Treatise of Artificial Magnets* [2-40] based partly on experiments of his own: 'The attraction and repulsion of magnets decreases, as the squares of the distances from the respective poles increase'. Michell was to know Priestley and Cavendish rather well in later years; among other things, he assisted Cavendish with some of his experiments. The definitive confirmation of Michell's semi-empirical law was left to Coulomb in the same series of experiments in which he confirmed Priestley's law of electrical repulsion with the torsion balance. (The torsion balance, in fact, was invented independently by Michell, though he died without making use of it.) In his second memoir of 1785 (the one in which he dealt with electrical attraction) he announced, on the basis of simple experiments with magnetized needles, that 'the magnetic fluid acts by attraction or repulsion in a ratio compounded directly of the density of the fluid and inversely of the square of the distance of its molecules' [2-41].

Coulomb's reference to a magnetic 'fluid' has a familiar ring. Not surprisingly, the contemporary preoccupation with, variously, one- or two-fluid theories of electricity provoked, before long, a spate of similar attempts at concocting theories of magnetism, by Æpinus and others. Æpinus' single magnetic fluid soon gave way to two imponderable fluids, termed *boreal* and *austral* as befitting polar fluids and similar in their properties to vitreous and resinous electric fluids. If nothing else, all this served to fuel the suspicion of some deep relationship between electricity and magnetism. Indeed, there were empirical hints of such a connection: lightning appeared to have the power of magnetizing steel needles, and Franklin had magnetized needles by discharging them through a battery of Leiden jars. However, a direct experimental search for such related phenomena only became practical with an experimental development of the highest importance in the spring of 1800, ushering a new epoch in electrical experimentation.

On March 20 1800 Alessandro Volta of Como, Italy, announced in a letter to Sir Joseph Banks, president of the Royal Society, his discovery of a method of producing a *continuous* current of electricity. Because England and France were at war, the letter, in two sections, was read before the Society in June of that year [2-42]. Volta had been educated by the Jesuits in classical subjects, but chose a secular life. From a young age he was greatly fascinated with all things scientific, but above all electricity. By the age of 16, he was corresponding with several known electricians, including Nollet in Paris. In time he was appointed professor of physics at Pavia, a post he kept until his retirement. The discovery of the new source of current was made at Pavia in the course of following up experimentally Luigi Galvani's work on 'animal electricity'. A controversy had arisen over the nature of Galvanism, as it also became known. Galvani himself maintained that the convulsions of the limbs of frogs when touched by scalpels or probes of dissimilar metals are caused by the transport of an internal electric fluid from the nerves to the muscles, the probe acting simply as a conductor. Volta, who had initially agreed with Galvani, soon changed his mind, suspecting that the frog is merely a sensitive detector, with the source of the electricity residing externally. In his opinion, the electricity causing the convulsions was generated by dissimilar metals in contact with a moist body, and he was engaged in magnifying the feeble effect with a number of bimetallic junctions joined together in the presence of a moist conductor when he came across his celebrated pile.

In the Voltaic pile, shown in figure 2.2, pairs of disks or strips of dissimilar metals (e.g. copper and zinc) are immersed in brine or a weak-acid electrolyte. It brought instant fame to Volta, and activated a rash of electrochemical experiments on both sides of the Atlantic. Volta himself demonstrated the action of the pile in 1801 before a distinguished Parisian audience headed by the First Council, Napoléon Bonaparte, who was a keen follower of the developing science of electricity [2-43]. Napoléon, who had driven the Austrians out of Italy in 1796 and subsequently proclaimed the Cisalpine (or Italian) Republic under France, saw to it that Volta received a special gold medal and pension, and retained him as his protégé. When eventually Volta requested his retirement from the Pavia chair with Italy still a Napoléonic regime, Napoléon refused, bestowing more honors on him, giving him a pension, and making him a count and senator of the kingdom of Italy. Despite all this attention, though, Volta practically disappeared from the scientific scene after the flurry of the pile, showing little interest in its electrochemical effects. With the Austrians back in power, he did retire in 1819, and died in 1827, leaving no school or followers of note.

Across the Channel, the chemist William Nicholson learned of Volta's communication from Banks. Assisted by the surgeon Anthony Carlisle, he at once constructed a pile with which he succeeded in the decomposition of water into hydrogen and oxygen. This seminal achievement, announced in Nicholson's own *Journal* [2-44], was followed up by a host of electrochemists. Among them, William Cruickshank decomposed solutions of metallic salts with an improved pile—the true Voltaic *battery*—consisting of rectangular zinc and copper plates in a resin-insulated wooden trough [2-45]. William Hyde Wollaston, a physician who abandoned his practice to devote himself to science (possibly because of his failure in a contest for the appointment of physician to Saint George's Hospital), demonstrated the identity of Voltaic currents with those in a discharge of frictional electricity, by showing that water could be decomposed by currents of either type [2-46].

Not to be outdone was Humphry Davy, a young Cornish apothecary's assistant already known for his discovery of nitrous oxide (laughing gas) and newly appointed by Count Rumford as lecturer in

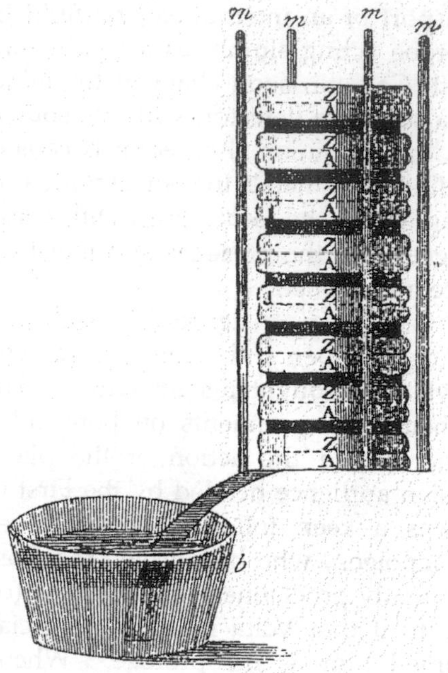

Figure 2.2. *Volta's first pile. A denotes silver, Z zinc. The dark layers are pieces of moistened cardboard.*

chemistry at the Royal Institution in London. Quick to appreciate the possibilities of the new source, Davy threw himself into a brilliant series of electrochemical investigations. In the short span of 10 years they propelled him into the limelight. He became Secretary of the Royal Society at the age of 29, and by 1810 was the mainstay of the Royal Institution, culminating in a knighthood in 1812 [2-47].

Davy's Bakerian Lecture for 1806 [2-48] contains the first clear exposition of the mechanism of electrolysis and of the Voltaic battery; Davy perceived the two effects to be basically the same, even though one requires a current and the other produces it [2-49]. The next year, with a powerful battery of 274 plates, he decomposed the so-called fixed alkalis, potash and soda, from which he extracted the new elements potassium and sodium [2-50]. When laid up with a serious illness for some months in 1808, Institution members took the occasion to provide funds by subscription for the construction of an even more powerful battery of 2000 pairs of plates; it was capable of throwing a spark from charcoal points through 4 inches of air. On recovering, Davy extracted additional new elements from the alkaline earths, namely, barium, strontium, calcium, and 'magnium' (magnesium).

On the Continent, the most active experimenters included Louis-Joseph Gay-Lussac and Louis-Jacques Thenard in France, Luigi G Brugnatelli in Italy, Auguste De La Rive in Switzerland, and the complicated Johannes Wilhelm Ritter in Jena, Germany. Ritter, co-discoverer with Nicholson and Carlisle of the electrolytic decomposition of water and inventor of the first storable or 'secondary' pile, is a curious figure in these unfolding developments. An experimenter of considerable skills [2-51], he was also an ardent disciple of the romantic movement of *Naturphilosophie* that, in Wilhelm Ostwald's words, 'ravaged Germany like a plague in the first years of the nineteenth century' [2-52]. Placing intuition above empiricism, this intellectual school stressed that 'all phenomena are correlated in one absolute and necessary law' [2-53]. Its leading proponent was Friedrich W J Schelling, who was especially fascinated by physical phenomena exhibiting polarity, such as electricity and magnetism. However, Ritter too was no slouch when it came to speculation. His search for the great unity even led him to see correlations between periods of maximum inclination of the ecliptic with outstanding discoveries in electricity, including Kleist's (and Musschenbroek's) Leiden jar and Volta's pile. He prophesied that the next great discovery would come in '1819 2/3 or 1820' in a letter to none other than Hans Christian Ørsted of Copenhagen [2-54]!

On a less lofty plane, an example of the experimental search for a relationship between Voltaic electricity and magnetism in the years 1800–1820 was the effort of Jean-Nicolas-Pierre Hachette and Charles-Bernard Desormes to determine whether an insulated Voltaic battery, resting on a float in a vat of water, is oriented by the earth's magnetic field. No orientation was observed, though a steel bar weighing nearly half of the battery and floated in the same manner appeared to indicate such an effect [2-55].

There was, in fact, little direct evidence for magnetic manifestations of electricity. Moreover, there was far from agreement among scientists that there *had* to be an *a priori* connection, aside from the insistence of the Nature Philosophers. Thomas Young, for one, maintained in his lectures that 'there is no reason to imagine any immediate connexion between magnetism and electricity' [2-56]. As late as 1819, Jean-Baptiste Biot, writing on magnetism in David Brewster's *Edinburgh Encyclopedia*, cautioned that, although 'there exists the most complete... analogy between the laws of the two magnetic principles, and those of the two electrical principles', nevertheless 'the independence which exists between their [magnetic] actions and the electric actions does not allow us to suppose them to be of the same nature as electricity' [2-57].

Yet, it nowadays seems odd that at least one particular line of investigation was hardly pursued at all: on the magnetic effects of an electric current. Save for the aforesaid effects of lightning on the compass needle, and some early experiments of Franklin, about the only experiment of relevance was one by Nicholas Gautherot in 1801 demonstrating that two wires, one attached to one terminal of a battery and the other to the other terminal, would adhere to each other when brought into close contact [2-58]. The same was independently observed by none other than Laplace (and Biot, for that matter), but they, too, failed to see its significance.

The *experimentum crucis*—and an extremely simple one at that—was performed at the University of Copenhagen in the spring of 1820: quite as Ritter had foreseen, one might say. Hans Christian Ørsted had graduated as a pharmacist at the University in 1797—not a bad scientific grounding, as we have already seen. The following year he joined the editorial staff of a new and short-lived Copenhagen journal defending Kantian philosophy against its many assailants. Contributing a paper on Kant's metaphysics, Ørsted elaborated it into a dissertation for which in 1799 he earned his doctorate. The turn of the century found him manager of an old and respected apothecary shop and 'adjunct' lecturer without salary in chemistry and physics at the University. That year, 1800, he was stirred by Volta's announcement of chemical electricity, and lost no time constructing a small Voltaic pile of his own; soon he was engrossed in experiments on acids and alkalis. The next year, however, he was off on the distracting but traditional Continental *Wanderjahr* under a grant from 'Cappel's Travelling Legacy' (carrying with him his little battery), in the course of which he ran into the leading German Nature Philosophers: Schelling, Friedrich and August Schlegel, and others. He also met Count Rumford, then serving the Elector of Bavaria, and particularly the disorganized Ritter—by training again a pharmacist. The close relationship struck up with Ritter at Oberweimar would prove to be a mixed blessing in times to come.

Ørsted returned to Copenhagen in 1804. As luck would have it, the University, weary of his enthusiasm for Nature Philosophy, turned down his application for a salaried appointment. Instead, he was entrusted charge of the royal cabinet of scientific instruments and lectured privately on physics and chemistry. The lectures proved highly popular, so much so that in 1806 the University relented and appointed him professor *extraordinaris* in physics, to be sure at a miserable salary. Only in 1817 did he realize full professorship (*ordinarius*). In the meanwhile, despite a heavy teaching load, he was a prolific writer of textbooks and pedagogical tracts, and found time for experiments and built up a chemical laboratory. Grandly named the Royal Chemical Laboratory, it was indeed augmented by the king's contribution of his own collection in 1815.

As noted, a recurring theme of the German romantic movement was the unity of nature, and, in particular, of electricity and magnetism. Although skeptical of the more extreme views of Schelling and his followers, Ørsted's own belief in this unity, reinforced by his *Wanderjahr* contacts, determined his own scientific program and led straight to the experiments of 1820. As he himself wrote,

> Throughout his literary career, he adhered to the opinion, that the magnetical effects are produced by the same powers as the electrical. He was not so much led to this, by the reasons commonly alleged for this opinion, as by the philosophical principle, that all phenomena are produced by the same original power. [2-59]

The years 1812–1813 saw him in Germany and France again. Though by now more critical of the philosophers, his guarded optimism in finding a relationship between electricity and magnetism by direct experiment surfaced at this time in an account of his own electrochemical system first published in Berlin [2-60] and extensively revised in a French version [2-61]:

> Magnetism exists in all the bodies of nature, as Bruckmann and Coulomb have proved. For this reason it is felt that magnetic forces are as general as electrical forces. One should test whether electricity in its most latent form has any action on the magnet as such. This experiment would offer some difficulty because electrical effects are always likely to be involved, making the observations very complicated.

As to the circumstances of the actual experiments in the spring of 1820, Ørsted has left us several accounts, one written the following spring.

> ...I was brought back to [experiments on the effects of an electric current on a magnet] through my lectures on electricity, galvanism, and magnetism in the spring of 1820. The auditors were mostly men already considerably advanced in science; so these lectures and the preparatory reflections led me on to deeper investigations than those which are admissible in ordinary lectures. Thus my former conviction of the identity of electrical and magnetic forces developed with new clarity, and I resolved to test my opinion by experiment. The preparations for this were made on a day in which I had to give a lecture the same evening. I there showed Canton's experiment on the influence of chemical effects on the magnetic state of iron. I called attention to the variations of the magnetic needle during a thunderstorm, and at the same time I set forth the conjecture that an electric discharge could act on a magnetic needle placed outside the galvanic circuit. I then resolved to make the experiment. Since I expected the greatest effect from a discharge associated with incandescence, I inserted in the circuit a very fine platinum wire above the place where the needle was located. The effect was certainly unmistakable, but still it seemed to me so confused that I postponed further investigation to a time when I hoped to have more leisure.* At the

> beginning of July these experiments were resumed and continued without interruption until I arrived at the results which have been published.
>
> *All my auditors are witness that I mentioned the result of the experiment beforehand. The discovery was therefore not made by accident, as Professor Gilbert [then editor of *Annalen der Physik*] has wished to conclude from the expressions I used in my first announcement. [2-62]

The four-page announcement of the discovery of electromagnetism appeared on July 21 1820 [2-63]. We know from a former student of Ørsted, Christopher Hansteen, that the renewed experiments on possible effects on a magnetic needle by a neighboring current-carrying wire were negative, since Ørsted placed the wire at right angles to the needle; however, it then occurred to him to place the wire parallel to the needle [2-64]. Behold! A pronounced deflection was observed. Pursuing the matter in a hectic series of additional experiments, Ørsted summarized 'the effect which takes place in [the] conductor and in the surrounding space, [which] we shall give the name of the *conflict of electricity*' [2-65] as follows:

> The electric conflict acts only on the magnetic particles of matter. All non-magnetic bodies appear penetrable by the electric conflict, while magnetic bodies, or rather their magnetic particles, resist the passage of this conflict. Hence they can be moved by the impetus of the contending powers.
> It is sufficiently evident from the preceding facts that the electric conflict is not confined to the conductor, but dispersed pretty widely in the circumjacent space.
> From the preceding facts we may likewise collect that this conflict performs circles; for without this condition, it seems impossible that the one part of the uniting wire, when placed below the magnetic pole, should drive it towards the east, and when placed above it towards the west; for it is the nature of a circle that the motions in opposite parts should have an opposite direction. [2-66]

That is, a wire carrying an electric current affects an adjacent magnetic needle by causing it to swerve to a position perpendicular to the direction of the wire. When the current is reversed, the needle swings diametrically about. When the wire is placed under the needle, it points in the opposite direction, indicating the presence of a circular magnetic field around the current-carrying conductor.

In the very journal that carried Ørsted's account of the first experiments in the spring of 1820 appeared his own description of additional experiments suggesting complete symmetry between the effects of a current on a magnet and that of a magnet on a current, or, as he put it: 'As a body cannot put another in motion without being moved in its turn, when it possesses the requisite mobility, it is easy to foresee that the galvanic arc must be moved by the magnet' [2-67]. This he verified by showing that, just as a current exerts a force on a magnet, a magnet causes a closed, current-carrying loop to turn, and that such a loop has a north-seeking end and a south-seeking end.

Less than a month after the publication of Ørsted's announcement, news of his discovery was described at a meeting of the French Academy on September 11 1820 by the professor of analytic geometry at the École Polytechnique, Dominique-François Jean Arago, who had just returned from abroad. Ørsted's experiments were repeated without delay, and the first quantitative analysis of the effect involved was presented by the initially skeptical Biot and by Felix Savart at a meeting of the Academy on October 30. They expressed the 'actions exerted on a molecule of austral or boreal magnetism when placed at any distance from a thin cylindrical wire of indefinite length rendered magnetic by the Voltaic current' [2-68] as follows:

> From the point where the molecule is, draw a perpendicular to the axis of the wire; the force which acts on the molecule is perpendicular to this line and to the axis of the wire. Its intensity is inversely as the distance. The nature of its action is the same as that of a magnetized needle, which is placed on the contour of the wire in a direction determined and always fixed with respect to the direction of the Voltaic current; so that a molecule of boreal magnetism and a molecule of austral magnetism will always be acted on in opposite senses, although always along the straight line determined by the preceding construction. [2-69]

Nobody could any longer doubt that a magnetic field was just as readily produced by a current-carrying wire as by a magnet. Indeed, Arago soon showed that a piece of iron, surrounded by a coil, was briefly magnetized by passing current through the wire [2-70].

One who heard Arago's report at the Academy's meeting on September 11 was André-Marie Ampère, professor of mathematics at the École Polytechnique. He plunged into a series of experiments of his own. At the next meeting of the Academy, just one week later, he presented a paper showing that two parallel wires carrying currents attract each other if the currents flow in the same direction, and repel each other if they flow in opposite directions [2-71].

Ampère continued his researches on *electrodynamics* (a term coined by him to distinguish it from *electrostatics*, also his own term) over the next three years, the results of which he published in his celebrated memoirs of 1825 [2-72]—results characterized by Maxwell, half a century later, as 'one of the most brilliant achievements in science. The whole, theory and experiment, seems as if it had leaped, full-grown and full-armed, from the brain of the "Newton of electricity"' [2-73]. Ampère determined that the force of attraction or repulsion between current-carrying parallel wires is directly proportional to the strength of the current and inversely proportional to the square of the distance between the wires. On a more basic level, all magnetism is essentially electromagnetic in nature: it is the tiny closed 'Ampèrian' currents circulating within the molecules of a lodestone that give rise to its magnetic properties.

Ampère was quick to appreciate the practical possibilities of electromagnetism, in particular that the response of a magnetic needle to a remote current-carrying wire could form the basis of an electromagnetic telegraph transmitting signals over an electric wire. To be sure, the application of Voltaic batteries in electro*chemical* telegraphs had occupied many between 1800 and 1820 with some success, but Ampère's was a new, more elegant approach. All that was necessary was to identify magnetized needles with the letters of the alphabet, and by energizing the wire adjacent to a particular needle, the corresponding letter could be so indicated. In the event, the telegraph proved impractical because of the complexity of wires involved; instead, the first practical application of electrodynamics (also suggested by Ampère) was the galvano-magnetic 'multiplier' or galvanometer (Ampère's term) of Johann S C Schweigger (who had been one of the earlier advocates of the electrochemical telegraph) in 1820. It consisted of a small insulated coil surrounding a magnetic needle which responded quantitatively to the magnetic force generated by the current flowing in the coil. Perhaps the first working prototype telegraph line was that of Gauss and Wilhelm Weber over the rooftops of Göttingen in 1834, $1\frac{1}{4}$ miles long, from Weber's laboratory to the astronomical observatory in charge of Gauss. The first practical telegraph line came ten years later across the Atlantic, in Samuel F B Morse's line connecting Washington and Baltimore.

Both Ampère and Ørsted led highly productive lives and attained great fame, although Ampère's later years were morose and unhappy whereas Ørsted's scientific and public life remained happy to the end. By contrast, Volta, the progenitor of electromagnetism, retired in relative obscurity. A fourth giant in the stable foundation of modern electrical science we have not yet met. Michael Faraday would complete the experimental unification of electricity and magnetism by demonstrating the generation of an electric current from magnetism. We will have ample to say about Faraday presently, but for now, lest we be seriously diverted from the real thread of our monograph, it is again necessary to backtrack and review another line of electrical experimentation hardly touched on: that of the electrical discharge in evacuated vessels.

Chapter 3

CATHODE RAYS TAKE CENTER STAGE

3.1. Wonders of the electrical fire

The earliest known form of an electrical discharge was that of lightning—a prominent subject in Benjamin Franklin's voluminous correspondence with Collison and John Mitchel in London. By 1749, Franklin was in full agreement with the rather commonplace opinion that clouds are electrified and that the lightning discharge is a rapid release of electric fluid [3-1]. In November of that year, he concluded that since the 'electric fluid is attracted by points', we might find out 'whether this property is in lightning... Let the experiment be made' [3-2]. Nevertheless, he wisely stalled, warning his readers that hills, trees, towers, spires, masts and chimneys would act 'as so many prominences and points', and would 'draw the electrical fire'.

Not until July of 1750 did he propose a definite test, his famous and dangerous sentry box experiment which he himself refrained from attempting. It was duly performed in May 1752 at Marly-la-Ville outside Paris, by Thomas François Dalibard, naturalist and translator of the French edition of Franklin's text, in the company of Buffon and De Lor. The experiment called for a box large enough to house a man, mounted on a tall building—say a church steeple. A long, pointed iron rod fixed to an insulated stand protruded through the door, extending some 9 m into the surrounding air. The hapless subject, perched on the insulated stand, held in his hand a wax handle affixed to a wire loop, whose purpose was to draw sparks from the rod. Dalibard and his co-workers had no more wish than Franklin to expose themselves to the electrical fire. Instead, the role of sentinel was offered to a retired dragoon who fortunately survived the experiment [3-3]. Not so lucky was G W Richtmann in Saint Petersburg who, some years later, repeated the experiment with fatal results.

A short time after the Marly experiment, Franklin himself communicated his simpler and even more famous kite experiment to Collinson; in fact, he appears to have flown his electrical kite prior to learning of the successful French test. Be that as it may, these experiments brought instant public fame to Franklin—not merely the attention of scholars.

For more controlled experiments on the electrical discharge we turn to Francis Hauksbee, whom we have briefly met. Hauksbee's background is rather obscure. He first attracted notice in an experimental demonstration before the Royal Society at its meeting on December 15 1703; when mercury was rushed into the evacuated glass of an air pump of his, a sparkling light was given off—'a Shower of Fire descending all round the Sides of the Glasses' [3-4]. From 1704 until his final illness in 1713, he served as curator of experiments at the Society, and in November 1705 was elected Fellow for even more noteworthy experiments.

The new experiments followed two years of lackluster repetition, before the Society, of experiments by Robert Boyle and others. However, late in 1705 Hauksbee returned to his first subject, 'mercurial phosphorus' [3-5]. In this he was actually following up on earlier observations by Johann I Bernoulli, among others, that a luminosity is produced by shaking mercury in an exhausted glass vessel as in a barometer. Hauksbee's experiments in the fall of 1705 explored in greater depth the mercurial light, clearly a result of friction between mercury and glass. To ascertain whether the phenomenon was peculiar to mercury, he devised a series of experiments with various sources of frictional electricity introduced in the receiver of his pump. Beads of amber rubbed against wool cloth produced light 'without Intermission, . . . discernible at three or four foot distance', while if 'the same Motion and the same Attrition was given the Amber in the Open Air, . . . little Light did ensue, in Comparison to the Appearance of it *in Vacuo*' [3-6]. Replacing the beads with a small glass globe rotated swiftly by a spindle against woolen cloth 'commonly sold for Gartering' secured to a brass spring, gave better results: 'a fine Purple Light ensued, the included *Apparatus* being distinguishable by it' [3-7]. Even replacing the wool with 'flat shells of oysters well dry'd' produced a light 'resembling a fierce flaming Spark' [3-8]. Finally, in rubbing a glass globe against glass tubes attached to brass springs,

> . . . a considerable Light was exhibited: The whole Included *Apparatus* became perfectly distinguishable by it, and would have been much more so, had not the Daylight prevented: It then being but very little after 5. P.M. a clear Horizon, and in an open Room, when the Experiment was made. [3-9]

Hauksbee steadily improved his experiments, the results of which he presented at regular intervals to the Royal Society presided over in these years by Isaac Newton. (Newton, keen on re-introducing the former practice of having experiments demonstrated at Society meetings, may have invited Hauksbee to the Society in the first place.) Aware that other materials might produce light when rubbed *in vacuo*, the glass nevertheless increasingly occupied Hauksbee's attention, and he spoke in terms of Gilbert's effluvia lodged in the glass and dislodged by rubbing.

The first quantitative investigations of these phenomena were probably those of Gottfried Heinrich Grummert and William Watson. Watson, whom we have also met, was admitted to the Royal Society on the strength of his botanical contacts and contributions to natural history, but his reputation was largely earned by his electrical researches that led him to formulate his one-fluid model. His thunder as an electrician was stolen by Dalibard's successful test of Franklin's theory of lightning, and he eventually phased out these researches in favor of what became a prestigious medical career which earned him a knighthood. While still an electrician, however, he pursued the discharge of electricity *in vacuo*, among other things. In one experiment, he passed a current produced by an electrical machine through an evacuated glass tube three feet long.

> It was a most delightful spectacle, when the room was darkened, to see the electricity in its passage: to be able to observe not, as in the open air, its brushes or pencils of rays an inch or two in length, but here the coruscations were of the whole length of the tube between the plates, that is to say, thirty-two inches. [3-10]

Watson speculated that the electricity 'is seen, without any preternatural force, pushing itself on through the vacuum by its own elasticity, in order to maintain the equilibrium in the machine' [3-11].

A different explanation was offered by Nollet, who experimented with the discharge in rarefied air about the same time as Watson. Nollet interpreted the luminosity in terms of his own ill-fated theory of effluent and affluent currents of electricity. In his opinion, the particles of the effluent stream collided with those of the oppositely moving affluent stream, and as a result were excited to a state of luminous emission [3-12].

Almost a century passed before more progress was made on the continuous discharge in evacuated vessels. Before bridging that gap, it may be helpful in understanding subsequent developments if we depart temporarily from the historical thread and review from a later perspective the phenomena accompanying the transmission of electricity through gases at low pressure, and the apparatus involved. In doing so, we rely entirely on the excellent discussion of the subject by Chalmers [3-13].

The apparatus, to be sure of more recent vintage, consists of a discharge tube, a source of high-voltage electricity, and a vacuum pump and associated equipment, as depicted in figure 3.1. The tube is of

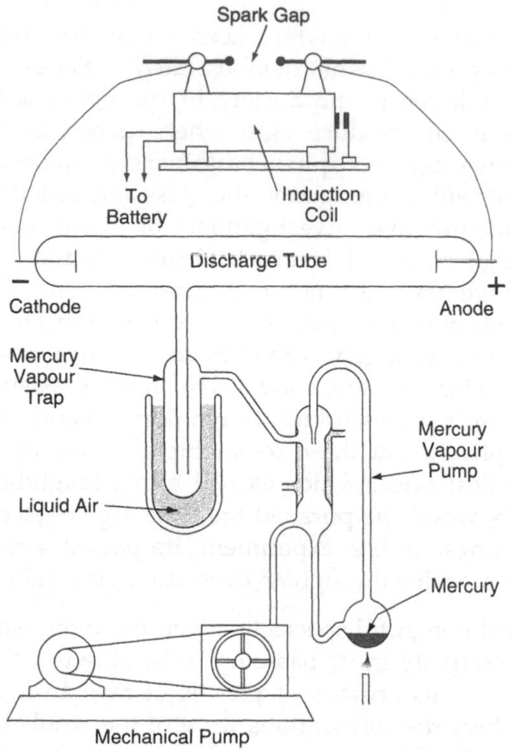

Figure 3.1. *Discharge tube, high voltage supply, and vacuum apparatus.*

glass, typically an inch or two in diameter and of any length up to, say, a yard, depending on the strength of the electrical source. Each end is sealed with a metallic electrode, the cathode and the anode. The source of electricity is depicted as an induction coil connected to a battery. The fact that an induction coil produces an alternating current is not of importance, since the output consists of a series of short, high-voltage pulses of one polarity alternating with lower-voltage pulses of the opposite polarity; typically the resistance offered by the gas in the tube prevents the latter pulses from passing. With the variable spark gap, two alternative paths are provided for the high-voltage current: through the external air across the gap, or through the gas in the tube. The vacuum apparatus depicted consists of a mercury vapor pump backed by a mechanical pump; the mercury trap prevents mercury vapor from diffusing into the tube by condensation on a liquid air-cooled surface.

Imagine now the tube filled with air, with the pump at work. At first, the spark discharge is confined to the external gap and the interior of the tube shows no effect, but at some reduced pressure, typically 10 mm Hg, a flickering of light appears in the tube and the discharge across the spark gap stops. Conditions are now such that the longer, rarefied path through the tube offers less resistance than the air at atmospheric pressure in the spark gap. As pumping continues, a uniform pinkish glow fills the whole tube between cathode and anode. Next, a violet glow appears at the cathode, growing in length and seemingly pushing the pink glow towards the anode, but never touching it; the violet glow is separated from the pink glow (or 'positive column') by a dark region—the 'Faraday dark space'. The air pressure at this stage is about 1 mm Hg.

As the reduction in pressure continues, the receding positive column breaks up into a number of regularly spaced, curved striations with their convex surfaces facing the cathode. At this point, the air resistance in the tube has reached a minimum value and is now rising. It is necessary to lengthen the spark gap to maintain the discharge in the tube. Soon the violet glow splits into two parts, one seemingly adhering to the cathode (the 'cathode glow'), and a second, a detached column (the 'negative glow'), extending towards the anode. The two parts are separated by the 'Crookes dark space' which gradually lengthens, appearing to push the negative glow, the Faraday dark space, and the striated positive column toward the anode. Again the spark gap must be lengthened.

As the Crookes dark space continues to grow, the positive column fades away except as a short-lived film at the surface of the anode (the 'anode glow'). The anode glow soon disappears, as does the Faraday dark space, with the negative glow extending all the way to the anode. Finally, the cathode glow and the negative glow disappear as well. Now the Crookes dark space completely fills the tube. The discharge is completely non-luminous, but a quite new phenomenon sets in: the walls of the glass tube are bathed in a yellowish-green fluorescence. At this stage, the pressure in the tube is roughly 0.01 mm Hg. If, finally, the tube were to be exhausted to the state of a perfect vacuum, the discharge would cease altogether.

In returning to our chronology of discovery, we skip the aforesaid unproductive period from the time of Nollet to the apogee in Faraday's electrical researches in the late 1830s. To be sure, prodigious and fruitful researches on the conduction of electricity through liquids followed Volta and his pile, as we have sketched earlier—researches employing relatively simple apparatus, uncovering electrochemical effects which to a considerable extent could be readily explained by prevailing chemical concepts and doctrines. The same could not be said for the concurrent studies of the transmission of electricity through gases. The reason was manifold: an over-abundance of experimental facts clouding relevant effects from irrelevant minutiæ and sensitivity, not only to the degree of purity and condition of the gases under study, but also to all possible aspects of the experimental arrangement (the size of the discharge tube, the shape and nature of the electrodes, and the strength of the current). The most serious experimental deficiency was the lack of an efficient vacuum pump. The pump still in use in the middle of the 19th century was essentially one described by Francis Hauksbee in 1709, a two-cylinder pump with pistons worked by rack and pinion. Its basic design, in fact, was simply a modification of Otto von Guericke's original pump dating from about 1650. The best Hauksbee's version could do was a vacuum within about 1 in Hg of a perfect vacuum—about as expected because of its water-saturated leather washers [3-14].

At least three modes of electrical transmission or discharge were recognized by the time of Davy and Faraday—indeed largely due to Faraday. Each was presumed to be characteristic of a distinct state of matter [3-15]. The 'conductive' mode was exhibited by an electric current passing invisibly through a metallic wire without causing any permanent material change; the 'electrolytic' mode, if a liquid replaced the wire, was accompanied by easily visible chemical decompositions; the 'disruptive' mode if the intervening medium were a gas such as air. The latter discharge tended to be highly visible, audible, and odorous, with the term 'disruptive' suggesting the electricity forcibly thrusting the gas molecules aside. As such, Chalmers reproves the term as an unfortunate one. 'It committed physicists to a rigid picture of the mechanism by which the mode of discharge was effected. Responsibility for it rests squarely on Faraday, the very man who, earlier, when investigating the transmission of electricity through liquids, had gone to great pains to invent and use terms avowedly designed to be completely non-committal' [3-16]. With that, the time has come to deal with Faraday himself.

3.2. Faraday the electrician

Two unconnected incidents were responsible for bringing Michael Faraday, a bookbinder's apprentice, to the Royal Institution where he, even more than Sir Humphry Davy, would leave his mark on the Institution and profoundly influence physical science generally. In late October 1812, while engaged in experiments to determine the properties of the unstable substance nitrogen chloride, an explosion injured one of Davy's eyes [3-17]. As he was temporarily blinded, Faraday was recommended as his amanuensis and Davy, taking an immediate liking to the young man, took him on. Unfortunately, with a laboratory assistant, William Payne, already on the Institution payroll, Davy was unable to offer Faraday a permanent position. By a quirk of fate, however, earlier in the same year the Librarian and Superintendent of the House, Mr Harris, had responded to a commotion in the lecture hall, finding that Mr Payne had come to blows with Mr Newman, the instrument maker, over some trifling matter [3-18]. In February 1813, Payne was dismissed and Davy immediately sent for Faraday, who on March 1 was appointed to Payne's position.

Michael Faraday was born in 1791 in Newington Butts, a small village near London; he was the third of four sons of a blacksmith in declining health and often out of work [3-19]. At the age of 13, he was engaged as errand boy to a bookseller and stationer, Mr George Riebau, in order to assist his mother in providing for the family. The following year he was apprenticed to his employer to learn the art of bookbinding—an apprenticeship that would prove enormously significant in the intellectual history of mankind. The young Faraday consumed the variegated material in the many books and manuscripts that came his way with ravenous appetite, encouraged by the enlightened Mr Riebau. His particular interest in science was kindled by the article 'Electricity' in a copy of the *Encyclopaedia Britannica* he was rebinding, written by James Tytler, a debt-ridden eccentric who almost single-handedly revised the original edition of the *Britannica*. It was not easy for a young man to further his education in London of 1810, but, as luck would have it, that year he fell in with a group of similar enthusiasts who had jointly formed a study group named the City Philosophical Society. The group was led by a goldsmith, John Tatum, at whose house they met on a weekly basis. On Mondays, Tatum would lecture on some scientific subject. On Wednesdays, a member of the CPS would lecture on a subject of his choice, and the ensuing discussion ranged widely over every conceivable aspect of elementary science.

Further stimulation was provided by Jane Marcet's lively *Conversations on Chemistry*. The climax came in the winter of 1812 when one of Mr Riebau's customers offered Faraday tickets to Davy's lectures on chemistry at the Royal Institution. Faraday took meticulous notes at these lectures, the last ones delivered by Davy as professor—notes which Faraday illuminated and bound himself. He wrote to the president of the Royal Society, Joseph Banks, for advice in obtaining any kind of post in science, receiving no reply. Shortly thereafter, however, occurred the aforesaid providential accident of Davy. While temporarily engaged with Davy, Faraday, on Mr Riebau's advice, shared his consummate lecture notes with him—notes which undoubtedly Davy remembered when, in due course, the Institution sought a replacement for Payne.

The duties of the young assistant included assisting Davy in his researches, attending to the needs of the various lecturers including Professor William Thomas Brande, and caring for the apparatus. No sooner had Faraday settled in at the Institution quarters on Albemarle Street, however, than Davy set out on a one-and-a-half year grand tour of scientific inquiry through the Continent, accompanied by Lady Davy, her maid, and Faraday who eagerly consented to come along. The party departed London in October 1813 at a time when England and France were still in a state of war; consequently, they entered France on a special dispensation of the Emperor himself. In Paris, Davy and Faraday were cordially received by many of the leading scientists of the day, including

Figure 3.2. *Michael Faraday.*

Ampère who presented Davy with a strange substance derived from seaweed. Within days, with his portable chest of chemical apparatus [3-20], and assisted by Faraday, Davy determined it to be a new element for which he coined the name 'iodine'. From France, the party journeyed to Genoa, and thence to Florence, Rome, Naples, and Milan, all the while Davy and Faraday performing experiments at every stop. Naturally, Faraday was thrilled.

In Geneva, there occurred the first of various rifts which would occasionally strain the personal relationship between Faraday and his mentor. Professor de la Rive the elder, upon learning that Faraday was Davy's scientific assistant, not simply his valet (who had chosen to remain behind in London), invited Faraday to dine with himself and the Davys. Davy himself (or actually Lady Davy) objected, and Faraday was served in a separate room [3-21]. Returning to Rome, relationships calmed and they wintered in harmony, busily engaged in scientific pursuits. This lull was cut short by the news of Napoléon's escape from Elba. Hastening northward, the party reached England in April 1815.

Returning to the Royal Institution Faraday, re-appointed, threw himself into experimental activities with boundless energy and singleness of purpose, buoyed by a penchant for orderliness and organization—qualities foreign to Davy. At the start, they were mainly small chores in support of Davy's program or modest chemical experiments suggested by the work of Davy and Professor Brande (Davy's successor as professor of chemistry). Before long, however, Faraday's own reputation spread, and he, too, became an indispensable mainstay at the Institution. One reason was his growing prowess as a lecturer, skills initially honed before the City Philosophical Society soon after the European tour. Before long his reputation in the lecture theater rivalled that of Davy and far overshadowed the lackluster lectures of the straight-laced Brande. In about 1925, evening meetings of the Institution Members were begun on a regular basis, organized around a lecture often accompanied by an experimental demonstration. This was the origin of the celebrated Friday Evening Discourses, of which Faraday gave over a hundred and which are still given today. The same year, 1825, the Managers appointed Faraday director of the laboratory on the urging of Davy—one of Davy's last official acts: this in spite of Davy's unsuccessful opposition, the year before, to Faraday's election to Fellowship in the Royal Society. In 1833, Faraday was elected to the newly founded Fullerian Professorship of Chemistry at the Institution, having already turned down a chair in chemistry at the University of London [3-22].

It is not possible nor desirable to dwell here on Faraday's variegated researches, though his electrochemical and electromagnetic experiments warrant some measure of attention in our chronicle. After all, the former experiments contain the seeds for the concept of the electron, and the latter presage the mature concept of an electric field.

Faraday's electromagnetic experiments had their origin in Ørsted's discovery of the magnetic effects of a current in 1820, news of which reached the Royal Institution only in October of that year. The next year William Wollaston and Davy, who were on close terms, undertook an inconclusive experiment at the Royal Institution involving a magnet and a wire. Pondering their attempt to make a current-carrying wire rotate in a magnetic field prompted Faraday to conduct a series of experiments of his own with similar apparatus. They culminated in his demonstration, in the fall of 1821, of electromagnetic rotation. A bar magnet was placed upright in a beaker of mercury, secured at the bottom but with the opposite pole protruding above the liquid surface. A wire dipping into the mercury, but free to rotate, rotated around the magnetic pole when current was passed through the mercury–wire circuit. If the wire was secured and the magnet allowed to move, the current caused the magnet to rotate. Sadly, Faraday's published paper [3-23] involved him in another rift with his mentor, who rightly accused him of failing to acknowledge his indebtedness to Wollaston and himself [3-24].

Faraday returned to electromagnetism off and on in the ensuing decade, suspecting, in light of the experiments of Arago and Ampère, that it ought to be possible to induce electricity in a coil by action in a neighboring coil without any metallic connection between the two. For a long time, experimental success eluded him. Finally, at the end of August 1831, acting on a clue from experiments of the American physicist Joseph Henry, he wound coils of insulated wire on opposite sides of an iron ring, the ring replacing wood and glass rings utilized in his earlier negative experiments. One coil was connected to a Voltaic battery, the other to a galvanometer. Behold, the improved coupling by the iron ring (something Faraday was in no position to appreciate) gave a sensible effect on 'making or breaking' the current in the circuit.

> The galvanometer was immediately affected, and to a degree far beyond what had been described when with a battery of tenfold power helices *without iron* were used; but though the contact was continued, the effect was not permanent, for the needle soon came to rest in its natural position, as if quite indifferent to the attached electro-magnetic arrangement. Upon breaking the contact with the battery, the needle was again powerfully deflected, but in the contrary direction to that induced in the first instance. [3-25]

Hard on the heels of this epic experiment, a scant three weeks later, Faraday further demonstrated that a bar magnet thrust into or rapidly withdrawn from a solenoid generated an electric current. His long-sought goal, *electromagnetic induction*, was no longer in doubt.

Having demonstrated the reverse of Ørsted's effect, Faraday proceeded to establish, through a series of beautiful experiments, the 'identity of electricities' from electrostatic generators, Voltaic cells, and electromagnetic induction, again confirming his long-held conviction 'that the various forms under which the forces of matter are made manifest have one common origin' [3-26]. In the course of these experiments, he was sidetracked into quantitative experiments on the chemical action of the electric current, from which he deduced his two laws of electrochemistry: (i) the chemical power of a current of electricity is in direct proportion to the absolute quantity of electricity which passes and (ii) the amounts of different substances deposited or dissolved by the same quantity of electricity are proportional to their chemical combining weights. That is, each element has its 'electrochemical equivalent', or, in Faraday's definition, the mass of an element (relative to that of hydrogen) deposited during a fixed time by a fixed amount of current [3-27].

These relationships imply that Faraday anticipated a fundamental unit of electrical charge, as suggested by Johnstone Stoney in his report to the British Association in 1874, introducing the concept of the electron, and subsequently by Hermann von Helmholtz in his Faraday Memorial Lecture of 1881 [3-28]. Alas, Faraday failed to anticipate a definite quantity of electricity associated with each material atom, and that during electrolysis each atom received or lost a definite quantity of electricity. The closest he came to a quantum of electricity was his finding that a given quantity of electricity deposited produced masses of different substances proportional to their equivalent values, or relative atomic weights. Unfortunately, for oxygen and heavier elements, Faraday used atomic weights too light by a factor of two, since he associated electrochemical equivalents simply with relative atomic weights, not with relative atomic weights divided by the valence—a concept that was developed somewhat later [3-29].

If we jump ahead and denote Stoney's or Helmholtz's 'atom of electricity' by e, then the amount of electricity required to deposit a gram ion (number of grams equal to the atomic weight) of material on an electrode is $F = N_0 e$ Faradays, where N_0 is Avogadro's number (number of atoms in a gram atom for monatomic ions). By Stoney's time, F and N_0 were reasonably well known, N_0 from kinetic theory and F from experiment.

A characteristic trait of Faraday in these electrolytic researches was a distrust of prevailing phraseology which implied certain preconceived but untested notions, such as the term 'pole' and its connotation of attraction and repulsion, or 'atoms' for the constituents in electrochemical decompositions. To guard himself against 'confusion and circumlocution' he coined, with the concurrence and assistance of the Rev W Whewell of Trinity College, Cambridge, a number of more non-committal terms such as 'electrode' (substituted for 'pole'), 'cathode' and 'anode' (for the negative and positive electrodes), and 'cations' and 'anions' (substituting for the overused term 'atom'). In so doing, Faraday also discarded the contemporary view of electromagnetic phenomena in terms of electrical 'fluids' and their mutual forces. Instead, he introduced his own idea of lines of force, whatever their precise nature might be and whether or not ponderable matter is essential to their existence, producing a state of molecular strain or vibration in current-carrying wires and the surrounding medium—a state he referred to as 'the electrotonic state'. It was this notion of strain, acting over some distance from its point of origin, that had been his guiding principle in searching for electromagnetic induction in the first place; indeed, it was the thread linking all his researches on electricity and magnetism [3-30].

Faraday returned to induction, this time *electrostatic* as distinct from electromagnetic, in 1835. Among other things, he wished to confirm, once and for all, what Franklin and Coulomb, in his opinion, had not convincingly demonstrated, that the electricity of a charged body resides on its surface alone, not inside or dispersed through the bodily material. He had a chamber constructed in the lecture theater, in the form of a metalized cube measuring 12 ft on a side, insulated from the floor and connected to an electrical machine by a wire in a glass tube extending into the interior of the cube.

> I went into the cube and lived in it, and using lighted candles, electrometers, and all other tests of electrical states, I could not find the least influence upon them, or indication of anything particular given by them, though all the time the outside of the cube was powerfully charged, and large sparks and brushes were darting off from every part of its outer surface [3-31].

A positive experiment, if there ever was one! At least as important, Faraday's three year preoccupation with electrostatic induction resulted in his concepts of 'dialectrics' between charged bodies and the corresponding 'specific inductive capacities'.

In 1838, Faraday turned his attention to phenomena of more immediate relevance to our ongoing chronology; namely, those accompanying the electrical discharge through gases in the form of a spark or electric arc. Ostensibly, he sought to determine the resistance offered to a spark discharge by different gases at atmospheric pressure, as well as to characterize the ability of a given gas to pass a discharge in terms of the pressure of the gas. In both problems, he had preliminary experiments by others to guide him. Already in 1821, Davy had drawn a luminous arc between charged electrodes by gradually separating them from initial contact. By exhausting the glass container, initial contact was no longer necessary, and the arc could be drawn across a gap to a length of half a foot or more. In 1834 more quantitative results were published by the electrician William Snow Harris, involving an electrical machine and a battery separated by a miniature Leiden jar serving as a spark gap. For a given air pressure, the amount of electricity required to initiate a spark discharge varied in exact proportion to the gap spacing; with a constant gap, the initiating charge varied in exact proportion to the air pressure [3-32].

Abroad at about the same time, the Dutchman Martin (Martinus) Van Marum, natural philosopher, physician, botanist, and sometimes electrician, was occupied in studies of spark discharges in different gases at atmospheric pressure. In hydrogen the discharge exhibited a fine crimson color; in nitrogen bluish or purple, accompanied by quite audible sound; in carbon dioxide greenish; in coal gas a highly variable color—sometimes green, sometimes red or even a mixture of the two, and sometimes showing distinct black bands [3-33]. Back in London, Charles Wheatstone of Kings College concentrated on *brush* discharges observed at the end of a singly charged conductor, in contradistinction to spark discharges between conductors. He showed that the brush discharge is not, as it appears to be, a continuous discharge to the surrounding medium, but a rapid succession of intermittent discharges [3-34].

Faraday picked up on Wheatstone's brush discharge, able to resolve, by particular motions of the eyeball, the brush visually into a succession of elementary brushes between the conductor and the surrounding air particles. The next logical step in this particular inquiry was to study the behavior of the brush discharge in different gases and at reduced gas pressure. To this end, he had the instrument maker construct a glass vessel with cathode and anode of variable separation. When the separation was sufficient to thwart a spark discharge between the two electrodes, a brush discharge was observed at the anode. The color and shape of the discharge, and the ease with which it could be initiated, depended strongly on the particular gas utilized. As the pressure of the gas was reduced, the brush gradually disappeared and was replaced by a more or less steady glow extending from the anode to the cathode. When the pressure reached a sufficiently low value, the glow was interrupted roughly halfway between the electrodes by a dark region which today is known as the 'Faraday dark space' [3-35].

This was as far as Faraday got in his study of the electrical discharge *in vacuo*, and about as far as he could be expected to advance with vacuum equipment then available. In addition, he carried an increasing intellectual burden in several other areas: he had his regular lectures and routine administrative duties at the Institution, served a part-time appointment as lecturer at the Royal Military Academy at Woolwich and as Advisor to Trinity House, was a member of the Scientific Advising Committee to the Admiralty, and acted as consultant to various private and public parties. During 1838–1839, he began to feel the strain of overwork, complaining of a loss of memory. In the summer of 1840, he suffered a serious breakdown and for over a year he ceased work entirely. Although he resumed his regular Institutional duties at the end of 1841, only in 1844 did he return full-time to research. At that point, he turned his attention to entirely new subjects, the action of magnets on light and diamagnetism. To be sure, these researches had repercussions as profound as any of his other investigations; indeed, they planted the seed for Clerk Maxwell's electromagnetic theory of light. However, with his newfound devotion Faraday never studiously returned to the nature of the gaseous discharge. Though he performed certain additional experiments on the subject in the 1850s with the Rühmkorff coil and Geissler tube then available, he never wrote anything for publication.

3.3. Plücker's light and Hittorf's shadow

The next serious experimental advance in characterizing the electrical discharge in rarefied gases, following Faraday with his negative glow and his 'dark space', was made fully two decades later. And it was made by a mathematician at that—hence, by an early forerunner of J J Thomson, the mathematician turned experimenter. Julius Plücker had graduated from the Gymnasium in Düsseldorf, after which he studied, in turn, at the universities of Bonn, Heidelberg, Berlin, and Paris. As if these were not enough universities, in 1824 he earned his doctorate from the University of Warburg (to be sure, *in absentia*) [3-36]. Not yet ready to settle down, he then rose from *Privatdozent* to extraordinary professor (*extraordinarius*) at Bonn, continued as extraordinary professor at Berlin while also teaching at the Friedrich Wilhelm Gymnasium there, and served briefly as ordinary professor (*ordinarius*) at Halle. He finally did find a permanent niche at the University of Bonn, as full professor of mathematics from 1836 to 1847, and of physics from 1847 to 1868.

Plücker's mathematical expertise lay in analytic geometry, his pioneering contributions involving algebraic line geometry that he utilized in describing different shapes of cubic surfaces and other figures; in the process, he created a new subfield of geometry occupying numerous researchers after him.

In 1847 he turned to physics—curiously, to experimental, not theoretical physics. (He did have one last fling in mathematics late in life, in the company of Felix Klein, the Göttingen mathematician.) For a starter, he investigated the anomalous magnetic behavior of uniaxial crystals in a magnetic field, an investigation that proved highly opportune for Faraday in his ongoing researches in diamagnetism just then. Plücker had communicated his preliminary results via letter to Faraday in the fall of 1847, but Faraday, unable to read German, had ignored it. However, Faraday's interest in Plücker's work was awakened when Plücker called on him at the Royal Institution the following summer while in England to attend the meeting of the British Association at Swansea. Their meeting provided Faraday with the clue to the anomalous magnetic behavior of bismuth which had puzzled him for some time, and led to a lengthy correspondence between the two [3-37]. Before long, Plücker took up a second subject also of major interest to Faraday, his scientific role model: the electrical discharge *in vacuo*. Not much more had been learned about this phenomenon since Faraday had tackled it in 1838. Faraday, as it happened, had resumed his own researches on this subject at about the same time that Plücker had started, but not with his earlier energy. Still, he wrote Plücker as follows:

> . . . Then again the question of transmission of the discharge across a perfect vacuum or whether a vacuum exists or not ? is to me a continual thought and seems to be connected with the hypothesis of the ether. What a pity one cannot get hold of these points by some *experiments* more close and searching than any we have yet devised. [3-38]

The chief reason for the hiatus in 'close and searching experiments', in fact, was the inadequacy of Hauksbee's vacuum pump—a device basically unaltered for 150 years. Equally lackluster in performance was the vacuum tube itself then in use—the so-called 'electrical egg'. The egg, a glass bulb with a negative and positive electrode protruding through the wall, crafted by Rühmkorff and Jean Quet, would shortly be replaced by a tube of superior design. Rühmkorff's induction coil (which we have encountered earlier), on the other hand, was something else; it would prove indispensable in the experiments now under way.

Plücker owed his forthcoming success in the electrical discharge experiments in large measure to his instrument maker, the skilled glassblower and mechanic Johann Heinrich Wilhelm Geissler [3-39]. Geissler descended from a long line of craftsmen in the Thüringer Wald and in Böhmen. He learned the art of glassblowing in the duchy of Saxe-Meiningen, following which he pursued his vocation at several German universities; he also spent some years in The Hague. He finally settled down as an instrument maker in a workshop of his own at the University of Bonn in 1852. Here he produced a succession of first rate chemical and physical instruments for various Bonn faculty members and stocked the university's 'physical cabinet'. Being equally adept in grasping the underlying principles of his instruments, he soon made himself an indispensable participant in the experiments utilizing them. He was awarded an honorary doctorate in 1868 on the occasion of the 50th Anniversary of the University of Bonn [3-40].

Geissler's first collaboration with Plücker, and his first known activity at Bonn, involved the construction of a high-precision, standard thermometer in 1852. His celebrated 'Geissler tubes', as Plücker called them, date from 1857. Basically, they were thin-walled glass tubes of various shapes; generally cylindrical and often widened in the middle in the form of an ellipsoid and/or provided with bulbous ends. Electrodes (usually of platinum) were fused directly into the glass wall at either end (not simply sealed with plugs, which was one reason for the poor vacuum attainable in earlier tubes). They were filled with rarefied elementary or compound gases or vapors, such as air or vapors of æthereal oil, hydrogen, or phosphorus. Many were dispatched by Geissler to Rühmkorff in Paris, replacing his simpler egg, and to Henry Bence Jones in London, then Secretary of the Royal Institution. Before long, few university physical cabinets were without these tubes. Some of them can still be admired in the historical cabinet of the Physics Institute on the Nusallee in Bonn.

Two additional pieces of apparatus were needed for producing conditions favorable to the electrical discharge: a source of high-voltage electrical current and a vacuum pump better than Hauksbee's. Prior to the mid-1830s, the only way of generating an electrical current of high voltage was by means of bulky batteries or by ponderously cranking a friction machine. However, during 1836–1837 an early version of the induction coil was devised independently by Nicholas J Callan, a priest and physics teacher at Maynooth College in County Kildare, Ireland, and by Charles Grafton Page, a physician with a bent for electrical experimentation in Salem, MA [3-41]. The induction coil is essentially an iron core wrapped with an inner primary, insulated winding of a few heavy turns and an outer secondary winding of many more fine

Figure 3.3. *Rühmkorff induction coil, for powering discharge tubes.*

turns. Periodic interruption of a direct current in the primary winding (by a manually cranked contact breaker in the early coils) induces (by the collapsing magnetic field) an alternating current of high voltage in the secondary. The more secondary turns, the higher the voltage.

Features of the induction coil (mainly iron core, coil construction, and contact breaker) were gradually refined during the 1840s and, like Geissler's tube somewhat later, it became a staple item in nearly every physical cabinet. Its definitive form took shape in the 1850s, principally in the hands of Heinrich Daniel Rühmkorff. After serving an apprenticeship to a German mechanic in Hannover, Rühmkorff had spent a few years knocking about Germany and Paris before, in 1824, obtaining a position in England with Joseph Brahmah, reputed to be the inventor of the hydraulic press. He returned to the Continent in 1827, soon settling in Paris. Here, he was in due course hired by Charles Chevalier, the well known maker of optical instruments. In 1855, he established his own shop in Paris. His coils, one of which is shown in figure 3.3, brought him much fame, including (in 1864) a 50 000 franc prize established by Emperor Napoléon III for the most important discovery in the application of electricity [3-42].

The problem of obtaining a better vacuum was solved by Geissler in his own pump, one utilizing a mercury column instead of pistons as in Hauksbee's pump. Developed in 1855, it relied on the principle of the barometer discovered by Evangelista Torricelli in 1644 (even before Guericke) and, consequently, needed no wet gaskets, the limitation in the Hauksbee–Guericke method. Its essential element was a bulb connected by a flexible tube to an open reservoir of mercury. The bulb was provided with a tap whereby it could be connected either to the external air or to the vessel to be evacuated [3-43]. With the tap turned so as to admit air, the reservoir was raised until all the air had been expelled from the bulb; the tap was then turned to make connection with the vessel, and the reservoir was lowered. The process was repeated again and again; it was slow but effective, and good for a vacuum of about 1 mm Hg.

Plücker's first experiments with Geissler's tubes were performed in early 1857, and communicated to the Rheinische Gesellschaft der Naturforscher und Aerzte that July, as reported in *Könische Zeitung* for July 22 1857 [3-44]. Specifically, he was interested in ascertaining what would happen to the gaseous discharge with the tube mounted between the two poles of a magnet. That the stream of electric current would be deflected in some manner by the magnetic field could not be doubted. After all, had not Davy observed that his luminous arc (p 48), drawn between carbon points by means of a powerful battery, was diverted by a magnet? Such a diversion could also be foretold from Ørsted's discovery of the mutual relationship between a magnetic field and an electric current. However,

> . . . on the actual performance of the experiment, in addition to the phaenomena which were looked for, certain unexpected ones presented themselves; namely, the division of the light-stream, its decomposition at the negative electrode into an undulating flickering light, and the extension of the stream from the positive electrode into a brilliantly illuminating fine point. [3-45]

Plücker dwells in great detail and discourses elegantly on the delicate and beautiful hues exhibited by Geissler tubes filled with different residual gases, and on the stratified light and the light enveloping the negative electrode—the two separated by the (Crookes) dark space. After this introduction, he reaches the crux of the matter: the effect of a magnetic field on these already highly complex phenomena. His Geissler tubes were typically 250 mm long, 10 mm wide and provided with bulbs at each end enclosing the sealed-in electrodes. The magnet was a large upright horseshoe magnet, to the two limbs of which were fastened two heavy armatures facing each other with conical pole pieces forming a short gap. In some of the experiments, a brass keeper was inserted in the gap and the Geissler tubes were attached to the armatures. In some runs, the tubes were mounted such that the bulb enclosing the cathode was situated in the poletip gap.

Prior to exciting the magnet 'diffuse violet light was spread through the bulb, surrounded by a pale green light, which appeared to form a thin coating immediately upon the surface of the glass bulb' [3-46].

> After the magnet was excited, the diffuse violet light collected to a horizontal, semilune-shaped, bright and uniformly luminous disc, bounded towards the tube by an almost circular well-defined concave arch, whose middle passed through the point of the platinum electrode. This disc was upon the opposite side enveloped by a narrow strip of beautiful bright green light, which

> followed the curvature of the glass. Moreover, the light (according to what has been already described), which was red at a distance from the pole, but became bright violet towards the pole, and showed the dark intervals beautifully, descended towards the bottom of the bulb on entering it from the tube, and terminated in a point. [3-47]

Apparently, then, the fluorescence, 'the beautiful green light, whose appearance is so enigmatic' [3-48], was not uniformly spread over the wall of the tube, as is usually the case in a modern discharge tube when the Crookes dark space attains a maximum. Instead, probably due to some quirk in tube construction in Plücker's experiments, it was concentrated in patches near the cathode. But for this fortuitous quirk, Plücker would not have discovered that the position and shape of the fluorescent patches are altered by a magnetic field [3-49]. Equally interesting, though, were his observations on reducing the cathode to a fine point.

> On leading an electrical discharge through a Geissler's tube provided at its extremities with bulbs to the centres of which the electrodes penetrate, such magnetic light radiates from every point of the negative electrode, and spreads through the whole interior of the bulb containing this electrode. If all the negative electrode except its extremity be isolated by a fused coating of glass, then the above radiation is confined to the free point. The rays proceeding from this point collect in one single line of light, which coincides with the magnetic curve passing through the end of the negative electrode, and which by its luminosity renders such magnetic curve visible.
>
> Thus every ray which is bent in this magnetic curve, forming a portion of the arc of light, behaves exactly as if it consisted of little magnetic elements placed with their attracting poles in contact. In other words, such a ray behaves as a magnetic thread of perfect suppleness, and which accordingly, on being rigidly held in any one point (the extremity of the negative electrode), assumes the form of the magnetic curve passing through this point, or, what comes to the same thing, as an electrical current twisted in an infinitely thin spiral. [3-50]

All this is highly qualitative, surprisingly so coming from an experimentalist with a flair for mathematics. (Plücker's propensity as a geometer is perhaps revealed in his preoccupation with the complex geometrical configurations of the luminous and dark regions of the discharge and of the 'equipolar magnetic surfaces' in a magnetic field.) Moreover one of his few quantitative findings led him to a quite erroneous conclusion. He observed that, when the cathode was of platinum, 'small particles of platinum [were] generally torn off . . . and deposited upon the internal surface of the glass bulb enclosing the electrode'. Consequently, 'this bulb becomes gradually blackened, and after a long-continued action, the bulb . . . becomes coated internally with a beautiful metallic mirror' [3-51]. Chemical analysis did indeed reveal traces of platinum in the deposit, from which he surmised that 'it is clearly most natural to imagine the magnetic light to be formed by the incandescence of these platinum particles as they are torn from the negative electrode' [3-52]. True enough, considerable sputtering takes place in the course of the discharge, something first observed by W R Grove in 1852 (who was the first in England to employ a Rühmkorff coil in the production of a glow discharge), but this is a highly secondary phenomenon. Anticipating subsequent developments relative to this point, measurements would eventually show that the properties of the cathode rays are largely independent of the nature of the cathode material, and that the sputtering is caused by positive ions of the discharge striking the cathode and ejecting cathode material.

Plücker's lasting contributions lay undoubtedly in his mathematics. Nevertheless, he made his mark in several other areas of physics as well, including magnetic phenomena in tourmaline crystals and particularly in spectroscopy. (He anticipated Bunsen and Kirchhoff in the discovery of the first three Balmer lines in hydrogen.) Paid scant recognition by his countrymen, he was honored by the British who awarded him the Copley Medal—the highest award of the Royal Society—in the last year of his life, 1868. (Ironically, that was the year, we recall, in which Geissler *was* honored by the University of Bonn.) Still, the last word had not been heard on the electrical discharge. The next significant discovery was made within a year of Plücker's death by Johann Hittorf, a student of his.

Johann Wilhelm Hittorf was born in Bonn in 1824. After studying in Bonn and Berlin, he obtained his doctorate in Bonn in 1846. The next year he accepted an offer to become *Privatdozent* at the Royal Academy of Münster in Prussia. In 1852, when the academy became a university, he was appointed professor of chemistry and physics there. In 1876, a re-organization at the university allowed him to drop chemistry, and he stayed on as professor of physics until his retirement in 1890.

At Münster Hittorf undertook, between 1853 and 1859, a long series of electrolytic investigations of ions under an electric field. His studies of the changes in concentration of electrolyzed solutions and of the relative carrying capacities of different ions were extended some years later by F W Kohlrausch, and would eventually influence S A Arrhenius in developing (in 1865) his electrolytic dissociation theory. Subsequently, Hittorf collaborated with Plücker in the discovery of both band spectra and bright line spectra in the gaseous discharge at low pressures [3-53].

In 1869, Hittorf took up the electrical discharge in rarefied gases on his own [3-54]. He found not only that the thickness of the glow sheathing the cathode grows rapidly with diminishing gas pressure, but also that, for a given pressure, the glow stretches out further from the cathode as its surface is reduced; at the same time the fluorescence of the glass gradually becomes more vivid. This is most noticeable when the cathode, in the form of a wire, is drawn down to essentially a point cross section. But his most important observation was the following:

> Any solid or fluid body, whether an insulator or a conductor, which is placed in front of the cathode cuts off the glow, which lies between it and the cathode; no bendings out of straight lines occur.
>
> ... We see most clearly the rectilinear transmission of the glow if it goes out from a point cathode and, through a great length of the tube, brings the surface of the glass to fluorescence. If, in such conditions, any object is set in the space filled with the glow, it casts a sharp shadow on the fluorescing wall by cutting off from its cone of light which goes out from the cathode as an origin. [3-55]

The fact that the negative glow consists of rays that spread out conically from source points on the cathode surface, to strike the surrounding glass wall and causing it to fluoresce, was really nothing new, but something Plücker had essentially observed with his platinum wire 'warmth-pole' (cathode) 'covered with glass for isolation' into single points [3-56]. What was new was the shadow-casting properties of Hittorf's glow rays ('Glimmstrahlen') and, by implication, their rectilinear propagation.

Seven years later, Hittorf's discovery was followed up by Eugen Goldstein who was then employed at the University of Berlin under Helmholtz. Goldstein, born in 1850 at Gleiwitz, now Gliwice in Poland, attended Ratibor Gymnasium, after which he studied for a year at the University and then joined Helmholtz in Berlin in 1871 [3-57]. That was the year Helmholtz succeeded Gustav Magnus in the prestigious chair of physics in Berlin. Helmholtz and Gustav R Kirchhoff had been the principal candidates for the chair, left vacant on Magnus' death in 1870. The Berlin philosophical faculty opted for Kirchhoff as the better teacher of the two, but when he declined, unsure of his health and quite comfortable at Heidelberg, the chair went to Helmholtz. On his insistence the appointment was coupled with authorization for the construction of a new physics institute in the bargain, including instrument collection, and one under his sole control at that. The Prussian ministry of culture concurred, though building the new institute had to be put off because of the Franco–Prussian War that began just then. Still, Helmholtz was sufficiently convinced that the institute would be speedily constructed once conditions returned to normal, and accepted the appointment. The institute finally opened in 1879 (and lasted until

1945 when it was destroyed in the waning days of the Second World War during the battle for Berlin).

Goldstein received his doctorate under Helmholtz in 1881. He was destined to spend most of his long professional career as a physicist at the Potsdam Observatory outside Berlin, becoming head of its astrophysical section at the ripe age of 77 in 1927.

In his experiments of 1876, Golstein placed a small object in front of a cathode of considerable area, in contrast to Hittorf's point cathode. Also in this case a shadow was cast on the wall of the discharge tube, though not as sharply defined [3-58]. Moreover, he demonstrated that the rays, for which he now coined the term *'Kathodenstrahlen'* [3-59], are not emitted in all directions from portions of the cathode surface, but only in directions normal to the surface [3-60]. Despite the apparent distinction thereby implied between the manner in which the cathode rays are emitted from the cathode surface and that of light from the surface of a filament, Goldstein, Hittorf, and most of their German colleagues were persuaded from their shadow-casting properties, whether from a point source or from extended objects, that the rays consisted of some kind of electromagnetic wave in the æther. Goldstein would have much more to contribute on cathode rays in the years to come; the subject would occupy him for essentially all of his career. However, with his and Hittorf's aforesaid papers the seed was sown for the great forthcoming debate over the nature of the rays—a classic debate in the annals of physics, waged initially between the German professional physicists and a waning clique of English amateur physicists of the old school [3-61].

Chapter 4

THE ENGLISH GET GOING

4.1. The Victorian amateurs

The English did not remain idle during the pathbreaking German experiments. In contrast to their professional German colleagues, they tended to be members of a lingering class of amateur scientists constituting one segment of the late-Victorian scientific establishment: in short, men pursuing experimental physics as a hobby in their own private laboratories and at their own expense. They were the last generation of a peculiarly British old-style genre: some with university backgrounds but schooled in areas other than experimental science (perhaps law, medicine, often pure mathematics); some inheritors of commercial enterprises; some landowners and country gentlemen. To them, natural science was not necessarily best pursued as a vocation, but as an avocation. Four of these gentlemen, ardent students of the electrical discharge as well, warrant noting in our chronicle. They are: Warren De La Rue, head of a prominent stationery-manufacturing firm; William Spottiswoode, head of an equally prosperous printing business; J Fletcher Moulton, barrister; and John Peter Gassiot, wine merchant.

De La Rue was born in Guernsey, the Channel Island, in 1815; he was educated in Paris and studied science privately. Though he entered his father's stationery business at an early age, a budding interest in electrochemistry soon gave way to astronomy. Before long an amateur astronomer of some renown, he became prolific in astrophotography, specializing in lunar photography. Later he invented the photoheliograph which he put to good use taking remarkable telescopic photographs of solar prominences during a total eclipse. In 1873 he retired from astronomy, devoting himself to the family business and turning to physics. Collaborating with Hugo Müller in experiments on the striated discharge, De La Rue's photographic prowess resulted in photographic plates of the discharge of exceptional beauty—images widely disseminated. His Friday Evening Discourse at the Royal Institution on these experiments 'was on a heroic scale' [4-1]. Preparation for the experiments consumed 9 months, and the battery enlisted utilized 18 000 cells.

Spottiswoode was born in London in 1825, educated first at Harrow and then at Balliol where he lectured in mathematics. In 1846, he succeeded his father as head of the printing house of Eyre and Spottiswoode. He also found time for original work in polarization, and wrote a mathematical treatise on determinants. Mainly, though, his scientific forte lay in his collaboration on the electrical discharge with J Fletcher Moulton, Senior Wrangler then preparing for admission to the Bar. The fruit of their teamwork was two 'very long and soporific' papers in the *Philosophical Transactions* for 1879 and 1880 [4-2]. Despite his subsequent brilliant work as a barrister, perhaps Moulton's greatest societal contribution would be the four years he devoted with singleness of purpose in organizing the supply of high explosives on a grand scale to British troops during World War I.

The starting point for the present round of English experiments had been a rather casual paper in 1852 by William R Grove, an earlier barrister also turned physicist [4-3], who followed up on Faraday's still earlier work on the electrical discharge ca. 1838. Grove reported on 'the dark space in the discharge to which Faraday has called attention', finding it far from devoid of subtle detail.

> In a well-exhausted receiver containing a small piece of phosphorus, the discharge is throughout its course striated by transverse non-luminous bands, presenting a very beautiful effect, and a yellow deposit, which, as far as I have yet examined it seems to be allotropic phosphorus, is deposited on the plate of the air-pump and on the neighbouring substances... [4-4]

An explanation of the discharge was not long in coming in terms of Faraday's model of electrolytic decomposition.

> Faraday observed, Experimental Researches, 1164, 'in an electrolyte induction is the first state, and decomposition the second'. My present experiments show, I believe, that in induction across gaseous dielectrics there is a commencement, so to speak, or decomposition, a polar arrangement not merely of the molecules, irrespective of their chemical characters, but a chemical alternation of their forces, the electro-negative element being determined or directed, though *not travelling* in one direction, and the electro-positive in the opposite direction. [4-5]

As noted in chapter 3, Faraday himself had taken renewed interest in the gaseous discharge in 1856, now in the company of John Peter Gassiot and for much the same reason as Plücker—the availability of Geissler's tube and Rühmkorff's coil. In addition, however, reconciling the discharge phenomena then being disentangled, first by Grove and then by the Continental electricians, with his own theory of electrical action became for him a matter of some importance, if not his highest scientific priority in these years.

The actual experiments were conducted by Gassiot, a wealthy wine merchant and at the time Vice-President of the Royal Society [4-6]. Gassiot began his investigations of Voltaic electricity in the late 1830s, becoming highly adept at drawing sparks from banks of Leiden jars or batteries equipped with circuit breakers and transformers. His interest in the glow discharge was stimulated by Grove's paper of 1852. The discharge tubes initially utilized by him were prepared under the supervision of a Mr Casella in the laboratory of the Royal Society, and a Rühmkorff coil for powering them was constructed for him by a Mr Ladd. Shortly after presenting his first account of his experiments on this subject to the Royal Society, he learned of Geissler's superior tubes and acquired several of them without delay from Geissler through the intermediacy of Bence Jones, Secretary of the Society. Initially, he showed that both a friction machine and a Rühmkorff coil powered with a Grove cell produced a striated discharge, and that 'the striations are very powerfully affected by a magnet'.

> When the discharge is made from wire to wire,... if a horseshoe magnet is passed along the tube so as alternately to present the poles to differently contiguous positions of the discharge, it will assume the form of ~ in consequence of its tendency to rotate round the poles in opposite direction, as the magnet in this position is moved up and down the side of the tube. [4-7]
> If [moreover] the tube is placed in a horizontal position *over the pole of a magnet*, the stratifications evince a tendency to rotate as a whole in the direction of the well-known law of magnetic rotation; but when the discharge is made from [tinfoil] coating, or from one wire to one coating, an entirely new phenomenon arises: the stratifications have no longer a tendency to rotate as a whole, but are *divided*. [4-8]

Gassiot was not above showmanship as well, using all sorts of arrangements of Geissler tubes mounted together on a stand, twirling them around as in 'Gassiot's revolving star' and, generally, enchanting his captive audience with luminous displays. Among those enchanted was his friend and colleague Faraday. Visiting Gassiot's home laboratory on one occasion in 1858 to take in the striated bands, he confided in his diary: '*At Mr Gassiot's*. Saw his fine experiments on the luminous striae of the Electric discharge, especially with his tubes, and above all with the four tubes one over the other' [4-9].

During the next few years, Gassiot kept up his serious work on the gaseous discharge, concentrating on conditions favorable for producing the striations. At first he availed himself of Geissler's tubes, of which he possessed no fewer than 60 in number. For all their delicacy, they were not entirely to his liking, however.

> As I was desirous, during the progress of the experimental research I had entered on, to know the exact conditions under which each vacuum was obtained, and found, by comparison, that there was some uncertainty in the description of those I had received from Mr Geissler, I reluctantly laid them aside, and for all the experiments I have to describe in this communication, each tube was charged and exhausted by myself or in my presence. The tubes were constructed by Mr Casella . . . [4-10]

He showed that Rühmkorff's coil is not strictly necessary for producing the striations; a powerful battery will also do [4-11]. The striations were shown to exist within a narrower range of pressure and temperature than necessary for producing the discharge itself, and thus but a limiting feature of the luminous discharge [4-12] that is altered by changes in the electrical resistance of the external circuit. In a final round of experiments, Gassiot demonstrated that in a circuit containing two discharge tubes in series, if one discharge is interrupted by a magnet, the electrical current in both tubes is interrupted [4-13]. He also found that there is a mechanical disruption of the metal in the negative electrode [4-14], something already observed by Grove in 1852 [4-15]. In 1863, Gassiot received the Gold Medal of the Royal Society in recognition of his work on Voltaic electricity and on the gaseous discharge of electricity.

Despite these painstaking efforts, which provided marginal scientific insight, a conceptually more important advance by the English amateur clique was made by the colorful electrical engineer and spiritualist Cromwell Fleetwood Varley. Varley claimed to descend from Oliver Cromwell and his general, Charles Fleetwood. He was a telegraph engineer who had worked actively on the development of submarine cables and ran an engineering consulting practice in partnership with Lord Kelvin (who himself spent much of his time during 1856–1866 on laying of the Atlantic Cable). In early 1870, Varley found himself semi-retired with time on his hands, when the government in effect nationalized the private telegraph companies [4-16].

That summer, a year after Hittorf's discovery of the shadow-casting properties of the luminous rays spreading from the cathode, Varley too took up the electrical discharge in rarefied gases, admitting that

> after the labours of Mr Gassiot, one approaches this subject with diffidence, lest he should appear to be attempting to appropriate the glory which so justly belongs to that gentleman and to Professor Grove. The nature of the action inside the tube is at present involved in considerable mystery, but some light is thrown upon the subject by the following experiments. [4-17]

Plücker, Varley reminds us, had shown that when a current-carrying exhausted tube of Geissler's design was subjected to a magnetic field, the luminous glow could be distorted into an arch coinciding with the course of the magnetic rays threading the tube. To further ascertain the nature of this arch, Varley inserted into his Geissler tube 'a piece of talc, bent into the form U, [with] a fibre of silk stretched across it; on this fibre of silk was cemented a thin strip of talc, 1 inch in length, 1/10 inch broad, weighing about 1/10 of a grain' [4-18]. The tube was sealed and exhausted, with carbonic acid and potash used to produce the high vacuum. The battery utilized was a Daniell battery, each cell of which had a resistance of 50–100 Ω; a U-shaped glass tube containing glycerine and water, inserted into the external circuit, served as a variable resistance.

> When the magnet was not magnetized, the passage of the current from wire to wire did not affect the piece of talc. When the magnet was charged, and the luminous arch was made to play upon the lower portion of the talc, it repelled it, no matter which way the electric current was passing.
> When the tube was shifted over the poles of the magnet so as to project the luminous arch against the upper part of the talc, the upper end of the talc was repelled in all instances; the arch, when projected against the lower part of the talc, being near the magnet, was more concentrated, and the angle of deviation of the talc was as much as 20°. [4-19]

Not only that, Varley was not averse to speculating on the physical nature of the luminous cloud. He may be fairly said to be the first to explicitly suggest that the rays are corpuscular streams, and that it is by virtue of their negative electrical charge that the particles are deflected by a magnetic field.

> This experiment, in the author's opinion, indicates that this arch is composed of *attenuated particles of matter projected* from the negative pole by electricity in all directions, but that the magnet controls their course, and these particles seem to be thrown by momentum on each side of the negative pole, beyond the limit of the electric current. [4-20]

The cylindrical arch from the cathode, on being projected against the 'little talc tell-tale', 'cut out the light, and a corresponding dark space existed throughout the remainder of the course of the arch'. Where the arch struck the talc, a small luminous cloud appeared, 'as though the attenuated luminous vapour were condensed by this material obstruction' [4-21].

Not to be minimized in the ongoing researches were further improvements on Geissler's pump by two other German instrument

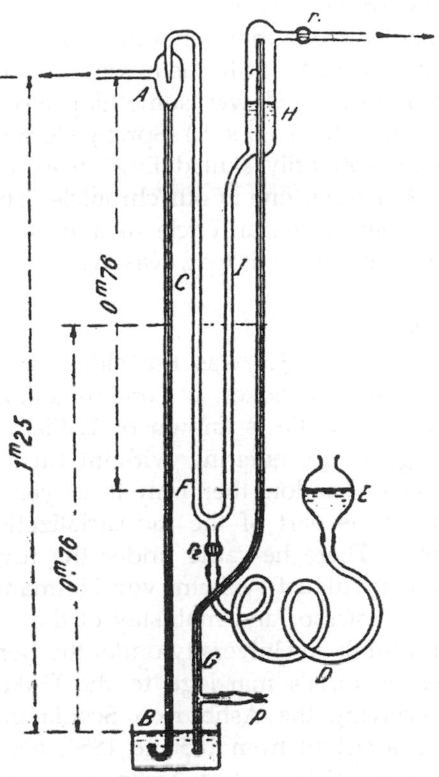

Figure 4.1. *Sprengel's vacuum pump. Mercury droplets, acting as pistons, drop in the tube AB. The right-hand apparatus is a system for restoring the mercury level.*

makers. The first was Töpler, about whom little else seems to be known. In 1862 he substituted for Geissler's tap (p 52) an arrangement of tubes; they provided passage for the air to be expelled. In addition, the vessel to be evacuated was automatically connected to the Torricellian vacuum at the appropriate position of the mercury level [4-22]. The other German was Hermann Johann Philipp Sprengel, an instrument maker with a background in chemistry who had spent some time under Bunsen in Heidelberg. As a young man, he gravitated to London and Oxford to further his technical career and pass on the experimental techniques developed under Bunsen, and he wound up a permanent English resident [4-23]. Among his other contributions, he too devised, in 1865 while engaged in the laboratory of London's short-lived Royal College of Chemistry, an improved mercury pump—one based on the continuous application of Torricelli's method of evacuation. In it (figure 4.1), the mercury was raised automatically to a reservoir from which droplets of mercury dripped steadily. They acted as pistons, trapping the air in the discharge tube and carrying it away.

These pumps enabled the routine generation of vacua in the 10^{-3}–10^{-4} Torr range—pressures verifiable with the invention of the McLeod gauge in 1874. (Their pumping *speed* was something else; Sprengel–Töpler pumps did hardly better than 1 l s^{-1}.) Sprengel's pump, moreover, in the hands of an extraordinarily skilled English experimentalist, made possible the next major milestone in our chronicle. That individual, the grand master of our late-Victorian circle of amateur scientists and an uncommonly interesting person at that, was William Crookes.

4.2. William Crookes

William Crookes, born in 1832, was the eldest son of no less than sixteen children of Joseph Crookes, a London tailor, and his second wife, Mary Scott [4-24]. Little is known of William's childhood. He left school at the age of 15, entering without further ado the Royal College of Chemistry in London, then only three years old and opened by the Prince Consort as part of the industrialization under way in England at the time. There he came under the tutelage of the well known organic chemist August Wilhelm von Hofmann, former student of Justus Liebig and Professor and mainstay of the college. (Hofmann had been imported from Bonn University under the Germanophile policy surrounding Queen Victoria's marriage to the Duke of Saxe-Coburg in 1840.) After receiving the Ashburton Scholarship, he served as Hofmann's personal assistant from 1850 to 1854, receiving a thorough grounding in experimental laboratory techniques under his guidance. Though ignorant in mathematics and without any formal schooling in physics, he picked up such physics as was needed for his modest researches in *inorganic* chemistry (in contradistinction to his fellow pupils who adhered to organic chemistry in deference to Hofmann). Crookes' lack of mathematics, his consummate experimental skill and intuition, superb organization, and great lecturing ability, make him an obvious ringer for Faraday; indeed, there are grounds for believing Crookes consciously modeled himself on him [4-25], somewhat like Plücker. Faraday, in fact, significantly abetted Crookes' early career by introducing him to Charles Wheatstone and George Gabriel Stokes.

Stokes took great interest in the budding scientist, whom he was able to steer away from traditional chemical problems into physical chemistry. The two became lifelong friends. Crookes' ignorance in mathematical physics would often be obscured by Stokes' helpful advice on theoretical questions and kind assistance in solving particular mathematical problems that cropped up on occasion [4-26]. Wheatstone's influence was more immediate; namely, in securing a position for Crookes in 1854 as superintendent of the Meteorological Department of the Radcliffe Observatory at Oxford. The next year he taught chemistry at the College of Science at Chester, and in 1856, on inheriting a fairly

Figure 4.2. *Sir William Crookes in 1889.*

sizable legacy from the prudent savings and investments of his father, he settled in London. There he took up a chemical consultationship, using his private home laboratory on Mornington Road, and edited several photographic and scientific journals. The most important of these was *Chemical News*, which he founded in 1859 and remained its proprietor and editor until 1906. (Spottiswoode (p 59) was initially its printer.) He married Ellen Humphrey, who bore him ten children. He was knighted in 1897, received the Order of Merit in 1910, and served as President over the Royal Society from 1913 to 1915.

Crookes proved an amazingly prolific experimentalist. His researches ranged from photography and spectroscopy up to ca. 1870, including the discovery of thallium, to his radiometer and the electrical discharge in rarefied gases after 1874. He studied electrical lighting and a variety of problems in technical chemistry, including fractionation of the rare earths, textile dyeing, disinfectants, and diamond formation, and he was a spokesman for the need for producing fertilizers from atmospheric nitrogen. After 1900, still smarting from missing out in the discovery of x-rays due to a period in South Africa in 1895, where he had gone to furnish evidence in a patent litigation, he threw himself with all his energy into the fledgling science of radioactivity. He established the continuous formation of uranium x from uranium, and invented the spinthariscope in 1903—a device which, suitably modified by Erich Regener, would be of decisive importance for Ernest Rutherford's work on atomic structure, based on a fluorescent screen for observing the scintillations of α-particles emitted by a radioactive substance.

Unlike Faraday, Crookes was not at all averse to making money in commercial endeavors, such as gold extraction by the sodium amalgamation process or the utilization of sewage as saleable manure through treatment with alum, blood, charcoal, and clay; these ventures, coupled with his legacy, enabled him to live comfortably, though he never made any great financial killing. As his reputation spread, and with it concomitant responsibilities to sundry scientific and other societies and causes, he was able to leave his strictly laboratory researches increasingly in the capable hands of assistants, notably Charles H Gimingham and James H Gardiner. At the same time, his publications ceased to be purely experimental, embracing more and more quasi-theoretical pronouncements and opinions; views seemingly plausible but often questionable and culled from notions of associates and prevailing schools of thought.

Superficially in the vein of Faraday the Sandemanian Christian, there was another side of Crookes' personality, one not out of line with the fairly common dual belief among Victorian scientists in both science and religion—indeed a common view from Newton onward. For Crookes, schooled in the religious belief in an afterlife, a critical period ensued not long after the death of a beloved brother at sea in 1867, when he was persuaded by Cromwell Fleetwood Varley to attend séances in an attempt to communicate with his brother. The upshot was that he devoted the period between his painstaking researches on thallium and on the radiometer, or from 1870 to 1874, mainly on investigations into what he called the 'psychic force' or spiritualism which excited Victorian society in these years. Though intolerant of the fraudulent practices frequently exposed in the production of psychic phenomena—tilting of tables, audible effects, and the like—he singled out a few mediums as genuine; in particular the Scottish–American Daniel Dunglas Home who electrified London audiences with his famous spiritualistic séances, and in broad daylight at that. Lady Crookes, caught up in this as well, is quoted to the effect that

> ...at these séances they had been sitting round a particular table for some days, then they put it aside and took another. The first table came out from its corner apparently to attack the other, which leaped on to the sofa, was pursued by the first and they had a fight there. Crookes, when appealed to, said he knew nothing about motives, but corroborated the facts [!]. [4-27]

Crookes subjected Home's psychic phenomena to a number of stringent tests, tests conducted in their specially modified home dining room. The simplest and most important of these tests was designed to verify that Home could alter the weight of a plank attached to a spring balance at one end and resting on a knife edge at the other end. Crookes communicated several papers on these experiments to the Royal Society, which were summarily rejected; however, he published essentially the same papers in his own *Quarterly Journal of Science* and *Chemical News*. They were sharply criticized, though Crookes countered with much vigor. Although he returned from metaphysics to physics rather abruptly in 1874 and threw himself into science with renewed energy, as if to atone for this dark period in his career, he retained his faith in the psychic force the rest of his life. Thus, after the death of Lady Crookes in 1916, he attempted to communicate with her through a medium. In his Presidential Address before the British Association at Bristol in 1898, he declared 'I have nothing to retract; I adhere to my already published statements. Indeed I might add much thereto' [4-28]. Despite Crookes' adamancy, however, Brock is probably correct in describing him, not as a spiritualist but as an occultist—'a man for whom traditional science left huge areas of creation unexplored' [4-29].

Crookes' introduction to cathode rays came by a circuitous route leading back to his early chemical researches on the selenocyanides under Hofmann ca. 1850, and to his work in photography in the ensuing decade inspired by Wheatstone. Concentrating on the spectral sensitivity of photographic collodion processes, Bunsen and Kirchhoff's breakthroughs in spectrum analysis in 1860 instilled in him an immediate and unbounded enthusiasm for spectroscopy. He soon had a spectroscope going, examining, among other things, his old selenium residues from 1850. In early March of 1861, he observed an anomalous bright green line in these residues that, by the end of the month, he identified with a new element. He named it 'thallium' after *thallos*, the Greek for a green twig. On invitation, he exhibited small samples of thallium at the International Exhibition in South Kensington in 1862, where to his chagrin he learned that the element had been simultaneously and independently isolated on a much larger scale by the French chemist C A Lamy. Not only that; when the list of awards and prizes at the Exhibition was made public, Lamy was found to have been awarded a medal for thallium, while Crookes was not even mentioned! Fortunately Hofmann intervened, as a result of which the award was revised, with Crookes receiving a medal for the *discovery* of thallium, and Lamy (who had observed the green line later than Crookes) a medal for its production as a metallic ingot. Nevertheless, a bitter controversy over priority in the discovery arose. Though Hofmann attempted to mollify the views of the protagonists, the feud dragged on, highlighted in the *Times*, the *Lancet*, and the *Athenaeum*; eventually it faded away. Still, the years 1862–1864 were not easy for

Crookes. Troubles with the *Chemical News* of various kinds, illness and anxiety at home, a reputation for an excessively quarrelsome stance as technical witness in courts of law—all conspired to keep him on edge. Happily thallium brought him a Fellowship in the Royal Society in 1863. (Among the signatories of the proposal for his admission were Gassiot and Faraday.)

Throughout the 1860s Crookes labored steadfastly on determining the atomic weight of his new element with the highest possible precision. In the course of this tedious investigation, utilizing a sensitive Oertling vacuum balance, he observed a weak dependence on the temperature of the bulb enclosing the thallium sample: a warmer sample appeared to be lighter than a colder one. Since his vacuum was rather good, disturbance of the scale by convection currents could be ruled out, as could condensation of vapor onto the colder body. Suspecting a link between heat and gravitation, perhaps a consequence of the 'psychic force' he himself was investigating as well just then, he launched a further investigation with the assistance of Gimingham (who had taken part in the Home séances). They found indications that a heavy metallic body attracted a small weight inside an evacuated tube when colder than the weight, and repelled it if warmer. Next, they suspended two pith balls on a pivoted horizontal rod in a glass tube attached to a mercury pump. Sure enough, when hot water flowed through a pipe in close proximity, the balls were repelled, while ice placed next to the tube attracted them. Since the effect was enhanced by lowering the residual pressure, Crookes concluded that the pressure of light was at work—an effect implied by the corpuscular theory of light, to be sure a theory then in disfavor, but predicted by Maxwell's electromagnetic theory. As it happened, Maxwell's *Treatise* [4-30] was published in 1873, the very year of Crookes' latest experiment. Thus both were working on the same problem, albeit Maxwell mathematically, Crookes purely empirically without the mathematical ability to follow Maxwell.

Not resting there, Crookes now modified his device into what he first called a 'Light Mill' and has gone down in history as the radiometer—his term as well [4-31]. A contraption that took scientists and the public by storm during 1874–1875, it contained a horizontal cross suspended on a steel point resting in a glass cup, enabling the cross to rotate freely. Four thin mica discs, blackened with lamp-black on one side and left bright on the other side, were attached to the extremity of each arm of the cross, the blackened sides all facing in the same direction. The device was mounted in a glass tube which was exhausted and sealed. When exposed to bright light the vanes rotated around the pivot in such a direction to suggest that the blackened sides were repelled or blown by the light. Much of the credit for its success was due to the indefatigable Gimingham, whose laboratory efforts had their bad days as well. Thus, a notebook entry of his for March 7 1876 reads as follows:

> Making radiometer for exhibition. German-silver arms answer well. Discs of rice pith.
> All went on well till putting in bulb when all came to grief. Put it in two bulbs. In the first it stuck to the side. Had to be taken out, unsoldered, put in another, when the cup took a piece out of the disc. The radiometer part is still good.
> This sort of thing makes me wretched. The more I tried the worse it got. I have tried hard to finish it, but find it impossible to-day. [4-32]

On seeing the note, Crookes added under it, 'Cheer up!'.

The radiometer became the rage of toymakers, shopkeepers, chemists and opticians, and the general public, and remains popular to this day. Its operating principle proved far from obvious, however, and became the subject of a considerable controversy involving Osborne Reynolds, Arthur Schuster, Peter Tait, Johnstone Stoney, and James Dewar. For one thing, it was soon clear that the impact from Maxwell's light pressure was far too feeble to account for the observed rotation. Moreover, the rotation from light pressure would be in the opposite direction. A photon impinging on the blackened surface would be absorbed and transfer its momentum to the vane, whereas it would be reflected and gain momentum from the bright surface equal and opposite to that it possessed before striking. Thus, reflection from the vane would double the momentum it would receive from absorbing a photon, and the bright surface would be repelled more than the dark surface.

Osborne Reynolds at Owens College, Manchester, among others, had another explanation: that the rotation was caused by residual molecules in the evacuated bulb. The dark surface would absorb more heat than the light one, and molecules striking it would rebound with greater velocity than from the colder light surface [4-33]. Stokes, too, hit on the alternate explanation, without downplaying the importance nevertheless of the radiometer.

> I confess that now the evidence to my mind is very strong, almost overwhelmingly strong, that the rotation of the radiometer is due to an action of heat between the radiometer (or more correctly the fly, the 'radiometer' being the instrument as a whole) and the case, through the intervention of the residual gas. But the action is none the less a perfectly new one. No one, so far as I know, had made the slightest approach to discovering it experimentally. No one had dreamt of its occurrence as a matter of theoretical prediction. And even now its theoretical explanation is not an application of well-ascertained laws, but the following out of a certain speculation as to the ultimate constitution of matter and the nature of heat; and your discovery, from the thorough novelty

of the action, cannot but exercise an important influence on the progress of our knowledge. [4-34]

A test to settle the matter was duly performed by Schuster, then a colleague of Reynolds at Manchester. The radiometer was suspended by a fine thread. If Crookes were right, the bulb should be carried around by the rotating vanes, whereas if Reynolds were right, by the principle of action and reaction the bulb should rotate in the direction opposite to that of the vanes. J J Thomson, then Reynolds' student, recalled the excitement and anxiety with which he awaited the verdict, and his relief on hearing that his teacher had been vindicated [4-35]. Crookes himself was quite unperturbed at the disproval of his own theory and eagerly embraced his opponent's—a personality trait that often served him well in science and in public life.

In 1878, Crookes decided to undertake a general investigation of 'lines of molecular pressure', suspecting 'that the dark space [separating the cathode from the cathode glow in Geissler's tubes] was in some way related to the layer of molecular pressure causing movement in the radiometer' [4-36]. To this effect the radiometer was modified into the 'electric radiometer'—a discharge tube in which the 'little windmill' acted as the cathode. The dark space was found to extend further from the blackened side of the vanes, but the vanes rotated only when the pressure in the tube was reduced to the point where the dark space (now 'Crookes' dark space', and not to be confused with Faraday's dark space) stretched all the way to the wall of the radiometer tube and there caused the glass to glow. This, indeed, suggested that the discharge was an illumination of the lines of molecular pressure. Moreover, since the thickness of the dark space grew as the pressure was reduced, the dark space could be construed to be simply a visible manifestation of the molecular mean free path.

Before long the radiometer was modified once more, as always in the capable hands of Gimingham—a first-rate glass blower and dextrous instrument maker who served Crookes as Geissler served Plücker. The modified tubes were, in fact, simply more efficient versions of Geissler's tubes, and now became known as 'Crookes' tubes'. With Sprengel's pump, also modified by Gimingham, the residual gas pressure slowly dropped (reaching a minimum value in the neighborhood of 3×10^{-3} mm Hg) and the dark space grew, stretching across the tube with the phosphorescent spot growing larger and more brilliant. The color of the spot depended on the kind of glass used. German glass produced a greenish-yellow spot, English lead glass a blue spot, and uranium glass a green one. A diamond became brilliantly blue. A superb showman, almost every year Crookes delighted his spellbound audience in the Royal Institution's packed lecture theatre, carrying on the Friday Evening Discourses initiated by Faraday.

In his Bakerian Lecture before the Royal Society in 1879, Crookes resurrected Faraday's notion of a fourth state of matter, radiant or ultra-gaseous matter, in which lines of molecular pressure were interpreted as streams of radiant matter (a term coined by Faraday) in a state so rarefied that 'the [molecular] free path may be made so long that the hits in a given time are negligible in comparison to the misses'. Moreover, if the mean free path becomes comparable to the dimensions of the vessel, 'the properties which constitute gaseity are reduced to a minimum, and the matter becomes exhalted to an ultra-gaseous or molecular state, in which the very decided but hitherto masked properties now under investigation come into play' [4-37].

> The phenomena in these exhausted tubes reveal to physical science a new world—a world where matter may exist in a fourth state, where the corpuscular theory of light may be true, and where light does not always move in straight lines, but where we can never enter, and with which we must be content to observe and experiment from the outside. [4-38]

Alice Bird, an old friend and literary adviser, was moved to write him as follows:

> My dear Mr Crookes,
> Of course you were discussed at a thousand breakfasts this morning, and as you could not hear us talk of your wonderful experiments and of the lovely way in which you acquitted yourself, George [her brother] says I must send you a line. He says whatever you may have suffered inwardly, your outward calm and self-possession were perfect. I hope you are not more thin to-day, although I daresay you lost a pound or two more last night. But I am so glad you are going to Brighton, and I hope on your return you will show yourself here in much better plight.
> I saw Ellen looking nervous, I thought, when she entered, but she must have been well pleased with your triumph.
> I feel I only understood in a glimmering way, you seemed to me like the magician of the Future before whom no secrets are hid.
> Love to you both, and best congratulations.
> Yours always,
> Lallah. [4-39]

Going even further, in his exultant lecture before the British Association at its annual meeting at Sheffield that August, Crookes declared

> In studying this Fourth state of Matter we seem at length to have within our grasp and obedient to our control the little indivisible particles which with good warrant are supposed to constitute the physical basis of the Universe... We have actually touched the border land where Matter and Force seem to merge into one

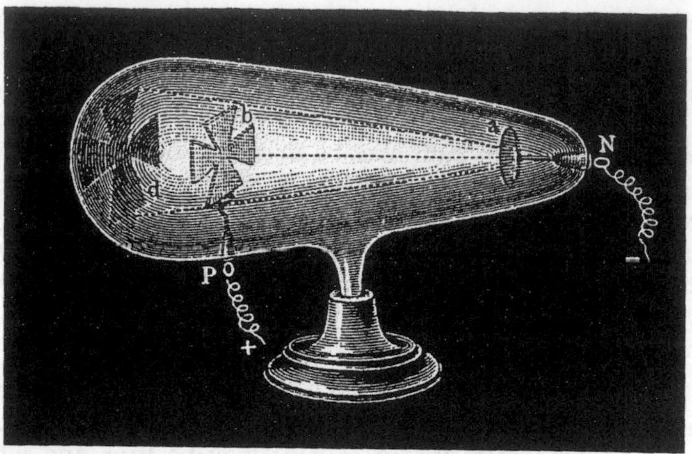

Figure 4.3. *Discharge tube with Maltese Cross, demonstrating rectilinear trajectories of cathode rays.*

> another, the shadowy realm between Known and Unknown which for me has always had peculiar temptations. I venture to think that the greatest scientific problems of the future will find their solution in this Border Land, and even beyond; here, it seems to me, lie Ultimate Realities, subtle, far-reaching, wonderful.
>
> Yet all these were, when no Man did them know, Yet have from wisest Ages hidden beene;
> And later Times things more unknowne shall show, Why then should witlesse Man so much misweene,
> That nothing is, but that which he hath seene. [4-40]

On a less lofty note, in the same Bakerian Lecture that year Crookes described the results of a series of quantitative experiments on cathode rays. He not only repeated with greater precision Hittorf's experiments ten years earlier, but also uncovered several new results [4-41]. For a starter, he employed a discharge tube in which Gimingham had mounted a Maltese cross of mica in the path of the ray, as in figure 4.3. The cross had been constructed with a hinge, allowing it to be positioned vertically or tilted in the horizontal plane. While the cathode was a small disc mounted on the axis of the tube, the anode was placed out of the way in a side branch of the tube. With the cross held vertically, it cast a sharply defined shadow on the end wall of the tube; only outside the shadow did the glass fluoresce. Whereas in moderately evacuated vacuum tubes the luminous rays would seek the anode through the most contorted paths introduced by curves or angles blown in the tube by a skilled glass blower such as Crookes or Gimingham, the present experiment showed that under significantly lower pressures the position of the anode was of no importance.

> The molecular ray which gives birth to green light absolutely refuses to turn a corner, and radiates from the negative pole in straight lines, casting strong and sharply-defined shadows of anything which happens to be in its path. In a U tube with poles at each end, one leg will be bright green and the other almost dark, the light being cut off sharply by the bend of the glass, a shadow being projected on the curvature. [4-42]

Another experiment was designed to demonstrate the mechanical effects to be expected from the molecules striking a solid surface, such as the glass wall of the tube. An egg-shaped tube (figure 4.4) was supplied with an anode *a* and cathode *b* in their normal relative positions. At *d* is a tiny 'radiometer fly' with clear mica vanes; it is supported on a glass cup and steel needle, and the whole small contraption is suspended from jointed glass fibers. Between the cathode and radiometer fly is an aluminum screen *e*, mounted slightly to one side of the axis *ab*. By tilting the tube, the radiometer fly could be fully exposed to the stream of 'projected molecules', or partially or entirely screened from the stream, and, indeed, the little fly behaved quite as expected [4-43]. When fully screened, no movement was seen. When half of it was exposed, it rotated rapidly—presumably under the impact of the rays striking one side only. When fully exposed, the rotation stopped; since no vanes were blackened in this particular apparatus, there was no tendency for the radiometer fly to turn one way or the other. If the other half of the fly was exposed, it rotated in the opposite direction.

In another version of this experiment, depicted in figure 4.5, the discharge tube had a little glass railway running along its length, with a radiometer fly traveling on it. The electrodes were mounted above the level of the rails, in order that the upper half of the fly would lie within the path of the rays. Under the influence of the rays the fly rolled merrily along the rails from the cathode end towards the anode. When the electrode connections were reversed, the fly rolled back again [4-44].

Here we may anticipate the thread of events by noting that, despite Crookes' precaution of *not* having alternate sides of the radiometer vanes blackened and polished, in due course it would be shown (by J J Thomson) that the momentum of the rays was quite insufficient to impart the observed motion. The cause of the movement of the radiometer fly in these experiments was, in fact, simply heating of one side of the vanes—ergo, the same effect responsible for the rotation of the original radiometer.

In the same bountiful sequence of experiments, Crookes also repeated Plücker's and Varley's experiments on the magnetic deflection of the cathode rays, though with greater experimental sophistication. First, an electromagnet *s* was placed beneath the anode end of the egg-shaped

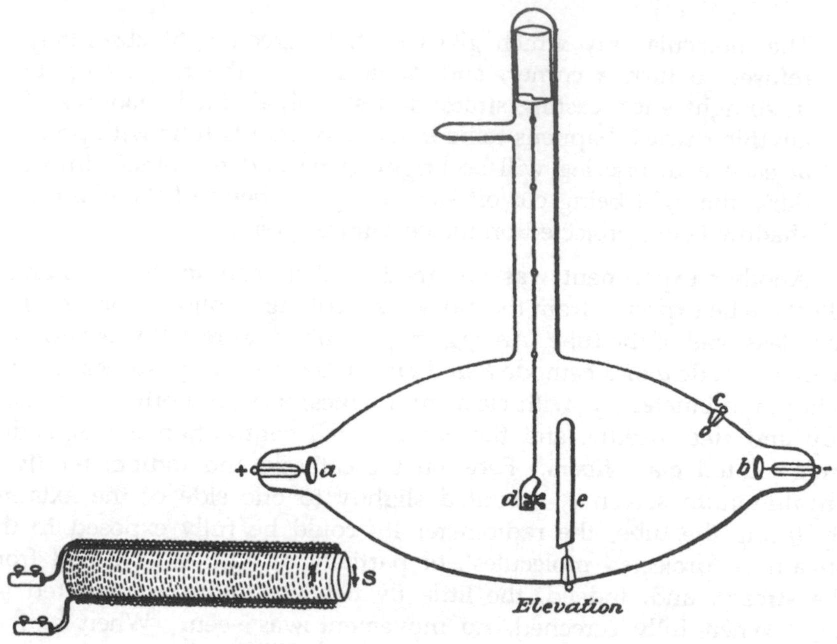

Figure 4.4. *Crookes' 'radiometer fly', purportedly demonstrating mechanical effects from 'Molecular pressure'.*

Figure 4.5. *Crookes' traveling radiometer fly.*

bulb used in the previous experiment (figure 4.4). With current flowing in the tube but without the electromagnet energized, the shadow of the screen *e* was projected as indicated by the cross-hatched region in figure 4.6, that shows a plan view of the tube. Exciting the electromagnet by gradually increasing the number of Grove's battery cells, the edge of the shadow was progressively deflected to positions 1, 2, 3 . . . .

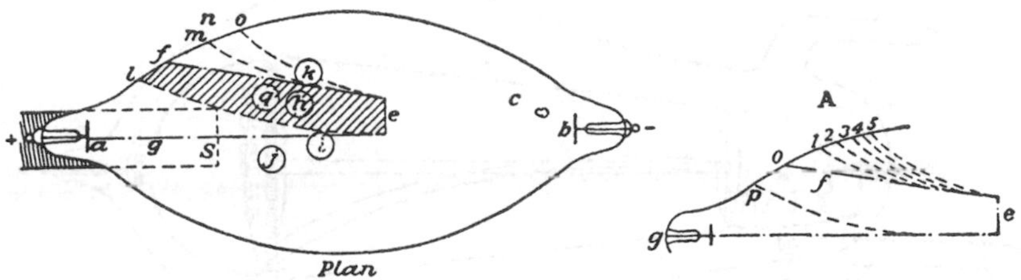

Figure 4.6. *Magnetic deflection by William Crookes.*

> I have spoken of shadows being deflected by the magnet as a convenient way of describing the phenomena observed; but it will be understood that what is really deflected is the path of the molecules driven from the negative pole and whose impact on the phosphorescent surface causes light. The shadows are the effect of a material obstacle in the way of the molecules. [4-45]

If the 'molecules' were projected from the cathode with different velocities under a constant magnetic deflection, would not the higher velocities show flatter trajectories? 'Though the deflecting force might be expected to increase as the velocity [of the molecules]', wrote Stokes, 'it would have to increase as the square of the velocity in order that the deflection should be as great at high as at low velocities' [4-46]. To verify this, Crookes next utilized the tube depicted in figure 4.7, one on which a ray was monitored instead of a shadow cast by it. Here *a* is the cathode with a notch cut in it, and *b* the anode. A mica screen pierced with a small hole is located at *c*. The glass plate *d* has a vertical scale engraved on it; it enabled a measurement of the deflection of a ray proceeding from *a* through *c* to *d*. The horizontal scale *e* indicated the variation in thickness of the dark space as the pressure was reduced. A vertical screen of mica in the plane of the movement of the rays, *ff*, was coated with a phosphorescent powder; it revealed the continuous path of a ray from *c* to *d*. A permanent magnet is also shown. Sure enough, with the magnet in place, the trajectory flattened with lower pressure, presumably indicating higher molecular velocities as a result of less frequent collisions.

That the rays, if indeed a molecular stream, should produce heating by bombardment, was verified somewhat painfully with another tube in which the cathode was a curved aluminum cup, focusing the ray against some spot on the glass wall. Touching the tube at that spot, Crookes immediately raised a blister.

The most important result of these experiments, however, was the direction of magnetic deflection, as shown in figure 4.7. With the permanent magnet in place, the rays were bent downward under the

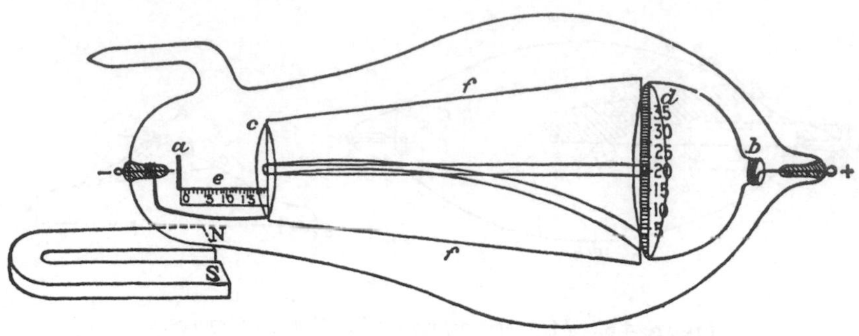

Figure 4.7. *Magnetic deflection as a function of velocity.*

influence of a particular polarity, and upward when the polarity was reversed. The direction of deflection was that expected from the force which, according to Faraday's rule, would be exerted by the magnetic field on a stream of molecules behaving as a perfectly flexible conductor joining the negative and positive poles. Such a current should, according to contemporary wisdom, consist of a flow of negative electricity from the negative cathode to the positive anode, and an equal flow of positive electricity in the opposite direction. Of the negative 'molecular torrent' there could be no doubt, but where was the positive counterflow? 'The extra velocity', Crookes argued, 'with which the molecules rebound from the excited negative pole keeps back the more slowly moving molecules which are advancing toward that pole. The conflict occurs at the boundary of the dark space, where the luminous margin bears witness to the energy of the collisions' [4-47].

Be that as it may, Crookes' shaky argument came under fire from different quarters the following year, 1880 [4-48]. That year Peter Tait of Edinburgh argued that, if the rays were indeed a torrent of particles moving with great velocity, it should be possible to measure their velocity by means of the Doppler effect exhibited by the luminous vibrations excited by them.

> The recent exhibition by Professor G Forbes of some of the latest of Mr Crookes' experiments, together with what I had read or heard about their explanation, led me to infer that I might determine directly the velocity of the luminous particles near the negative pole (and perhaps other parts) of a vacuum tube by means of observations of the spectrum made in directions perpendicular to, and parallel to, the lines of motion of the incandescent particles of gas.
>
> I made the attempt on some charcoal–bromine vacuum tubes, for which I have to thank Professor Dewar, but I found the light to be so feeble that it was impossible to employ an eight-prism

> spectroscope. A one-prism spectroscope, when the spectra taken in and perpendicular to the direction of motion of the particles were placed side by side, showed merely that the velocity could not amount to anything like 90 miles per second. There did seem to be a very slight shifting of the former spectrum towards the violet, but this appearance was probably due to the fact that its light had been weakened by two reflections, while that of the other was taken direct. [4-49]

The possibility of exploiting the Doppler shift was, in fact, negated by Crookes' own argument that the particles only become luminous on colliding with other particles, in the process of which they lose part of their velocity. In any case, another skeptic was Eugen Goldstein in Berlin, whom we have previously met. As may be recalled, Goldstein was persuaded from their shadow-casting properties that the rays were some kind of electromagnetic radiation in the all-pervasive æther. He, too, decided to subject the discharge to the Doppler test.

> It has long been known that, in accordance with Doppler's principle, the velocity of translation of the luminous gas-molecules in the discharge must influence the spectrum of the gas. As it appeared to me that the experimental treatment of this point might furnish a new criterion whether the discharge consists of a convective transport of electricity by the gas-molecules or not, I was glad to avail myself of the opportunity offered me by Professor Helmholtz to make use of a spectroscope of powerful dispersion. [4-50]

Alas, neither did his experiment, conducted the same year as Tait's, reveal any Doppler shift, from which he inferred that the velocity of the so-called torrent would have to flow at less than 10 miles s^{-1} (16 km s^{-1}), a quite modest speed. In addition, calculation and experiment convinced him that the mean free path for molecules in the most highly exhausted tube then possible in the Physical Institute in Berlin with a mercury pump (approximately $\frac{1}{125}$ mm Hg), or approximately 6 mm, was far shorter than that demanded by Crookes' new state of matter (6 cm); and yet the rays easily traveled 150 times as far as the mean free path [4-51]. There was little doubt in his mind that the *Kathodenstrahlen* were just that: ætheral electromagnetic rays. In this he was not alone. In particular, Heinrich Hertz, fellow student of Helmholtz, had grounds for sharing his view on the nature of the rays: this despite the atomistic views of Helmholtz himself just then.

Chapter 5

MEANWHILE, BACK IN BERLIN

5.1. From Hertz to Lenard

At this point in time, the scene shifts back to the Physical Institute of the University of Berlin, that we last visited in the company of Eugen Goldstein under Helmholtz's professorship. Goldstein's and Hittorf's cathode ray experiments constituted the basis for the German wave interpretation of the rays, despite the embarrassing fact that they were deflected by a magnetic field. Now another student of Helmholtz jumped on the wave party bandwagon, namely, Heinrich Hertz.

Heinrich Rudolf Hertz, the son of Gustav F Hertz, a barrister and later senator, was born into a cultured Hanseatic family in Hamburg in 1857 [5-1]. Following his boyhood education in the private school of Richard Lange, he entered the Johanneum Gymnasium at 15. With his *Abitur* (qualifying examination for university entrance) behind him, he nevertheless put off matriculating at a university. Instead, he spent a year preparing for a career in engineering by gaining practical experience with several construction bureaus in Frankfurt. There followed an obligatory year of military service with the First Railway Guards Regiment in Berlin. Vacillating between engineering and science, he finally opted for the latter course. In 1877, he matriculated at the University of Munich, and spent a year dividing his time between studying mathematics and mechanics on his own and receiving a first-class grounding in experimental physics—both under Philipp Johann Gustav von Jolly at the University and in F Wilhelm Beetz's laboratory at Munich's Technische Hochschule. Impatient to move on, he then migrated to the University of Berlin where he immediately came under the tutelage of Helmholtz, and his career was thereby assured. He now threw himself into research with singleness of purpose, winning within a year of arrival a prize competition offered by the Berlin Philosophical Faculty (on the inertial mass of an electric charge moving in a conductor), obtaining a first publication in Wiedemann's *Annalen*, and earning Helmholtz's considerable respect [5-2].

In February 1880, Hertz passed his doctoral dissertation on electromagnetic induction in rotating conductors, earning a *magna cum laude* for it—a distinction rarely conferred in Berlin. He then began a salaried appointment as Helmholtz's assistant in the Physics Institute, a post he held for three years. He supervised practice exercises in general physics, and did research on his own, both theoretical and experimental, on electrodynamics and elasticity. For him to obtain a regular faculty appointment, however, necessitated a period as *Privatdozent*. There were already too many *Privatdozenten* in Berlin, and Hertz was again in a frame of mind to move on. As luck would have it, a call came through from Kiel just then for a *Privatdozent* in mathematical physics, and on Helmholtz's and Kirchhoff's recommendation Hertz moved there in 1883. Though off to a good start academically, he soon had second thoughts about purely theoretical work, Kiel lacking a physics laboratory. He set up a laboratory at home, at his own expense, but that proved a poor substitute for a bona fide university laboratory. He lasted but two years at Kiel, increasingly frustrated and restless. He moved on, first to the Karlsruhe Physical Institute where he succeeded Karl Ferdinand Braun, and in 1889, on the invitation of the Prussian *Kulturministerium*, he became professor of physics at Bonn. Here, tragically, his premature demise from chronic blood poisoning, just short of 37, cut short what was already a brilliant career [5-3].

Hertz's research reputation had, in fact, an auspicious start during his three year tenure under Helmholtz, in which time he completed research for no less than 15 publications—the majority on electrical phenomena, including electromagnetic induction, the aforesaid inertia of electricity, and diamagnetism. The most lasting of these researches, however, was on cathode rays, a subject Hertz took up in the autumn of 1882. With neither a Geissler nor Gimingham to assist him, his brief exposure to engineering practice, coupled with his penchant from early boyhood for crafting things on a workbench, now proved invaluable. He blew his own glass tubes and patiently constructed a battery of a thousand cells—albeit one prone to problematic performance. The results of this particular series of experiments, begun in September, were submitted to *Annalen der Physik und Chemie* in late spring of the following year while Hertz was at Kiel [5-4]. Mindful of the work of the English and their particulate interpretation of cathode rays, he was naturally inclined towards the opposing view engendered by his compatriots Plücker, Hittorf, and above all his companion under Helmholtz, Goldstein, whose advice and expertise he duly acknowledges in his paper of 1883.

Hertz's first order of business, then, was to challenge the view of the English school originating with Gassiot, and subsequently propounded in turn by Varley, De la Rue, and Spottiswoode, that the luminous rays are simply a disruptive train of electrical charges. On the contrary, his own results indicated that the rays are continuous in nature [5-5], a view reinforced by his evidence that the course of the rays had no direct connection with the path of the current. Rather, the rays appeared to be a secondary effect produced by the current in the gas or in the glass.

> In certain spots the line of the current is perpendicular to the course of the cathode rays. Certain portions of the gaseous volume are brightly illuminated by the cathode rays, even though the intensity of the current is vanishingly small. The current flows [approximately] from pole to pole in a distribution similar to that in a solid or liquid conductor. From this it follows that the cathode rays have no direct connection with the path of the current. [5-6]

All of this convinced Hertz, as it had Goldstein and Eilhard Wiedemann before him, that the rays were not so much an electrical current as an ætheral disturbance in the sense of the Maxwellian doctrine then making the rounds. To be sure, their deflection by a magnetic field was not easily reconciled with such an interpretation, something Hertz vaguely brushed aside as analogous to the rotation of the plane of polarization [5-7]. In any case, the rays appeared quite unaffected by an electrostatic field between a pair of electrically charged plates [5-8]. Had they been electrically charged particles, they should have been repelled by the plate carrying like charge and attracted to that carrying an unlike charge. This crucial experimental result would turn out to be incorrect, as we shall see in due course.

Apparently his arguments were sufficiently convincing to sway Helmholtz, who had hitherto been inclined to go along with the particle school. In his Faraday Lecture before the Chemical Society in London the year before, Helmholtz had expounded on his own particulate hypothesis: 'If we accept the hypothesis that the elementary substances are composed of atoms, we cannot avoid concluding that electricity also, positive as well as negative, is divided into definite elementary quanta, which behave like atoms of electricity' [5-9]. He is also said, by a contemporary whose opinion counted, to have urged Goldstein, for one, to adopt the corpuscular hypothesis [5-10]. In light of Hertz's results, however, and as always eager to encourage his students, he was moved to respond favorably: 'I cannot refrain from wiring to say bravo!' [5-11].

Next to be heard from was, in fact, Goldstein, while he, too, remained under Helmholtz's supervision. In 1886, he resolved the nagging problem of the missing positive counterflow, by arguing that the detection of any positive rays flowing from the anode to the cathode might be obscured by the simultaneous presence of the cathode rays in the same space between anode and cathode. What better way to detect the positive rays than to pierce the cathode with holes, allowing the rays to pass through them into the free space behind the cathode? To this end,

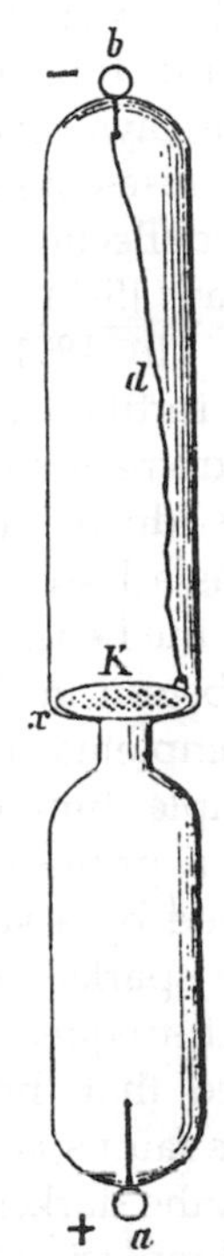

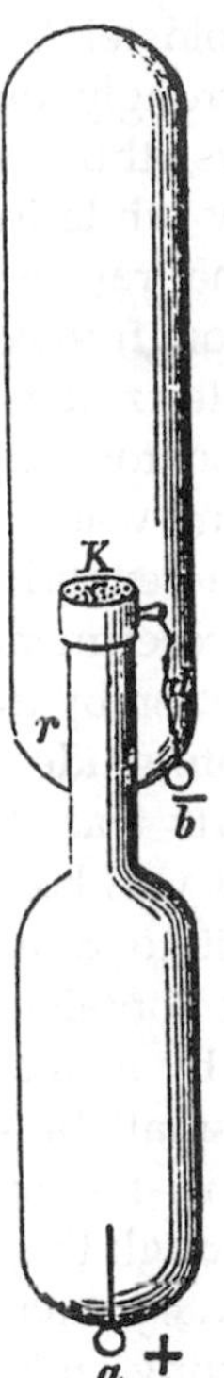

Figure 5.1. *Two versions of Goldstein's positive ray tube, incorporating a pierced cathode K. In the left-hand tube, the cathode is a flat plate K, connected to the induction coil via the wire d. In the right-hand tube, the cathode forms the bottom of a small pillbox fitting tightly over the glass tube r; again the negative charge is carried through the wire d. In either version, the anode consists of the wire a.*

he utilized a cathode in the form of a plate pierced with holes roughly 1 mm in diameter (figure 5.1). Behold!

> The result was ... the appearance of ... yellow light without any perceptible admixture of the blue cathode light ...
> The yellow light consists of regular rays which travel in straight lines. From every opening in the cathode there arises a straight, bright, slightly divergent yellow beam of rays. The separate bright beams are surrounded with a widely distributed cloud which is very weak in light, but in general has the same color as the beams. Until a suitable name has been found, these rays, which we now cannot distinguish any longer by their color, which changes from gas to gas, may be known as 'canal rays' [*Kanalstrahlen*]. [5-12]

The canal rays differed from the cathode rays in several important respects. They were self-luminous and visible, unlike the cathode rays which were visible only on encountering a fluorescent body. Moreover, as noted in Goldstein's citation, the color of the canal rays depended much more strongly on the residual gas in the tube than did that of cathode rays; thus, in rarefied air the canal rays assumed a vivid yellowish glow while in hydrogen they had a rose-colored tint [5-13]. Finally, the canal rays were quite insensitive to deflection in a magnetic field strong enough to readily deflect cathode rays [5-14].

Nor had Hertz left the subject for good. In 1891, now settled in Bonn, he too took up cathode rays again, fortified by his soaring reputation in the wake of his detection of electromagnetic waves three years earlier—an experiment that calls for a brief diversion. Preparation for that great experiment, originally proposed as a lesser experiment in a prize competition by the Prussian Academy at the behest of Helmholtz with his favorite student in mind [5-15], had begun in the autumn of 1886 while Hertz was still at Karlsruhe. He completed it in 1888 [5-3]. His experiment was based on a *resonance* principle, tuning a secondary electrical circuit to receive waves radiated by a primary circuit. The primary circuit contained a brass rod interrupted by a spark gap, and was energized by a small induction coil. When sparks crossed the gap between two small brass spheres, violent high-frequency oscillations were induced in the rod. Hertz demonstrated that the waves were transmitted through the air by detecting them as faint sparks induced in a similar secondary circuit some meters away in the darkened Karlsruhe lecture hall. Acting on 'friendly advice', he tried replacing the secondary spark gap with a frog's leg prepared for detecting currents. However, 'this arrangement which is so delicate under other conditions does not seem to be adapted for these purposes' [5-16]. By moving the secondary spark gap detector to and fro in the hall, he measured the wavelength of the waves. With its value and the calculated frequency of his oscillatory circuit, he determined the speed of propagation of the waves as none other than the speed of light. He showed, furthermore, that they are transverse vibrations of electric and magnetic forces by conducting experiments of his own on their reflection, refraction, and polarization [5-17].

Hertz's newfound reputation was enormous and world-wide. It was also responsible for his final relocation. In the early fall of 1888, the University of Giessen tempted him with an academic appointment. The Prussian *Kulturministerium* countered by urging him to consider instead an appointment as mathematical physicist and successor to Kirchhoff at the University of Berlin. Hertz demurred, feeling he was not yet prepared for such a lofty position and weary of a purely theoretical post in light of his experience at Kiel. Helmholtz carefully refrained from influencing him but offered him laboratory space in his own Physical Institute on the Reichstagsufer. Hertz still hesitated. Finally, in December the *Kulturministerium* offered him instead the professorship in physics at Bonn, and he accepted at once [5-18].

Still convinced that the cathode rays were electromagnetic waves in the æther similar to the 'Hertzian waves', as the latter were now commonly termed, he resolved to demonstrate the similarity by experiment. His approach was to pass them through a barrier impervious to particles of atomic size, particles to which Crookes attributed the rays. Accordingly, he undertook a new series of experiments utilizing various metallic foils. Initially, a diaphragm of pure gold leaf blocked the path of the rays [5-19]. At first the diaphragm simply threw a shadow on the glass wall, but when the vacuum reached a sufficiently low pressure the shadow faded and the glass on the far side of the diaphragm began to fluoresce. The ray appeared 'as a faint veil upon the glass, scarcely recognizable except by its general shape and the lines of its small wrinkles' [5-20]. Some portion of the cathode rays had evidently succeeded in penetrating the diaphragm. Other metal foils—silver, aluminum, tin, and copper alloys—worked equally well, as did silver or platinum precipitated by discharge within the tube. Hertz ruled out by experiment that the rays were simply percolating through pores invariably present in the thin metal leaf [5-21], this despite the fact that some materials prevented the rays from penetrating; thus a thin mica plate cast through the gold leaf a black shadow on the glass. More puzzling, while some metal leafs passed the rays while opaque to light, other metals transparent to light 'offered an invincible resistance to the penetration of cathode rays' [5-22]. Be that as it may, if the rays really were particles, they had to be far smaller than molecules or even atoms in order to penetrate the space between the closely packed atoms in the metallic barrier.

This would be Hertz's last experimental investigation; time was running out. Already at Karlsruhe he had complained of toothaches, and no sooner had he settled in Bonn than he had all his teeth extracted. In the summer of 1891, he was forced to hand over his ongoing investigation to his newly arrived assistant, Philipp Lenard. The following summer he was in such agony in his nose and throat that he had to stop work altogether. He visited spas to no avail, and underwent several unsuccessful operations by his physicians who failed to diagnose his condition correctly. He resumed lecturing in the spring of 1893, while working feverishly on the manuscript of his book *Prinzipien der Mechanik*. On December 3 1893, he fowarded his largely completed manuscript to the publisher, and gave his last lecture four days later. He passed away on the first day of 1894, at the age of 36.

Hertz's successor was Heinrich Kayser, the spectroscopist, though it was his young assistant who inherited his cathode ray program. Philipp Eduard Anton Lenard, to give him his full name, was the son of a wealthy wine merchant. He was born in Pressburg, Austria–Hungary (now Bratislava, Czechoslovakia), in 1862 [5-23]. His education began in the cathedral school in Pressburg, followed by the normal

Figure 5.2. *Philipp Lenard.*

schooling in the city's Realschule. As a boy Philipp showed great flair for mathematics and physics, studying both subjects and conducting experiments in chemistry and physics—all on his own. His father wanted him to carry on the wine business, but reluctantly consented to him continuing his studies, provided he did so at the Technische Hochschulen in Vienna and Budapest, and on the stipulation that he concentrate on the chemistry of wine. At the end of his studies, Lenard dutifully joined his father's business. However, having thereby earned some money, Lenard could afford a journey to Heidelberg. There he was entranced by the lectures of Robert Bunsen, the guru of chemistry and his private hero. The upshot was that he matriculated at the University of Heidelberg during the winter of 1883. Following a stint at Heidelberg, Lenard, like Hertz, gravitated to Berlin. There he began his dissertation on a topic suggested by Helmholtz: the oscillation of precipitated water droplets. He completed his doctorate *summa cum laude* back in Heidelberg in 1886, whereupon he became assistant to Georg Hermann Quincke, professor of physics on the Heidelberg faculty.

With three years in Quincke's employ under his belt, Lenard spent a half year in London in the electromagnetic and engineering laboratories of the City and Guilds of the London Central Institution, followed by a few months at Breslau (now Wroclaw, Poland). On April 1 1891, he arrived in Bonn to become assistant to Hertz, someone he greatly admired [5-24].

Lenard was not new to cathode rays. While still under Quincke at Heidelberg, he had performed an experiment in an attempt to bring the cathode rays in a rarefied tube out into the external air. This necessitated an airtight seal in the wall of the tube that would allow the rays to penetrate. As one school of thought (mainly that of Wiedemann) maintained that the rays were a form of ultraviolet light, a quartz window seemed like a good choice. In the event, the test was unsuccessful [5-25]. Now, four years later, the opportunity arose again. Shortly after joining Hertz, Hertz called him over and showed him a discovery he had just made, that a piece of uranium glass covered with aluminum foil and placed inside the discharge tube became luminous beneath the foil when struck by the cathode ray.

> We ought to separate two chambers with aluminum leaf [said Hertz], and produce the rays as usual [i.e. with a slight residual air pressure] in one of the chambers. It should then be possible to observe the rays in the other chamber more purely than has been done so far and even though the difference in air pressure between the two chambers is low because of the softness of the leaf, it might be possible to completely evacuate the observation chamber and see whether this impedes the spread of the cathode rays—in other words find out whether the rays are phenomena in matter or phenomena in æther. [5-26]

Here was Lenard's opportunity, Hertz proclaiming himself too busy to carry out the actual experiment.

5.2. Lenard's window

Lenard wasted little time in following up on Hertz's cathode ray experiments; here was his chance to repeat his unsuccessful experiment under Quincke at Heidelberg. Undaunted by the softness and porosity of Hertz's aluminum leaf (soft enough for use in bookbinding), he piled layers on top of layers in the tube; ten to 15 layers still transmitted the rays. Aluminum foil of comparable thickness worked equally well, and withstood the considerable air pressure, provided its surface area was sufficiently small. On a hunch, he replaced his quartz window in the end of the tube—a holdover from the Heidelberg experiment—with a metal plate pierced with a small hole sealed with aluminum foil. On the outside surface of the foil, he spread a few grains of alkaline-earth phosphor (the tube being oriented vertically with the end on top). 'Lo and behold', on exciting the tube 'the grains glowed brightly!' [5-27] Next, he mounted the grains slightly above the aluminum window, and still the glow persisted. Not only had the rays passed out of the hitherto confining discharge tube, but, even more remarkably, they had passed some distance through the external air at atmospheric pressure.

Lenard's next cathode ray tube, the handiwork of Glastechniker Müller-Unkel in Braunschweig, is shown in figure 5.3 [5-28]. In it, the 'production chamber' with its aluminum anode disk *A* and hollow brass cathode *C* was separated from the external room by the seal *mm* and the window beyond it, as shown. For observing the rays, he used

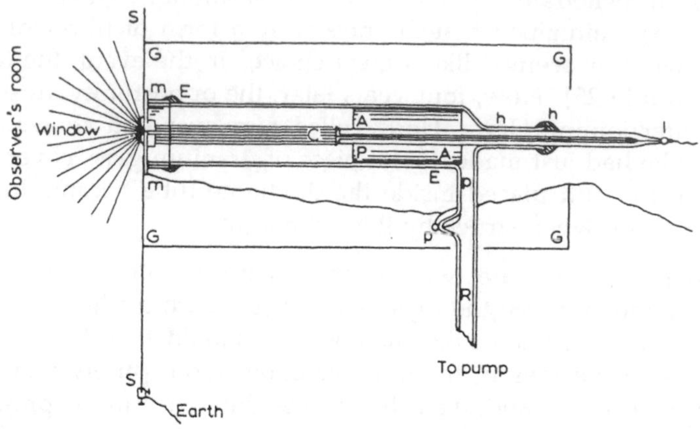

Figure 5.3. *Lenard's cathode ray tube, featuring a 'window' between the production chamber and the observer's room.*

sheets of paper coated with ketone, platinum cyanide, or an alkaline-earth phosphor, these being phosphorescent screens routinely used to render ultraviolet rays visible.

> When we use the phosphorescent screen, we find it glowing brightly close to the window; as the distance from the window increases, the intensity of the rays progressively diminishes until at a distance of about 8 cm the screen remains quite dark. Apparently air at full atmospheric pressure is not very permeable to cathode rays, certainly far less permeable than it is to light. But it was far more interesting to find that air is even a turbid medium for these rays, just as milk is for light. If we place an impermeable wall with a hole in it a suitable distance from the window and put this view [figure 5.4]. Here the dotted lines indicate the narrow pencil of rays that we should expect in the case of rectilinear propagation; but it is the broad bent bunch of rays that we really see on the screen in the open air, just as if we had passed light through the same hole into a tank containing slightly diluted milk. What clouds the air? In milk it is numerous small suspended fat particles that make it turbid to light. Pure air on the other hand contains nothing except molecules of the gases contained in it, suspended in the æther. These molecules are extremely small, 10,000 times smaller than the fat particles, far too small to act individually on light. But, as we see, the cathode rays are hindered by each of these molecules. Thus these rays must be extremely fine [short], so fine that the molecular structure of matter, which is minute compared with the very fine light waves, becomes pronounced in comparison with them. It may then be

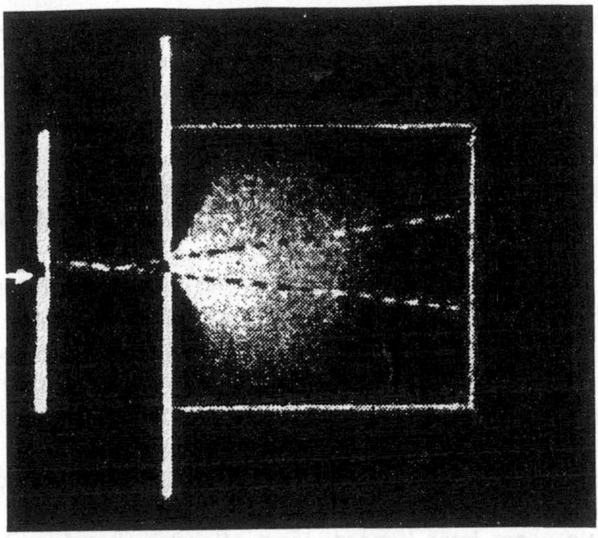

Figure 5.4. *The business end of Lenard's tube: an early clue to molecular structure and the nature of cathode rays.*

possible to obtain data by means of these rays concerning the nature of molecules and atoms. [5-29]

These results pointed to the importance of subjecting the rays to a wide variety of metallic foils. By interposing foils of different materials between the window and detection screen, it soon became 'abundantly clear that the permeability or impermeability of a material to light is not even slightly related to its behavior in relation to cathode rays' [5-30]. Ergo, Hertz's puzzle was laid to rest. Nor was it simply a question of the permeability or impermeability of the material to cathode rays. What really mattered was the degree of absorption, or of the foil's *absorptive power*. This, in turn, boiled down to the *weight* of the material (in Lenard's terminology; that is, its density). Less dense materials admitted more cathode rays, more dense materials admitted fewer.

Finally, here was the means at hand to settle once and for all the paramount question of particles versus waves—a question awkwardly, or guardedly, phrased by both Hertz and Lenard in terms of whether the rays are 'phenomena that take place in matter or in the air' [5-31]. Analogously, since a ray of sound is unable to pass through a highly evacuated chamber, it represents 'a phenomenon in matter' whereas light or electromagnetic forces are clearly phenomena in the æther. The same question concerning cathode rays could not be resolved with 'ordinary discharge tubes because [true enough] once all the air is removed the production of the rays in such a tube ceases' [5-32]. Instead, the observation chamber outside the window was evacuated as well as possible with a Töpler–Hagen pump. As a result, all absorption and turbidity due to any residual gas molecules ceased, and the extracted rays traveled several meters with rectilinear sharpness akin to a beam of light. There could no longer be any doubt. The rays were surely not radiant matter nor emitted gas molecules, as maintained by the English, but an ætheral propagation of some kind, just as Wiedemann, Goldstein, and Hertz had argued all along [5-33]. But *what sort* of propagation remained an open question.

At this critical juncture, Lenard's experiment was interrupted by the untimely death of Hertz. It fell on Lenard to take charge of the publication of Hertz's *Gesammelte Werke* [5-34]—no small task. In particular, volume 3, the *Prinzipien der Mechanik*, though complete, required considerable editing, a chore Hertz had entrusted Lenard on his deathbed. At the same time, Lenard was appointed to a theoretical professorship and, on top of that, was even obliged to assume the acting directorship of the laboratory.

Thirty-five years later Lenard, long after he abandoned a brilliant experimental career and became simply an ideological crank, would publish a tribute to Hertz in his *Grosse Naturforscher*, a volume containing biographies of great scientists from Pythagoras onward [5-35]. In it, he was compelled to reveal that, contrary to his contemporary opinion in 1894 [5-36], he had detected a split personality in Hertz, whose father was Jewish. (Heinrich Hertz himself was Lutheran.) Hertz's theoretical work, personified in his *Prinzipien der Mechanik*, was something Lenard, himself weak in theory, declared to be a product of Hertz's Jewish legacy.

In 1894, however, though highly sensitive to the brilliance of others, above all Hertz, Lenard himself remained a pertinaciously enthusiastic experimentalist with several important researches on his agenda. In particular, in addition to the cathode ray work, he was also extending the work of Hertz in another area—one highly germane to the nature of the rays: the photoelectric effect [5-37]. That effect, the emission of electricity from a metal plate illuminated by ultraviolet light, was discovered in 1887 by Hertz at Karlsruhe in the course of preparing his experiments on electromagnetic waves. Hertz had noticed that the spark in the detector circuit was longer when it was exposed to the light of the spark in the primary circuit. True to form, he undertook a major investigation, interposing different materials between the two circuits, and concluded that ultraviolet light was the sole cause of the effect [5-38].

The photoelectric effect soon caught the attention of, among others, Wilhelm Hallwachs in Germany, Augusto Righi in Italy, and Alexandr Stoletov in Russia. They concluded that the ultraviolet radiation liberates negative electricity from a metallic plate whether charged with negative electricity or not [5-39]. The precise nature of the photocurrent remained unclear, but it was generally thought that the ultraviolet rays caused molecular dissociation. Ambient gaseous molecules are separated into charged atomic constituents, with the subsequently ejected charges being repelled atoms of negative polarity [5-40].

Lenard, too, took an interest in the photoelectric effect during his Heidelberg stint under Quincke, and in the spring of 1889 undertook some experiments on it with the future astronomer Max Wolf. Lenard's skepticism of the prevailing notion that the photocurrent consisted of charged gaseous atoms seemed at first to be vindicated, as their experiments indicated that ultraviolet light roughened or pulverized metal plates. Hence, it seemed likely that metallic particles carried the negative charge off the illuminated surface [5-41]. Subsequent experiments cast doubt on this, however. Illuminating a sodium amalgam cathode should have passed a detectable amount of sodium to a platinum wire anode, though a fluorescent screen failed to detect any rays in the vacuum adjacent to the anode. Nor, on removing the anode wire from the bulb, was any trace of sodium found spectroscopically. Only with a more sensitive electrometer was the existence of negative rays revealed. When it was found, furthermore, that the rays were also deflected by a magnetic field, the conclusion that cathode rays are produced by ultraviolet light seemed inescapable [5-42]. Not only that; the hypothesis which formed the basis of Hertz's and his own interpretation of cathode rays was now in jeopardy [5-43]. It seemed increasingly likely that the cathode rays, too, were charged particles, a conclusion which Lenard kept to himself for nearly a decade. Only in 1899 did he publish a detailed account of these researches [5-44]. Alas, by that time the brunt of the cathode ray investigations had passed on to others, leaving Lenard more or less in the lurch. It would not be the only time his work was anticipated by others, a habitual situation undoubtedly fueling his pathological crankiness in later years.

A further setback, and a more dramatic one at that, was indeed not long in coming—one we have touched on in chapter 1. Even before his distraction by Hertz's passing away, Lenard had devised a particular discharge tube in which his window was mounted in the end of a cylindrical platinum extension fused into the main body of the tube (figure 5.5). As it happened, he would later claim, this tube would prove especially effective in generating x-rays the following year in the hands of Röntgen, with the cathode rays impinging on the large area of platinum metal [5-45]. With it, Lenard soon found that the electrical conductivity induced in the air outside the tube was considerably enhanced, but when the tube failed early in the experiment he set it aside. Meanwhile, Röntgen wrote him in search of a window foil suitable for his own attempt at repeating Lenard's cathode ray experiments. Lenard responded at once, even sending him pieces of his own precious aluminum foil.

The source for thin Aluminum sheets presents great difficulties

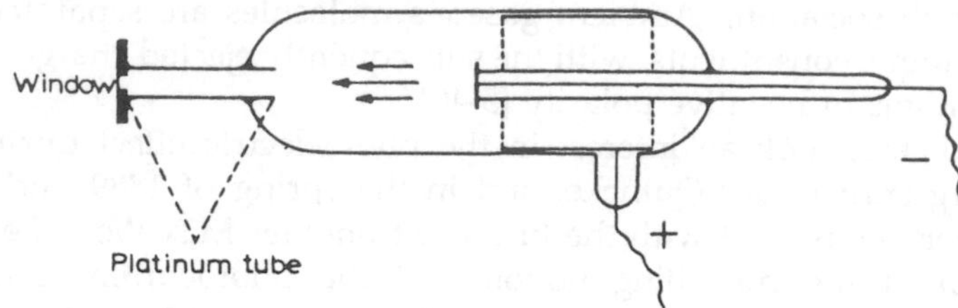

Figure 5.5. *Lenard's discharge tube, employed with great success by Röntgen.*

> for me too, since the manufacturers hesitate to give up unusual sizes (thicknesses) or they take little care with small parts, so that the sheets turn out full of holes. I too am lacking a good source of Al. I permit myself to send you two sheets from my supply. The thickness is about 0.005 mm. By the way I have heard lately from Mr. Müller-Unckel that he is manufacturing apparatus with a complete window and not evacuated. [5-46]

On Lenard's advice, Röntgen had a similar tube constructed by the Braunschweig glassblower Müller-Unckel who routinely blew Lenard's tubes [5-47]. Though it is unclear whether Röntgen did, in fact, deploy this particular tube when he made his discovery of x-rays, Lenard, soon smarting in missing out in the discovery, was convinced that he did. Given the tube, the fact that Lenard's experiments had shifted attention from the interior to the exterior of the tube, his use of phosphorescent screens—even his suggestion of covering the tube with black paper—made the discovery all but a certainty, or so it seemed to him in retrospect [5-48]. Failing to make the discovery exasperated Lenard's already bitter scar, even though for a period he was friendly toward Röntgen and even basked in the limelight accorded him.

> For your kind letter that I was happy to receive, I thank you many times ... I was particularly happy to know for sure, what I had never any reason to doubt, that you are friendly toward me. I was often afraid it could have been otherwise and I would have been sorry for that. However, I am completely innocent of any remarks which could have caused this to be. I never took part in any of the slightest Polemics. Because your great discovery caused such swift attention in the farthest circles my modest work also came into the limelight, which was of particular luck for me, and I am doubly glad to have had your friendly participation. [5-49]

The question of why Lenard did not discover x-rays [5-50] is one thing. More to our point, the embarrassing stance he exhibited as the discovery was announced must be attributed to his failure to distinguish x-rays from cathode rays. Both, he believed at the time, were electromagnetic waves, only x-rays were of much shorter wavelength than the cathode rays. Since, as he had amply demonstrated, cathode rays could traverse metallic bodies, there was no reason to doubt that the effects observed by Röntgen were produced simply by some escaping cathode rays as well. Thus, he was at first inclined to deny the novelty of Röntgen's discovery while at the same time claiming substantial credit for his own role in the discovery in his correspondence with Röntgen [5-51]. His claim in this, as it made its rounds among physicists, was supported by few beyond his immediate colleagues, however. Recalls Arthur Schuster, who would enter the fray as well, in his retrospective lectures in Calcutta:

> I remember in this connection an interesting discussion on board an excursion steamer on the occasion of the Kelvin Jubilee. A German Professor stood up for Lenard's priority in the discovery of the new radiation. The whole thing, he said, was in Lenard's mind, when he carried out his researches. 'Ah', said Sir George Stokes with a characteristic smile, 'Lenard may have had Roentgen rays in his own brain, but Roentgen got them into other people's bones'. Nobody enjoyed the remark more, or subsequently repeated it oftener, than the eminent Professor who had called it forth. [5-52]

Röntgen, for his part, being quite aware of Lenard's priority claim and his haughty attitude, took some pains in distinguishing the x-rays from Lenard's extracted cathode rays in his paper announcing the discovery. Though, like cathode rays, 'neither refraction nor regular reflection takes place to any sensible degree' with the x-rays, 'air absorbs a far smaller fraction of the x-rays than of the cathode rays... Other substances behave in general like air; they are more transparent to x-rays than to cathode rays' [5-53].

> A further difference, and a most important one, between the behavior of cathode rays and of X-rays lies in the fact that I have not succeeded, in spite of many attempts, in obtaining a deflection of the X-rays by a magnet, even in very intense fields.
> I therefore reach the conclusion that the X-rays are not identical with the cathode rays, but that they are produced by the cathode rays at the glass wall of the discharge-apparatus. [5-54]

If Lenard was bitter about the x-ray business, he soon had cause for more bitterness. This time his nemesis was not a fellow German scientist but one of the English, albeit not one of the Victorian amateur scientists. Two English physicists, quite as professional as Hertz and Lenard, were looming as effectual advocates of the particle school of interpretation of cathode rays. One was Arthur Schuster, who began his researches in this area shortly after Hertz first visited the subject during 1882–1883.

Chapter 6

THE ENGLISH KEEP GOING

6.1. Arthur Schuster; a discovery narrowly missed

Arthur Schuster, born in 1851, was one of three sons of Francis Joseph Schuster, the head of a prominent Jewish family in Frankfurt am Main. The family records, though sketchy, can be traced back to 1607 [6-1]. In the mid-1700s, the family started a business in cotton goods, trading mainly with Great Britain. After the Seven Weeks War between Austria and Prussia, when Frankfurt was annexed by Prussia, the elder Schuster moved the family business to Manchester, England. Young Arthur was educated privately and at the Frankfurt Gymnasium, where he received his first instructions in physics and mathematics from Harald Schütz. In 1868, he was sent to the Geneva Academy where he remained until joining his parents at Manchester in the summer of 1870. At Geneva, in addition to studying French, he studied chemistry under Jean Charles Galissard de Marignac, physics under L Soret, and astronomy under Plantamour.

Schuster was 19 years old when he arrived in Manchester, old enough to be apprenticed to the firm of Schuster Brothers. His parents, especially his sympathetic mother who was blind from Graves' disease, soon sensed his discontent in the business milieu. They consulted with Henry Roscoe, then professor of chemistry at Manchester's Owens College, whose evening classes Schuster was attending. Roscoe, who was tireless in his effort to spread the importance of science in public schools, academia, industry, and government alike, was a mainstay in building Owens from a wretched school of minimal resources to a college as good as any outside London. He arranged for Schuster to enroll as a day student at the college in October 1871. With Schuster's grounding in science at Geneva, and further inspired to go into physics by Roscoe's highly popular and elementary textbook on spectrum analysis, he studied physics under Balfour Stewart and pure and applied mathematics under Thomas Barker—excellent teachers who left their mark on the slightly younger J J Thomson at about the same time. A nasty railroad accident had nearly cost Stewart his life soon after his appointment at Owens in 1870, but in the fall of 1871 he was able to resume his teaching load. Though he was a teacher second to none, instruction at Owens nevertheless left something to be desired, predating by a century today's all-too-common complaint among university undergraduate students.

> The main part of the instruction was handed over to teachers, whose knowledge was tempered by sufficient ignorance to allow them to be dogmatic without carrying conviction. In consequence we—the students—not being able to distinguish between bluff and solid teaching, took nothing on trust and started investigations of our own, to confirm or disprove our instructors. [6-2]

Schuster's heart lay in laboratory research. Spectroscopy was then a hot topic for both physicists and chemists [6-3], and under Roscoe's guidance he took up spectrum analysis—his first serious experimental investigation. Within a year of enrolling at Owens, he published his first paper, 'On the spectrum of nitrogen', albeit a not very successful one [6-4]. Schuster himself decided that he had taken up research too soon, but Roscoe would have none of that. Instead, he encouraged his protégé to spend a *Wanderjahr* under Kirchhoff and Bunsen at Heidelberg—the birthplace of spectrum analysis—as he himself had done. Schuster agreed, and enrolled at Heidelberg under Kirchhoff, where he spent two years in the company of fellow visitors Kamerlingh Onnes and Gabriel Lippmann. His research there, on the reflection of light from metallic surfaces, was rather lackluster but good enough to earn him his PhD in 1873.

On returning to Manchester, he was appointed honorary (unpaid) demonstrator in physics, the physics laboratory having relocated, during his absence, from its miserable quarters in Quay Street to a comparatively palatial building on Oxford Road [6-5]. Schuster now became preoccupied with Wilhelm Weber's diamagnetism based on circulating electric currents, and with Maxwell's recently published *Treatise* [6-6]. His wanderlust proved too strong, however, and in the summer of 1874 he was off to Germany again. He first spent a few months at Göttingen under Weber, who was then 70 years old, and his assistant Eduard Riecke, by then extraordinary professor in Weber's institute. There he began his protracted and controversial work on an experimental test of Ohm's law for rapidly alternating currents, as noted earlier [1-51]. Like everybody else when afforded the opportunity, he then journeyed to Helmholtz's laboratory in Berlin, with an introduction from Roscoe. Among the promising students under Helmholtz at the time was Eugen Goldstein.

> Goldstein was working at his electric discharges in high vacua, and trying to explain effects he discovered near the kathode by a theory which I know to have been distasteful to Helmholtz; yet

> no word did he ever say to discourage the purely experimental side of those experiments. [6-7]

Schuster had intended to spend the winter of 1874–1875 in Berlin. However, while on Christmas vacation in Manchester, as luck would have it, he received an invitation from Norman Lockyer, the well known astronomer [6-8], to join a forthcoming solar eclipse expedition to the coast of Siam—a temptation he could ill afford to turn down. In the event, Lockyer himself was unable to accompany the expedition. Much to his pride, Schuster was appointed head of the expedition by its sponsor, the Royal Society. It was, on the whole, successfully carried out the following April, but attempts to photograph the spectrum of the solar corona failed, due to the slow speed of the wet photographic plates then in use [6-9]. On his return journey he toured the Himalayas, hiking from Simla to Leh, and returned via Calcutta.

Back in Manchester by November 1875, Schuster lasted only one semester at Owens, lecturing on the Siamese expedition and on Maxwell's *Treatise*. Among the students attending the latter lectures, all of three, in fact, was J J Thomson, who otherwise appears not to have interacted much with Schuster at Owens. Schuster's experiments on the radiometer, which we have touched on earlier, were also conducted during this session. At the end of the semester, he joined Maxwell as a researcher at the Cavendish Laboratory; here he would reside off and on for five years. What brought him to the Cavendish was his aforementioned and unfortunate experiment begun at Göttingen under Weber and continued at Owens: a test of Ohm's law for alternating currents. Since it was discussed briefly in section 1.4, we need not dwell on it here. It suffices to say that his results seemed to indicate a slight deviation from Ohm's law—something that caused Maxwell to take notice, since his electrodynamics gave no reason for such a deviation (nor for the law itself, for that matter). The upshot was that, on Maxwell's behest, the anomaly was investigated further by his laboratory assistant, George Chrystal, and found to be a dud [6-10]. However, the matter had raised Maxwell's interest in Schuster, and he responded favorably when Schuster applied for a position at the Cavendish.

> It would do us all great good if you were to come and work at the Cavendish Laboratory. The very prospect of your coming has caused all our pulses to beat about one per minute quicker.
> The Schuster effect and the anti-Schuster effect have long been the objects of our regard, but we look forward to the time when these terms will have lost their significance as applied to the phenomena of electric conduction, and when every department of physics will have a recognized Schusterismus . . . [6-11]

> I was well received by Maxwell [recalls Schuster], but otherwise was looked upon with friendly curiosity by a University that was

> not accustomed to see anyone coming to it merely for the purposes of research, without previously availing themselves of the benefits of degree courses and the prospect of the usual University rewards. After a little trouble and delay I was allowed to enter one of the colleges. My position was irregular, if not illegal, for though inscribed as an undergraduate I was allowed to stay five years without passing the 'Little Go' [entrance scholarship examination], though subject, of course, to the discipline of the Proctor. I remember meeting that formidable officer of the University at dinner, and walking home with him unchallenged and not fined, though I was without cap and gown. [6-12]

With the 'Schuster effect' put to rest, Schuster needed something else to tackle. The opportunity came when Rayleigh succeeded Maxwell at the Cavendish in 1880. No sooner had Rayleigh arrived in Cambridge than he instituted a renewed determination of the absolute ohm in collaboration with Schuster and his sister-in-law Mrs Henry Sidgwick, Principal of Newnham College for women in Cambridge. The first determination had been made in 1864 under the direction of a committee of the British Association; the experiments were performed by Maxwell while he was at King's College, London. Later experiments by others cast doubts on the professed accuracy of that determination. Hence, it was repeated under Rayleigh at the Cavendish by no less than three independent measurements which 'gave very concordant results' [6-13]. These arduous experiments consumed three years; Schuster only participated in their earliest phase.

In 1881, a professorship in applied mathematics was established at Owens College, and Schuster applied for it, as did J J Thomson and Oliver Lodge. Curiously, the chair went to Schuster, an experimentalist at heart, and not to Thomson who was manifestly a theorist. (Lodge, an experimentalist, accepted the first professorship in physics at Liverpool University the same year [6-14].) Schuster probably owed his success to Roscoe's effort on his behalf to convince certain skeptical electors to the Manchester chair of his mathematical prowess [6-15].

Settling back in Manchester, Schuster once again cast about for a new subject of research. What better than his original field of spectroscopy? Wasting no time, he promptly took G J Stoney to task for his 'rule' explaining spectral lines in terms of simple harmonic series. Schuster himself found that, on the contrary, spectra of certain elementary gases appeared to conform to a random distribution, though he suspected that some rule—however complex—was probably at work [6-16]. Within a year, however, he abandoned the spectroscopic problem as too formidable to handle. Since the spectra in question were produced by glow discharges in evacuated tubes, he decided a more fruitful tack was concentrating on the discharge mechanism itself. In this he was delayed by another solar eclipse the same year, 1882—this time to Egypt (and this time, he met with success in capturing the spectrum of the solar corona on specially prepared, faster dry photographic plates).

Schuster now capitalized on the dubious notions of his friend Lockyer. Lockyer held that chemical elements dissociate into subatoms under the influence of sources of heat, and that spectra of both elementary and compound bodies vary with varying degrees of heat [6-17].

> The spectra vary as we pass from the induced current with the jar to the spark without the jar [wrote Lockyer], to the Voltaic arc, or to the highest temperature produced by combustion. The change is always in the same direction; and here, again, the spectrum we obtain from elements in a state of vapour (a spectrum characterized by spaces and bands) is similar to that we obtain from vapours of which the compound nature is unquestioned.
> At high temperatures, produced by combustion, the vapours of some elements (which give us neither line- nor channelled space-spectra at those temperatures, although we undoubtedly get line-spectra when electricity is employed, as before stated) give us a continuous spectrum at the more refrangible end, the less refrangible end being unaffected. [6-18]

Back from Egypt, Schuster began a series of investigations which he reported in his Bakerian Lecture before the Royal Society in June 1884, presented on the invitation of Sir George Stokes (who remained a Secretary of the Society for over three decades). Curiously, he makes no reference to Lockyer in this important report, though he acknowledges De La Rue and Müller, Crookes, Goldstein, and particularly Faraday and Hittorf [6-19]. He also acknowledges the help of Arthur T Stanton, his personal assistant. The gist of his argument was that the particles in the gaseous discharge are not charged molecules, as Crookes insisted, but *dissociated* molecules—that is, molecules broken up into their positively and negatively charged atomic constituents. (Only later does Schuster use the term 'ionization' (or 'ionisation'), a term not yet coined in 1884, though 'ions' were so named by Faraday.) This he sought to verify with mercury, whose vapor, according to the kinetic theory, consists of single atoms. Sufficiently freed of air by a Sprengel pump, mercury should exhibit simpler behavior than the striations, band spectra, and other phenomena characteristic of gases commonly used in glow discharge experiments (e.g. nitrogen, oxygen, carbonic oxide, or air). Indeed, the resulting glow showed no negative glow, no dark spaces, and no striations, merely a continuous glow uniting the positive and negative electrodes. Viewed through a spectroscope, a line spectrum dominated the band spectra typical of more complex molecules.

> While we thus have ample evidence of dissociation whenever an electric current passes through a gas, we also find that whatever

> increases independently the dissociation improves the conducting power of the gas. Thus we know that a flame is a good conductor, as Hittorf has, in a series of very interesting experiments, shown that if we heat up the electrodes to a white heat, an electromotive force of a few volts will send a current through the gas.
>
> The fact also discovered by Hittorf, that if a discharge is set up a small electromotive force is sufficient to pass a current across, is also easily explained by our theory, as the original discharge throws the molecules into that state of dissociation which favors the passage of the current. [6-20]

The glow surrounding the negative electrode, and the influence of the positive electrode on it, seemed particularly relevant to uncovering the mechanism at work, something Goldstein had appreciated before him. Schuster found the glow to be divided into three layers. As the residual pressure is reduced, the innermost layer changes: what was a golden layer exhibiting a spectrum rich in the hydrogen and sodium lines (that he ascribed to matter settled on or absorbed in the cathode) loses its golden glow and spectral lines save for the spectrum of the positive half of the tube. The second layer is the dark space, and the third the glow proper, separated from the positive part of the discharge by another non-luminous space which Schuster calls the 'dark interval' [6-21]. The fall of potential across the negative electrode suggested the action of a condenser.

> The rapid fall of potential in the neighbourhood of the negative electrode renders the presence of positively electrified particles in its neighbourhood necessary.
>
> If the discharge through which the condenser action takes place is sensible, the positively electrified particles will be acted upon by a neighbouring positive electrode.
>
> A steady state will be established in which the fall of potential along the normal from the surface will be everywhere the same.
>
> As however the flow is stronger away from the positive electrode, we must conclude that other forces besides electrical forces determine the flow.
>
> It is natural to assume that these are chemical forces: that in other words the positively electrified particles are the decomposed molecules, which by their presence assist the decomposition of others, and therefore the formation of the current.
>
> Unless a flaw is detected in this line of argument, I think that the conclusion must be granted, namely, that the decomposition of the molecules at the negative electrode is essential to the formation of the glow discharge. This is really all that I endeavor to support in this paper. [6-22]

Focusing on the dark space, 'the region through which the greatest number of atoms can freely pass' [6-23], he explained that

> ... Mr Crookes has given reasons why we should consider the region of the dark space as one in which directed motion prevails, and although Hittorf has raised serious objections against the arguments drawn from his radiometer experiments, which seem to be explained by secondary temperature effects, the general conclusion which he has drawn from his experiments is not thereby invalidated, for the rise of temperature itself requires explanation. [6-24]

Then comes what he considered his most important test of the theory.

> The most conclusive proof of our theory would be the demonstration of the fact that each particle of matter carries with it the same amount of electricity. We shall not, of course, be able to prove this for each single particle, but I propose to show how we can decide the point experimentally as far as the average amount is concerned. Suppose a small straight beam is cut out of the glow and placed in a [magnetic] field of uniform force, the lines of which cut the rays of the glow at right angles. The force being everywhere normal to the rays, these will curl up in a circle. This has been shown to be true experimentally by Hittorf. I think a careful measurement of the radii of such circles will give us important information, and I have already made preliminary experiments which have shown me that such a measurement is possible. [6-25]

Equating the magnetic force exerted on a beam of particles of mass m and charge e, moving in a field of intensity H, to the centripetal force experienced by the particles constrained by the field to move in a circle of radius r, or

$$Hev = mv^2/r$$

we obtain

$$e/m = v/Hr. \tag{6.1}$$

Thus, e/m, or the radius, is dependent on the velocity, a parameter then not readily measured other than qualitatively by the Doppler shift, and even that gave doubtful results, as we have seen. There *are* methods for measuring it quantitatively, but that development came later and not in Schuster's hands. Meanwhile, to get around this particular difficulty, Schuster argued as follows. If V is the fall of potential across the region in which the velocity is acquired (Schuster uses F for 'fall'), then the corresponding kinetic energy is related to the potential by

$$Ve = \tfrac{1}{2}mv^2$$

and hence

$$e/m = 2V/(H^2r^2). \tag{6.2}$$

Here V, H, and r are all measurable quantities. However, in his Bakerian Lecture, Schuster simply noted that r is, according to equation (6.1), proportional to v/e, and from equation (6.2) (and assuming e is a constant) proportional to the square root of the fall in potential near the cathode. Only in his second Bakerian Lecture of 1890 [6-26] did Schuster offer numerical upper and lower limits for e/m, by then a quantity appreciated in several quarters at home and abroad as of particular relevance in the final phase of the cathode ray studies. We have alluded to its importance in chapter 1, and will have much more to say about it later. For now, we are not amiss in anticipating Schuster's subsequent limits. His upper limit for nitrogen was 1.1×10^6 emu (later revised to 3.5×10^6 emu), assuming the particles cross the region of potential difference without any loss in energy, which is the assumption underlying equation (6.2). His lower limit, 10^3 emu, assumed the particle's velocity is reduced, by numerous collisions with residual molecules, to the extent that it is in statistical equilibrium with the surrounding gas; consequently v is the velocity of a gas particle at the temperature of the discharge, as given by the kinetic theory of gases. Since the lower value came very near that which is found in electrolysis—Schuster's model for the discharge phenomenon in the first place—whereas the upper value seemed unrealistic in view of the shaky assumption of no energy loss, Schuster settled for the lower value. In hindsight, had he 'had more faith in his own experiment', he might have concluded that the mass of the particles in the discharge (Schuster eschews the term 'cathode rays') was much smaller than that of electrolytic ions [6-27]. In the event, J J Thomson stole the spotlight in due course, though that is definitely getting too far ahead of our chronology.

6.2. Thomson and Schuster

We last met J J Thomson during his unexpected appointment to the Cavendish Professorship in December of 1884, about the same time that Schuster's first Bakerian Lecture went to press. (Schuster, as we recall, was one of numerous applicants for that prestigious post—the second time he ran against Thomson, having previously beat him to an Owens College chair.) Immediately following his election to the Professorship, Thomson (as he tells us), excited by the beautiful experiments of William Crookes on the electrical discharge in gases and with the intuitive conviction that here lay the experimental key to the structure of matter and the nature of electricity, began some experiments of his own in collaboration with his colleague and close friend Richard Threlfall [6-28]. Threlfall was very skillful with his hands, something Thomson was decidedly not, despite having blown off several fingers in a boyhood experiment with explosives. Threlfall had just taken a First Class in the Natural Sciences Tripos, and soon succeeded T C McConnel as Assistant Demonstrator at the Cavendish. He was diverted from this assignment almost at once, however, by an appointment as professor in physics in the University of Sydney, which he could hardly turn down. His post at Cambridge passed on to Hugh Frank Newall, the future astrophysicist.

As luck would have it, in time the Antipodes more than made up for snatching Threlfall (who eventually gave up his generously endowed chair at Sydney); they furnished Thomson with one of the first 'Empire Students', when a research program leading to a degree on the German model was instituted at Cambridge in 1895. The student, a farmer's son, came from Canterbury College, New Zealand, and his name was Ernest Rutherford.

> Although the senior members of the University accepted the new arrivals with enthusiasm, there were one or two demonstrators with the ancient prejudice that no good things can come to pass from the Colonies. Rutherford told the writer [A S Eve] that these men used to pass his door at the Laboratory with a snigger. He politely asked them in, and told them he was in some difficulties with his experiments [on Hertzian waves with a magnetic detector, something he began in New Zealand] and would be grateful for their help. They quickly realized that they had not the faintest idea of what he was doing... After that, they gave me no more trouble; they had got on my nerves a bit'. [6-29]

To be more specific on his choice of research, Thomson explains that he was drawn to the electrical discharge because of his wish 'to test the view that the passage of electricity through gases might be analogous to that through liquids, where the electricity is carried by charged particles called ions' [6-30]: in other words, precisely by what also motivated Schuster's work. In Thomson's view, the current results from molecules split into positively and negatively charged ions in the discharge tube. Experiments with iodine and bromine subjected to a spark discharge appeared to confirm this view; evidently vapor densities after the discharge were lower than before, suggesting some degree of molecular dissociation [6-31]. One of several other collaborators in these experiments besides Threlfall was H F Newall, distinguished as the first undergraduate to work under Maxwell at the Cavendish, and, as noted, Threlfall's successor. Newall had started out as a salaried private assistant to Thomson. He soon succeeded W N Shaw as full Demonstrator at the Laboratory, a post he retained until 1900 when he took charge of a 25 in refracting telescope his father had bequested to Cambridge University. With him, Thomson studied the dielectric breakdown in poorly conducting liquids such as benzene and paraffin oil [6-32], but without shedding much light on the discharge mechanism.

Figure 6.1. *J J Thomson.*

At about the same time, Thomson began experiments on electrolytic conduction—experiments necessitated by circumstance in the untimely departure for an engineering appointment in India of D S Sinclair, superintendent of the Laboratory workshops and an indispensable glass blower. Sinclair's departure caused a dearth in vacuum tubes for a period before his replacement came on board. Marking time, Thomson took up his electrolytic experiments somewhat casually—experiments nevertheless highly germane in light of the conduction analogy. However, attempts to determine whether molecules of electrolytes are, indeed, decomposed by the passage of a current, or whether the current simply redirects the motion of already dissociated atomic fragments, were unsuccessful; passage of the current through a salt solution produced no detectable change in osmotic or vapor pressure [6-33].

While this was going on, lectures and demonstrations could not be neglected. The same year as the electrolytic experiments, 1887, W N Shaw, who with Richard Glazebrook had organized practical physics instruction at the Laboratory in the first place, had to drop his duties as Demonstrator upon accepting a tutorship at Emmanuel College.

He was succeeded by Newall, though he agreed to continue his course of lectures in physics, as did Glazebrook. (Indeed, 'Glazebrook and Shaw' was for many years the premier textbook in practical physics in Britain.) In addition, physics was taught by L R Wilberforce who succeeded Newall as Assistant Demonstrator, and by Thomas Cecil Fitzpatrick who also was appointed Assistant Demonstrator in 1887. When Thomson instituted classes for medical students the following year, the attendance grew so fast that the Laboratory could no longer accommodate the students. An interim solution saved the day with an iron shed that had served as a dissecting room. 'It was not, to begin with, all that could be desired, for such a nauseating odour clung to it that Fitzpatrick and Wilberforce had to exhaust all the resources of physics and chemistry to dispel it' [6-34].

Thomson himself also lectured, on both physics and mathematics, and to both elementary and advanced students, accompanied by demonstrations.

> As a lecturer to elementary students, Thomson was hardly to be surpassed. He had that rare quality of not going too fast, or trying to cover too much ground: thus he avoided being over the heads of the slower students; at the same time he was never so slow as to bore the quicker ones. Though the main outlines of the subject were fully emphasised, he often introduced speculative ideas which reflected his own recent thoughts or historical matter from his own recent reading. He often showed one or two experiments, but these, though introduced in an appropriate place, were not allowed to occupy the forefront of the picture. The lectures were not an exhibition of showmanship. They were intended to teach serious students and were quite exceptionally stimulating to any intelligent pupil. [6-35]

One student who attended the various lectures in 1887, and, in particular, Thomson's own lectures and demonstrations the following year, was Miss Rose Elizabeth Paget, daughter of Sir George and Lady Paget and sometime student at Newnham, the women's college. (Sir Paget was Regis Professor of Physics at Cambridge.) The admission of Newnham and Girton women students to Cambridge was a heated issue in the Senate House at the time of Thomson's Cavendish appointment. Their admission was finally voted in the affirmative, though only much later were women entitled to a degree. Thomson himself was rather cool on the matter, as is evident in his letters to Threlfall in 1887.

> There is a great agitation going on to admit women to all the University privileges that men have and it seems a most ill-advised thing as it would involve their sitting on boards, etc. and it has divided the friends of the women nearly as much as Home Rule has divided the Liberals. Sedgwick [Reader in Animal

> Morphology at Cambridge] in his usual vigorous way, declares that unless they drop the agitation he will turn all the women out of his laboratory, and not allow them to attend his lectures. Is yours a mixed university?...[6-36]

Admitting nevertheless 'that it would [not] do the university very much harm' [6-37], Thomson sided with the Senate House vote at the turn of the year, and a good thing it was. Thomson always made it a point to look in on his students, no doubt including Miss Paget. In 1889 she undertook, on his suggestion, some laboratory work upstairs at the Cavendish on vibrating soap films—something Thomson soon decided needed his personal assistance. One day he came downstairs in a splendid mood, followed by Miss Paget with a flush on her cheek. It abruptly spelled the end of her experiments; within six weeks they were married [6-38].

Along with his marginal experiments on the electric discharge grew a change in Thomson's views on the nature of the electric current [6-39], from explaining it simply in terms of molecular dissociation to one in which the current was viewed as the dissolution of 'tubes of electrostatic induction' or 'Faraday tubes'—harking back to his earlier work on vortex filaments [6-40]. Such tubes must either start and end on opposite electric charges or form endless loops in free space devoid of electric fields. In a discharge tube they would stretch from the atoms of one electrode to those of the other, as they would between atomic constituents of polarized, residual gas molecules. Borrowing the Clausius–Williamson view of chemical combinations, a continuous molecular exchange of partners with the atoms of the electrode resulted in contractions or elongations of the tubes of force; the displacement energy released in the process constituted the electric current [6-41]. However, though the electric discharge arises in principle from an interaction between the electric field and molecules, the complications encountered in the discharge phenomena are due to the presence of the electrodes. Since 1886, Thomson had sought to eliminate these complications by producing an electrodeless discharge in an endless ring. He finally succeeded around 1890 with a clever scheme in which the path of the discharge was the secondary circuit of an alternating current transformer. The primary circuit was a few turns of wire wound around a highly rarefied vacuum bulb, excited by discharging a Leiden jar through it [6-42]. The electrodeless discharge proved to be very useful as a brilliant spectroscopic source; it was less successful in providing quantitative insight into the discharge mechanism because of the rapidity of the alternating current [6-43].

As noted, Schuster too was inspired by the analogy between the gaseous discharge and electrolytic conduction—a subject occupying some of his attention at Owens College since returning from the latest solar eclipse expedition in 1882. Though influenced by the views of his eclipse partner Norman Lockyer, his avowed guide was the electrolytic theory of Helmholtz [6-44], whom Schuster knew from his early sojourn at Helmholtz's laboratory in Berlin and whom he occasionally met while summering at Pontresina in Switzerland's Grabunden area. (Schuster was a keen mountaineer despite being plagued by poor health—one reason why he spent most of his summer vacations abroad.) According to Helmholtz, the molecules of the electrolyte are split into atoms carrying a definite quantity of electricity, one a negative and one a positive charge. Tests with monatomic mercury vapor appeared to show less glow, but were inconclusive in showing whether the atomic constituents carried one of Helmholtz's elementary units of electric charge. Another test, performed somewhat later, involved a discharge tube containing the usual pair of electrodes in one part and two auxiliary electrodes mounted elsewhere in the same tube. Provided a continuous current was maintained between the auxiliary electrodes, a small e.m.f. sufficed to initiate a steady current between the primary electrodes [6-45]. This Schuster explained in terms of ions produced in the main discharge diffusing through the vessel; influenced by the electric field between the auxiliary electrodes, they drift as a continuous current between the main electrodes.

> Supposing the difference of potential of two electrodes is gradually increased, a point will be reached at which a spark will pass, that is to say, the molecules will be broken up by electric forces, the positive ions diffusing towards the kathode will tend to form a polarising layer of finite thickness, increasing in width as the pressure diminishes. If the discharge becomes steady, the decompositions are continuously kept up at the kathode, the negative ions being projected with great velocity away from it. These ions will move through the so-called dark space without much loss of energy by impacts, but when, probably owing to sufficient diminution in the electric force, the impacts become more frequent, the translational energy becomes transformed into the luminous vibrations of the glow. The positive ions forming an atmosphere round the kathode must have a greater velocity the nearer the kathode, where their energy becomes visible in the first luminous layer. Whether decompositions take place only at the electrode or through a finite distance from it is at present an open question, nor can we decide as yet whether the negative molecules projected outwards are the main carriers of the current inside the dark space. [6-46]

A better test of Schuster's theory seemed to be a measurement of the charge-to-mass ratio of the particles with the aid of a magnetic field. As discussed in the previous section, only in his second Bakerian Lecture of 1890 was he able to offer a definite, numerical prediction for e/m and compare it with known electrochemical equivalents.

> In my former paper I described a method by means of which I hoped to be able to measure the charges carried by the ions, and thus directly to test the truth of the theory. It is clearly most desirable that this should be done, for if it could be shown that the molecular charges are the same as those carried by the atoms in electrolytes, all further doubt as to the correctness of the view which I advocate would vanish. I have met with very considerable difficulties in the attempt to carry out the measurements in a satisfactory manner, and have only hitherto succeeded in fixing somewhat wide limits between which the molecular charges must lie. [6-47]

His limits for e/m, as noted above, were 10^3 and 10^6 emu.

> The actual value of e/m is 10^4 for hydrogen and 0.7×10^3 for nitrogen, if we imagine each atom of nitrogen to carry the same charge as the atom of hydrogen in water; but, as nitrogen may unite with these atoms of nitrogen we must assume three charges at least to be carried, which would make e/m equal to 2×10^3. It thus appears that there is nothing in the actual facts which is in any way not in harmony with the theory. The lower limit for e/m comes very near the actually observed values, and it is not astonishing that the upper limit yields so great a value... I think I may take the experiments hitherto recorded as a confirmation of the theory. [6-48]

Two basic suppositions were involved here—suppositions quite at variance with the views of Schuster's British colleagues, and vigorously defended in the conclusion to his Bakerian Lecture. One was the possibility of a *volume electrification*; that is, electrification distributed throughout a volume, not simply confined to the surface of a conductor [6-49].

> The possibility of a volume electrification is denied by some of Maxwell's disciples, who look on a current of electricity as on a flow of an incompressible liquid in a closed circuit. But there is nothing, as far as I can see, in the conclusions I have drawn from the gas discharges which is inconsistent with the fundamental tenets of Maxwell's theory, however much they may disagree with the accessory embellishments with which that theory is occasionally adorned. There may be a volume electrification without interfering with the equation of continuity of an incompressible liquid as long as we admit the possibility of displacement currents and displacements in conductors, and I see nothing improbable in this. [6-50]

The second offense against the Maxwellians was couched in a reconciliatory tone.

> I have offended in another manner against so-called modern views of electricity, for I have spoken of positive and negative electricity as real substances possessing a separate existence. I have tried to place myself, however, under the shelter of recognised authority by quoting at the top of this lecture Helmholtz's saying that we have as much ground for the supposition that electricity has an atomic constitution as we have for the atomic constitution of matter. We must trust to the future to bring this view into harmony with the electromagnetic theory of light, which may be accepted now as an established fact. There is no real antagonism between the two views. If ever we are able to explain chemical and gravitational attraction by the stresses in a medium, we shall still find it convenient to speak of atoms and molecules; and in the same way the belief in an electric strain and stress is consistent with a belief in something in the atom from which the strain proceeds, and which may be taken as the elementary quantity of electricity. Even taking the extreme view that electric stress is due to vortex filaments in the ether, we need only assume all these filaments to have the same intensity, and some to end at the surface of atoms, in order to reconcile apparently antagonistic views. But there is no need to commit ourselves at present to any particular ideas. In some electric phenomena we shall find it most convenient to speak of electric strain and stress (displacement I think to be a misleading term, which, however, has come too much into use to be dispensed with); in other, and at present more numerous, cases, we shall still continue the old nomenclature, and speak of positive and negative electricity as real quantities. [6-51]

Schuster would remain somewhat distressed over the lack of recognition by the community of his contributions in illuminating the nature of the gaseous discharge. In particular, in his retrospective lectures on the history of the subject delivered at Calcutta in 1908, he complains that his major paper of 1890, containing his second Bakerian Lecture, was ignored by J J Thomson [6-52] in a paper on the same subject published by Thomson in the *Philosophical Magazine* for 1894. In it, Thomson obtains a value for e/m in agreement with the electrolytic value for hydrogen; the paper ends with the statement that 'I cannot find any quantitative experiments on the deflection of [cathode] rays by a magnet' [6-53]. A letter of complaint from Schuster to Thomson is no longer extant [6-54]. However, Thomson's reply appears as a letter to the editor in the *Philosophical Magazine* for 1896; it is reprinted in Rayleigh's *Life of J J Thomson*.

Dear Schuster,
I am exceedingly sorry to find that my papers have given you the impression that I wished to slight your work. I can assure you that nothing is or has been further from my intention. I will confess that I am not as well acquainted with your second Bakerian Lecture as I ought to be. It was published when I was very busy writing my *Recent Researches*. I had practically finished the part relating to the discharges through gases; when I read your paper it was with reference to the insertions it would require me to make and I no doubt passed over the part of your paper relating to parts of the subject which I had not introduced into the Chapter on Discharge through Gases too hurriedly. I cannot, however, plead having forgotten Stanton's experiments on the discharge from red hot copper for I had them in my mind when I was writing the paper in the Dec. number of the *Phil. Mag.* and I fully meant to have referred to them, it was a mere accident that I did not. I am writing by this post to the Editor of *Phil. Mag.* in which I refer to his experiments and point out that you more than five years ago stated that the negatively electrified atoms moved faster than the positive. If I had known of your results earlier, it would have saved me a great deal of time and worry, for when some of my experiments seemed directly to suggest it I had the greatest repugnance to the idea and fought against it much longer than I should have done if I had known it had commended itself to you. With reference to the third point I am not quite sure as to which paper you refer. The only reference I remember to have made on the subject of the deflection of the cathode rays by a magnet was at the end of a paper on the velocity of these rays when I made a rough calculation as to whether the deflection by a magnet was compatible with the velocity I had found. I certainly did not know until I received your letter that you had made a series of measurements of the deflection of the rays by a measured magnetic force, and I should not have thought of looking in your first Bakerian Lecture as I thought I did remember that paper, and that it was about the discharge through mercury vapour and the residual effects observed near the poles when the current was reversed.
I can only express my regret that my ignorance should have caused you any annoyance. [6-55]

The *Recent Researches* which Thomson refers to in the above was written as a sequel to Maxwell's *Treatise* [6-56]. At the request of the Clarendon Press, Thomson had edited the third edition of Maxwell's work, and found the need for adding extensive footnotes, mainly on account of the advances in electromagnetism since the *Treatise* originally went to press, but also in the interest of pedagogical clarity. In the end, he decided a supplementary volume was the wiser course. The volume contains chapters on Faraday tubes of force, on the discharge of electricity through gases, on conjugate functions, on electric waves, on the distribution of rapidly alternating currents, and on the electromotive intensity in moving bodies. Except for the chapter on the electrical discharge, which is largely experimental, the book is highly mathematical and (like Maxwell's *Treatise* for that matter) not easy going [6-57]. The experiments by Arthur Stanton, Schuster's assistant, appear in an appendix to Schuster's Bakerian paper of 1890, and show that the process of oxidation strongly affects the electrical discharge from red-hot copper or iron—a fact of importance for Schuster in showing that the discharge from hot bodies is accompanied by chemical action [6-58].

At about this time a quite separate development in physics ensued—one that effectively spelled the end of Schuster's work on the gaseous discharge, and that we dwelt on at the start of this volume. Returning to Manchester after a Christmas holiday in early January 1896, Schuster found an envelope in his office mailbox without a letter or clue to the sender, but carrying photographs containing shadow images of a human hand clearly showing the bone structure, and a copy of Röntgen's communication entitled 'Über eine neue Art von Strahlen' [6-59]. This was the first news of Röntgen's discovery of x-rays to reach England. As elsewhere, attempts to repeat Röntgen's sensational discovery were made at once in most English laboratories. Schuster was in a better position than most, having the requisite high-vacuum pumps and other apparatus in hand, notwithstanding the fact that English lead glass was less suited for generating and transmitting the rays than soft German glass. The medical significance of the radiation was obvious to one and all, not least the medical profession. (Indeed, within a year and a half of Röntgen's discovery, medical units accompanying the British expedition to Omdurman on the Nile had x-ray equipment for diagnosing wounds.) Schuster was inundated by physicians with patients in tow, requesting x-ray images of bodily parts containing needles, bullets, or other metallic objects.

> The discharge tubes had all to be prepared in the laboratory itself, and where a few seconds exposure is required now, half an hour had to be sacrificed owing to our ignorance of the best conditions for producing the rays. More difficult problems also arose, as when I had to travel to a small manufacturing town in the north of Lancashire, in order to locate a bullet in the skull of a poor dying woman who had been shot by her husband. My private assistant [A Stanton] completely broke down under the strain and excitement of all this work, and the experiments on the magnetic deflexion of kathode rays on which I was then engaged were seriously interfered with by this interruption. [6-60]

Except for a few desultory experiments, Schuster never seriously returned to cathode ray studies. He was, however, the first to suggest, if hesitatingly, that the x-rays are high-frequency *transverse* vibrations of the æther, contrary to Röntgen himself who tentatively assumed they were longitudinal vibrations [6-61]. Ironically, Röntgen's discovery would prove pivotal in J J Thomson's eventual elucidation of cathode rays.

Chapter 7

FROM PARIS TO THE SCOTTISH HIGHLANDS

7.1. France enters the fray: Thomson and Perrin

At Cambridge, meanwhile, Thomson concluded that the electrodeless discharge was not so promising after all in shedding light on the discharge, due to the oscillatory nature of the current and its high frequency (10^8 cycles s^{-1}). Instead, he took up an experiment first carried out by Adolphe Perrot in 1861 on the electrolysis of steam by the electric spark [7-1]. Perrot had found that hydrogen was preponderantly liberated at the negative electrode, and oxygen at the positive electrode. In addition, the amount of hydrogen liberated was equivalent to the amount of copper deposited in a voltameter inserted in the circuit.

Thomson repeated the experiment in 1893, while in the midst of bringing out the third edition of Maxwell's *Treatise* and the third volume of his own *Notes on Recent Researches*. He found that the excess of hydrogen was *not* always at the cathode, but under certain conditions at the anode [7-2]. All in all, this finding must have shaken his faith in the electrolytic analogy; at any rate, he never referred to the subject in his later books [7-3].

The year 1893 was also when Thomson instituted what was termed the 'Cavendish Physical Society'. This was not a society in the usual sense, but rather a regularly scheduled colloquium held every fortnight—the first of its kind in England, though a standard practice on the Continent. Thomson held sway in the still extant two-story Maxwell Theater room of the Cavendish, with the meeting open to anyone interested. Somebody, usually Thomson, would report on some ongoing work or interesting paper. Mrs Thomson served tea prior to the meeting, occasionally assisted by other ladies. To avoid its degenerating into a social affair, Thomson insisted on the plainest possible cups... and saucers! [7-4].

A year later, Thomson attempted a determination of the velocity of cathode rays. As we recall, both a fellow Scottish colleague (Tait) and Goldstein in Germany had attempted an indirect velocity determination from the Doppler shift back in 1880, but with inconclusive results. Thomson now tried a direct determination,

> ...as it seemed that a knowledge of this velocity would enable us to discriminate between the two views held as to the nature of the cathode-rays. If we take the view that the cathode-rays are ætherial waves, we should expect them to travel with a velocity comparable with that of light; while if the rays consist of molecular streams, the velocity of these rays will be the velocity of the molecules, which we should expect to be very much smaller than that of light. [7-5]

His experiment this time was a modified version of one he had used earlier in measuring the velocity of the positive luminosity traversing a 30 ft long vacuum tube, using a rotating mirror; the result was roughly half the speed of light [7-6]. The younger Rayleigh remembers the setup well. 'I saw the arrangement for this experiment on a visit to Cambridge about 1892, when I was taken round the laboratory by Glazebrook, but J J was not there at the moment, and I did not make his acquaintance till later' [7-7].

In his 1894 measurements, Thomson used a more modest vacuum tube covered with lampblack, except for two thin slits spaced 10 cm apart. The scheme involved observing the fluorescence of the glass at each slit as the rays passed them in succession—always passing the slit closest to the cathode first, as he soon ascertained. The light from the slits fell on the rotating mirror from his earlier experiment, placed nearby and driven by a large 'grammemachine (turntable)'. The images reflected from the mirror were observed through a telescope. When the mirror was at rest, the two images were brought into coincidence by interposing a prism between one of the slits and the mirror. The idea was to measure the velocity of the rays by hunting for that particular speed of rotation at which the two images remained in coincidence.

In the event, it proved very difficult to obtain sharp, distinctly separated images. After a number of unsuccessful attempts, Thomson obtained a velocity of $2 \times 10^7\ \mathrm{cm\,s^{-1}} = 200\ \mathrm{km\,s^{-1}}$—a result erroneously low by fully two orders of magnitude, as time would show (section 9.3). Though far slower than the previously measured velocity of the positive striations moving from the anode to the cathode, the velocity was clearly well below the velocity of light ($\sim 3 \times 10^{10}\ \mathrm{cm\,s^{-1}}$) demanded by Hertz and Lenard, yet above the molecular speed estimated by Goldstein ($16\ \mathrm{km\ s^{-1}}$, or 10 miles $\mathrm{s^{-1}}$).

> The velocity of the cathode-rays found from the preceding experiments agrees very nearly with the velocity which a

negatively electrified atom of hydrogen would acquire under the influence of the potential fall which occurs at the cathode. For, let v be the velocity acquired by the hydrogen atom under these circumstances, m the mass of the hydrogen atom, V the fall in potential at the cathode, e the charge of the atom; then we have, by the conservation of energy,

$$mv^2 = 2Ve.$$

Now e has the same value as in electrolytic phenomena, so that $e/m = 10^4$.
Warburg's experiments show that V is about 200 volts, or 2×10^{10} in absolute measure. Substituting this value, we find

$$v^2 = 4 \times 10^{14}$$

or

$$v = 2 \times 10^7 \text{cm/sec}.$$

A value almost identical with that found by experiment. The very small difference between the two is of course accidental, as the measurements of the displacement of the images on which the experimental value of v was founded could not be trusted to anything like 5 per cent.
The action of a magnetic force in deflecting these rays shows, assuming that the deflexion is due to the action of a magnet on a moving electrified body, that the velocity of the atom must be at least of the order we have found. [7-8]

For the next development, we switch to Paris and the first published research of Jean Baptiste Perrin, whom we met briefly in section 1.2. Perrin was born in Lille, France, on September 30 1870. He had been raised by his mother, along with two sisters, under somewhat trying circumstances after his father died of wounds received in the battle of Saint-Privat in the Franco-Prussian war [7-9]. He was educated at the Lycée in Lyons and at the Lycée Janson-de-Sailly in Paris. After serving a year in the army, he was admitted to the École Normale Supérieure in 1891 on the strength of his mathematical aptitude; here he became *agrégé-préparateur* in 1895. His teacher was Marcel-Louis Brillouin, who had been *maître de conferences* (lecturer) at the École Normale since 1887. Brillouin was one of the few outspoken adherents of the kinetic theory of gases and statistical mechanics in France at the time, and equally outspoken in his rejection of Ostwald's and Mach's 'energetics'. He was also sympathetic toward the use of mechanical models in scientific methodology *à la* Kelvin 'without being excommunicated from the scientific community, or being accused of backwardness' [7-10].

Brillouin no doubt saw to it that Perrin received a thorough grounding in Maxwell's electrodynamics and Boltzmann's statistical mechanics; under his tutelage Perrin became a staunch atomist, despite the prevailing anti-atomic stance in French educational circles at the time [7-11].

Indeed, Perrin's matriculation at the École Normale occurred shortly after a period when anti-atomic issues were most hotly argued among faculty members in Paris and elsewhere in France. By 1890 Henri Saint-Claire Deville, chemist at Besançon and an ardent anti-atomist, was no longer around. Other influential skepticists, notably the physicist Pierre Duhem and chemists Henri Le Chatelier and Marcellin Berthelot, were still around, though it was increasingly acceptable to conjecture about atoms [7-12]. Moreover though arguments as to the molecular reality had by no means ceased, the experiments of Crookes and J J Thomson fueled the debate. Looking back, Perrin mused that the goal of his scientific career from its inception was establishing the reality of invisible atoms [7-13]—atoms, in fact, not altogether invisible if one accepted the British interpretation of luminous cathode ray phenomena. Given his mentor's enthusiasm for corpuscles, what better experiment to launch his career than one on cathode rays?

Perrin's first piece of serious research, then, was on cathode rays, conducted in the laboratory of Brillouin and the physicist Jules Violle. His paper was read before the French Academy on December 30 1895 [7-14].

> Two hypotheses have been presented to explain the properties of the cathode rays.
>
> Some with Goldstein, Hertz, or Lenard, think that this phenomenon, like light, results from vibrations of the ether, or even that it is light of short wave length. We then easily see that these rays might have a straight trajectory, excite phosphorescence, and act upon photographic plates.
>
> Others, with Crookes or J J Thomson, think that these rays are formed of matter negatively charged and moving with great velocity. We then can easily understand their mechanical properties and also the way in which they bend in a magnetic field.
>
> This latter hypothesis suggested to me some experiments which I shall present without troubling myself for the moment to consider if the hypothesis accounts for all the facts which are at present known, and if it alone can account for them.
>
> Its partisans assume that the cathode rays are negatively charged; so far as I know, no one has demonstrated this electrification; I have therefore tried to determine if it exists or not. [7-15]

Inside his cathode ray tube, shown in figure 7.1, Perrin had mounted a coaxial insulated aluminum cylinder with a small opening in the end

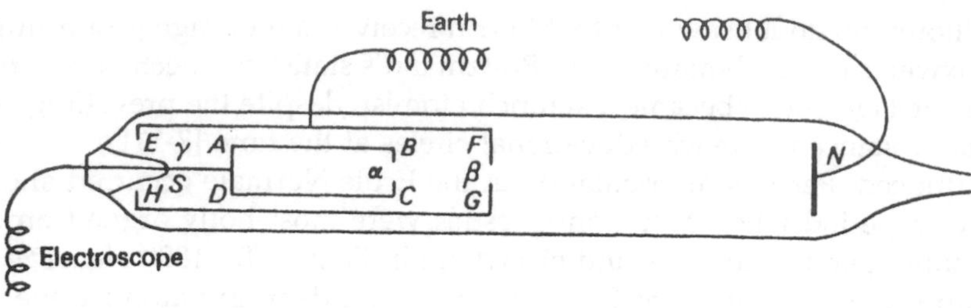

Figure 7.1. *Perrin's cathode ray tube, incorporating a Faraday cylinder for determining the charge of the cathode rays. N is the cathode, EFGH the anode in the form of a grounded cylinder, screening the Faraday cylinder within, ABCD.*

facing the negative electrode—a cylinder known as a Faraday cylinder or Faraday cup (after Faraday's much larger electrical cage used in his Friday evening lectures). A metallic wire sealed in the rear wall of the tube connected the cylinder to an electroscope. The cylinder was enclosed in a larger metallic cylinder which was grounded. It screened the Faraday cylinder from external electrical influences, and acted as an anode for attracting negatively charged particles from the cathode, if indeed they existed. When the tube was in operation, the electroscope indicated that the Faraday cylinder invariably became charged with negative electricity. The tube could also be placed between the poles of an electromagnet. When it was excited, the cathode rays were deflected and no longer entered the Faraday cylinder, which consequently received no charge.

It was not easy to reconcile these results with the supposition that the rays were some form of ætherial waves. However, they were readily understood if the rays consisted of negatively charged particles, in line with the British view.

At this point, we must deviate slightly from the strict historical chronology, and turn to a modification of Perrin's experiment by J J Thomson, one first described by Thomson before the Cambridge Philosophical Society on February 8 1897. Perrin's experiment was open to the objection that it did not conclusively prove that the cause of the negative electricity recorded in the electroscope had anything to do with the cathode rays—that is, that the negative electricity collected was, in fact, bound to the rays. Thomson's discharge tube differed from Perrin's in one vital respect: the Faraday cylinder was not coaxial with the undeflected cathode ray. The cathode, a plane metallic disc, was placed in a small side tube fused onto a much larger bulb. The rays passed through a hole in a grounded brass plug constituting the anode. Emerging into the larger bulb, the rays impinged on the opposite wall of the bulb at a spot glowing softly from bluish phosphorescence. In the absence of a magnetic field, this spot was well removed from the entrance to the collecting cylinder. The collector, like Perrin's, consisted of a pair of coaxial cylinders, the outer grounded and the inner insulated from it and connected to an electrometer. With an electromagnet straddling the apparatus, the rays could be deflected until the fluorescent spot vanished from view, showing that they fell squarely on the aperture of the collector; the electrometer then suddenly registered a large negative charge. When the rays were further deflected so as to overshoot the entrance to the collector, the charge fell to a very low value.

> This experiment [wrote Thomson] shows that however we twist and deflect the cathode rays by magnetic forces, the negative electrification follows the same path as the rays, and that this negative electrification is indissolubly connected with the cathode rays. [7-16]

The experiment showed something else, as well. If the rays were deflected so as to fall continuously on the aperture of the collector, the negative charge in the cylinder did not increase indefinitely, but reached a limiting value and then remained constant, even though the rays continued to pour into the cylinder. Thomson concluded that the residual gas in the bulb was rendered conducting by the rays passing through it, and that a steady state was reached when the collector lost as much negative electricity by gaseous conduction to the anode as it gained from the cathode rays. This he confirmed by giving the inner cylinder an initial negative charge. If this charge was less than a certain amount, the inflow of the cathode rays increased it; if above a certain amount, some of the initial charge leaked away until the loss by conduction equalled the gain from the rays received [7-17].

These results from Paris and Cambridge would be ruefully overshadowed by a momentous piece of scientific news shortly after the new year of 1896—news which, however, had particular bearings on the conductivity of gases exposed to radiation.

7.2. Preoccupation with Röntgen rays; research students

One month after Perrin's experiment, news of Röntgen's discovery of x-rays reached Arthur Schuster in Manchester—a discovery which 'relieved the gloom of January 1896' [7-18] and that would have tremendous ramifications for the cathode ray studies. Thomson must have heard of the discovery at just about the same time, since a translation of Röntgen's paper to the Würzburg Medical Society appeared in *Nature* on January 23. Indeed, reports of the rays must have reached Cambridge before then, since popular newspapers—even *Punch*—had picked up the news from German newspaper articles [7-19]. It appears, moreover, from Rutherford's letter to Mary Newton, his long-term fiancée, that Thomson must have lost no time in having a copy made of Röntgen's tube.

> *Cambridge, 25 January, 1896*
>
> The Professor has been very busy lately over the new method of photography discovered by Professor Röntgen and gives a paper on it on Monday to which of course I will go. I have seen all the photographs that have been got so far. One of a frog is very good. It outlines the general figure and shows all the bones inside very distinctly. The Professor of course is trying to find out the real cause and nature of the waves, and the great object is to find the theory of the matter before anyone else, for nearly every Professor in Europe is now on the warpath.... [7-20]

Thomson did indeed expound on Röntgen's paper before standing room only at the Cambridge Philosophical Society on January 28—the first meeting of the Society that year [7-21].

The first order of business was, as Rutherford noted, establishing the nature of the new-found radiation. Much was already known about the rays, thanks to Röntgen's perseverance. They were emitted from the region where the cathode rays struck the wall of the discharge tube; they traveled in rectilinear paths and threw regular shadows; their penetrating power in matter depended solely on the density of the absorber; in addition to activating barium platinocyanide and other fluorescent substances, they produced images on photographic plates that could be developed. They were neither refracted nor underwent regular reflection or polarization to any appreciable degree. Most significantly, they were not deflected by a magnetic field.

The photographic aspect naturally evoked the greatest interest in the rays, much to the annoyance of Röntgen himself; thus they were widely hailed as a breakthrough in photography, rather than in physics [7-22]. However, the self-evident medical importance of this aspect could hardly be ignored. Before long Thomson, like Schuster, was inundated with requests for photographs of patients brought by local surgeons not yet equipped for producing radiographs of their own. The early x-ray tubes at the Cavendish were made by Mr Ebenezer Everett, Thomson's lifelong assistant, who had his start as a boy under Professor G D Liveing in the Chemical Department. They were operated by W H Hayles, the lecture assistant and photographic expert at the laboratory. Hayle's photographs were sometimes unsettling to patient and surgeon alike, revealing broken bones poorly reset in the past.

> One of the patients was a prominent member of the University who could express himself strongly. He had broken his arm and, as it did not heal, he insisted on having it Röntgen-rayed and brought his doctor with him. My assistant came back after a shorter time than usual. I asked him why. He said, 'I came away because I thought the doctor would not like me to hear what Mr. — was saying to him after he saw the photograph'. [7-23]

A letter from Thomson to Oliver Lodge is telling, both for his earliest ideas on x-rays and for his attempting in vain to obtain photographs with a plate in an ebonite box mounted *inside* the x-ray tube.

> *6 Scroope Terrace. Jan. 19th, 1896.*
>
> I think the cause of the Röntgen photographs must be something not quite identical with the ordinary cathode rays as Röntgen finds that they are not affected by a magnet. I have been trying to get the photographs by putting the plate inside the vacuum tube so that the cathode rays may fall directly on a little ebonite box enclosing the plate and are so prevented from striking against the glass and causing phosphorescence. Under these circumstances I have never succeeded in getting the photographs. This looks as if phosphorescence as well as a cathode was necessary. That phosphorescence without a cathode is not sufficient is, I think, shown by some experiments I have tried with my bulbs without electrodes which though showing strong phosphorescence on the glass were quite inoperative as far as these photographs were concerned. It seems to me that both a cathode and phosphorescence are necessary.
>
> On the whole I incline to the opinion that they are due to waves so short that the wave length is comparable to a molecule. Whether these waves are transverse or longitudinal is, I think, at present an open question. The absence of refraction is, I think, not surprising for these very short waves. If the excess of the specific inductive capacity over unity is due to the molecules setting themselves under the electric field we should, I think, expect it to be very small for these small waves. For supposing the half wave lengths were just equal to the length of the molecule, the force on the positive atom would be equal and parallel to the force on the −atom and there would not be a couple tending to make the molecule set, but merely a force tending to push it along.
>
> I am trying to find whether there is any motion of the ether close to a cathode. [7-24]

Indeed, Thomson's intuition in regard to the absence of refraction is right on the mark; the extremely high frequency (short wavelength) of the vibrations was bound to defeat all contemporary attempts to refract the rays, due to the smallness of $1-\mu$, where μ is the index of refraction.

The very next day after his lecture before the Cambridge Philosophical Society, Thomson made an important discovery: the x-rays discharged electrified bodies, in the process rendering the surrounding gas temporarily conducting. In this work, Thomson was assisted by John A McClelland, one in the first crop of research students at Trinity College. Their result was communicated to the Royal Society at its meeting on February 13 [7-25]. The experiment went as follows. Everett's evacuated tube and the Rühmkorff coil used in charging it were mounted in a wooden box covered by tin plate—a precaution devised to screen their highly sensitive current-measuring electrometer—a mainstay device at the turn of the century physics laboratories—from electrostatic noise invariably generated by the coil [7-26]. A hole in the box allowed the x-rays to emerge. An electrified plate outside the hole was connected to one of the quadrants of the electrometer. The two pairs of quadrants were normally physically connected, but electrically insulated, and the insulation of plate and quadrants was good enough that the quadrants remained at the same potential. When the x-rays struck the plate, however, a violent leakage of electricity from it took place, promptly driving the spot of light reflected from the mirror of the electrometer off scale.

The leakage effect appeared to differ in several important respects from the photoelectric effect discovered by Hertz nearly a decade earlier (section 5.2) and, most recently, studied by the curiously inseparable but brilliant team of Julius Elster and Hans Geitel [7-27]. That effect, the liberation of negative electricity from a metallic plate illuminated by ultraviolet light, only involved the leakage of negative charge, whereas x-ray bombardment discharged either negative *or* positive electricity, depending on the initial charge of the plate. In addition, the photoelectric effect liberated negative electricity whether or not the metallic surface was negatively charged, though Thomson was unable to detect any leakage of charge from x-rays striking an *uncharged* plate. The discharge from x-rays appeared to be a much more sensitive indicator of the ray's existence than a photographic plate. The effect occurred even when the electrified plate was embedded in a solid insulating material, such as paraffin wax or ebonite.

> Hence [concluded Thomson], when the Röntgen rays pass through a dielectric they make it during the time of their passage a conductor of electricity, or that *all substances when transmitting these rays are conductors of electricity*. The passage of these rays through a substance seems thus to be accompanied by a splitting up of its molecules, which enables electricity to pass through it by a process resembling that by which a current passes through an electrolyte. By using a block of solid paraffin in which two pairs of electrodes are embedded, the line joining one pair being parallel, that joining the other pair perpendicular, to the Röntgen rays, which were kept passing through the block, I found that there is but little difference between the rate of leakage along and perpendicular to the rays. [7-28]

In the above, we have referred to J A McClelland. As noted, McClelland was one of a handful of young men initially responding to the milestone admission in 1895 of postgraduate students to Cambridge from other universities, whether at home or abroad, as 'Research Students'. After two years in residence, if they submitted a thesis outlining their research work, and providing the work was deemed sufficiently original, they were entitled to a Cambridge degree. At first the degree was simply a Master of Art, but in a few years it was replaced by a Doctor of Philosophy—a new Cambridge degree created specifically for the benefit of the Research Students. The PhD was instituted to overcome the disadvantage of Cambridge graduates in competing for teaching positions with graduates from German universities where the doctorate was routinely granted.

Three of the Research Students are of particular interest in the present context. First to arrive was Ernest Rutherford, who was followed by the second, John Townsend, within the hour.

Ernest Rutherford was a farmer's son, born in 1871 in Brightwater near Nelson on New Zealand's South Island [7-29]. He had six brothers and five sisters, three of whom died in childhood. At the age of ten, late in his preliminary schooling, Ernest was inspired by Balfour Stewart's popular text in physics. Two years later the family relocated to Pelorus Sound, where he attended secondary school, followed by a stint at Nelson College; there he easily won a scholarship covering a year's expenses. In 1889, on the basis of a second scholarship, he entered Canterbury University College in Christchurch.

No sooner had he settled in at Canterbury, than Rutherford was stimulated by Heinrich Hertz's detection, the year before, of electromagnetic waves predicted by Maxwell, and he began a research project on a device for detecting such waves by means of magnetized iron wire. The idea was to capitalize on the fact that a magnetized iron wire exposed to a high-frequency discharge loses some of its magnetism when placed inside a solenoid. Two papers of his resulted—both accepted for publication in the *Transactions of the New Zealand Institute*. With that accomplishment under his belt, and having avidly soaked up everything written by J J Thomson, in 1895 he applied for and handily won an 1851 Exhibition Scholarship to Cambridge [7-30]. For the moment the honor did him no good, as he was forced to borrow money for the long passage to England. He stopped en route at Adelaide University in South Australia, where he called on William H Bragg, who at the time was attempting to operate a Hertz oscillator. Rutherford's detector made a good impression on Bragg, who was slightly older than Rutherford and already a professor of physics at the ripe age of 24 [7-31]. In later years, Bragg's son William L Bragg would succeed Rutherford at both Manchester and Cambridge.

Rutherford arrived in London in September 1895, promptly coming down with a sore throat and a severe case of neuralgia from the city's unhealthy air. A letter from Thomson awaited him.

If you could spare the time to come to Cambridge for a few

> hours, I should be glad to talk matters over with you; so much depends upon the requirements and intentions of a student that a personal interview is generally more efficacious than even a long correspondence. In case you decide to visit Cambridge, if you will let me know when to expect you, I will arrange to be at the Laboratory; if equally convenient to you, Mondays, Wednesdays or Fridays are the days I should prefer, though just at present I could arrange for an interview on any day. I am much obliged to you for your paper, I hope to take an early opportunity of studying it. [7-32]

Recovered somewhat from his odorous affliction, in no way thanks to the stuff doled out by London's chemists, Rutherford reached Cambridge by railway express, and caught a cab to the Cavendish.

> I went to the Lab. [wrote Rutherford to Mary] and saw Thomson and had a good long talk with him. He is very pleasant in conversation and is not fossilized at all. As regards appearance he is a medium sized man, dark and quite youthful still: Shaves, very badly, and wears his hair rather long. His face is rather long and thin; has a good head and has a couple of vertical furrows just above his nose. We discussed matters in general and research work, and he seemed pleased with what I was going to do. He asked me up to lunch to Scroope Terrace where I saw his wife, a tall, dark woman, rather sallow in complexion, but very talkative and affable. Stayed an hour or so after dinner, and then went back to Town again. I have forgotten to mention *the* great thing I saw—the only boy in the house—3 1/2 years old—a sturdy youngster of Saxon appearance but the best little kid I have seen for looks and size. Prof. J. J. is very fond of him and played about with him during lunch while Mrs J. J. apologised for the informality. I like Mr and Mrs both very much. She tries to make me feel at home as much as possible, and he will talk about all sorts of subjects and not shop at all. [7-33]

The little kid, G P Thomson, would himself, like his father, one day receive the Nobel Prize in physics—for slightly different but related subjects, in contrast to Bragg and his son who shared the Prize for the same investigation [7-34].

Rutherford had barely moved into lodgings, found for him by Mrs Thomson, when he was invited to dinner at the Thomsons; there he met a fellow Research Student also just arrived: John Searly Edward Townsend. Townsend, born in 1868 in Galway, nowadays in the Republic of Ireland, was the second son of Edward Townsend, a professor of civil engineering at Queen's College in Galway [7-35]. He was educated at Corrig School and entered Trinity College, Dublin, in 1885. There he read mathematics, mathematical physics, and experimental science, and became an excellent student of George F FitzGerald. He received several scholarships in mathematics, and his BA in 1890. He was tempted to become a fellow of Trinity College, but was put off by the murderous examination in mathematics that lay in the way, though he was obviously no slouch as a mathematician. Instead, after a stint of lecturing in mathematics, he opted for the new opportunity offered for Research Students at Cambridge. Accepted and settling in at Cambridge, he and Rutherford struck it off well from the start, having no other close friends in town.

> Townsend, my particular friend at present [wrote Rutherford to Mary], is going up to his cousins at Yorkshire for Xmas. . . . I don't know whether I have described him to you. Imagine a middle-sized man, very fair hair rather scanty on top, very fair moustache and a true Irish complexion and a merry blue eye—rather good features and a very pleasant appearance altogether. He is a very fine mathematician and is a good deal of assistance to me in that way. I think it probable he and I will research together on abstruse subjects next term, for in some of the work I am doing, it is very difficult for me to do all without assistance. [7-36]

John Alexander McClelland was another Irishman who joined Rutherford and Townsend as a research student after one term. He was born in 1870, the son of an Ulster farmer of Dunallis, Coleraine. He received his early training at Coleraine Academical Institution, after which he entered Queen's College, Galway; there he obtained his MA degree in 1893. Recalls Rayleigh, he 'was of powerful build and grave demeanor, friendly and capable' [7-37]. All three, Rutherford, Townsend, and McClelland, would become indispensable in Thomson's forthcoming researches and go on to make their own marks of distinction in experimental physics—above all Rutherford.

7.3. Fire on the mountain

The experiments with McClelland continued through most of the month of February 1896, with Thomson submitting an interim letter to *Nature* on February 27 [7-38]. That the leakage of electricity through a dielectric substance was due to the condition of the dielectric, rather than the electrified metallic disk embedded in it, was shown by the fact that the leakage occurred whether the plane of the disk was parallel or at right angles to the rays. The electrified gas retained traces of conductivity for some time after the rays ceased passing through it. The leakage in air was, over a wide range of pressures, approximately proportional to the square root of the pressure. On the assumption that the air traversed by the rays was ionized, and in light of the well known law of ionic dissociation which holds that the number of ions is proportional to the square root of the pressure, their results showed that the rate of leakage was proportional to the number of ions.

In his letter to *Nature*, Thomson draws attention to two other ongoing investigations at the Cavendish pertinent to the subject at hand—i.e. the electrolytic properties of gases under the action of x-rays. One was by J Erskine Murray, who was working on the potential difference between metal plates in air traversed by x-rays. Murray confirmed that the plates behaved as if they were connected by a weak electrolyte. The other piece of research was by C T R Wilson, who was investigating the effect of x-rays on the condensation of clouds of water vapor formed by the expansion of air. He found that when the rays passed through the containment vessel, the cloud was much denser than when the rays were absent. Apparently, the rays increased the number of nuclei acting as centers of cloud condensation. Ions, with their electrical charges, turned out to be particularly effective condensation nuclei; thus, once again this favored the view that the rays turn the air into an electrolyte.

Charles Thomson Rees Wilson would loom large at the Cavendish in the years ahead, and is someone we will encounter again. He was known to his contemporaries as 'C T R', and also as 'cloud Wilson' to avoid confusion with Harold Albert Wilson of Trinity College. His fame rests on his invention of the cloud chamber, a device as important for nuclear physics in its infancy as was the discharge tube in the saga leading to the electron [7-39]. The youngest son of a sheep farmer and his second wife, he was born in 1869 in Glencorse parish in Midlothian, the Scottish county south of Edinburgh [7-40]. His father died when Charles was only four, whereupon Mrs Wilson moved to Manchester to be near her parents. Charles attended the private Greenhayes Collegiate School; it taught no science, but his interest in natural science was stimulated by the gift of a microscope from a family friend. Charles and a brother built a microtone for slicing botanical and zoological specimens, and eagerly immersed themselves in microscopy. Impressed with their scientific aptitude, an elder and prosperous stepbrother in Calcutta contributed money which enabled them both to enter Owens College, that familiar institution; Charles registered as a medical student, but with the intention of first taking a degree in science.

Young Wilson graduated with a first-class degree in zoology, but, as seems to have been the wont of Owens graduates, stayed on at Manchester for another year while preparing for an entrance scholarship to Cambridge. He was admitted to Sidney Sussex College, Cambridge, in 1888, where the chemist F H Neville became his tutor. By now, he had decided to take up physics instead of medicine—a decision no doubt spurred by his exposure to Balfour Stewart's excellent lectures on elementary physics at Owens College. He studied practical physics under Newall and chemistry under S Ruhemann, an organic chemist and former pupil of Hofmann who, we recall, had left his mark on

William Crookes. Ruhemann, like everybody else, had quarrelled with the cantankerous James Dewar, Jacksonian Professor of Chemistry at Cambridge, and, as a result, taught at Caius College—well out of range of Dewar's utterly irascible personality. Wilson also attended the lectures in physics by Glazebrook and Shaw, passed the Tripos in physics, and obtained his degree in 1892.

Upon graduating, Wilson briefly held a dual position as Demonstrator at the Cavendish and at the Caius Chemical Laboratory, followed by an appointment as Assistant Science Master at Bradford Grammar School. Finding no opportunity for research at Bradford, he returned to Cambridge where, as luck would have it, he was hired by Thomson to assist Fitzpatrick in the supervision of medical students doing laboratory work for their course in physics [7-41]. Fortunately, his duties now allowed time for some independent research as well, and this was in the nick of time, as his scientific career was suddenly and unexpectedly determined by an experience in early fall 1894 on the summit of Ben Nevis in the Scottish hills off Loch Linnhe—the highest mountain in Britain [7-42].

An observatory was maintained on lofty Ben Nevis by the Scottish Meteorological Society, having been declared open by Lord Abinger in 1883. It was the custom of the permanent observatory staff to offer a stint of mountain top observing to volunteers while staff members took their fortnightly summer vacation. Wilson volunteered in September 1894. Conditions proved rough up there, even in summer. Snow drifts from the previous winter were many feet deep. Atlantic gales howled; there was fog, snow, rain, and thunderstorms, occasionally relieved by crystal clear skies, with the outer isles visible 100 miles away [7-43].

Wilson began his solitary observations at the crack of dawn, when the sea of cumulus clouds was below the summit and shimmering in the light of the rising sun.

> The wonderful optical phenomena shown when the sun shone on the clouds surrounding the hill-top and especially the coloured rings surrounding the sun (coronas) or surrounding the shadow cast by the hill-top or observer on mist or cloud (glories) greatly excited my interest and made me wish to imitate them in the laboratory. [7-44]

Back in Cambridge, he lost no time in preparing to do just that. In essence, his project called for the production of artificial clouds. As it happened, it was known from the private experiments of the Scottish meteorologist John Aitken of Edinburgh that water vapor in the atmosphere will not condense to form fog or clouds unless dust particles are present which act as condensation nuclei. Aitken made the discovery while attempting to conjure a method for creating artificial clouds by suddenly expanding, and thus cooling, moist air in a flask.

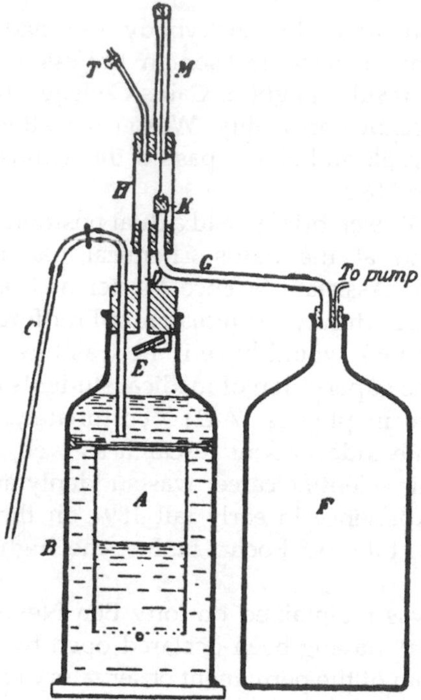

Figure 7.2. *C T R Wilson's initial cloud chamber, described in note 7-46.*

Wilson skillfully adapted Aitken's apparatus with great success, and was, indeed, able to reproduce many of the atmospheric effects he had observed on Ben Nevis—halos and glories, rainbows, aurorae, zodiacal light, and Saint Elmo's fire [7-45].

Wilson's experiments utilized a glass chamber with a moveable piston at one end, as shown in figure 7.2 [7-46]. When the pressure outside the piston was suddenly reduced, the piston rapidly pushed outward, thereby suddenly expanding the chamber volume. This adiabatic expansion cooled the enclosed air. Since, moreover, the air, being in contact with water, was saturated before the expansion and supersaturated after the expansion, the excess moisture condensed into droplets, providing dust nuclei were present. Wilson soon discovered that clouds were produced even in the absence of dust, provided a somewhat heftier expansion was resorted to.

When news of Röntgen's discovery of x-rays reached Cambridge in the fall of 1895, Wilson set out to investigate the effect of these rays, which by Thomson's experiments at the beginning of the new year were known to electrify air, on the phenomena in his cloud chamber. Behold, remembered Wilson in retrospect,

I can well recall my delight when I found at the first trial [in

February 1896] that while no drops were formed on expansion of the cloud chamber when exposed to X-rays if the expansion were less than 1.25, a fog which took many minutes to fall was produced when the expansion lay between the rainlike and the cloudlike limits; X-rays thus produced in large numbers nuclei of the same kind as were always being produced in very small numbers in the air within the cloud chamber. [7-47]

During the next two years, Wilson pursued his studies of the effects of x-rays, of the newly discovered radiation from uranium, of ultraviolet light, and of other sources, with singular dedication and unexcelled patience.

This work of C. T. R. Wilson [wrote Thomson], proceeding without haste and without rest [*ohne Rast ohne Hast*] since 1895, has rarely been equalled as an example of ingenuity, insight, skill in manipulation, unfailing patience and dogged determination. Those who were not working at the Cavendish Laboratory during its progress can hardly realise the amount of work it entailed. For many years he did all the glass-blowing himself, and only those who have tried it know how exasperating glass-blowing can be, and how often when the apparatus is all but finished it breaks and the work has to be begun again. This never seemed to disconcert Wilson; he would take up a fresh piece of glass, perhaps say 'Dear, dear', but never anything stronger, and begin again. Old research students when revisiting the Laboratory would say that many things had altered since they went away, but the thing that most vividly brought back old reminiscences was to see C. T. R. glass-blowing. [7-48]

Indeed, Thomson himself had foretold the electrical effect in his *Application of Dynamics to Physics and Chemistry* of 1888, the year of Wilson's admission to Cambridge, and in a later paper on the effects of electrification on a steam jet [7-49].

If the drop is electrified, the effect of the electrification is the opposite to that of surface tension [which promotes evaporation]. An electrified drop will not evaporate even if the air around it is not saturated, and the smaller the drop the smaller is its tendency to evaporate, and the greater the deposition of moisture from the air outside. For drops below a certain size the electrical effect will be greater than that of surface tension, and a drop, if formed, will increase until it reaches the critical size. Thus drops should be formed even in the absence of dust if they are electrified. [7-50]

Thomson had shown that the effects of electrification on a drop and its diminishing size could be compared quantitatively—calculations which Wilson repeated in analyzing his own observations on the role of x-rays in promoting condensation. Alas, according to Rayleigh,

> he did not publish [his calculations] because Lord Kelvin on a visit to the Cavendish Laboratory had urged him strongly to keep theory out of his paper. We may remark in passing that this was very characteristic of Lord Kelvin, who was enthusiastic about other people's experiments, but, at any rate in his mature years, was apt to consider their theories deplorable. [7-51]

Meanwhile, on March 9 Thomson presented the full report on the electrical measurements with McClelland to the Cambridge Philosophical Society. This time McClelland shares the masthead of the printed paper as coauthor [7-52]. The paper mainly amplifies Thomson's interim report to *Nature* a little over a week earlier. Not only was the electrical conductivity proportional to the number of ions, as reported earlier, but under pressure conditions such that the number of ions was small compared to the number of undissociated molecules, the conductivity was independent of the molecular mean free path. Another characteristic property of the conductivity produced by the x-rays was that the rate of leakage as a function of increasing potential difference soon reached a maximum value beyond which it was independent of the potential difference. Such a 'saturation' current could be understood in terms of the e.m.f. simply removing the ions as fast as they were generated by the incident radiation [7-53].

It must be kept in mind that, amidst these interesting experiments, Thomson, the 'Prof', had a laboratory full of students to administer as well. The very month of the above report to the Cambridge Philosophical Society, March 1896, a badly needed extension of the Cavendish was unveiled. We have noted in an earlier section the strain on adequate teaching space caused by the introduction of physics instruction for medical students in 1888. The University sorely lacked funds for any expansion, though it did provide a parcel of land adjacent to the Cavendish for that purpose. Only after eight years could Thomson himself, being by nature averse to asking outright for money [7-54], scrape together about £2000 from laboratory and lecture fees 'by a cheese-paring policy almost comparable with that practised by Lord Cromer when he was restoring the finances of Egypt' [7-55]. In addition, the University saw fit to augment Thomson's contribution with another £2000. The extension afforded by this modest sum included an underground cellar for experiments requiring a constant temperature, lecture rooms, and a private room for the Professor—the old room serving Maxwell and Rayleigh having long ago been relinquished for physics demonstrations. Most importantly, the design of the annex ensured that additional space could be incorporated when further funding became available.

> The opening of the new buildings in March 1896 [according to Rayleigh] was celebrated by a conversazione at the laboratory.

Figure 7.3. *A view of the old Cavendish Laboratory from Free School Lane. The Rayleigh Wing of 1906 can be seen in the distance.*

> The entrance from Free School Lane to the laboratory was covered in with an awning, carpets were put down on the stairs, and everything brightly lighted with electric light, temporary wiring having been put up by the laboratory staff. There were scientific exhibits, not only in physics, but in other subjects as well, Rutherford's magnetic detector for electric waves being perhaps the show piece. Mrs. Thomson received the guests holding a bouquet of flowers, the gift of the research students. Various workers in the laboratory, the present writer among them, acted as stewards, wearing badges. As Rutherford describes it in a contemporary letter: 'J. J. himself wandered round looking very happy and grinning at everybody and everything in his own inimitable way'. The party amounted to some seven or eight hundred guests, including most of the leading figures in the University. [7-56]

Thomson's parsimonious attitude with regard to laboratory space extended to acquisition of laboratory apparatus as well, in stark contrast to Rayleigh, who was an ardent fund raiser and even contributed the bulk of his Nobel Prize money to the Laboratory. This stringent policy created considerable frustration among research students.

> Naturally, this financial stringency and the rapidly increasing number of workers in the laboratory created a severe competition for such apparatus as there was. A few who could afford to do so provided things of their own. Naturally the scarcity led to the development of predatory habits, and it was said that when one was assembling the apparatus for a research, it was necessary to carry a drawn sword in his right hand and his apparatus in his left. Someone moved an amendment—someone else's apparatus in his left. [7-57]

For an update of life at the Cavendish ca. 1930, see Norman de Bruyne's 'A personal view of the Cavendish 1923–30' [7-58]. Except for a second addition during 1906–1908, the external appearance of the old Cavendish building remains unchanged on narrow Free School Lane behind Corpus Christi College, although the Laboratory vacated its former Victorian quarters in 1974. The Gothic ornamentation (figure 7.3) is much as it was when the building was formally opened 100 years earlier, including the obscure inscription over the gateway: *'Magna opera domini exquisita in omnes voluntares elus'* [7-59]. In the summer of 1974, when under the Cavendish Professorship of Sir Brian Pippard, the Laboratory (formally the Department of Physics in the University of Cambridge) relocated to its modern, spacious quarters in an open field off Madingley Road in West Cambridge. The old quarters on Free School Lane are nowadays used for other University purposes.

Chapter 8

FROM LIVERPOOL TO PRINCETON

8.1. Events of 1896

With his reputation steadily rising, both as experimentalist and as 'Prof', commitments from near and far gradually crowded Thomson's already busy schedule. As he ruminated on April 17 1896,

> I have got my hands full just as present. I have to give the Rede lecture this term, preside over Section A [of the British Association] in September, and lecture in America in October. [8-1]

The Rede lecture was sponsored by a Cambridge foundation dating from the time of Henry VIII [8-2]. This year's subject was the Röntgen rays, something preoccupying virtually everybody that spring—scientists and the public alike. Casually browsing through the issue of *Nature* for the week of Thomson's apprehensive musing about upcoming tasks says as much [8-3].

Oliver Lodge, in this particular issue of the journal, gets the discussion going by arguing that 'it has been asserted that the action of x-rays on a film is a photographic one, depending on the fluorescence of the glass backing. The truth is that a film on a ferrotype plate [8-4] is just about as rapid as a similar film on glass' [8-5]. V Novac and O Sulc of Prague conclude that the absorbing powers of the chemical elements depend on their atomic weights alone, and that the absorbing power of a compound depends only on the atomic weights of the constituent elements and not on the complexity of its molecules [8-6]. A Winkelmann and R Strand of Jena find optical refractions of the rays amounting to typically 1:0.0038 'referred to air' [8-7]. Augusto Righi of Bologna and the team of A Fontana and A Umani in Rome are said to have found the effect of x-rays in stopping the action of Crookes' radiometer to be purely electrostatic—that is, due to the electrification of the glass bulb housing the radiometer [8-8]. A number of correspondents, among them H Van Heurck of the Botanical Gardens in Antwerp, Mr Basilewski in addressing the Paris Academy, and Columbia University's Michael Pupin before the New York Academy, observe the reduction in exposure time achieved by replacing the glass backing of sensitive photographic plates with a fluorescent screen such as a celluloid backing [8-9]. Ferdinando Giazzi of Perugia confirms what Edison had communicated to Lord Kelvin via telegraph, namely, that calcium tungstate, suitably crystallized, exhibits fluorescent phenomena under the action of x-rays to a far more marked degree than barium platino-cyanide—the ubiquitous laboratory powder [8-10].

Both Ogden N Rood and Nikola Tesla are said to have found indications of regular reflections of the rays. So had Röntgen himself, as he reported in his original paper. Since he had nevertheless felt compelled to add that

> if we connect these facts with the observation that powders are quite as transparent as solid bodies, and that, moreover, bodies with rough surfaces are, in regard to the transmission of x-rays, as well in the experiment just described, the same as polished bodies, one comes to the conclusion that regular reflection, as already stated, does not exist, but that the bodies behave to the x-rays as muddy media do to light. [8-11]

Pupin concludes that 'Prof. Rood's and Mr. Tesla's experiments must be interpreted as a confirmation of Prof. Röntgen's results, and not as a demonstration of the existence of regular reflection' [8-12]. Nevertheless, better evidence for reflection was forthcoming from several sources, among them the measurements by FitzGerald and Frederick T Trouton at Trinity College, Dublin. In addition, experiments on the relative opacities of various substances to Röntgen rays were under way by W L Goodwin at Kingston's School of Mining, Canada, and by E Doelter of Graz, among others [8-13].

> Finally [ends the survey in *Nature*], attention may profitably be called in this summary to the April number of the *Proceedings* of the Physical Society. In the admirable collection of abstracts of physical papers there published, will be found concise descriptions of the scope and results of no less than forty papers concerned with Röntgen rays. [8-14]

Evidently, then, much was known about the rays by the time Thomson gave his Rede lecture on June 10 before a packed house in the Lecture Theatre of Anatomy and Physiology at Cambridge.

Thomson began his lecture with a review of the phenomena accompanying the discharge of electricity through gases as the pressure is gradually reduced, until at very low pressures the luminosity is confined to the glass wall of the discharge tube, with little or none in the gas itself; only then are Röntgen rays produced by the tube [8-15].

> [From experiments] we... conclude that we have something starting from the negative electrode, travelling in straight lines, and producing phosphorescence when it reaches the glass.... This something... is called the kathode rays: these rays are of great interest in relation to the subject of this lecture, for the kathode rays seem to be the parents of the Röntgen rays. [8-16]

Thomson stressed the striking difference between 'the parent—the kathode rays—and the child, the Röntgen rays', in terms of the effect of a magnet on them: the cathode rays are deflected by a magnet, while the Röntgen rays are not so affected.

Repeating before his enrapt audience some of the experiments by which Röntgen established the existence and properties of the new-found radiation, Thomson admitted that

> the experiments made on these rays have not led to any results absolutely decisive as to their nature, but we can profitably discuss the question whether the facts known about these rays oblige us to regard them as due to a new form of energy, or whether they are consistent with these rays being a variety of some form of energy already known to us; before calling in a new form, we ought to be quite sure that it is necessary to abandon the old. The rectilinear propagation of these rays, their powers of producing phosphorescence and of affecting a photographic plate, their insensibility to a magnet, suggest that of the old forms of energy light is the one to which these rays are most closely allied. We are acquainted with so many varieties of light (by light I mean transverse vibrations propagated with a definite velocity) with such widely different properties, that we can well contemplate the existence of other kinds with still different properties. [8-17]

Examples of different forms of radiant energy were, as noted, not hard to come by. There was ultraviolet light, transverse vibrations of very short wavelength, so thoroughly elucidated by Stokes in his classic researches. There was visible light, and infrared light of yet longer wavelength. Then there were waves emitted by vibrating electrical systems—waves 'so different in properties from ordinary light that it required the genius of Clerk-Maxwell to recognise them as light at all' [8-18], and most recently studied by Rutherford. Still more recently, only a little over three months earlier, news from abroad of yet another form of radiation had startled the Cambridge research community—news strongly supportive of Thomson's (and Röntgen's) interpretation of the radiation.

> So far [continued Thomson], I have confined myself to showing that there is nothing in the effects known to be due to these rays

inconsistent with their being a variety of light. I must now pass on to some evidence of a more positive character. Since the discovery of the Röntgen rays, Becquerel has discovered a new kind of light, which in its properties resembles the Röntgen rays more closely than any kind of light hitherto known. Becquerel found that certain uranium salts emitted, after being exposed to the sunlight, radiations which, like the Röntgen rays, could pass through plates of aluminum or of cardboard, and affect a photographic plate behind. I have here a photograph of a perforated piece of zinc, which has been taken by Becquerel's method by simply scattering over the zinc plate powdered uranium nitrate, and placing it over a photographic plate well protected from ordinary light. After a long exposure of from twenty to forty hours, the photograph now on the screen was taken. Becquerel has shown that the radiation from the uranium salts can be polarised, so that it is undoubtedly light; it can also be refracted. It forms a link between the Röntgen rays and ordinary light, it resembles the Röntgen rays in its photographic action in power of penetrating substances opaque to ordinary light, and in the characteristic electrical effect, while it resembles ordinary light in its capacity for polarisation, in its liability to refraction. [8-19]

From this discovery of Becquerel, we may conclude that besides the vibrations emitted by luminous bodies with which we have hitherto been acquainted, there are others having a much greater frequency and, it may be, arising in a different way.
To sum up, we may say that though there is no direct evidence that the Röntgen rays are a kind of light, there is no known property of these rays which is not possessed by one or other of the forms of light. [8-20]

Thomson's Presidential Address before Section A on Mathematical and Physical Science of the British Association, which met at Liverpool in September, was also mainly on x-rays [8-21]. He spoke in the University's Arts Theatre on the morning of September 17. His audience included a host of prominent Association members, among them Lord Kelvin, Stoney, Lodge, and FitzGerald. Guests from abroad included Friedrich Kohlrauch (Director of the Physikalisch-Technische Reichsanstalt at Berlin-Charlottenburg), Lenard (then at the Technische Hochschule in Aachen, but scheduled to leave for Heidelberg upon his return), the Norwegian meteorologist Vilhelm Bjerknes [8-22], the American astrophysicist James E Keeler [8-23], Max Wolf of Heidelberg, sometime collaborator of Lenard [8-24], and the inseparable team of Elster and Geitel [8-25].

There is a melancholy reminiscence connected with this meeting of our Section [began Thomson], for when the British Association last

> met in Liverpool the chair in Section A was occupied by Clerk-Maxwell. In the quarter of a century which has elapsed since that meeting, one of the most important advances made in our science has been the researches which, inspired by Maxwell's view of electrical action, confirmed that view, and revolutionised our conception of the processes occurring in the electro-magnetic field. When the Association last met in Liverpool Maxwell's view was almost without supporters; to-day its opponents are fewer than its supporters then. Maxwell's theory, which is the development and extension of Faraday's, has not only affected our way of regarding the older phenomena of electricity, it has, in the hands of Hertz and others, led to the discovery of whole regions of phenomena previously undreamt of. It is sad to think that his premature death prevented him from reaping the harvest he had sown. His writings are, however, with us, and are a storehouse to which we continually turn, and never, I think, without finding something valuable and suggestive.
> 'Thus ye teach us day by day,
> Wisdom, though now far away.' [8-26]

By way of opening remarks, Thomson noted several milestones in physics in the year gone by: the jubilee of Lord Kelvin's tenure of the Professorship of Natural Philosophy at Glasgow, and the passing away of William Grove, famous for his cell, and Stoletov of Moscow (best known for his researches on the electrical properties of gases illuminated by ultraviolet light). Perhaps because Section A was less commonly a forum for airing questions of science education than Section B (Chemistry and Mineralogy), Thomson also dwelt on the teaching of physics in schools and universities, admonishing teachers in his audience that

> I think... that in the teaching of physics at our universities there is perhaps a tendency to make the course too complex and too complete. I refer especially to the training of those students who intend to become physicists. I think that after a student has been trained to take accurate observations, to be alive to those pitfalls and errors to which all experiments are liable, mischief may in some cases be done if... he is kept performing elaborate experiments, the results of which are well known. It is not given to many to wear a load of learning lightly as a flower. With many students a load of learning, especially if it takes a long time to acquire, is apt to crush enthusiasm. Now, there is, I think, hardly any quality more essential to success in physical investigations than enthusiasm. Any investigation in experimental physics requires a large expenditure of both time and patience; the apparatus seldom, if ever, begins by behaving as it ought; there are times when all the forces of nature, all the properties of matter,

> seem to be fighting against us: the instruments behave in the most capricious way, and we appreciate Coutts Trotter's saying, that the doctrine of the constancy of nature could never have been discovered in a laboratory. These difficulties have to be overcome, but it may take weeks or months to do so, and, unless the student is enthusiastic, he is apt to retire disheartened from the contest. I think, therefore, that the preservation of youthful enthusiasm is one of the most important points for consideration in the training of physicists. In my opinion this can best be done by allowing the student, even before he is supposed to be acquainted with the whole of physics, to begin some original research of a simple kind under the guidance of a teacher who will encourage him and assist in the removal of difficulties. If the student once tastes the delights of the successful completion of an investigation, he is not likely to go back, and will be better equipped for investigating the secrets of nature than if, like the White Knight of 'Alice in Wonderland', he commences his career knowing how to measure or weigh every physical quantity under the sun, but with little desire or enthusiasm to have anything to do with any of them. [8-27]

Turning, finally, to the Röntgen rays, he began his subject proper by acknowledging Lenard's illumination of a coated screen, darkening of a photographic plate, or discharging of an electrified body, when screen, plate, or body was placed outside the window of his discharge tube. The acknowledgment brought hearty applause for Lenard, seated in the audience. Lenard's rays responsible for these external effects, like the cathode rays inside the tube, were deflected by a magnet, whereas Röntgen's rays were *not* deflected, as far as could be determined. Thomson ventured that the cathode rays

> are particles of gas carrying charges of negative electricity, and moving with great velocities which they have acquired as they travel through the intense electric field which exists in the neighbourhood of the negative cathode.... This view of the constitution of the cathode rays explains in a simple way the deflection of those rays in a magnetic field, and it has lately received strong confirmation from the results of an experiment made by Perrin. [8-28]

On the other hand, with regard to Röntgen rays,

> we are not yet acquainted with any crucial experiment which shows unmistakably that these rays are waves of transverse vibration of the ether, or that they are waves of normal vibration, or indeed that they are vibrations at all. As a working hypothesis, however, it may be worth while considering the question whether

> there is any property known to be possessed by these rays which is not possessed by some form or other of light. The many forms of light have in the last few months received a noteworthy addition by the discovery of M. Becquerel of an invisible radiation, possessing many of the properties of the Röntgen rays, which is emitted by many fluorescent substances, and to an especially marked extent by the uranium salts. By means of this radiation, which, since it can be polarised, is unquestionably light, photographs through opaque substances similar to, though not so beautiful as, those obtained by means of Röntgen rays can be taken, and, like the Röntgen rays, they cause an electrified body on which they shine to lose its charge, whether this be positive or negative. [8-29]

As to the apparent absence of polarization in the case of Röntgen's rays, based on numerous unsuccessful attempts with crossed and parallel tourmaline plates—a strong objection, along with the aforesaid absence of refraction, to the rays being electromagnetic vibrations in the æther—Thomson opined that

> though the structure of the tourmaline is fine enough to polarise the visible rays, it may be much too coarse to polarise the Röntgen rays if these have exceedingly small wave-lengths. As far as our knowledge of these rays extends, I think we may say that though there is no direct evidence that they are a kind of light, there are no properties of the rays which are not possessed by some variety of light. [8-30]

Following Thomson's address, the Section adjourned from the Arts Theatre to the Physics Theatre, where the report of the Committee on the Establishment of a National Physical Laboratory was presented by Sir Douglas Strutt Galton, President of the British Association the previous year, when it met at Ipswich. The Committee had been appointed the year before to follow up on the suggestion of Oliver Lodge in his presidential address to Section A that the Kew Observatory at Richmond be expanded to include a nucleus for the Laboratory, organized along the lines of the highly successful *Reichsanstalt* in Berlin [8-31]. The ensuing discussion ranged from stellar evolution by Isaac Roberts and periodic orbits (G H Darwin) to physical applications of the phonograph (McKendrick).

The next day, Friday, saw a joint discussion with Section B on 'Röntgen rays and allied phenomena' before a crowded audience in the large lecture theatre of University College. The subject was introduced by none other than Lenard, who, as we have recounted earlier, had reasons for becoming rather bitter about priorities surrounding the discovery of x-rays. Though Lenard would turn overtly hostile toward Thomson (and Röntgen) by the time of World War I, he attended the Liverpool meeting on Thomson's express invitation. That he was still on good terms with his host that year is underscored by his subsequent civil letter to Thomson:

> Dear Prof. Thomson,
> Let me thank you most heartily for your kind letter and invitation to Cambridge. This award [of the Rumford Medal] by the Royal Society makes me of course very happy, I think it is the greatest event that has happened in my life, but to have so many kind words from you on this occasion makes me still happier. To come to England and to enjoy personal intercourse would be an increased delight to me now.... I hope for future occasion to have the pleasure of meeting you. I shall never forget the pleasant meeting at Liverpool and how much of the delight in it was due to you. Please present my best compliments to Mrs. Thomson, and believe me always,
> Yours very sincerely, P. Lenard [8-32]

In any case, Lenard described his researches on cathode rays and his views as to their nature. Though they were strongly attenuated in air under atmospheric pressure, they traveled much further if the vacuum tube was extended beyond the aluminum window and the air pressure in the second chamber was reduced. 'This', argued Lenard, 'favors the view that they are not due to projected matter, but are of the nature of ether-waves' [8-33].

In the discussion that followed, George Stokes took issue with the wave interpretation of the thin-foil experiments, since it was not necessary to suppose the rays passed through the foil. One could as easily argue that if the inside of the foil is bombarded by gaseous molecules or by particles discharged from the electrode, the bombardment gives rise to a process in which molecules are projected from the far side of the foil. FitzGerald congratulated Lenard for his experiments, noting that 'whereas Röntgen's experiments had soon been repeated by hundreds of observers, Lenard's earlier experiments were of such a difficult nature that no one had since repeated them' [8-34]. Nevertheless, FitzGerald too supported the English particulate view; the deflection of cathode rays by a magnet appeared to him analogous to the Hall effect, since 'the Hall effect only occurs when matter is present' [8-35].

Next, Thomson gave an account of experiments by himself and Ernest Rutherford on the conduction of electricity through gases ionized by x-rays. He had referred to the subject in advance in his presidential address, when he noted that

> Mr. Rutherford and I have lately found that the conductivity is destroyed if a current of electricity is sent through the Röntgenised gas. The gas in this state behaves in this respect like a very dilute solution of an electrolyte. Such a solution would cease to conduct

> after enough electricity had been sent through it to electrolyse all the molecules of the electrolyte. When a current is passing through a gas exposed to the rays, the current destroys and the rays produce the structure which gives conductivity to the gas; when things have reached a steady state the rate of destruction by the current must equal the rate of production by the rays. [8-36]

This brief statement says virtually as much as does their elaborate—albeit definitive—paper published in November—a paper filled with measurements and tabulated results in terms of electromagnetic forces applied and leakage currents observed and calculated, for various gases [8-37]. The paper bears the stamp of a style one identifies with the 'mature' Rutherford in his papers and notebooks—particularly his habit of marking certain tabular readings with an asterisk, these being what he judged to be 'good' results [8-38]. It is clear from the paper that Rutherford performed most of the measurements while Thomson supplied the theory and the discussion. From the opening paragraph and from Thomson's advance remarks one gathers that the bulk of these researches occupied Rutherford and his mentor much of the summer months and that they had a preliminary conclusion in hand by the beginning of September. At the same time, Rutherford continued his work on his magnetic detector with Thomson's approval—a device he was invited to demonstrate on Monday of the Liverpool meeting. From Arthur S Eve's official biography of Rutherford we can follow his nervous exhilaration as the Liverpool meeting draws closer, in letters to Mary Newton and his mother (to both of whom he wrote regularly).

> *Cambridge, 6 June*
>
> I have got some news to tell you about... J. J. has asked me to give an experimental lecture at the British Association meeting in Liverpool on September 6. I expect you know the British Association meets every year and all the scientific men of the country turn up to it, so it is rather an honour to give a paper before it... My paper would deal with my work on the wave detectors and I think I could make it interesting to any who turned up. [8-39]
>
> *Cambridge, July*
>
> I have been working pretty steadily with Professor J. J. Thomson on the X-rays and find it pretty interesting. Everett, who is the professor's assistant, makes the bulbs which give out the X-rays. You know one can see the bones of the hand and arm, and coins inside with the naked eye. [8-40]
>
> *Cambridge, 12 August*
>
> The work the professor and I have done this long vac. will probably be published in the Philosophical Magazine. We will also give an account of it at the British Association meeting at

Liverpool. I am therefore steadily progressing to my desired aim, a lectureship or a fellowship. I believe the Cambridge degree will be of considerable assistance to me in the end, for it counts a great deal with the outside world. [8-41]

Cambridge, 27 August

I have been very busy today running over my experiments for the British Association. Pye, the workshop man, has made some things very nicely for me and I will be very well equipped for my apparatus, and I only hope things will pass off successfully. Tomorrow will be my last day in the Lab. for work till October. J. J. is going for a short spell before he goes to the British Association and then to America. I have still got to write up my paper for the British Association but am putting it off till the last minute . [8-42]

We tried [reads the Thomson/Rutherford paper] whether the conductivity of the gas would be destroyed by heating the gas during its passage from the place where it was exposed to the [Röntgen] rays to the place where its conductivity was tested... The gas after coming through [a porcelain tube raised to a white heat] was so hot that it could hardly be borne by the hand; the conductivity, however, did not seem to be at all impaired ... A very suggestive result is the effect of passing a current of electricity through the gas... The current produces the same effect on the gas as it would produce on a very weak solution of an electrolyte [i.e. ionization]... We shall find that the analogy between a dilute solution of an electrolyte and gas exposed to the Röntgen rays holds through a wide range of phenomena, and we have found it of great use in explaining many of the characteristic properties of conduction through gases. [8-43]

In commenting on this paper in his biography of Rutherford, David Wilson relates an often-repeated story about Rutherford's indignation at someone's suggestion that ions might not exist—a reaction indicative of 'how intensely the man flung his whole person into his research'. Rutherford vociferously insisted that ions *do* indeed exist; they are 'jolly little beggars, so real that I can almost see them' [8-44].

Thomson's presentation of his work with Rutherford provoked considerable discussion by, among others, William E Ayrton of Central Technical College, South Kensington, Frederick T Trouton, a pupil of FitzGerald, and Lord Kelvin; all agreed that experiments of their own tended to confirm the electrolytic analogy. The last paper of the session was by Silvanus P Thompson, professor of physics and principal of the City and Guilds Technical College, Finsbury (where Ayrton also served on the faculty for a period). His paper dealt with the relationship between cathode rays, x-rays, and Becquerel's rays [8-45]. In Thompson's experiments, various screens or obstacles were interposed between the cathode and wall of a Crookes tube, and the resulting shadow or emerging radiation studied as a function of the degree of exhaustion of the tube. The discussion of Thompson's paper was deferred until the next day, Saturday, when Vilhelm Bjerknes—a man of some intellectual weight—argued in favor of Lenard's contention that the cathode rays are due to ætheral vibrations. Bjerknes' argument was based on certain vague observations of the line spectrum of cathode rays undergoing magnetic deflection by his countryman Kristian Birkeland of Kristiania (also known as Christiania, and later renamed Oslo, its original name).

On Monday September 21, the Section met again, in two separate sessions. One, on electromagnetic experiments, met in the Physics Theatre. (The second, on meteorology, met in the physics class room but is not of particular relevance to our chronology.) Kelvin opened the session with an account of experiments by himself, Maclean, and Galt on the communication of electricity from electrified steam to air. He was followed by Rutherford's demonstration of his detector, repeating essentially what he had presented to the Royal Society the previous June. He reported that he had received signals at a distance of half a mile, or from the Cavendish to his lodgings in town through intervening houses. Eve, in his biography, recalls a story that when Rutherford was giving his demonstration, his apparatus at first refused to work despite Pye's diligence. Calmly, Rutherford looked up and said: 'Something has gone wrong! If you would all like to go for a stroll and a smoke for five minutes, it will be working on your return'. And so it was [8-46].

Twenty-seven years later (in 1923) the British Association again met at Liverpool, at which time Rutherford himself was president. On that occasion he naturally recalled the 1896 meeting.

> The visit to your city in 1896 was for me a memorable occasion for it was here that I first attended a meeting of this Association, and here that I read my first scientific paper... a paper which I had the honour to read, on a new magnetic detector of electrical waves. [8-47]

In the adjourned discussion, William H Preece, the electrical engineer in the Post Office telegraphic system, announced that Signor Marconi and a Mr Kempe had outdone Rutherford in having transmitted and detected electrical waves across a distance of one and a quarter miles on Salisbury Plain. The Post Office encouraged Marconi, even to the extent of providing facilities for testing his apparatus.

Within a month after the Liverpool meeting, Thomson and Mrs Thomson journeyed by steamer to America, where Thomson was to attend the celebration of the Sesquicentenary of Princeton University and give a course of lectures on the conduction of electricity through gases. Stopping first in Baltimore where he was hosted by Johns Hopkins

University, 'the first university [world-wide] to be founded where the primary consideration was research and not teaching', as he tactfully put it [8-48], Thomson thoroughly enjoyed a game of baseball. Indeed, he was exceptionally well informed about all forms of sport.

> The idea is the same as that of the English game of rounders in that the batsman, armed with a broomstick, tries to hit the ball so far that he has time to run to a base before a fieldsman can gather the ball, throw it at him and hit him; but there is as much difference between this and a good baseball match as between cricket played by boys in a side street... and the match between Gentlemen and Players at Lord's. The fielding, i.e. the throwing and catching of a first-class baseball team, is superb and reaches a standard but rarely attained in English cricket. The pitcher, who corresponds to the bowler at cricket, throws, not bowls, the ball full pitch at the batsman, and by putting spin on the ball can make it swerve in the air either to the right or to the left, or up or down... [8-49]

At Princeton University, 'of all the American universities I have visited,... the one most reminiscent of Cambridge' [8-50], his lectures were well attended [8-51]. Thomson found many aspects of Princeton University interesting, among them the dubious role of the University Clubs (today's Fraternities). In promoting student activities which would ensure their election to a Club—activities rarely of a scholarly nature or work related—they effectively lowered academic standards. This trend Woodrow Wilson countered initially by the appointment of preceptors who gave individual attention to students, like Cambridge's tutors, and later by replacing the Clubs with Colleges, again in the vein of Oxford and Cambridge. Thomson also paid a visit to the Military Academy at West Point, where he was greatly impressed by the rigorous training program in force. Despite the method of learning by rote, it produced cadets that 'certainly did not show any of the usual symptoms of overstudy, but were a remarkably vigorous, well-set-up body of men' [8-52].

The proceedings of the Sesquicentenary lasted three days. Thomson did his part on the first day, offering congratulations on behalf of European universities and learned societies. 'The second day began with the recitation of the Academic Ode by Dr. Henry van Dyke representing the Clio Debating Society, while the oration was delivered by Professor Woodrow Wilson representing the rival debating society, the American Whig Society' [8-53]. In the afternoon there was a football match between Princeton and the University of Virginia—a game Thomson compares in some detail with rugby football in England [8-54]. He added that 'the players are much more padded than they ever were over here, and look rather like the advertisements for Michelin tyres' [8-55].

The third day was spent conferring honorary degrees. Thomson ventures that it was the first time doctor's robes had been worn in America.

> Two at least of the American Professors had, with characteristic thirst for improvement, borrowed, from President Gilman of Johns Hopkins, gowns which he had brought home from Europe, where he had received doctors' degrees from several universities. They took these to their wives' dressmaker and told her to select the best features of each and produce something stylish. The result was remarkable: they were the only doctors' gowns I ever saw that had anything that could be called a good fit. These, however, went in at the waist and then billowed out into something very like a skirt in a very coquettish and most unacademical fashion... [8-56]

8.2. Radioactivity

As we perhaps recall, on January 1 1896, Röntgen mailed out preprints of his fact-filled announcement of x-rays, along with photographs, to colleagues and individuals at home and abroad—among them Arthur Schuster in Manchester. Another prominent recipient was Henri Poincaré, professor of mathematics and mathematical physics at the University of Paris and member of the Académie des Sciences. The Academy's regular Monday meeting on January 20 was largely devoted to x-rays. There Poincaré presented x-ray pictures of the human hand communicated by two French physicians, Paul Oudin and Toussaint Barthélémy [8-57]. In so doing, he took the opportunity of reporting on Röntgen's own paper. In the subsequent discussion, Henri Becquerel inquired about the place of origin of the rays in the discharge tube. Poincaré replied that they apparently originated in the region of fluorescence in the tube wall struck by the cathode rays. Looking back on the discussion in 1903, Becquerel recalls that

> I thought immediately of investigating whether the new emission could not be a manifestation of the vibratory movement which gave rise to the phosphorescence and whether all phosphorescent bodies could not emit similar rays. I communicated this idea to Monsieur Poincaré and the very next day I began a series of experiments along this line of thought ... [8-58]

Antoine Henri Becquerel was the second son of Alexandre-Edmond Becquerel and grandson of Antoine-César Becquerel, both renowned physicists, members of the Academy, and each in turn professor of physics at the Musée d'Histoire Naturelle in Paris [8-59]. Since Henri Becquerel's son, Jean, also became a distinguished physicist, this family dynasty, spanning four generations of physicists, is worth a brief digression. To top it off, there was also Paul Becquerel, nephew of Henri and a prolific biologist.

Antoine-César Becquerel was one of the first *polytechniciens*—that is, one of the first to graduate from the École Polytechnique, the engineering school founded in 1794 under the leadership of Lazare Carnot and Gaspard Monge. He fought and was wounded in the Napoléonic War in Spain during 1810–1812; after the fall of Napoléon he resigned from the army. Before long he became professor of physics at the Museum of Natural History, and later its director. He wrote about 530 papers and six multi-volume textbooks. His son Alexandre-Edmond Becquerel also attended the École Polytechnique, but skipped military service; instead, he became his father's assistant at the Museum. In due course, he too became professor there—a professorship destined to be passed on to Edmond's son Henri, and from Henri to his son Jean. Among the research topics of Edmond is one of particular interest to our chronicle: the fluorescence of uranium exposed to light of different wavelengths.

Henri Becquerel was born in Paris in 1852. He was schooled at the Lycée Louis-le-Grand, after which he attended the École Polytechnique as a matter of course. He then entered the École des Ponts et Chaussées where he received further training in engineering. Subsequently he was appointed to the Administration of Bridges and Highways in 1875 with the rank of *Ingénieur*, in which post he rose steadily until becoming *Ingénieur en chef* in 1894—having in the meanwhile been promoted to *Ingénieur de premiére classe* in the Ponts et Chaussées. The years 1875–1876 also marked the start of Becquerel's academic career. He commenced his private researches in 1875, and his teaching career the following year as *répétiteur* at the Polytechnique. The year 1878 was personally tumultuous for Becquerel: his grandfather died, as did his wife Lucie-Zoe-Marie following the birth of their son Jean; moreover, he succeeded his father to the post of *aide-naturaliste* at the Museum of History. He obtained his doctorate from the Faculty of Sciences of Paris in 1888 on the strength of his early optical researches embracing polarization of light in magnetic fields, infrared illumination of phosphorescent crystals, and absorption of light in crystals. The next year he was elected to the Academy of Sciences, at the tender age of 37.

With his doctorate in hand, Becquerel's active researches declined sharply [8-60]. In 1890 he married again, and the following year, upon the death of his father, he succeeded to his father's two chairs of physics, at the Conservatoire National des Arts et Métiers and at the Museum. The same year he took up teaching at the École Polytechnique, where he was appointed professor of physics in 1895. He thus held no fewer than three simultaneous chairs in physics!

There remains to take notice of Becquerel's only descendant, Jean Becquerel, who, like his father before him, began as assistant to his father, collaborated and published jointly with him, and succeeded him at the

Figure 8.1. *Henri A Becquerel.*

Museum after Henri's death in 1908. Jean's mature researches centered mainly on magneto-optics—much of it carried out in collaboration with Kamerlingh Onnes at Leiden. His retirement from the Museum in 1948 marked the end of the Becquerels' century-long hereditary occupation of the chair of physics; similarly, on his death in 1953 the lineal membership in the Academy ceased to be.

We will have more to say about Jean Becquerel and certain unfortunate developments plaguing French physics during his formative years. Returning for now to January of 1896, the first French contribution to x-ray studies was communicated on January 27 by Jean Perrin while still at the École Normale under Brillouin [8-61]. He had repeated Röntgen's preliminary experiments, of which he had only vague notions culled from the daily press, and came to the conclusion that if the radiation were periodic, its wavelength had to be much shorter than that of green light [8-62].

Meanwhile, Becquerel had wasted no time in testing his conjecture of some relationship between x-rays and fluorescence—that is, that all fluorescent bodies emit x-rays. The very next day after the meeting of the Academy on January 20 he took up the problem in earnest. Alas, his first experiments gave negative results, failing to reveal any x-rays emitted from the first fluorescent substances tested. Nevertheless, on January 30 an article published by Poincaré repeated Becquerel's conjecture without extending due credit to him: 'Can it not be asked then whether all bodies whose fluorescence is sufficiently intense would not emit, apart from the luminous rays, the x-rays of Röntgen, *whatever the cause of this fluorescence*?' [8-63]. The article caused a flurry of communications purporting to report some evidence for the elusive relationship, or other effects [8-64].

For his part, Becquerel stubbornly continued his search, now turning to uranium salts. In particular, he concentrated on bisulfate of uranium and potassium, a substance which his father had studied and of which Becquerel himself had prepared samples some 15 years earlier in the form of thin, transparent crystal lamellas. On February 24, he reported to the Academy, in a paper even skimpier than Röntgen's legendary report of the previous December, as follows:

> A *Lumière* photographic plate [one sensitive to visible light] having bromide emulsion was wrapped with two sheets of thick black paper, so thick that the plate was not clouded by exposure to the sun for a whole day. Externally, over the paper sheet, was placed a piece of the phosphorescent substance, and all were exposed to the sun for many hours. Upon developing the photographic plate I recognized the silhouette of the phosphorescent substance in black on the negative. If a coin or metallic screen with an openwork design were placed between the phosphorescent substance and the paper, the image of these objects appeared on the negative.
> The same experiment can be repeated by interposing between the phosphorescent substance and the paper a thin glass plate; this excludes the possibility of a chemical action resulting from vapors that could be emitted by the substance upon being heated by the sun's rays.
> From these experiments it may be concluded that the phosphorescent substance emits radiations which penetrate paper that is opaque to light and reduce silver salts in a photographic plate. [8-65]

It would thus appear that x-rays were emitted from the uranium salt while fluorescing. However, to be sure on this point, Becquerel decided to repeat the experiment. Unfortunately, on February 26 and 27 the sun appeared only intermittently in Paris, as is usual at that time of year—not long enough for adequate exposures. He therefore set the experiment aside, placing the wrapped plates in a cabinet drawer with the uranium salt samples left in place on top of the wrapping. As he reported to the Academy on March 2:

The sun did not appear on the following days and I developed the plates on March 1st, expecting to find only very faint images. The silhouettes appeared, on the contrary, with great intensity. I thought therefore that the action must continue in the dark... [8-66]

Ergo, the new phenomenon was not connected with sunlight, nor was it caused by luminous radiation emitted by phosphorescence! Rather, wrote Becquerel succinctly,

a hypothesis presents itself very naturally to explain these radiations (by noting that the effects bear great resemblance to the effects caused by the radiations studied by M. Lenard and M. Roentgen), as invisible radiation emitted by phosphorescence, but whose persistence is infinitely greater than the duration of the luminous radiation emitted by these bodies. The present experiments, although not contrary to this hypothesis, do not lead to any formulae [i.e. to definite conclusions]. The experiments which I intend to do from now on, I hope, will clarify this new kind of phenomena. [8-67]

In contrast to Röntgen's solitary experiment, several eyewitnesses were present during Becquerel's epochal discovery. One was his son Jean, then 18. Another was the venerable William Crookes, who was to throw himself energetically into radioactivity as compensation for missing out in the discovery of x-rays, due to his ill-timed visit to South Africa in 1895. Crookes was in the habit of visiting Becquerel's laboratory now and then, and even assisted him with an occasional experiment. He was struck by the remarkable facility with which his host extemporized experimental apparatus.

Card, gummed paper, glass plates, sealing wax, copper wire, rapidly and almost spontaneously seemed to grow before one's eyes into just the combination suitable to settle the point under investigation. The answer once obtained, the materials were put aside or modified so as to constitute a second interrogation of nature. [8-68]

Crookes has left a record of what exactly took place one memorable morning in Paris.

Becquerel was working on the phosphorescence of uranium compounds after isolation; starting with the discovery that sun-excited uranium nitrate gave out rays capable of penetrating opaque paper and then acting photographically; he had devised another experiment in which, between the plate and the uranium salt, he interposed a sheet of black paper and a small cross of thin copper. On bringing the apparatus into daylight the sun

> had gone in, so it was put back in the dark cupboard and there left for another opportunity of insolation. But the sun persistently kept behind clouds for several days, and, tired of waiting (or with the unconscious prevision of genius) Becquerel developed the plate. To his astonishment, instead of a blank, as expected, the plate had darkened as strongly as if the uranium had been previously exposed to sunlight, the image of the copper cross shining out white against the black background. This was the foundation of the long series of experiments which led to the remarkable discoveries which have made 'Becquerel rays' a standard expression in Science. [8-69]

By March 9, Becquerel had ascertained that not only did the radiation emitted by uranium blacken the covered photographic plate, but it discharged an electroscope by ionizing gases traversed by the radiation [8-70]. Thus it was also possible to measure the activity of a sample by measuring the ionization with a gold-leaf electroscope.

Alas, his paper on March 9 contains a serious error which would befuddle other researchers for some time to come: Becquerel felt that he had demonstrated the refraction and polarization of the rays, which would imply they were transverse waves in the æther [8-71].

Doggedly pursuing the subject, Becquerel reported on May 18 that

> several months ago I showed that uranium salts emit radiations hitherto unknown, and that the radiations exhibit remarkable properties, some of which are not unlike the effects studied by M. Roentgen. The radiations are emitted not only when the salts are exposed to light, but also when they are kept in the dark, and for two months the same salts continued to emit, without noticeable decrease in amount, these new radiations. From March 3d to May 3d the salts were kept in a closed opaque box. Since May 3d they have been in a lead-walled box, which was kept in the darkroom. By means of a simple arrangement, a photographic plate can be placed under a black paper at the bottom of the box, on which rest the salts under test, without exposing them to any radiation not passing through the lead.
>
> Under these circumstances the salts continued to emit active radiation. [8-72]

Following descriptions of experiments on different uranium compounds, Becquerel correctly surmised that, since all the uranium salts studied, whether phosphorescent from light or not, whether as crystals or in solution, yielded the same results,

> I have thus been led to the conclusion that the effect is due to the presence of the element uranium in these salts, and that the metal should give more noticeable effects than its compounds.

> An experiment performed several weeks ago [with a disk of pure uranium] confirmed this belief; the effect on photographic plates is much greater for the element than that produced by one of the salts, particularly by the bisulphate of uranium and potassium. [8-73]

This *experimentum crucis* marks the completion of Becquerel's discovery of radioactivity (a term coined subsequently by the Curies), though he continued his researches through the remainder of 1896—mainly ionization studies. Becquerel himself looked upon his discovery, which he modestly characterized as the first observation of phosphorescence in a metal, as destined to occur. Wrote Jean Becquerel on the 50th anniversary of the discovery

> He [Henri Becquerel] said the investigations, which during sixty years had followed one another in this same laboratory, formed a chain which, at the propitious hour, were ineluctably to end up with radioactivity. [8-74]

to which may be added Poincaré's remarks, pronounced the very year after the discovery, that

> it seemed that the path which was followed by the Becquerels was destined to lead to a dead end. Far from it, one can think today that it will open for us an access to a new world which no one suspected. [8-75]

Still, the discovery went largely unnoticed in the scientific community at the time, certainly compared with the hoopla over x-rays, and probably due to the nearly universal preoccupation with Röntgen's discovery. Becquerel himself published seven papers on radioactivity in 1896, two in 1897, and none the year after. Others contributed perhaps a dozen additional papers on the topic in the same years. In contrast, over 1000 papers and books on x-rays were published in 1896 alone!

The paucity of interest in radioactivity lasted several years. Even Becquerel soon abandoned the subject, turning instead to another newly discovered phenomenon: the Zeeman effect, which we touched on in section 1.2, and take up again later. Only in 1898 was interest in Becquerel's radiation rekindled, when George Cecil Jaffé and Marie Curie independently showed that it was emitted not only from uranium but also from thorium. When, later the same year, Pierre and Marie Curie, with Gustave Bémont, announced the discovery of two more powerful radioactive elements, polonium and radium, the scientific community finally took serious notice. One who had taken notice much earlier was Ernest Rutherford, still at Cambridge but not for long. He took up the subject even before the discovery of thorium's activity [8-76]. In light of his definitive paper with Thomson on the conduction of electricity in gases ionized by x-rays, it is hardly surprising that he was intrigued by Becquerel's rays. He apparently got started in the Lent term of 1897—certainly by June, when he was required to report on his work to the Commissioners for the Exhibition of 1851 Scholarship. The Commissioners were sure to take notice of his expanding research interests [8-77].

The year 1897 was one of distractions for Rutherford, now 26, with Mary Newton arriving from New Zealand that spring for a half-year visit. She spent a week with the Thomsons at Cambridge; she and Ernest attended the races on the Cam, and they visited Killarney, Ireland, in the company of some friends. Yet the year has been characterized by a former student of his as 'marking the beginning of the first great period of Rutherford's achievement as an experimenter' [8-78]. In short order, he successively tackled the electrical effects produced by x-rays, ultraviolet light, and Becquerel's rays from uranium. However, the next year, just as he was in the thick of his work on the uranium rays, an even greater distraction befell him, prompted by the announcement of a vacancy in physics at McGill University in Montreal, Canada.

Hugh Longbourne Callendar, a student of Thomson in the mid-1880s, had subsequently served as professor of physics, first at the Royal Holloway College, Egham, and then at McGill University. In 1898, he accepted an appointment as Quain Professor of Physics at University College, London. Rutherford was not that keen on leaving Cambridge just then, especially for a colonial university far away in Canada. On the other hand, the offer of a fellowship at Trinity College seemed not in the offing, yet he was desirous of monetary security to marry Mary Newton. And no doubt, the Canadian professorship intrigued the adventurous New Zealander. He therefore entered the competition, with testimonies from a number of Cambridge dons, among them Thomson. Thomson wrote that

> I have never had a student with more enthusiasm or ability for original research than Mr Rutherford, and I am sure that if elected he would establish a distinguished school of Physics at Montreal...I should consider any Institution fortunate that secured the service of Mr Rutherford as a Professor of Physics [8-79].

That did the trick. The Principal of McGill University and the chairman of the department of physics (himself a Trinity man) pounced on Rutherford, who sailed for Canada on the *Yorkshire* in September 1898.

At McGill, Rutherford was warmly received in what was probably the finest physics laboratory in Canada—perhaps anywhere at the time—financed and generously supported by the tobacco millionaire Sir William Macdonald. Before long the department chairman, John Cox, a mathematician who was a fine teacher but had less flair for research, offered to assume some of Rutherford's teaching duties, leaving his young colleague largely free for research. Cox had been impressed by Rutherford when he and the Principal of McGill interviewed him at Cambridge the previous July.

Though Rutherford's research at McGill soon made him the undisputed experimenter *par excellence* on radioactivity, his pioneering paper in this field was actually submitted to the *Philosophical Magazine* a few days before he sailed for Canada. This classic paper appeared in print in January 1899 [8-80]. It is a carefully articulated paper, chock full of facts, and some 54 pages long. As such, its tone gives no hint of any preoccupation with the imminent upheaval in its author's life, nor does his notebook from this period [8-81]. His bottom line was that, like x-rays, the uranium rays rendered gases traversed by them electrically conducting by ionization. Moreover, he concluded that the refraction and polarization that Becquerel purported to have found was erroneous; in point of fact, again like x-rays, there was neither refraction nor polarization. Though the characteristics of the two types of ray appeared very similar, they differed in at least one respect: the uranium rays consisted of *two* components. Absorption experiments with thin foils showed that

> ... the uranium radiation is complex and that there are present at least two distinct types of radiation—one that is very readily absorbed, which will be termed for convenience the α-radiation, and the other a more penetrating character, which will be termed the β-radiation. [8-82]

The nature of the β-rays would be elucidated before very long as a natural step in the ongoing cathode ray studies, as we shall see presently. The nature of the α-rays took much longer to ascertain. Rutherford's untiring pursuit of these rays—less interesting to many physicists because of their poor penetrating power [8-83]—would bear fruit nine years hence when their true nature was established. Though his work at McGill would earn him the Nobel Prize the very same year, 1908, his utilization of the α-rays in 1911 resulted in the greatest discovery of his career: the atomic nucleus. However, once again, that development takes us far ahead of our chronicle.

Chapter 9

THE RACE FOR *e/m*

9.1. Prelude to April 1897

The year 1897 began with at least two crucial questions in physics not yet resolved: (i) the nature of the ubiquitous cathode rays and (ii) ditto for x-rays which had so taken society by storm. Since the latter were produced by the former, investigation of either type of ray invariably threw light on the other. However, though the cathode rays had been studied for well-nigh half a century, agreement among the two contending schools as to their basic constitution—waves or particles—was still far from unanimous. In contrast, it was generally agreed from the time of their discovery that x-rays, studied for scarcely a year, constituted some kind of vibration in the æther; the only question was whether transverse or longitudinal, pulsed or periodic. Röntgen himself, in light of the apparent absence of refraction and interference but possibly swayed by the belief of his countrymen that cathode rays were some form of wave motion, tentatively assumed that his rays were the long-sought longitudinal, compressive vibrations also thought to propagate in the æther.

> There seems to exist some kind of relationship between the new rays and light rays; at least this is indicated by the formation of shadows, the fluorescence and the chemical action produced by them both. Now, we have known for a long time that there can be in the ether longitudinal vibrations besides the transverse light-vibrations; and, according to the views of different physicists, these vibrations must exist. Their existence, it is true, has not been proved up to the present, and consequently their properties have not been investigated by experiment.
>
> Ought not, therefore, the new rays to be ascribed to longitudinal vibrations in the ether?
>
> I must confess that in the course of the investigation I have become more and more confident of the correctness of this idea, and so,

therefore permit myself to announce this conjecture, although I am perfectly aware that the explanation given still needs further confirmation. [9-1]

Röntgen was not alone in his conjecture. The Austrian theorist Ludwig Boltzmann, of the University of Vienna, shared his views in a lecture on x-rays in January 1896, as did Oliver Lodge—then at University College, Liverpool. J J Thomson was at first noncommittal on the matter, declaring that 'whether these waves are transverse or longitudinal is... at present an open question' [9-2]. Lorentz, too, refrained from committing himself (though see Whittaker: note 1-19, pp 360–1). As noted earlier, perhaps Schuster was the first to get it right, suggesting in early 1896 that the rays were transverse vibrations of very short wavelengths. One who definitely agreed with Schuster was Emil Wiechert, of whom a brief background is not amiss.

Wiechert, born in 1861, came from a merchant family in Tilsit, an East Prussian city on the Memel River. He received his doctorate in physics in 1889 under Paul Volkmann at the University of Köningsberg. In 1896, when we first encounter him, he was *Privatdozent* at Köningsberg. A year later he left for Göttingen, where he started out as assistant to Ernst Schering at the observatory, charged with geodesy, terrestrial magnetism, and astrophysics. He soon became an associate professor at Göttingen, and in due course was responsible for a new and ultimately world famous geophysical institute there. The Göttingen school of geophysics produced some of the most far-reaching contemporary results on the internal structure of the earth. The fields of seismology generally, and of geophysical prospecting by seismic methods, were placed on a firm foundation under Wiechert's leadership [9-3]. We need not remind the reader at home in physics, however, that Wiechert's geophysical preoccupation did not, at least initially, prevent him from pursuing his interests in electrodynamics. His work on the 'retarded potentials' that bear his name and that of Liénard dates from 1900.

Confining ourselves to 1896, however, we note that in April of that year Wiechert suggested that 'Röntgen rays are most probably light rays of the ordinary transverse kind, but with shorter wave length, or with quite irregular, impulse-like oscillations' [9-4]. The same suggestion was put forth later in the year by Sir George Stokes. Stokes argued, as did Wiechert, that if the cathode stream consists of a large number of negatively electrified particles moving at a high speed, the sudden stopping of these particles must generate an irregular succession of pulses in the æther. The suggestion of Wiechert and Stokes echoes that of Schuster, who spoke of 'an impulsive wave... generated by the impact of a cathode ray' [9-5]. As to the apparent absence of refraction of x-rays, we have noted (p 117) that it simply stems from the very high vibrational frequency of these rays—the most characteristic property distinguishing x-rays from ordinary white light [9-6].

The next development of importance for our chronicle occurred in the opening days of 1897. By way of setting the stage for the unfolding events of that memorable year, let us divert our attention briefly to the prevailing atmosphere at the Cavendish just then, relying mainly on Rayleigh's *Life of Thomson*.

We have, earlier, touched on the opening of the new addition to the Cavendish next to Free School Lane in the Lent (spring) term of 1896, not withstanding Thomson's parsimonious attitude toward the need for laboratory space and equipment. (Hardly had it opened its doors than another academic development at Cambridge rendered the new accommodations insufficient: the admission (in 1896) of graduates of other universities as research students, making the demand for space and apparatus even more acute.) Not that this financial stringency was borne lightly by Thomson, who, notes Rayleigh, denied himself the commercially produced, professional instruments he deserved [9-7]. What seems to have saved the day was the narrow line between what the instrument makers could produce and what the resourcefulness of the researchers accomplished with string and sealing wax.

The 'Prof.' made a habit of spending an hour each morning, between his lectures and lunch time, looking in on the research students, encouraging when necessary. 'If the work had got stuck he would sit on a stool alongside, push his glasses up on his forehead, remain silent for a short while, and then shoot out a suggestion for a modification of the apparatus or something new to try' [9-8]. His interaction with the workers on the ground floor (virtually all of its 100 feet × 40 feet being devoted to laboratory space) was not limited to his regular round in the morning, however, and while his council was eagerly sought, Thomson *could* be stymied. Rayleigh recalls one instance when Thomson ducked a question on some detail in Henry Rowland's experiment on the magnetic effects of electrical convection [9-9]. He was even known to blurt out some silly answer if overworked or deeply preoccupied. Thus, when Frank Horton was required to accurately measure the diameter of a 0.001 cm diameter fiber, Thomson offered the unhelpful suggestion of measuring the length of n turns of a thread wrapped around the fiber! [9-10].

Rayleigh has left a vivid account of the day-to-day routine in the refurbished laboratory.

> The rooms on the ground floor all opened into one another, and their occupants wandered to and fro as they felt inclined. J.J. occupied the room at the end, and for a time his assistant, Everett, tried to establish the convention that it was private, indeed I think there was a notice to this effect on the door. But in practice little attention was paid to it, and when Rutherford, McLennan and others were established to work there as well as Everett, the game

> was up. Rutherford and J.J. talked on and off at all times, discussing the papers in the latest number of Wiedemann's *Annalen*, or the *Philosophical Magazine*. Later, after Rutherford had left, H.A. Wilson became his chief confidant. I myself began work upstairs, but when I had acquired some seniority among the research workers, I petitioned to be removed downstairs, where the heating was less miserably inadequate. The copper hot-water pipes which had been put in by Maxwell were altogether too small. This material was no doubt chosen to avoid magnetic disturbance, but its high cost had probably led to too small a size. [9-11]
>
> The picture that remains of the laboratory in those days is of a score of individuals scattered about in various rooms, two or three in a larger room—but each working at his own particular problem, for there was no team work then. Glass work of very varying quality was usually conspicuous at bench level, with the ubiquitous Töpler pump attached, and a maze of wires overhead: at least they should have been overhead, though I remember making a friendly protest to Townsend, who worked in the same room as I did, against his stretching wires in a position which threatened me with decapitation. The research workers had their own habits for commencing work and finishing it at times suited to their own individual temperaments. Some arrived at noon or even lunch time. There were no strict regulations in force (though this may have been tightened up later) and some I believe even got the key from the Porter's Lodge and came in on Sundays. [9-12]

The highlight of the day seems to have been the tea hour—that is, the daily tea and biscuit break, as distinct from the normal ritual presided over by Mrs Thomson before the meetings of the Cavendish Physical Society. Here Thomson invariably held forth on virtually any subject *sans* physics, be it the Boer War or some upcoming cricket match.

As in Cambridge, the establishment of a new physics building in Germany was a protracted affair—witness Göttingen, where a new institute was only opened in 1905 after Woldemar Voight and Eduard Riecke had labored for some 22 years on its creation [9-13]. On the other hand, research, not only teaching, became the academic mission in German university circles long before it became so across the English Channel (where the universities belatedly followed Cambridge's example). The establishment of the Physikalisch-Technische Reichsanstalt in Berlin in 1897 (serving as a model for England's subsequent National Physical Laboratory) was another example of the desire of the German authorities to provide physicists with an institution devoted, exclusively in this instance, to research—partly as a result of the political stability following the establishment of the German Reich in 1871 and in response to a spreading need for technological innovations in the wake of growing German nationalism and self-confidence [9-14]. As early as 1866, the redoubtable Wilhelm Weber, installed at Göttingen University 35 years earlier by Carl Friedrich Gauss, informed the curator of the university that the purpose of the physics institute was to enable research to be carried out by teachers and students alike. From then the number of students of mathematics and the natural sciences grew steadily until around 1881 when a decline in enrolment set in, bottoming out in about 1892. After then it swelled dramatically [9-15].

In mathematical physics, at any rate (and certainly in mathematics), Göttingen was second to none with its schools of electrodynamics stretching from Weber to Riecke and electro-optics from Voigt (and Franz Neumann) to Paul Drude. The electrodynamics effort would peak in Göttingen's interdisciplanary seminar of 1905 [9-16]. One of the contributing seminar docents was Emil Wiechert, whom we have already met.

On January 7 1897, in a demonstration lecture before the Physical and Economic Society of Köningsberg, Wiechert announced his view, based on cathode ray experiments of his own, that 'we are not dealing with the atoms known from chemistry, because the mass of the moving particles turned out to be 2000 to 4000 times smaller than the one of hydrogen atoms, the lightest of the known chemical atoms' [9-17]. Wiechert's experimental procedure did not differ significantly from that in other contemporary cathode ray experiments: the crux of his experiment lay in his analysis of the results. The rays were deflected in a magnetic field H normal to the plane defined by the direction of ray propagation and the velocity vector; thus they followed a curved path whose radius of curvature r is determined by equating the magnitude of the force exerted by the magnetic field on the rays of charge e and velocity v, Hev/c, to the resultant centripetal acceleration experienced by the rays, mv^2/r, where m is the mass of the rays, or

$$\frac{mv^2}{r} = \frac{Hev}{c}. \tag{9.1}$$

(The velocity of light, c, enters when the magnetic force is expressed in Gaussian units. We have discussed this relationship before, but in view of its timeliness for the events of 1897, it bears repeating.)

Equation (9.1) gave Wiechert an upper limit for the ratio m/e if it could be shown that v is larger than some minimum velocity. Moreover, if V is the potential difference between the electrodes in the cathode ray tube, the kinetic energy of the rays is given by

$$\frac{mv^2}{2} \leqslant eV \tag{9.2}$$

where strict equality only holds if there is no loss of energy by collisions with residual molecules in the evacuated tube. Combining these two expressions, we have

$$\frac{e}{m} \leqslant \frac{2Vc^2}{H^2r^2}$$

in which V, H, and r are all experimentally measurable.

To determine the minimum velocity, Wiechert attempted to compare the transit time of the cathode rays through the tube with the period of a Hertzian oscillator. Unable to obtain oscillations sufficiently rapid, however, he had to settle for an order-of-magnitude value of one-tenth of the velocity of light—an estimate right on the mark, as we shall see. Assuming, moreover, that the electrical charge e is a constant, namely the unit of charge identified in electrolysis—a quite *ad hoc* assumption, to be sure—he obtained his aforesaid mass limit for m.

Wiechert's experiment is not very different from that reported seven years earlier by Schuster in his second Bakerian Lecture of 1890 (section 6.1), as are his supporting assumptions, with one exception. Whereas Wiechert assumed that the cathode rays lost practically none of the kinetic energy acquired from the electric field in the tube, Schuster, as luck would have it, deliberately discarded that possibility, on a hunch, in favor of assuming that the rays lost nearly all of their energy. Schuster, like Wiechert, also made the unwarranted assumption with regard to the charge e. The capital point, however, was Wiechert's faster ray velocity, which led him to propose for the first time in print that the rays are particles of *subatomic* masses, and thus probably smaller than atoms as well.

Nevertheless, a truly convincing experiment would require not simply a measurement of mv/e, which strictly speaking is all Wiechert or Schuster could claim, but a direct, independent determination of the velocity v, in order to isolate m/e. Such a measurement was not long in the making. Wiechert himself measured the velocity in 1898—alas, well after J J Thomson had beaten him to it by an elegant experimental step quite overlooked by all his competitors.

9.2. April 30 1897

If there is a month in the annals of science that might be singled out as of decisive importance in the history of elementary particle physics, it is April 1897. That month marks the final resolution of the debate between the German and British (or Anglo-French) camps that has preempted our attention for so long: the physical nature of the cathode rays—waves or particles. Curiously, the resolution in favor of the largely British particulate view was presaged by the researches of two young German physicists within months of each other. One, Emil Wiechert, we have met. Wiechert was on the right track, but he drew correct conclusions from a less than convincing experiment. Perhaps closer to the mark was Walter Kaufmann, who, unfortunately, hedged on the importance of his excellent results. Possibly of significance is the nearly reverse order of their institutional affiliations. Wiechert performed his marginal experiment at Köningsberg, adding a convincing extension at Göttingen where he subsequently completed a brilliant career in geophysics. Kaufmann's initial work was done at Berlin—work which he significantly elaborated upon at Göttingen before moving on, settling for good at Köningsberg.

Walter Kaufmann was born in 1871 in Elberfeld, part of Wuppertal in northwestern Germany, where he obtained his early schooling. He studied at Berlin and then Munich, where he received his doctorate in 1894 [9-18]. In 1896, he became an assistant at the Physics Institute at Berlin, and three years later he accepted a similar position at Göttingen. He became an associate professor at Bonn in 1903, and full professor and director of the Physics Institute of Köningsberg in 1908. He retired as guest professor at Freiberg, where he remained until his death at 76 in 1947.

Kaufmann undertook his determination of e/m for cathode rays while serving his assistantship in Berlin. Like Wiechert, he based his measurements on magnetic deflection of the rays and relied on the expression for their kinetic energy, $\frac{1}{2}mv^2 = eV$, where V is the potential between the electrodes in the discharge tube. Kaufmann, too (unlike Schuster) assumed no kinetic energy was lost by collisions with residual molecules; the assumption also ignores the possibility that, at moderate pressures, ions may not start from the cathode itself, or, if they do, some work is expended in tearing ions out of the metal [9-19]. His initial paper in *Annalen der Physik und Chemie* was completed in April 1897. The apparatus is depicted in figure 9.1. His primary finding was that the magnetic deflection varied inversely with the square root of the potential difference, independent of the nature of the residual gas in the tube, whether air, hydrogen, or carbon dioxide, from which it also follows that e/m is independent of the residual gas [9-20].

Two aspects concerning this important finding bothered Kaufmann, however. For one thing, 'this assumption [of constant e/m] is physically hard to interpret; for if one makes the most plausible assumption that the moving particles are ions then e/m should have a different value for each gas' [9-21]. Moreover, his order-of-magnitude value for e/m was about 10^7 emu g^{-1}, 'while for a hydrogen ion [e/m] equals only 10^4' [9-22]. In view of these misgivings, he professed reservations as to whether 'the hypothesis of cathode rays as emitted [*abgeschleuderte*] particles [might be] by itself inadequate for a satisfactory explanation of the regularities I have observed' [9-23].

We will have more to say presently about Kaufmann's e/m program. For now, we note that during his subsequent Göttingen period, 1899–1902, he concentrated on e/m values for Becquerel rays—β-rays, in particular. Shortly before, a controversy had arisen over whether the

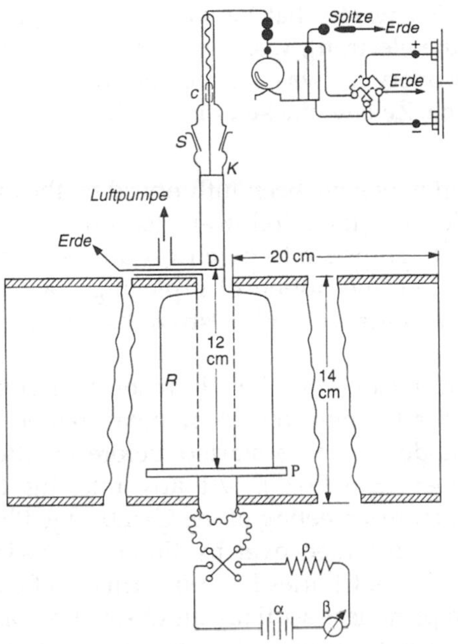

Figure 9.1. *Kaufmann's e/m apparatus of 1897. The bell-shaped tube (R) is cemented to a glass plate (P). At the top, a glass extension with a cathode (K) and wire anode (D) is interrupted by a ground-glass joint (S). The lead-in to the cathode is a copper spiral terminating on the outside in a mercury bath. An iron cylinder (c) allows displacement of the cathode by an external magnetic field. The shadow of the anode, projected onto the graduated surface of the plate glass, allows measurement of the deviation of the cathode ray beam in a magnetic field, facilitated by a layer of chalk on the glass surface that fluoresces when bombarded. The deflecting field is produced by a Helmholtz coil of insulated copper wire, wound on the 20 cm long cylindrical formers depicted. The coil is powered by a battery (α) in conjunction with a rheostat (ρ) and precision Siemens/Halske ammeter (β).*

electron (an accepted term to all but, possibly, J J Thomson by then) could possess an apparent 'electromagnetic mass' gained from the interaction of a moving charge with its own field, in addition to its normal 'material' mass. Such a suggestion, it may be recalled, had been put forth as early as 1881 by J J Thomson. The question seemed particularly relevant for β-rays, some of which had velocities approaching that of light. However, for now, Kaufmann's lingering unease over the 'particle hypothesis' in April 1897 may reflect the last gasps of the German wave school, possibly coupled with a conservative outlook at the Institute in Berlin that discouraged unwarranted speculation. Thus George Jaffé, who went to Cambridge for a postdoctoral stint after receiving his doctorate under Ostwald, recalls John Zeleny, who followed him to Cambridge from Berlin, looking back as follows:

> I heard John Zeleny say that he was in Berlin [at the time of the discovery of the electron], working in the laboratory of Warburg. When the discovery... was announced, nobody in Berlin would believe it. So he, Zeleny, packed his trunks to go to Cambridge... [9-24]

Finally, Kaufmann may have been influenced by the anti-atomistic views of Mach, Ostwald, and their followers (section 1.2) that regarded it is as scientifically inappropriate to dwell on unobservable entities such as atoms [9-25]. Indeed, the notion of replacing mass by electromagnetic energy was not at variance with Ostwald's and Helm's penchant for energetics [9-26].

At Cambridge, meanwhile, J J Thomson was not idle either. It is not certain whether his new round of measurements on the magnetic deflection of cathode rays was started before or after his return from America in October 1896 [9-27]. At any rate, his first report on the subject was read before a meeting of the Cambridge Philosophical Society on February 8 1897, presided over by Francis Darwin, president of the Society, and third son of Charles Darwin. This brief communication, less than two pages in print, is essentially an abstract of various experimental results which he would soon elaborate on in his forthcoming lecture before the Royal Institution. (His modest report on the occasion was followed by a very much longer one on a related subject by his research student J S E Townsend, as we shall relate in the next chapter.) One result in Thomson's communication warrants particular notice; namely, that 'the magnetic deflection of the cathode rays in air, hydrogen, carbonic acid gas and methyl iodide is the same provided the mean potential difference between the cathode and the anode is the same' [9-28]. Here was a fresh result, indeed—perhaps the first indication of the independence of the cathode ray behavior of the constitution of the residual gas, well ahead of Kaufmann [9-29]. Hitherto, the guiding principle in cathode ray experimentation had been to maintain constant *gas pressure*, not constant tube voltage—a more difficult procedure—in which case different ion species invariably gave different deflections.

The February communication contains no hint, however, whether Thomson was yet prepared to declare the 'negatively electrified particles', as he then calls them, as anything other than chemical atoms or ions—that is, as *sub*atomic entities. Apparently Thomson was unaware of Wiechert's lecture at Köningsberg just a month earlier—a lecture which may not have appeared in print before Thomson's brief report [9-30]. Nor does the main text of the printed version of Thomson's Princeton lectures offer a clue, if we ignore the summary of his Royal Institution lecture subsequently appended to the volume as it went to the printer late in 1897 [9-31]. In his *Recollections and Reflections*, Thomson is no more helpful, stating merely that he 'made the first announcement of the existence of these corpuscles in a Friday Evening Discourse at the Royal

Institution on April 29, 1897' [9-32], and we have to leave it at that. Or almost at that. Curiously, Thomson is off by a day. The epochal Discourse was indeed delivered on Friday—which fell on April 30.

Thomson's Discourse was reprinted in the *Electrician* for May 21 [9-33], though this classic paper is often ignored in favor of the fuller article published later that year in the October issue of the *Philosophical Magazine*. The latter paper is rightly considered *one* of Thomson's definitive memoirs on the discovery of the electron, but only one of three—the second and third appearing in 1898 and 1899, respectively. Nevertheless, despite the fame attached to the October memoir, it is instructive to dwell on its April precursor; doing so gives a feeling for the state of his experiments at that time and lets us place the relative importance of the two papers in perspective.

The weekly evening lecture at the Royal Institution on Friday April 30 was chaired by Sir Frederick Bramwell, Secretary of the Institution (who was succeeded in that position in 1900 by Sir William Crookes). After some preliminaries on cathode rays from Plücker's time onward, Thomson launched his lecture proper with a review of more recent investigations of chemical effects of the rays, starting with Goldstein's observation that substances exposed to the rays changed color, something Wiedemann and Ebert attributed to the formation of a subchloride (e.g. chloride of alkaline metals). More recently, Elster and Geitel had shown that these substances, after exposure to cathode rays, also acquired the power of discharging negative electricity under the action of light—that is, they became photoelectric. A more widespread manifestation of the effects produced by the impact of the rays was that of *thermoluminescence*. If bodies are exposed to the rays for some time, to all appearances they revert to their original condition when the bombardment stops. However, if they subsequently are heated up, they become luminous at a temperature well below the onset of incandescence.

Next, Thomson turned to his own investigations then occupying him—a renewed study of the magnetic deflection of the rays. In his discharge tube, crafted by Everett, a narrow beam was formed by sending the cathode rays through a tube containing a plug with a slit in it, the plug serving as anode and grounded. Under the action of a uniform magnetic field, the beam spread out into a fan-shaped luminosity in the residual gas; like the luminosity, the phosphorescence produced when the rays struck the glass wall was not uniformly distributed, but spread out into a luminous patch crossed by bright and dark bands—a pattern of bands first reported the year before by the Norwegian Kristian Birkeland [9-34]. As Thomson had already observed in his February report,

> ...in a given magnetic field, with a given mean potential difference between the terminals, the path of the rays is independent of the nature of the gas; photographs were taken

> of the discharge in hydrogen, air, carbonic acid, methyl iodide, i.e. in gases whose densities range from 1 to 70, and yet not only were the paths of the most deflected rays the same in all cases, but even the details, such as the distribution of the bright and dark spaces, were the same; in fact, the photographs could hardly be distinguished from each other. [9-35]

Not only that,

> The curves described by the cathode rays in a uniform magnetic field are, very approximately at any rate, circular for a large part of their course; this is the path which would be described if the cathode rays marked the path of negatively electrified particles projected with great velocities from the neighbourhood of the negative electrode. Indeed, all the effects produced by a magnet on these rays, and some of these are complicated, as, for example, when the rays are curled up into spirals under the action of a magnetic force, are in exact agreement with the consequences of this view. [9-36]

To clinch his argument that the rays are negatively electrified particles, Thomson also repeated Perrin's elegant experiment performed in Paris two years earlier, in which the rays were accumulated and their aggregate charge measured directly—but with an important modification. We have dwelt on Perrin's experiment, and on Thomson's repetition of it, at some length in section 7.1; thus we need not do so here. Briefly, Perrin collected the rays in a coaxial, insulated pair of metallic cylinders—a so-called Faraday cup—in which the inner cylinder, connected to an electroscope, was screened by the outer, grounded cylinder which acted as the anode. The problem with the experiment, as designed by Perrin, was that the Faraday cup was coaxial with the undeflected beam from the cathode.

> Though the experiment shows [wrote Thomson] that negatively electrified bodies are projected normally from the cathode, and are deflected by a magnet, it does not show that when the cathode rays are deflected by a magnet the path of the electrified particles coincides with the path of the cathode rays. The supporters of the theory that these rays are waves in the ether might say, and indeed have said, that while they did not deny that electrified particles might be shot off from the cathode, these particles were, in their opinion, merely accidental accompaniments of the rays, and were no more to do with the rays than the bullet has with the flash of a rifle. [9-37]

In Thomson's version of the experiment, as outlined earlier, the Faraday cup was not in the direct path of the undeflected rays, but placed in an extension of the tube protruding at an angle to the cathode–anode axis

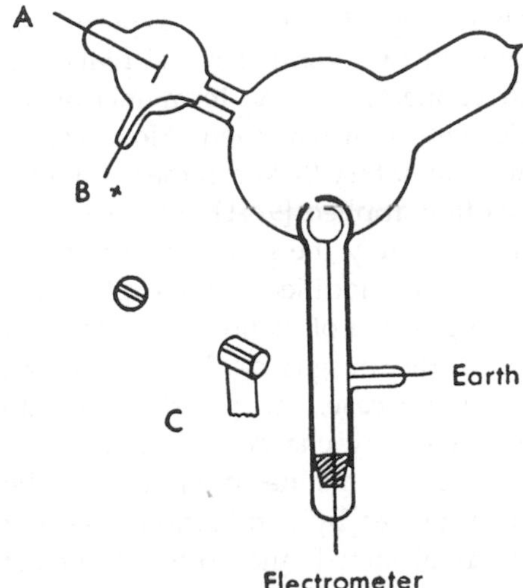

Figure 9.2. *Thomson's tube for measuring the cathode ray charge. The rays from the cathode A pass through a slit in a metal plug (the anode B), and, when deflected by a magnet, enter a Faraday cup c connected to an electrometer.*

(figure 9.2). With this clever tube arrangement, as always the handiwork of Mr Everett, the cathode rays could not enter the Faraday cup *except* when suitably deflected by a magnet [9-38]. If now the collector registers a negative charge, we may agree with Thomson that 'the stream of negatively-electrified particles is an invariable accompaniment of the cathode rays' [9-39].

Even so, one serious obstacle still lay in Thomson's path. If the cathode rays were truly charged particles, why were they not deflected in an *electric* field—something Hertz had emphatically been unable to do. Hertz, an experimenter *par excellence*, had passed the rays between a pair of plates connected to a battery and thus maintained at a known potential difference, without the slightest hint of any deflection. The clue to Hertz's negative experiment, realized Thomson, lay in an old experiment initially carried out by Eugen Goldstein and improved upon by Crookes—one in which two cathode ray beams generated simultaneously in one discharge tube appeared to repel each other by electrostatic action between the two beams [9-40]. This experiment led Crookes to conclude that the two beams were composed of radiant matter carrying electrical charges of the same polarity. A further experiment, this time an improvement on an old one by Plücker involving the magnetic deflection of a single cathode ray, convinced him that the 'molecules' in the beams carried *negative* charges.

Thomson had given the matter considerable thought the previous fall, and dwells on it in his Princeton lectures. No sooner had he returned from America than he repeated Hertz's experiment, passing a beam of cathode rays between two parallel metal plates mounted inside the discharge tube and connected to the terminals of a battery, and once more no convincing deflection was seen. However, as he relates in his *Recollections*, a faint, momentary flicker in the beam was discernible when the electric field was first applied [9-41].

Thomson saw immediately the problem. From the saturation of the accumulated charge in his modified version of Perrin's experiment, he knew (p 115) that the passage of cathode rays through the residual gas in a discharge tube made the gas *conducting* by ionization, just as x-rays did. That being the case, the positive ions thus produced will be attracted to the negatively charged plate and the negative ions to the positive plate, leading to a state of equilibrium between the charge collected and the charge lost by conduction. As a result, the charges on the plates will be neutralized, and so will the electric force between them.

What to do? The obvious solution was easier said than done: reducing the neutralizing action of the ions by pumping on the gas for all the equipment was worth, exhausting the tube to an even higher state of rarefaction than hitherto believed possible. The necessity of reducing the amount of gas condensed on the interior surfaces of the discharge tube by prolonged baking was not yet appreciated as good practice in vacuum experiments. However, it was painfully obvious that during a discharge run, gas was liberated and the vacuum deteriorated too fast for available pumps to cope with. Here Thomson's competitor Kaufmann had a potential advantage, vacuum experimentation being his strong card. In fact, Kaufmann would invent the first rotary high-vacuum pump which he deployed with great success in his further e/m program [9-42]. At any rate,

> by running the discharge through the tube day after day without introducing fresh gas, the gas on the walls and electrodes got driven off and it was possible to get a much better vacuum. The deflection of the cathode rays by electric forces became quite marked, and its direction indicated that the particles forming the cathode rays were negatively electrified. [9-43]

In his April lecture Thomson offered another way out, which he demonstrated in the Royal Institution's well equipped lecture theater, assisted by G W C Kaye (with apparatus prepared at Cambridge and transported to London by Everett): mounting the plates in the dark space next to the cathode—the region most highly exhausted. Connecting either plate to ground in turn gave a decided deflection of the cathode rays in one or the other direction. Thomson 'ascribed the deflection to the gas in the dark space either not being a conductor at all, or if a conductor, a poor one compared to the gas in the main body of the tube' [9-44].

After reviewing the relationship between the distance traveled by Lenard rays from their 'window' (oiled silk did about as well as a window as did very thin—and scarce—aluminum) and their absorption, Thomson's lecture continues as follows.

> We see that though the densities and the coefficients of absorption vary enormously, yet the ratio of the two varies very little, and the results justify, I think, Lenard's conclusion that the distance through which these rays travel only depends on the density of the substance—that is, the mass of matter per unit volume, and not upon the nature of the matter.
>
> These numbers raise a question which I have not yet touched upon, and that is the size of the carriers of the electric charge. Are they or are they not the dimensions of ordinary matter?
>
> We see from Lenard's table that a cathode ray can travel through air at atmospheric pressure a distance of about half a centimetre before the brightness of the phosphorescence falls to about one-half of its original value. Now the mean free path of the molecule of air at this pressure is about 10^{-8} cm, and if a molecule of air were projected it would lose half its momentum in a space comparable with the mean free path. Even if we suppose that it is not the same molecule that it carried, the effect of the obliquity of the collisions would reduce the momentum to one-half in a short multiple of that path.
>
> Thus, from Lenard's experiments on the absorption of the rays outside the tube, it follows on the hypothesis that the cathode rays are charged particles moving with high velocities, *that the size of the carriers must be small compared with the dimensions of ordinary atoms or molecules* [italics added]. The assumption of a state of matter more finely subdivided than the atom of an element is a somewhat startling one; but a hypothesis that would involve somewhat similar consequences—*viz.* that the so-called elements are compounds of some primordial element—has been put forward from time to time by various chemists. Thus Prout believed that the atoms of all the elements were built up of atoms of hydrogen, and Mr Norman Lockyer has advanced weighty arguments, founded on spectroscopic consideration, in favor of the composite nature of the elements.
>
> Let us trace the consequences of supposing that the atoms of the elements are aggregations of very small particles, all similar to each other; we shall call such particles corpuscles, so that the atoms of the ordinary elements are made up of corpuscles and holes, the holes being predominant. Let us suppose that at the cathode some of the molecules of the gas get split up into these

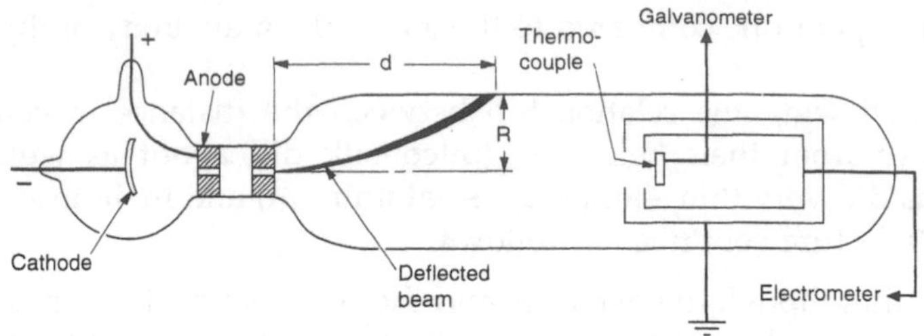

Figure 9.3. *Thomson's e/m tube of April 1897, employing a Faraday cup supplied with an iron–copper thermocouple for measuring the kinetic energy of the incident corpuscles, deposited in the form of heat. Not shown is a pair of Helmholtz coils for measuring the deflection of the rays.*

> corpuscles, and that these, charged with negative electricity and moving at high velocity, form the cathode rays. The distance these rays would travel before losing a given fraction of their momentum would be proportional to the mean free path of the corpuscles. Now, the things these corpuscles strike against are other corpuscles, and not against the molecules as a whole; they are supposed to be able to thread their way between the interstices in the molecule. Thus the mean free path would be proportional to the number of these corpuscles; and, therefore, since each corpuscle has the same mass to the mass of unit volume—that is, to the density of the substance, whatever be its chemical nature or physical state. Thus the mean free path, and therefore the coefficient of absorption, would depend only on the density; this is precisely Lenard's result. [9-45]

Thomson's discussion has brought us to his *experimentum crucis*, his e/m measurements reported in April. Thomson did not depict in print his particular discharge tube deployed for this purpose, though from his description its features are readily reconstructed, as in figure 9.3. The rays from the cathode were collimated into a narrow beam by traversing a pair of slotted metal plugs, the first of which was maintained at a high positive potential and served as the anode. Emerging from the second plug, the beam enters the usual Faraday cup with the outer cylinder grounded and the inner cylinder insulated and connected to an electrometer. Just inside the entrance to the collector is a small iron–copper thermocouple with leads passing through and insulated from both cylinders and glass wall; it is connected to a low-resistance galvanometer.

Assuming all the 'corpuscles' struck the thermocouple, they deposited their kinetic energy in the form of heat. Knowing the weights and specific heats of the metals in the thermocouple, and, from the galvanometer deflection, the rise in temperature of the thermoelectric junction,

Thomson obtained the mechanical equivalent of heat W imparted to the couple in unit time. Equating W to the kinetic energy of the particles, we have

$$W = \tfrac{1}{2}Nmv^2 \tag{9.3}$$

here N is the number of particles, m is the mass of each particle, and v is their mean velocity. The deflection of the electrometer gave the total charge Q acquired by the collector in the same time,

$$Q = Ne \tag{9.4}$$

where e is the charge of each particle [9-46]. Eliminating N from the two equations gives

$$W/Q = 1/2(m/e)v^2. \tag{9.5}$$

One more measurement was necessary; namely, the velocity v in terms of measurable quantities. After recording the galvanometer and electrometer readings, the tube, still running, was placed in a uniform magnetic field generated by a pair of Helmholtz coils (split coils, separated by a distance equal to their radius), in order to measure the deflection of the beam in a known magnetic field. Referring to figure 9.3, let R be the radius of the discharge tube and d the distance from the second plug to a spot on the tube wall made fluorescent by the deflected beam. The radius of curvature of the path of the deflected beam, r, is obtained from the relationship between the chord, radius, and height of a segment of a circle, and is given by [9-47]

$$2r = d^2/R + R.$$

The radius of curvature is also given, from equation (9.1), by

$$r = (mv)/(He).$$

Eliminating v from equation (9.5) gives

$$e/m = (2W)/(Qr^2H^2). \tag{9.6}$$

With an exposure of about 1 s, the accumulated charge raised the capacitor, 1.5 μF, to a potential of 16 V, giving a charge $Q = 2.4 \times 10^{-6}$ emu. The thermocouple, with a heat capacity of 5×10^{-3} cal °C^{-1}, was raised by 3.3 °C by the impact of the rays. Using 1 cal = 4.2×10^7 erg, we then have

$$W = 6.3 \times 10^5 \text{ erg}.$$

Inserting, finally, a value for the product Hr, Thomson obtained

$$e/m = 0.62 \times 10^7 \text{ emu g}^{-1}$$

or

$$m/e = 1.6 \times 10^{-7}\,\text{g emu}^{-1}.$$

> This [concludes Thomson] is very small compared with the value 10^{-4} for the ratio of the mass of an atom of hydrogen to the charge carried by it. If the result stood by itself we might think that it was probable that e was greater than the atomic charge of atom rather than that m was less than the mass of a hydrogen atom. Taken, however, in conjunction with Lenard's results for the absorption of the cathode rays, these numbers seem to favor the hypothesis that the carriers of the charges are smaller than the atoms of hydrogen. [9-48]

His value for e/m (actually m/e) for cathode rays, adds Thomson, agrees with Zeeman's order-of-magnitude value, 10^{-7}, deduced from the effect of a magnetic field on the period of the sodium light—a matter touched on in section 1.2, and to which we return in some detail shortly.

As Thomson recalls, not everybody in his audience in the historic lecture theater on London's Albemarle Street was convinced of the existence of bodies smaller than atoms. 'I was even told long afterwards by a distinguished physicist who had been present... that he thought I had been "pulling their legs". ... There were, however, a few, I think Professor G. F. FitzGerald was one, who thought I had made out a good case' [9-49]. Nor does Robert Strutt, the fourth Lord Rayleigh and then an undergraduate at Cambridge, think his lecture caused much of an immediate stir.

> I do not think that I myself heard anything about it at the time, and only heard the conclusion he had reached some weeks later at Cambridge. The probabilities are that few of the audience really took in Thomson's argument which, after all, requires the assembling of a good many lines of reasoning which were not then familiar. However, they no doubt realised that he was saying that he had found bodies smaller than hydrogen atoms, a statement which, in the then condition of science, was thought to be paradoxical, or even self-contradictory, an atom being (it was said) the smallest portion of matter that did or could exist [9-50].

The first the young Rayleigh heard about it was, in fact, from Thomson himself, in the summer term of 1897 as nearly as he can recall.

> Chancing to meet on King's Parade he began to unfold to me what he had been doing—telling me that the cathode rays had now 'turned out' to be particles, and particles quite different from atoms. My rooms were at that time in Whewell's Court, but I did not want to interrupt the tale he was unfolding by stopping there,

Figure 9.4. *Thomson giving a lecture demonstration of the Braun e/m tube.*

> and walked on with him past St John's and the Round Church to the other entrance of Whewell's Court in Sidney Street, where he left me, after standing talking for a few minutes. [9-51]

Not long after his encounter with Thomson, the rest of the Cavendish students had the opportunity of hearing a first-hand account of the work by the Professor in a lecture demonstration, albeit poorly announced, before the Cavendish Physical Society. True to form, Thomson enthralled his young audience, as he had at the Royal Institution, with plenty of apparatus including the cathode ray tubes—to be sure, not evacuated. The cumbersome Töpler pumps still required several days of strenuous pumping—an effort not warranted for the informal gatherings of the Physical Society.

9.3. October 1897

Thomson's full paper on cathode rays was submitted to the *Philosophical Magazine* on August 7 1897, and appears in the October issue of that journal [9-52]. The first page of Thomson's ink-written manuscript of this historic paper is reproduced here as figure 9.5. Among other things, the paper contains a fuller account of his *e/m* determinations based on magnetic (but not electrostatic) deflection and the ratio of heat to charge deposited—measurements first reported in April's Royal Institution lecture. Cathode ray tubes of three different types were used in this round of investigations. The first, reproduced in figure 9.3, was the one described but, oddly, is not depicted in his printed lecture. For some reason, he omitted it in an early stage of his manuscript [9-53]. It needs no further description here. Its chief drawback, notes Thomson, was an annoying tendency, in consequence of the charging up of the glass tube, for a secondary discharge to sheath the collecting cylinders with a luminous glow—a problem normally cured by further pumping and then letting the tube remain quiescent for a period. Results obtained with this tube are listed under 'tube 1' in table 9.1 [9-54].

The second tube ('tube 2') was a bell jar with the cathode mounted in a small extension tube protruding from the side wall of the jar. In the aperture between extension and bell jar was located the usual slotted, metallic plug serving as anode and ray collimator. The bell jar could be placed between a pair of Helmholtz coils producing a nearly homogeneous magnetic field perpendicular to the axis of the jar and to the path of the rays. Under its action, the rays spread out into a fan-shaped pattern ('Birkeland magnetic spectrum' as it was commonly called) in a plane perpendicular to the field direction; the luminous rays in the rarefied gas could be photographed against a glass plate ruled into small squares. As in the first tube, a cylindrical collector equipped with a thermoelectric junction was placed in the 'line of fire'. The troublesome glow plaguing tube 1 was eliminated by lining the inside of the bell jar with a grounded copper gauze. A drawback of tube 2 was the need for lots of sealing wax (used in seating the electrodes), which limited the range of operating pressures due to outgassing.

Tube 3 was similar to the first, except for the slits in the two plugs being replaced by small holes; this (which could effectively have been accomplished simply by rotating one plug through 90°) reduced the accumulated charge. The reduction was compensated for by reducing the capacity of the inner Faraday cylinder (thereby increasing the potential for a given charge) and by increasing the sensitivity of the quadrant electrometer.

The results are shown for air, hydrogen, and carbonic acid (carbon dioxide) in table 9.1. Here W/Q is calculated from equation (9.5), I is the 'magnetic rigidity' Hr in Gaussian units, m/e is obtained from equation (9.6), and the mean velocity v ($\mathrm{cm\,s^{-1}}$) results from eliminating $m/e = rH/v$ from equation (9.5). That is,

$$v = 2W/QHr. \tag{9.7}$$

Values for v are seen to vary over nearly an order of magnitude, from typically 23 000 $\mathrm{km\,s^{-1}}$ (about $\frac{1}{15}$ the velocity of light) to as high as 130 000 $\mathrm{km\,s^{-1}}$. On average, then, the ray velocity was just about what Wiechert had concluded from his own experiments announced early in the year (section 9.1). These velocities are, in any case, not particularly interesting, except that the highest values for air and carbon dioxide (approaching the speed of light) are either misprints or

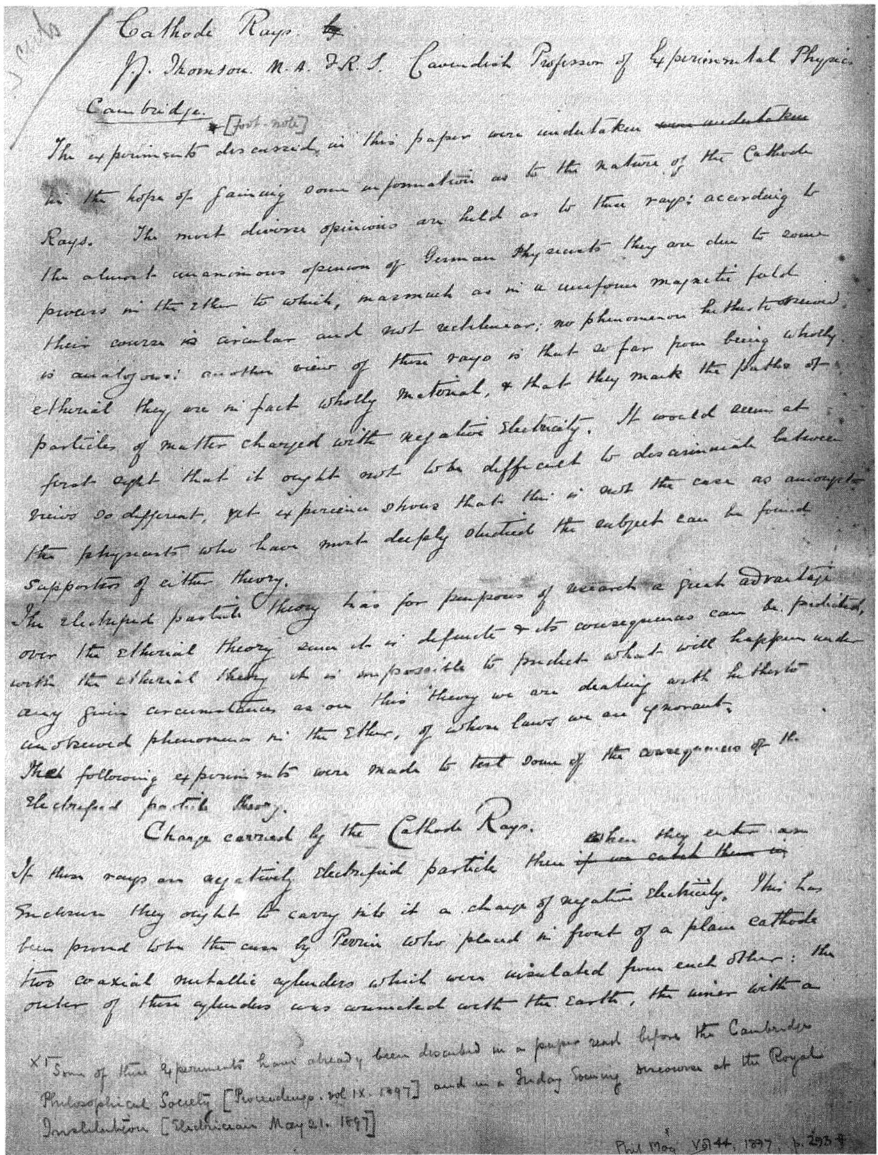

Cathode Rays.

J.J. Thomson. M.A. F.R.S. Cavendish Professor of Experimental Physics Cambridge.

*[foot-note]

The experiments discussed in this paper were undertaken ~~were undertaken~~ in the hope of gaining some information as to the nature of the Cathode Rays. The most diverse opinions are held as to these rays; according to the almost unanimous opinion of German physicists they are due to some process in the ether to which, inasmuch as in a uniform magnetic field their course is circular and not rectilinear, no phenomenon hitherto observed is analogous: another view of these rays is that so far from being wholly etherial they are in fact wholly material, & that they mark the paths of particles of matter charged with negative electricity. It would seem at first sight that it ought not to be difficult to discriminate between views so different, yet experience shows that this is not the case as amongst the physicists who have most deeply studied the subject can be found supporters of either theory.

The electrified particle theory has for purposes of research a great advantage over the etherial theory since it is definite & its consequences can be predicted; with the etherial theory it is impossible to predict what will happen under any given circumstances as on this theory we are dealing with hitherto unobserved phenomena in the ether, of whose laws we are ignorant.

The following experiments were made to test some of the consequences of the electrified particle theory.

Charge carried by the Cathode Rays.

If these rays are negatively electrified particles then when they enter an enclosure they ought to carry into it a charge of negative electricity. This has been proved to be the case by Perrin who placed in front of a plane cathode two coaxial metallic cylinders which were insulated from each other: the outer of these cylinders was connected with the earth, the inner with a

* Some of these experiments have already been described in a paper read before the Cambridge Philosophical Society [Proceedings. vol IX. 1897] and in a Friday Evening discourse at the Royal Institution [Electrician May 21. 1897]

Phil Mag Vol 44, 1897, p. 293 ff

Figure 9.5. *The first page of Thomson's manuscript on the discovery of the electron.*

spurious. Definitely more interesting are the values for e/m (or m/e), which also vary over a considerable range. Before dwelling on these results, however, Thomson introduces another, independent method for measuring e/m, making use of his recent finding that the cathode rays can be deflected by an electrostatic field as well as by a magnetic field. This

Table 9.1.

Gas	Value of W/Q	I	m/e	v
Tube 1				
Air	4.6×10^{11}	230	0.57×10^{-7}	4×10^{9}
Air	1.8×10^{12}	350	0.34×10^{-7}	1×10^{10}
Air	6.1×10^{11}	230	0.43×10^{-7}	5.4×10^{9}
Air	2.5×10^{12}	400	0.32×10^{-7}	1.2×10^{10}
Air	5.5×10^{11}	230	0.48×10^{-7}	4.8×10^{9}
Air	1×10^{12}	285	0.4×10^{-7}	7×10^{9}
Air	1×10^{12}	285	0.4×10^{-7}	7×10^{9}
Hydrogen	6×10^{12}	205	0.35×10^{-7}	6×10^{9}
Hydrogen	2.1×10^{12}	460	0.5×10^{-7}	9.2×10^{9}
Carbonic acid	8.4×10^{11}	260	0.4×10^{-7}	7.5×10^{9}
Carbonic acid	1.47×10^{12}	340	0.4×10^{-7}	8.5×10^{9}
Carbonic acid	3×10^{12}	480	0.39×10^{-7}	1.3×10^{10}
Tube 2				
Air	2.8×10^{11}	175	0.53×10^{-7}	3.3×10^{9}
Air	4.4×10^{11}	195	0.47×10^{-7}	4.1×10^{9}
Air	3.5×10^{11}	181	0.47×10^{-7}	3.8×10^{9}
Hydrogen	2.8×10^{11}	175	0.53×10^{-7}	3.3×10^{9}
Air	2.5×10^{11}	160	0.51×10^{-7}	3.1×10^{9}
Carbonic acid	2×10^{11}	148	0.54×10^{-7}	2.5×10^{9}
Air	1.8×10^{11}	151	0.63×10^{-7}	2.3×10^{9}
Hydrogen	2.8×10^{11}	175	0.53×10^{-7}	3.3×10^{9}
Hydrogen	4.4×10^{11}	201	0.46×10^{-7}	4.4×10^{9}
Air	2.5×10^{11}	176	0.61×10^{-7}	2.8×10^{9}
Air	4.2×10^{11}	200	0.48×10^{-7}	4.1×10^{9}
Tube 3				
Air	2.5×10^{11}	220	0.9×10^{-7}	2.4×10^{9}
Air	3.5×10^{11}	225	0.7×10^{-7}	3.2×10^{9}
Hydrogen	3×10^{11}	250	1.0×10^{-7}	2.5×10^{9}

experiment has gained prominence quite out of proportion to its intrinsic importance compared to the thermocouple experiments. It is largely because of its fame that the publication date of the journal article, October 1897, is frequently cited as the birth date for the electron. At any rate, the new experiment was simpler, and potentially more accurate, than its forerunners, though Thomson himself appears not to have attached any special virtue to it. (However, his views on the results from tube 1 and tube 2 may indicate some prejudice in favor of the latter experiment, as we shall see.) In light of its unquestioned elegance and classic reputation, the new experiment warrants description, if for no other reason than as an interesting alternative e/m experiment.

The original discharge tube, shown in figure 9.6, is one of many pieces of apparatus on display in the Cavendish Museum in the modern Cavendish Laboratory in west Cambridge [9-55]. It is also depicted

Figure 9.6. *Thomson's e/m tube of October 1897, employing crossed electric and magnetic fields.*

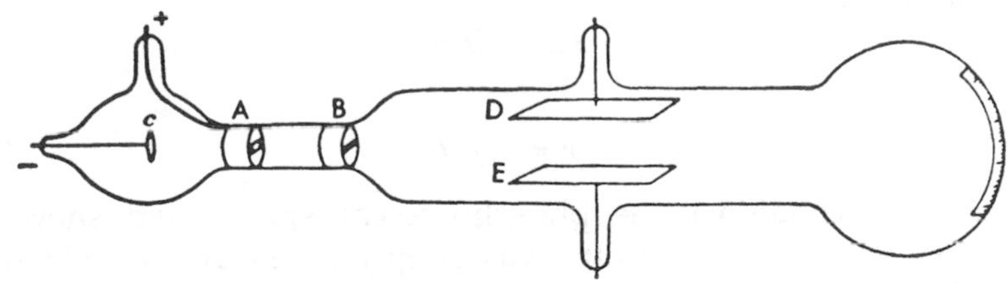

Figure 9.7. *A schematic rendering of Thomson's crossed-field tube of October 1897. The slotted plug A is the anode, c is the cathode, and D and E are the electric deflection plates.*

schematically in figure 9.7. As in the tube in figure 9.3, the rays from the cathode exit through a pair of slotted plugs into the main chamber, which now contains a pair of parallel aluminum plates connected to the terminals of a battery or storage cell. Externally, a pair of Helmholtz coils (not shown), whose diameter is equal to the length of the plates, produces a magnetic field extending over the same length of the tube as the electrostatic field between the plates. The plates and split coils were energized in such a manner that a negatively charged particle passing between the plates would be deflected vertically *downward* by the electrostatic field and vertically *upward* by the magnetic field. The deflection of the phosphorescent patch at the end of the tube was measured with a scale pasted to the tube, as shown in figure 9.7. Since it was necessary to darken the room to see the phosphorescent patch, a needle coated with luminous paint was moved up or down by Thomson or Everett, his assistant (whom Thomson acknowledges as 'Mr. Everitt'), by a screw until it coincided with the patch. In this way, the deflection could be read off when the light was switched back on.

The magnitude of the force exerted by an electrostatic field of strength E on a particle of charge e moving across the field with a velocity v is

$$F_e = eE \tag{9.8}$$

where both E and e are measured in any consistent set of units; let us assume in electromagnetic units. Assuming other forces are negligible, the particle undergoes uniformly accelerated motion in the direction of the field. The magnitude of the force exerted on the same particle by a magnetic field of strength H is, as we know,

$$F_m = Hev \tag{9.9}$$

and is directed perpendicular to both H and v. In this case, the particle experiences a change in its *direction* but not in *speed*; the magnetic force causes the particle to move in a circular path in a plane perpendicular to the field direction. In practice, purely for convenience, Thomson adjusted the two field strengths such that the two opposing forces cancelled each other, in which case

$$eE = Hev$$

or

$$v = E/H. \tag{9.10}$$

Under these conditions the phosphorescent spot would show zero deflection, and the ray velocity was simply given by the ratio of the two field strengths.

The next step was to measure the deflection of the phosphorescent patch on the scale with the magnetic field acting alone; this, as noted before, gave the radius of curvature of the path followed by the rays in the field H:

$$r = \frac{mv}{He}.$$

Inserting the velocity, as given by equation (9.10), we then have

$$\frac{e}{m} = \frac{v}{Hr} = \frac{E}{H^2 r} \tag{9.11}$$

where $2r = d^2R + R$ as discussed in section 9.2.

The new results are listed in table 9.2. In all cases, the current through the Helmholtz coils was adjusted so that the electrostatic deflection was equal to the magnetic deflection. In table 9.2, l (essentially the length of the plates) is the length (cm) over which the rays experienced a uniform electric field, and θ is the angle of deflection upon entering a region free from electric force [9-56]. Note that Thomson uses F, not E, for the electric field intensity.

Though Thomson notes that, in principle, the latter method of determining v, and hence e/m, 'is much less laborious and probably more accurate than the former method' [9-57], table 9.2 does not bear this out. The values for e/m, ranging from 0.67×10^7 to 0.91×10^7 emu g^{-1}, or $(e/m)_{av} = 0.78 \times 10^7$ emu g^{-1}, are systematically low compared with the modern value for the electron (1.75×10^7 emu g^{-1}), whereas the

Table 9.2.

Gas	θ	H	F	l	m/e	v
Air	8/110	5.5	1.5×10^{10}	5	1.3×10^{-7}	2.8×10^{9}
Air	9.5/110	5.4	1.5×10^{10}	5	1.1×10^{-7}	2.8×10^{9}
Air	13/110	6.6	1.5×10^{10}	5	1.2×10^{-7}	2.3×10^{9}
Hydrogen	9/110	6.3	1.5×10^{10}	5	1.5×10^{-7}	2.5×10^{9}
Carbonic acid	11/110	6.9	1.5×10^{10}	5	1.5×10^{-7}	2.2×10^{9}
Air	6/110	5	1.8×10^{10}	5	1.3×10^{-7}	3.6×10^{9}
Air	7/110	3.6	$1, \times 10^{10}$	5	1.1×10^{-7}	2.8×10^{9}

values in table 9.1 range from 1.0×10^7 to 3.12×10^7 emu g^{-1} with $(e/m)_{av} = 1.96 \times 10^7$ emu g^{-1}. If we ignore the results for tube 3, the former measurements give even better agreement with the modern value, $(e/m)_{av} = 2.14 \times 10^7$ emu g^{-1}, whereas tube 3 alone gives a result too low, like table 9.2, $(e/m)_{av} = 1.15 \times 10^7$ emu g^{-1}. Thomson himself states that 'I am of the opinion that the value of m/e [e/m] got from Tube 1 and 2 are too small [large]' [9-58]. Perhaps in evaluating his initial round of e/m (thermocouple) experiments, he was swayed by the lower values subsequently obtained in the latter crossed-field experiments. This may be the reason why he altered the published value of $m/e = 1.6 \times 10^{-7}$ in the margin of his personal copy of this paper in the *Electrician* for May of that year to 1.5×10^{-7} (note 9-33, above) [9-59].

It appears that the last row of values (for air) in table 9.2 was added in pencil by Thomson to his ink-written manuscript before it was mailed off to the *Philosophical Magazine*. A pencilled addition to the paragraph following table 9.2 explains why: 'In the last experiment Sir William Crookes' method of getting rid of the mercury vapour by inserting tubes of pounded sulphur, sulphur iodide, and copper filings between the bulb and the pump was adopted' [9-60]. In fact, the inclusion of a seventh value for e/m makes negligible difference in $(e/m)_{av}$ deduced from the crossed-field experiments, or from table 9.2.

Be that as it may, the chief conclusion speaks for itself.

> From these determinations we see that the value of m/e is independent of the nature of the gas, and that its value 10^{-7} is very small compared with the value 10^{-4}, which is the smallest value of this quantity previously known, and which is the value for the hydrogen ion in electrolysis.
>
> Thus for the carriers of the electricity in the cathode rays m/e is very small compared with its value in electrolysis. The smallness of m/e may be due to the smallness of m or the largeness of e, or to a combination of these two. That the carriers of the charges in the cathode rays are small compared with ordinary molecules is shown, I think, by Lenard's results as to the rate at

which the brightness of the phosphorescence produced by these rays diminishes with the length of path travelled by the ray. If we regard this phosphorescence as due to the impact of the charged particles, the distance through which the rays must travel before the phosphorescence fades to a given fraction (say $1/e$, where $e = 2.71$) of its original intensity, will be some moderate multiple of the mean free path. Now Lenard found that this distance depends solely upon the density of the medium, and not upon its chemical nature or physical state. In air at atmospheric pressure the distance was about half a centimetre, and this must be comparable with the mean free path of the carriers through air at atmospheric pressure. But the mean free path of the molecules of air is a quantity of quite a different order [that is, much shorter]. The carrier, then, must be small compared with ordinary molecules [9-61].

Chapter 10

THE CHARGE AND THE MASS

10.1. The charge

In his classic cathode ray paper of October 1897, as in his Royal Institution Discourse earlier that year, Thomson had declared that 'the smallness of m/e [for cathode rays] may be due to the smallness of m or the largeness of e'. He put it more strongly in the October paper with the added statement that 'the smallness of the value of m/e is... due to the largeness of e *as well as* the smallness of m [italics added]'—something he justified vaguely in terms of the high specific inductive capacity of gaseous molecules as an 'additive quantity' [10-1]. Before long he abandoned this opinion, however. As he relates in his *Recollections*,

> After long consideration of the experiments it seemed to me that there was no escape from the following conclusions:
> (1) That atoms are not indivisible, for negatively electrified particles can be torn from them by the action of electrical forces, impact of rapidly moving atoms, ultraviolet light or heat.
> (2) That these particles are all of the same mass, and carry the same charge of negative electricity from whatever kind of atom they may be derived, and are a constituent of all atoms.
> (3) That the mass of these particles is less than one-thousandth part of the mass of an atom of hydrogen. [10-2]

That the high value for the specific charge e/m is solely due to the smallness of m follows from the first conclusion, that atoms are not indivisible. If an electrically *neutral* atom contains one or more particles, each of mass m and negative charge e, then tearing a particle from the atom leaves an ion with an unbalanced positive charge of magnitude e. If n particles are torn off, the residual ion will be left with an unbalanced positive charge ne. Thus, the unbalanced positive ionic charge may be greater than or equal to the negative charge of each of its constituent particles, but the latter charge cannot exceed the residual ionic charge [10-3].

The next step, clearly, was a direct determination of the absolute value of either *m* or *e*. As it turned out, measurement of *e* proved the way to proceed, though not a simple task. That task fell initially on one of Thomson's first research students, J S E Townsend, who, we recall, arrived in Cambridge in 1895 on the very same day as Rutherford. On Thomson's suggestion, Townsend began his research on the magnetic properties of the salts of iron of various types, measuring their susceptibility with a sensitive induction balance. This work only occupied him for a short period, however. He soon switched from magnetic measurements to a study of electrified gases—in particular, gases released by electrolysis. In so doing, he trod on territory already visited by C T R Wilson and, for that matter, by Thomson himself much earlier.

It was already known from the work of Enright in 1890, and indeed familiar to Laplace and Lavoisier, that hydrogen liberated by electrolysis (say, by dissolving iron in sulfuric acid) carries an electric charge. After repeating some of Enright's experiments, Townsend embarked on a systematic study of the electrification of gases liberated in the electrolysis of liquids—a subject which aroused his interest in a field which would occupy him for the greater part of his career. Specifically, he wished to ascertain whether the charge resided on the gas or on the dust or 'spray' it may contain, something Enright had been unable to determine. He soon found that under certain conditions the liberated gases were highly charged. In the electrolysis of dilute sulfuric acid (H_2SO_4), the oxygen evolved carried a positive charge, whereas oxygen liberated during the electrolysis of potassium hydroxide (caustic potash or KOH) was negatively charged. In both cases the oxygen released came off at the anode of the cell, though in that case one would have expected only a negative charge. This charged gas caused the spontaneous formation of a fine mist of water drops when the gas came in contact with air saturated with water vapor. Apparently, the charges acted as nuclei on which the droplets forming the cloud condensed. The *origin* of the charge carriers, clearly larger than the ordinary free ions in gases, proved problematic for Townsend at the time; however, he devised an elegant method for measuring the magnitude of their charges—a technique which 'included practically all the ideas which were later used in accurate measurements of the [electronic] charge' [10-4].

Townsend measured the total charge per unit volume of the cloud with a quadrant electrometer, and its total weight by bubbling it through a series of tubes containing a drying agent (sulfuric acid); the mass of water in the cloud was found from the increase in weight of the tubes. The average weight of single droplets was obtained by observing, photographically, their rate of descent under gravity through the moist air. The top layer of the cloud was illuminated with an arc light, and was thus reasonably well defined. Assuming spherical drops of radius r falling freely in a viscous medium, their limiting or terminal velocity is given by an expression entering into the derivation of Stokes' law [10-5]:

$$v = \frac{2}{9}\frac{gr^2}{\eta}(\rho - \rho'). \tag{10.1}$$

Here ρ is the density of a drop, ρ' the density of the medium, η the coefficient of viscosity of the medium, and g the acceleration of gravity. The unknown quantity was r, which in turn gave the mass per drop. (Stokes himself, though apparently not personally responsible for inspiring the investigation, was still very much around at the Cavendish; among other things, he was then Secretary of the Royal Society. Rayleigh recalls helping him to find some piece of apparatus by chance located in the very room in which Townsend was working [10-6]). Dividing the weight of the cloud by the average weight of a droplet gave the number of droplets. Assuming every particle or gaseous ion had a drop condensed on it, he obtained the average charge per ion (in electrostatic units) as 2.8×10^{-10} for positively charged oxygen ions and 3.1×10^{-10} for negatively charged ions.

Townsend's results were reported to the meeting of the Cambridge Philosophical Society on February 8 1897 [10-7]. That, in fact, was the very meeting which Thomson opened with his own preliminary communication on cathode rays (section 9.2, above)—a report under two pages, compared to Townsend's hefty 14 pages in the published proceedings. In light of his experimental uncertainties, Townsend added that

> . . . the numbers given above as the charges on each carrier are not more unequal than the errors of the experiments would account for, so that there is nothing in this result to disprove that the oxygen carriers have the actual atomic charges. [10-8]

Hence, he concluded that the fundamental ('atomic') unit of the positive and negative charges was equal, with a value of $e = 3 \times 10^{-10}$ esu—presumably the same for different gases. To be sure, his uncertainties were large, as was his concomitant range of values for e. His novel photographic method, he admits, was only good to within 8%. The uncertainties include the assumed lack of evaporation of the drops (a subject J J Thomson had treated in his *Applications of Dynamics to Physics and Chemistry* in 1888), the applicability of Stokes' law for very small drops, and above all the assumption of a single charge per drop [10-9]. Indeed, Townsend subsequently showed that, in ozone, it was possible to produce clouds with uncharged nuclei! In light of the uncertainties, his experiment cannot be construed as unambiguously establishing a unique value for the fundamental unit of electrical charge. Credit for that belongs to J J Thomson.

The next year, 1898, Thomson, in fact, took matters into his own hands, using a combination of Townsend's falling cloud method in conjunction with C T R Wilson's expanding cloud apparatus. Wilson had not been idle in the intervening period. He first produced artificial clouds at the Cavendish in early 1895 with John Aitken's method of expanding moist air. Late in that year came news of Röntgen's discovery of x-rays. With Thomson losing no time in attacking the conductivity of air exposed to the rays, neither did Wilson. He borrowed one of the tubes crafted for Thomson's purpose by Ebenezer Everett, hoping to see whether x-rays might affect the air in his own expansion apparatus. Sure enough, he soon found that dust-free gas, ionized by exposing it to x-rays or, it later became apparent, to ultraviolet light, uranium rays, or, for that matter, point electrical discharges, produced cloud-forming nuclei if subjected to the same expansion. The nuclei had to be positively or negatively charged ions, just as Thomson had predicted, since no cloud was formed when they were swept away by an electric field. Also, the negative ions were more effective in this respect than the positive ions. Wilson found that it was possible to regulate the expansion such that condensation occurred around the negative ions, but not around the positive ions—a development that ultimately (in 1911) led him to his ubiquitous 'cloud chamber'. A preliminary account of this work was communicated to the Royal Society in March 1896, with a full report on the subject following in two installments—one published the next year, and the second a year later [10-10].

Thomson's own approach to the magnitude of e was to expose a known volume of gas uniformly to x-rays. Measuring the current passing through the gas rendered conducting by the exposure, and acted upon simultaneously by a known potential gradient, gave him the product nev, where n is the number of ions in unit volume, e the charge per ion, and v the mean velocity of positive and negative ions dragged by the same electric force. The velocity was known from ionic measurements already completed by Rutherford. With it, the current through the gas gave the product ne. Then, if n could be determined, so could e.

The critical parameter, obviously, was n, which Thomson took to be equal to the number of drops in Wilson's chamber adjusted to produce condensation on negative ions only. His method for obtaining n was Townsend's indirect falling cloud method (Thomson having given up on a direct optical method in which light from an arc lamp, after passing through the cloud, produced diffraction rings on a screen which could, in principle, be measured to yield the drop size). The rate of fall of the water drops, once formed in Wilson's chamber, gave their size, and hence their mass. From the degree of expansion needed in condensing the water vapor into fog particles, he knew, on thermodynamic grounds, the total

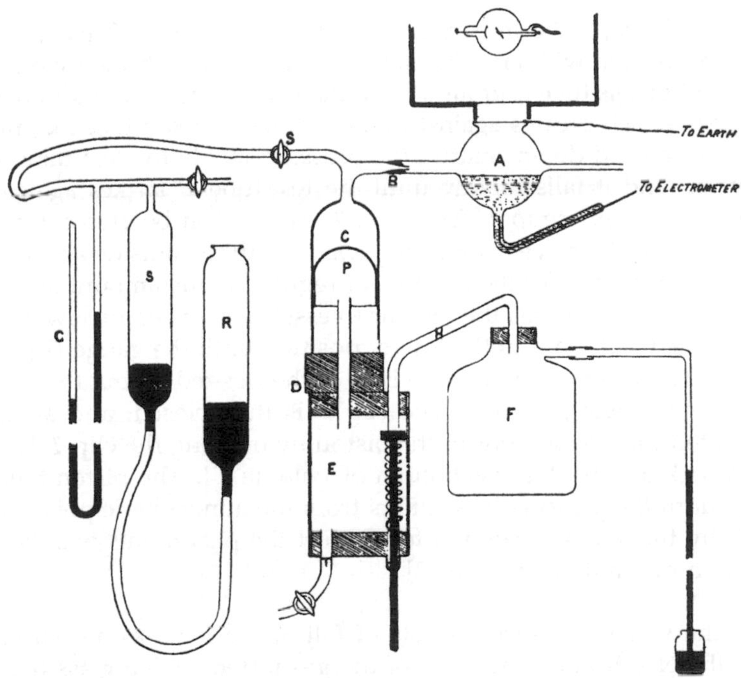

Figure 10.1. *Thomson's (and Wilson's) cloud expansion apparatus for measuring the value of e.*

weight of vapor condensed. Dividing the total weight by the mass per particle gave him *n*, the number of fog particles. Knowing *n*, he knew the ionic charge *e*. Curiously, in Thomson's published account in the *Philosophical Magazine* for 1898 no mention is made of Townsend, though he does acknowledge Townsend in this connection in his *Recollections*. [10-11]

The apparatus used in producing the cloud and for measuring the expansion, apparatus used earlier by Wilson, is shown in figure 10.1.

> The gas which is exposed to the rays [explains Thomson] is contained in the vessel *A*; this vessel communicates by the tube *B* with the vertical tube *C*, the lower end of this tube is carefully ground so as to be in a plane perpendicular to its axis, and is fastened down to the indiarubber stopper *D*. Inside this tube there is an inverted thin-walled test-tube, *P*, with the lip removed and the open end ground so as to be in a plane perpendicular to the axis of the tube. The test-tube slides freely up and down the larger tube and serves as a piston. Its lower end is always below the surface of the water which fills the lower part of the outer tube. A glass tube passing through the indiarubber stopper puts the in-

side of the test-tube in connexion with the space *E*. This space is in connexion with an exhausted vessel, *F*, through the tube *H*. The end of this tube is ground flat and is closed by an indiarubber stopper which presses against it; the stopper is fixed to a rod, by pulling the rod down smartly the pressure inside the test-tube is lowered and it falls rapidly until the test-tube *P* strikes against the indiarubber stopper. The tube *T*, which can be closed by a stop-cock, puts the vessel *E* in connexion with the outside air. The tubes *R* and *S* are for the purpose of regulating the amount of the expansion. To do this, the mercury-vessel *R* is raised or lowered when the test-tube is in the lowest position until the gauge *G* [far left] indicates that the pressure in *A* is the desired amount below the atmospheric pressure. The clip *S* is then closed, and air is admitted into the interior of the piston by opening the clip *T* [i.e. stop-cock in tube *T* at the bottom of volume *E*]. The piston then rises until the pressure in *A* differs from the atmospheric pressure only by the amount required to support the piston, this is only a fraction of a millimetre. [10-12]

A is the vessel in which the rate of fall of the fog was measured and the electrical conductivity of the gas tested. It is a glass tube about 36 millim. in diameter covered with an aluminum plate; a piece of wet blotting-paper is placed on the lower side of the plate and the current of electricity passed from the blotting-paper to the horizontal surface of the water in this vessel. The blotting-paper was placed over the aluminum plate to avoid the abnormal ionization which occurs near the surface of a metal against which Röntgen rays strike normally. M. Langevin has shown that this abnormal ionization is practically absent when the surfaces are wet. [10-13]

The coil and focus-bulb producing the rays were placed in a large iron tank elevated on supports; in the bottom of the tank a hole was cut and closed by an aluminum window. The vessel *A* was placed underneath this window and the bulb giving out the rays some distance behind [above] it, so that the beam of rays escaping from the tank were not very divergent. [10-14]

The tank and the aluminum plate at the top of *A* were connected with earth and with one pair of quadrants of an electrometer. The other pair of quadrants were connected with the water-surface *B* [in *A*]; this surface was [temporarily] charged up . . . with a battery [for purposes of testing for any leakage of charge with the x-ray tube inactive]. [10-15]

With the insulation passing muster

> ... the rays were turned on ... By measuring the rate of leak, the quantity of electricity crossing in one second the gas exposed to the rays can be determined if the capacity of the system is known. [10-16]

The charge flowing was determined as follows.

> If, when the rays are on, the movement of the spot of light [reflected from the mirror of the electrometer] indicates a change in the potential equal to V per second, then the quantity of electricity flowing in that time across the cross-section of the vessel exposed to the rays is CV [where C is the capacity of a condenser connected with a quadrant of the electrometer]. But if n is the number of ions, both positive and negative, per cubic centimetre of the gas, u_0 the mean velocity of the positive and negative ions under unit potential gradient, A the area of the plates, E the potential-gradient, this quantity of electricity is also equal to neu_0EA, hence we have
>
> $$CV = neu_0EA;$$
>
> so that if we know n and u_0 [the latter from Rutherford's data] we can from this equation deduce the value of e. [10-17]

Thomson's value for e was 6.5×10^{-10} esu, with a spread of about 5–8 $\times 10^{-10}$ esu (the accepted value is 4.8×10^{-10})—a value 'equal to, or at any rate of the same order, as the charge carried by the hydrogen ion in electrolysis'. [10-18]

> In connection with this result [Thomson ends his report], it is interesting to find that Professor H. A. Lorentz ... has shown that the charge on the ion whose motion causes those lines in the spectrum which are affected by the Zeeman effect is of the same order as the charge on a hydrogen ion in electrolysis. [10-19]

We turn to Zeeman and Lorentz presently. For the moment, we simply note Thomson's repetition of his falling drop experiment in 1901–1903, incorporating a more constant source of radiation (a radium salt) and, most important, a faster expansion. On further examination of the old experiment, the number of drops in an expansion judged sufficient to catch both negative and positive ions appeared to be less than twice what it was with an expansion only sufficient for negative ions, leading him to suspect that the earlier technique did not catch all the ions in the expansion. With a modified apparatus, the expected doubling of drops was indeed observed, and his new value for e was considerably lower: 3.4×10^{-10} esu. We will come back to this result in due course.

10.2. The mass

By 1898, then, Thomson had in hand independent values for the charge-to-mass ratio, e/m, from magnetic and electric deflection, and for the charge e from the cloud method. There was no longer any doubt that the large value of e/m was due to the smallness of m and not to the magnitude of the charge. Just the same, Thomson had not yet succeeded in any explicit determination of what he really sought, the mass m, or what amounted to the same thing, a simultaneous determination of e/m *and* e by the same method. However, in the spring of 1899 he conjured a method for doing just that, with the aid of yet another means of making the gas conductive: utilizing the negative electrification discharged by ultraviolet light.

The photoelectric effect, then ten years old, had most recently been studied abroad by the singular team of Elster and Geitel, and in his own laboratory, first by John Zeleny and then by Rutherford—by the latter as a logical extension of his work with Thomson on the ionization by x-rays. Rutherford, while still at Cambridge, had shown that a gas ionized by ultraviolet light behaves in much the same way as a gas exposed to x-rays (or to uranium rays). That is, the velocity of the charged carriers was the same for all metals illuminated, and agreed with the velocity of the negative ions produced in the same gas by x-rays [10-20].

This was, strictly speaking, not something new, as Rutherford grudgingly admitted. 'Most of these papers [by the German investigators] have dealt with the general character of the discharge' he allowed, 'but the subject of the nature of the conduction and of the carrier that discharges the electrification has not been specially attacked' [10-21]. In particular, the experiments of Lenard and others were 'capable of other interpretations' [that the loss of electricity was not necessarily due to the disintegration of a metal by the action of ultraviolet light, as indicated by Lenard's experiment, but due to ionization of the surrounding gas].

Thomson, too, had followed the German experiments with great interest. He was particularly struck by the asymmetry between the positive and negative electricity revealed in them. Thus, in Wilhelm Hallwach's experiment of 1887, a thoroughly cleaned zinc plate illuminated by ultraviolet light from an arc lamp caused a bountiful electrical discharge if negatively charged, but hardly a trace if positive [10-22]. This asymmetry was even more glaring in cathode ray tubes generally, in the difference in behavior between cathode rays and Goldstein's positive canal rays. Another matter caught Thomson's attention. Elster and Geitel had more recently (1890) shown that the discharge of negative electricity was diminished by a magnetic field if the lines of magnetic force were directed at right angles to the lines of electric force—i.e. were parallel to the illuminated metal surface [10-23]. As Rayleigh recalls it, neither Elster and Geitel, nor Thomson, speculated on the causes of this directional effect at the time. However, in 1897, when giving a 'verbal account' of his work on cathode rays at the Cavendish, Thomson observed that 'reasonable guesses' for the strengths of the magnetic and electric fields involved in their experiments would imply that the carriers of the negative electrification from the zinc surface must be more like the cathode ray corpuscles than like charged atoms or the disintegrated zinc dust supposed by Lenard and his countrymen [10-24].

By this time, Thomson had determined the charge carried by the ions begotten by x-rays, and realized that the same ought to be possible for the carriers of negative electricity liberated by illuminating zinc with ultraviolet light. Not only could e be determined, but, in fact, e/m by crossed electrostatic and magnetic fields in the very same experiment, thus promising a true measure of the evanescent mass m. (In the case of x-rays, the resulting charged carriers could not be collimated sufficiently to form a beam amenable to e/m measurements.) When news got around at the Cavendish of the Professor's latest experimental proposal, someone remarked to Rayleigh that the experiment was 'tempting providence'. In due course, with the experiment successfully completed, Rayleigh repeated the remark to Thomson, who failed to see the humor in it, noting simply that he had felt pretty confident [10-25].

In the experiment of Elster and Geitel, and as adopted by Thomson, an electrostatic force drove the negatively charged particles from the illuminated zinc surface, and at the same time a magnetic force, applied at right angles to the electric force, deflected the particles from their rectilinear path. In this arrangement, however, the magnetic deflection was more complicated than in Thomson's cathode ray experiment of 1897. In the older arrangement, the electric field imparting a uniform acceleration to the particles acted only over a finite length (between the parallel plates in figure 9.7), beyond which only the deflecting magnetic field was operative. In the present experiment, the electric field continued to accelerate the particles in the initial direction while simultaneously the magnetic field deflected them away from that direction. Under these conditions, Thomson showed, the effect of the magnetic force is obtained as follows [10-26].

Let the particle start from rest from the plane $x = 0$ at the time $t = 0$, acted upon by a uniform electric field E parallel to the x-axis and a uniform magnetic field H parallel to the z-axis. The equations of motion for such a particle of mass m and charge e are

$$m\frac{\mathrm{d}^2x}{\mathrm{d}t^2} = Ee - He\frac{\mathrm{d}y}{\mathrm{d}t}$$

and

$$m\frac{\mathrm{d}^2y}{\mathrm{d}t^2} = He\frac{\mathrm{d}x}{\mathrm{d}t}.$$

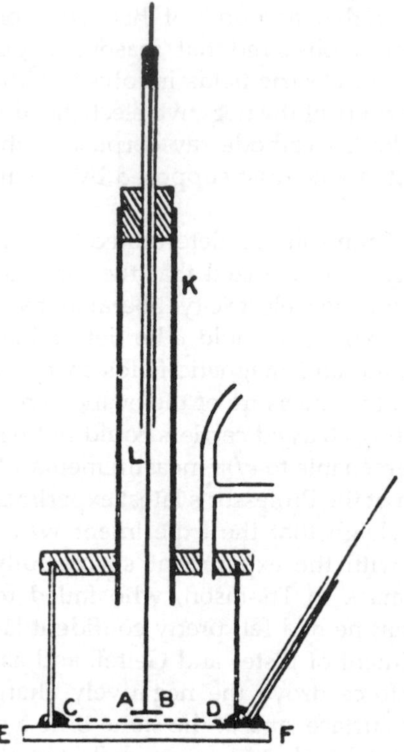

Figure 10.2. *Thomson's e/m apparatus based on the photoelectric effect.*

Their solution gives the coordinates of the particle at some time t,

$$x = \frac{m}{e}\frac{E}{H^2}\left[1 - \cos\left(\frac{e}{m}Ht\right)\right] \tag{10.2}$$

and

$$y = \frac{m}{e}\frac{E}{H^2}\left[\frac{e}{m}Ht - \sin\left(\frac{e}{m}Ht\right)\right] \tag{10.3}$$

which, in contrast to the circular path performed in the cathode ray experiment, describes a cycloidal path. The utility of such a complicated path, for Thomson's purpose, will be evident in what follows.

Thomson's new experimental arrangement is shown in figure 10.2. Between a zinc plate AB and a quartz plate EF is a gauze or mesh of fine wires CD, insulated and soldered to a ring of metal. The quartz plate forms the base of an evacuated glass tube, as shown. The plate AB is connected to a battery via the rod L passing through a sealing-wax stopper in the tube K, and the gauze CD to an electrometer by the indicated connection. Ultraviolet light from a source below the apparatus, not shown, enters the quartz plate, passes through the gauze, and illuminates the zinc plate. In the absence of a magnetic field, and supposing the gauze to be maintained at a higher potential than the plate AB, negatively charged particles dislodged from AB will be driven to CD by the electric force. If now a uniform magnetic field parallel to the plate is activated by an external electromagnet, the paths of the particles are bent. If the field is low, or if the separation between AB and CD is less than $2Em/eH^2$, the so-called generating circle of the cycloid determined by equation (10.2) and equation (10.3), the particles will still reach the gauze (provided the gauze extends over a sufficiently large area) [10-27]. However, at a certain critical field strength, for a given electric field, or equivalently, if the separation between AB and CD exceeds the value $2Em/eH^2$, particles starting from AB are turned back before reaching CD, and the rate at which CD receives a negative charge is sharply reduced (if not zero). Hence, a measurement of the distance between AB and CD at the onset of a drop in the charging current determines the value of $2Em/eH^2$. Knowing E and H, e/m is determined as well.

In practice, the gauze was kept at a fixed distance from the plate, the magnetic field was also held constant, and the potential difference between the plate and gauze was diminished until a deflection of the electrometer was observed, signaling the critical voltage. As one might expect, the transition from full leakage current to zero current was actually gradual, attributable to any number of causes; among them, a range of fringing magnetic fields for which the current is diminished but not extinguished, and the possibility of ions originating in a finite boundary sheath adjacent to the zinc plate.

The mean value for e/m deduced by Thomson in this experiment was 7.3×10^6 emu g^{-1}. This is in excellent agreement with his value for cathode rays—say, 7.8×10^6, depending on precisely which value in the 1897 data is picked. At any rate, Thomson's primary conclusion was not to be denied: 'The value of e/m in the case of ultra-violet light is of the same order as in the case of the cathode rays, and is very different from the value of e/m in the case of the hydrogen ions in ordinary electrolysis... ' [10-28]. We may add that Thomson's latest experimental procedure does not differ, in principle, from that used originally by Schuster and later refined by Wiechert and Kaufmann, utilizing magnetic deflection and the potential difference between electrodes in cathode ray tubes. Letting the electric and magnetic fields act *simultaneously*, as in the ultraviolet light experiment, is perhaps simpler experimentally but complicates the analysis. We may, with Rayleigh, 'consider this method as being in theory a rather complex compromise in spite of its experimental simplicity' [10-29].

Having measured e/m for the ions, there remained to be determined their specific charge e. For this purpose, Thomson availed himself of the falling cloud method so successfully employed earlier in measuring the charge of ions produced by the action of x-rays. C T R Wilson had already shown that ions produced by ultraviolet light act like x-ray ions in forming nuclei around which supersaturated water can condense from dust-free air. Again it boiled down to finding *n*, the number of ions produced in unit volume. As with x-ray ions, air was cooled by a sudden expansion until the supersaturation formed a cloud around the ions, and the problem was reduced to finding the number of drops per unit volume in the cloud. As usual, this was obtained from the volume of water deposited (known from the expansion) divided by the volume per drop (from Stokes' law). The total charge crossing unit area in unit time under a known electric force, *nev*, was obtained by allowing the ions to impinge on a plate of known electrical capacity and measuring the rate of fall in potential. Previous knowledge of *v* from Rutherford's earlier measurements gave *ne*, and finally *e*.

The great advantage of working with ions from ultraviolet light lay in the feasibility of operating in vacuum *or* at atmospheric pressure. Obviously a good vacuum was a prerequisite for the e/m determination, which demanded a pressure low enough to ensure an ionic mean free path long compared to the spacing between the electrodes. On the other hand, the falling cloud method necessitated working at atmospheric pressure. In addition, however, forming a cloud from ultraviolet light, instead of from x-rays, proved possible only in the presence of an electric field; without a field the ions remained close to the zinc surface without diffusing into the body of the chamber, and no cloud materialized.

The falling cloud apparatus is shown in figure 10.3. Here *AB* is a glass plate and *CD* is a quartz plate. On top of *CD* is a layer of water, and above it the illuminated zinc plate. As before, the ultraviolet light was produced by an arc lamp placed below the quartz plate. The space between the zinc plate and the water surface was illuminated by an arc light to allow accurately observing the rate of fall of the drops. The tube *LK* leads to the same apparatus as used in the experiments of 1898 (figure 10.1).

Skipping the details of the actual determination, it suffices to note that Thomson's new value for *e*, 6.8×10^{-10} esu, was practically the same value as found for x-ray ions (6.5×10^{-10}). Before dwelling on this result, however, one further measurements in the 1899 experiments warrants noting, one following up on another effect observed by Elster and Geitel. It had been common knowledge since the time of du Fay and Nollet (pp 21–22) that incandescent bodies were not under all circumstances able to retain an electrical charge. Quantitatively, Elster and Geitel had found that a white-hot filament in an atmosphere of hydrogen, or in a high vacuum, discharges negative ions at will to a neighboring body if the filament is negatively charged and the body is positive, while there is no leakage current if the charges are reversed—a condition reminiscent of that for emission from the illuminated zinc plate. Moreover, the negative discharge is diminished by a transverse magnetic field—again a familiar

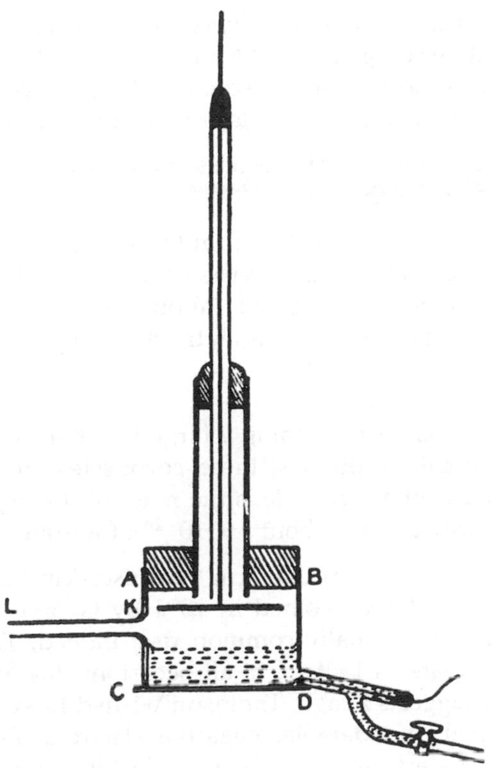

Figure 10.3. *Thomson's falling cloud apparatus for measuring the value e from electrification discharged by ultraviolet light.*

effect, hinting at a mechanism at work similar to that operative in the case of ions discharged by zinc under the action of ultraviolet light. An e/m experiment very similar to that already described showed this to be the case. It gave an e/m value for negative ions from an incandescent wire of 8.7×10^6 emu g^{-1}, in agreement within the experimental errors with the values found for cathode rays and ions from ultraviolet light.

Thomson concluded that all three carriers of negative electricity, whether ions in cathode rays, ions liberated by ultraviolet light, or ions emitted from incandescent carbon, were 'of the same nature' [10-30], characterized by the same value for e/m. Of the three kinds of ion, to be sure, only for those from ultraviolet light had he been able to make a direct measurement of the actual charge e. Since this value for the charge agreed with the charge for x-ray ions in air, however,

> ...the experiments just described, taken in conjunction with previous ones on the value of m/e for the cathode rays..., show that in gases at low pressures negative electrification, though it may be produced by very different means, is made up of units

each having a charge of electricity of a definite size; the magnitude of this negative charge is about 6×10^{-10} electrostatic units, and [in light of Townsend's results] is equal to the positive charge carried by the hydrogen atom in the electrolysis of solutions. [10-31]

With the charge and charge-to-mass ratio both in hand, so was a quantitative value for the mass.

In gases at low pressures these units of negative electric charge are always associated with carriers of a definite mass. This mass is exceedingly small, being only about 1.4×10^{-3} of that of the hydrogen ion, the smallest mass hitherto recognized as capable of a separate existence. [10-32]

I regard the atom as containing *n* large number of smaller bodies which I will call corpuscles; these corpuscles are equal to each other; the mass of a corpuscle is the mass of the negative ion in a gas at low pressure, i.e. about 3×10^{-26} of a gramme. [10-33]

Significantly, Thomson uses the term 'corpuscle' in favor over the term 'electron' which had been coined in 1891 by George Johnstone Stoney and was by now in virtually common use; indeed, Thomson shunned Stoney's term as late as 1906 in, of all occasions, his Nobel Lecture that year. Perhaps, suggests Kragh, Thomson wished to avoid any confusion between the 'real', ponderable, negative electron, alias his corpuscle, and the positive electron which (at least in his view) remained purely hypothetical ca. 1900 [10-34].

Thomson's results were read before the meeting of the British Association at Dover in September of 1899, in a joint session of Section A, Mathematics and Physics, with members of L'Association Française pour l'Advancement des Sciences. The French Association was meeting at the same time at Boulogne across the Channel. A station for wireless telegraphy by Hertzian waves had been installed for the occasion in the Dover Town Hall, crowned with a lofty pole mounted atop the crenellated battlements; direct transmission of messages to Wimereux outside Boulogne was thereby possible. This particular session, chaired by J H Poynting, began with a hearty welcome to the French visitors by Poynting, acknowledged by Justin-Mirande René Benoit as president of the Physical Section of the French Association.

Thomson delivered the keynote address of the joint session, 'On the existence of masses smaller than the atoms' [10-35], an abbreviated version of his full paper published subsequently in the *Philosophical Magazine* [10-36]. The recurring theme in his talk was that the mass of an atom is not invariable and that electrification (ionization) of a gas involves the detachment from atoms of negative corpuscles—all of unit charge and mass a small fraction of the atomic mass. Thus, a neutral atom consists of a large number of such corpuscles. 'The negative effect is balanced by something which causes the space through which the corpuscles are spread to act as if it had a charge of positive electricity equal in amount to the sum of the negative charges on the corpuscles' [10-37].

Not much remains on record of the ensuing discussion. Apparently, and oddly, Poynting, who became a student at the Cavendish in 1876 and was close to Thomson from his student days at Manchester, had not heard of the 'corpuscles' or had little to say on the subject, judging by his opening, ponderous address at Dover. In it, he dwells on the 'atomic and molecular hypothesis of matter' *vis-à-vis* the intervening æthereal medium. 'The sole difference in the atoms are differences of position and motion. Where there are whirls, we call the fluid matter; where there are no whirls, we call it ether' [10-38]. Norman Lockyer, however, rose and expressed his interest in Thomson's complex atom, something well-nigh demanded by line spectra and spectra of hot stars. So did A W Rücker, pointing to certain spectroscopic investigations by Schuster involving arc discharges. Oliver Lodge, too (who was next on the agenda of speakers), declared his enthusiastic support, hoping that Thomson's investigation might form a basis for an electron theory embracing mass and inertia—a matter to which Lodge himself had devoted considerable thought in these very years [10-39].

Less kindly disposed was Henry Edward Armstrong, the prolific chemist and outspoken pioneer in the heuristic school of chemical education [10-40], who dramatically unrolled an impromptu 'blackboard' in the form of a black cotton cloth covered with chemical symbols. He took strong exception to the tone of Thomson's remarks, stressing the unalterable mass and integrity of the chemical atom. Armstrong's attitude on the subject was all too well known. He was in the habit of saying 'that J J and his school were insufficiently instructed in chemistry and that but for this defect in their education they would have known better' [10-41].

The next paper on the agenda after Thomson's was one by Lodge on 'the controversy respecting the seat of Volta's contact force' [10-42]. Lodge is said to have later remarked that he had been so mesmerized by the preceding communication that he found it difficult to concentrate on his own presentation which, consequently, received scant attention [10-43]. Thomson himself, looking back on the Dover meeting, considered that it was an occasion when he 'made a good many converts' to his views [10–44].

Chapter 11

LEIDEN 1896

11.1. The Leiden town council flap

The Physical Laboratory of the University of Leiden effectively dates, as a renowned center of experimental physics, from November 11 1882. On that day, Heike Kamerlingh Onnes succeeded Pieter L Rijke as professor of experimental physics at Leiden, and was appointed director of the laboratory. In his inaugural address, Kamerlingh Onnes sketched out a program in quantitative research 'in establishing the universal laws of nature and increasing our insight into the unity of natural phenomena' [11-1]. Leading his audience from the laws of Boyle to Gay-Lussac, and onward to those of his countryman van der Waals, each step furnishing successive approximations to molecular behavior, he warned that

> . . . from approximate hypotheses concerning the molecule, theory can provide indispensable guides to experiment, but it is measurement that must decide their validity; this must secure research from being lulled to sleep by the seduction of a rounded off mathematical theory. It is measurement finally, that in the deviations provides the natural material for new hypotheses regarding the properties of molecules. [11-2]

Measurement, then, was the key to success, and was predicated on apparatus of first rank, in capable hands,

> . . . an indispensable part of a laboratory equipment, while at the same time a part of a modern physical laboratory must be modelled upon astronomical lines. It must be furnished with instruments, the individual peculiarities of which are completely familiar and properly recorded and with localities suitable for using these instruments to the best advantage. [11-3]

Door meten tot weten ('from measurement to knowledge') would indeed be the working dictum of Kamerlingh Onnes' famous laboratory, an establishment soon to rival the Cavendish and Royal Institution across the Channel in the breadth of its research program in the closing years of the 19th century. It was here that a discovery was made in 1896 on a par in importance with, and augmenting, J J Thomson's culminating elucidation that year of the nature of his corpuscles—a discovery which must now divert our attention. The Leiden result, by a quirk of fate, brought immediate attention to Kamerlingh Onnes' laboratory to a degree perhaps not altogether to his approval, if only for a short time, as we shall see. To do so, we need first to acquaint ourselves with the doyen of low-temperature physics and the background for his laboratory—a laboratory in many ways reminiscent of the Cavendish.

Kamerlingh Onnes, born in 1853, was the son of a prominent owner of a tile factory in Groningen. He was schooled at the Hoogere Burgerschool (Higher Citizen's School) at Groningen, where his scientific bent was strongly influenced by van Bemmelen, principal of the school and later Professor of Chemistry at Leiden. He received his 'candidaats' (bachelor's) degree with honors in physics and mathematics at the University of Groningen in 1871, winning the same year the gold medal in a prize competition in chemistry sponsored by the University of Utrecht. Intending to go into chemistry, he then secured a stipend under Robert Bunsen and Gustav Kirchhoff at Heidelberg. (Among his fellow students at Heidelberg was Arthur Schuster.) He finished his *Wanderjahr* of 1872 with experiments on Foucault's pendulum, the fruit of a *Seminarpreis* entitling him to an assistantship under Kirchhoff—a singular honor. This investigation became the starting point for his doctoral dissertation, 'New proofs of the axial rotation of the Earth', which he defended at Groningen in 1879.

The year before receiving his doctorate, Onnes had secured a position as assistant to Johannes Bosscha, director of the Polytechnic School at Delft. His four years under Bosscha proved an invaluable training period in designing and handling experimental apparatus. More important yet for the course of his future career were two papers published by Diderik van der Waals, by then professor at the University of Amsterdam and with whom Onnes was in close contact. Van der Waals had obtained his doctor's degree at Leiden under Bosscha based on a thesis introducing his famous equation of state for simple gases involving molecular forces. His published thesis, 'On the continuity of the gaseous and liquid states', not only caught the attention of Onnes (who would put the equation of state to good use when he undertook to liquefy helium at a later time), but also impressed Maxwell to the point of going to the trouble of learning Dutch to read it, as it did Thomas Andrews of Queen's College, Belfast—the father of isotherms characterizing 'real' gases. The second of van der Waals' papers to leave its mark on Kamerlingh Onnes was the one in which he introduced his law of 'corresponding states', a paper published two years before Kamerlingh Onnes' appointment to the chair at Leiden.

No sooner was Kamerlingh Onnes inaugurated at Leiden, in a professorship he would hold for 42 years, than he set about reorganizing the laboratory to suit his own liking. His experimental program was predicated mainly on two lines of research inspired by his two countrymen, van der Waals and H A Lorentz. Lorentz, who had also obtained his doctorate under Bosscha, occupied a newly created second chair of physics at Leiden. Experimental tests of the equation of state and of the law of corresponding states, in Kamerlingh Onnes' opinion, held the key to molecular physics and quantitative low-temperature experimentation. The second line of research was largely the legacy of Lorentz, his theoretical colleague (though Kamerlingh Onnes himself was no slouch at theorizing): investigations embracing magnetic, optical, and electrical properties of matter. The earliest volumes of Kamerlingh Onnes' unprecedented in-house laboratory journal, *Communications from the Physical Laboratory of the University of Leiden*, cover both areas with equal frequency, including magnetism, magneto-optics, electrical measurements, radioactivity, thermodynamics, and molecular phenomena, authored (mostly in English) by a brilliant roster of compatriots and visitors alike, among them C A Crommelin, Jacob Clay, H Braak, J Becquerel, Lenard, P Weiss, Perrier, Wander Johannes de Haas, Bengt and Anna Beckman, Ekko Oosterhuis, Weber, Mme Curie, Willem Hendrik Keesom, Gilles Holst, Gerard P Kuyper, H R Woltjer, Willem Tuyn, Vegard, Boks, and M Wolfke; the roster goes on and on [11-4]. Only later was this all-embracing program narrowed to one specializing in low-temperature physics exclusively, though undoubtedly low-temperature studies occupied center stage virtually from the start.

> Precision measurements at low temperature must be very attractive in my opinion. For this purpose, it was necessary to dispose of large apparatus with which the measurements could be made at a constant temperature, and it was indispensable to construct suitable temperature baths, baths which could also be used for numerous other investigations. [11-5]

With singleness of purpose the laboratory inventory of equipment grew systematically, from the acquisition of pumps and compressors to the installation of a complete liquefaction plant which, by 1893, was fully operational, providing a cascade of low temperatures extending down to the temperature of liquid oxygen. The next goal in the quest for absolute zero was liquid hydrogen, pursued in a closely contested race with Kamerlingh Onnes' nemesis at the Royal Institution, James Dewar. Hydrogen was first subjugated in London in 1896, if only as a jet of partially liquefied gas. Alas, in February of that year a serious crisis befell Kamerlingh Onnes' program. The crisis echoed a calamity in Leiden 89 years earlier—a calamity which, in a way, had proved crucial in the establishment of the Leiden laboratory in the first place.

In 1807, during the Napoléonic occupation of the Netherlands, a barge carrying gunpowder from Amsterdam to the armory in Delft tied up at the Rapenburg in the center of Leiden, not far from the university grounds. The cargo caught fire and the barge blew up, levelling a sizable portion of the city. Not before the end of the century was the last of the rubble cleared away; in the meantime the partially cleared space provided room for a large university building housing laboratories for anatomy, physiology, chemistry, and physics. By 1896, physics occupied virtually the entire building, except for a small portion still belonging to physiology to this day. In the February meeting of the Amsterdam Academy that year, the Leiden Town Council expressed grave concern over permitting continued work in the laboratory 'on the Ruin' with compressed hydrogen gas 'on account of the explosions which [it] might cause and the danger to which persons and buildings are exposed in consequence' [11-6]. A commission of experts, including van der Waals, was appointed to look into the matter. Both Dewar and Karol S Olszewski of Crakow, another contender for liquid hydrogen, testified on Kamerlingh Onnes' behalf in the ensuing inquiry. Fortunately, the Town Council accepted the recommendation of the commission and allowed the laboratory to resume operations, albeit with stringent safety measures instituted.

Hardly had the Town Council flap been laid to rest when there occurred another event, albeit more positive, which conspired to affect the Leiden program. An experiment in the laboratory by Pieter Zeeman, former student of Kamerlingh Onnes, caused a stir in scientific circles, the likes of which had not been seen since the hoopla in the wake of Röntgen's discovery of x-rays the year before. With Zeeman propelled into the scientific limelight, and almost at once to the University of Amsterdam as well, the scope of Kamerlingh Onnes' program was narrowed rather suddenly to solely encompass the low-temperature work. Zeeman, for his part, continued in the magneto-optic footsteps of Lorentz. A hint of bitterness is perhaps discernible in the utter lack of reference in print by Kamerlingh Onnes, then or any time later, to the Leiden Town Council incident. 'Perhaps', note C J Gorter and K W Taconis wryly, 'he did not wish the meddlesome Town Council to receive credit for the observation of the magnetic splitting of the sodium spectral lines by his collaborator Zeeman at the time of the incident' [11-7].

11.2. Pieter Zeeman and his effect

The inscription accompanying three stained glass commemorative windows in what is nowadays known as the Kamerlingh Onnes Laboratory reads as follows:

> These panels have been dedicated to Peter Zeeman; they commemorate the fortunate discovery which became known

> as the Zeeman effect and which, together with the ideas on light conceived by Lorentz, contributed to experimental and theoretical research and offered a glorious example of the power of cooperative efforts in lifting the spirit; presented and dedicated by H. Kamerlingh-Onnes on the same day, the eve of the kalends of November, on which twenty-five years earlier, these discoveries were published; H. Kamerlingh-Onnes caused these panels to be painted and installed in the very place where the cradle of the Zeeman effect stood. Leiden, October 31, 1921. [11-8]

Zeeman's experiment commemorated here took place, in its initial form, in late August 1896, following an unsuccessful attempt several years earlier with extemporized apparatus. The crucial component of the new apparatus was a Rowland grating recently acquired by the Leiden laboratory, one of a number distributed at cost to laboratories worldwide by Johns Hopkins University, home institution of the physicist Henry Rowland. Such a grating was far superior in resolving power to the prism spectrographs available in Faraday's time [11-9]. In Zeeman's experiment, the flame of a Bunsen burner was placed between the poles of a Rühmkorff electromagnet capable of about 10 kG. A piece of asbestos soaked in a salt solution was inserted in the flame, and the two prominent D-lines in the emission spectrum of sodium were seen as narrow, sharply defined linear images of an adjustable slit, as long as the electromagnet was not energized. When the magnetizing current was switched on, the two D-lines grew in width—to be sure, not very much; perhaps $\frac{1}{30}$ of the distance between the lines, or about 1 mm with the Rowland grating. The same was observed with a flame of coal-gas fed with oxygen replacing the Bunsen flame.

Could the widening be due simply to changes in density or temperature of the sodium vapor from the magnetic action on the flame, whose outline was, in fact, noticeably altered? To eliminate this possibility, Zeeman repeated the experiment without delay, using a closed porcelain tube mounted horizontally between the magnet poles. A pinch of sodium was placed in the tube whose central portion was heated to incandescence by a large Bunsen burner, and the light of an arc lamp was passed through the tube. To counteract the heating, the copper and glass end caps of the tube were cooled by little water-jackets, and the tube was continuously rotated to further avoid temperature variations. The same broadening of the (absorption) D-lines was seen on closing the magnetic circuit; hence, the broadening was most likely due to the altering of the frequency of the sodium light in the magnetic field, and not due to spurious causes, say from convection currents due to temperature gradients. To be doubly sure on this point, the experiment was repeated once more with a tube of smaller diameter rendered white hot by a blowpipe instead of the Bunsen burner; if anything, the magnetic effect was more striking.

These experiments were completed in time for submission to the October 31 meeting of the Amsterdam Academy [11-10]. As reported by *Nature* in its routine coverage of the meeting,

> Prof. Kamerlingh Onnes communicated two papers: (a) by Dr. Zeeman, on the influence of a magnetisation on the nature of the light emitted by a substance. Pursuing a hint given by Faraday, several experiments were tried. The principal was this: the light of the electric arc, being sent through a heated tube containing sodium vapour, is analysed by a Rowland's grating. The tube is placed between the poles of an electro-magnet. When acted on by the magnet, a slight broadening of the two sodium lines is seen, tending to show that forced vibrations are produced in the atoms by the action of magnetism; (b) by Dr. J. Verschaffelt... [11-11]

Not long afterward, Zeeman met with Lorentz as well, explaining to him the experiment and his own interpretation in terms of rotary motion in a magnetic field and a mechanical æther.

> Prof. Lorentz... at once kindly informed me of the manner in which, according to his theory, the motion of an ion in a magnetic field is to be calculated, and pointed out to me that, if the explanation following from his theory be true, the edges of the lines of the spectrum ought to be circularly polarized. The amount of widening might then be used to determine the ratio between charge and mass, to be attributed in this theory to a particle giving out the vibrations of light. [11-12]

The 'cooperative effort in lifting the spirit', as Kamerlingh Onnes was to poetically put it in characterizing the productive outcome of Zeeman's meeting with Lorentz, does indeed constitute a fine example of theory abetting experiment. Let us now review the immediate background for Zeeman's experiment, and the contributions of his mentor.

Pieter Zeeman, the son of a Lutheran clergyman, was born in the little village of Zonnemaire on the isle of Schouwen, Zeeland, in 1865 [11-13]. Following his elementary education in Zonnemaire, he attended the secondary school in nearby Zierikzee, and then studied Latin and Greek for two years at the Gymnasium in Delft to qualify for admission to the university. He wrote an account of the aurora borealis, readily visible in Zonnemaire, which was accepted for publication in *Nature* with praise by the journal's Editor of 'Professor Zeeman in his observatory at Zonnemaire'! Zeeman also impressed Kamerlingh Onnes, whom he met at Delft, with his grasp of Maxwell's treatise on heat.

Zeeman entered the University of Leiden in 1885, studying mainly under Kamerlingh Onnes and Lorentz. Five years later he was appointed assistant to Lorentz, a privilege opening up excellent research opportunities. Among them, a study of the Kerr magneto-optic effect sufficed for his doctoral dissertation. (That effect, discovered by the Scottish physicist John Kerr in 1875, involved the influence of magnetism on polarized light. Kerr, an associate of Kelvin, found that reflection of plane-polarized light from the polished pole face of a magnet rotated the plane of polarization.) Zeeman's careful measurement of the effect won him a gold medal from the Netherlands Scientific Society of Haarlem in 1892, and assured him the doctorate the following year. After a semester at the Kohlrausch Institute in Strasbourg, devoted to the propagation and absorption of electrical waves in fluids under E Cohn, he returned to Leiden in 1894 as *Privatdozent*.

Settled in Leiden, Zeeman preoccupied himself once more with the interaction between magnetism and light. Two magneto-optic effects were then known. One was Kerr's effect. Long before, fully 30 years earlier, Michael Faraday had discovered (in 1845) that the plane of polarization of a beam of light was rotated when passed through certain specimens of magnetizable glass placed in a strong magnetic field. 'Thus [had Faraday noted in his *Diary*] magnetic force and light were proved to have relation to each other. This fact will most likely prove exceedingly fertile and of great value in the investigation of both conditions of natural force' [11-14]. The theoretical basis for Faraday's effect was to be provided in Maxwell's electromagnetic theory of light. Despite Hertz's confirmation of that theory, it was in some respects insufficient, being devoid of any assumption of atoms at work. It remained for Lorentz to ground the theory on electrically charged particles in atoms or molecules vibrating at particular frequencies corresponding to the colors of emitted light. Since moving charges constitute an electric current, their motion, too, should be affected by a magnetic field, and, since the light from a sodium flame was known to emit, not a continuous spectrum like the Sun, but light of discrete frequencies, a sodium flame was obviously the source to try.

Zeeman placed his flame, colored with sodium, between the poles of the strongest Rühmkorff electromagnet he could lay his hands on. The vibrations of the hypothetical charges giving rise to the two prominent yellow–orange D-lines should, he reasoned, be affected by the magnetic field by an amount hopefully observable as a broadening of the lines, this despite Maxwell's own declaration at a meeting of the British Association at Liverpool in September 1870, in reference to the light-radiating particles in a flame, that 'no force in nature can alter even very slightly either their mass or their period of oscillation' [11-15]. Sure enough, Zeeman's improvised experiment, interrupting his ongoing work on the Kerr effect, agreed nicely; no hint of a change in the sodium spectrum was observed. About a year later, however, his attention was drawn to an entry in Bence Jones' *Life and Letters of Faraday* which was published in 1870, and may also have been what provoked Maxwell's blunt remark at Liverpool the same year:

> 1862 was the last year of experimental research. Steinheil's apparatus for producing the spectrum of different substances gave a new method by which the action of magnetic poles upon light could be tried. In January he made himself familiar with the apparatus, and then he tried the action of the great magnet on the spectrum of chloride of strontium, and chloride of lithium. [11-16]

The actual experiment, Faraday's very last, was performed in his laboratory at the Royal Institution on March 12 1862. On that day, Faraday wrote in his diary as follows.

> Apparatus as on last day (January 28), but only ten pairs of voltaic battery for the electromagnet.
>
> The colourless gas-flame ascended between the poles of the magnet, and the salts of sodium, lithium, &c. were used to give colour. A Nicol's polarizer was placed just before the intense magnetic field, and an analyser at the other extreme of the apparatus. Then the electromagnet was made, and unmade, but not the slightest trace of effect on or change in the lines of the spectrum was observed in any position of polarizer and analyser.
>
> Two other pierced poles were adjusted at the magnet, the coloured flame established between them, and only that ray taken up by the optic apparatus which came to it along the axis of the poles, *i.e.* in the magnetic axis, or line of magnetic force. Then the electromagnet was excited and rendered neutral, but not the slightest effect on the polarized or unpolarized ray was observed. [11-17]

Surely, reckoned Zeeman, if the great Faraday had thought the experiment worth doing, it warranted repeating with newer spectroscopic equipment in hand at Leiden, especially with Rowland's grating possessing a diameter of 10 ft and ruled with no fewer than 14 938 lines in^{-1}; the grating had arrived while Zeeman was at Strasbourg. The rest is history, as the saying goes. In any case, that much said of Zeeman and the background for his effect, his mentor warrants equal time.

Hendrik Antoon Lorentz was born in Arnheim in 1853, as the son of a nurseryman [11-18]. He attended primary and secondary schools in Arnheim, always the first in his class, in no small measure due to his parents' enlightened attitude and enthusiasm for learning. In 1870 he matriculated at the University of Leiden, intending to study physics and mathematics. Two years later, with his candidate's examination behind him, he returned to Arnheim, where he prepared at home for his doctoral thesis while teaching at an evening high school (Hoogere Burgerschool)—an interesting feature of the Dutch school system. In 1875, he obtained his doctorate based on physical optics treated from the viewpoint of Maxwell's electromagnetic theory of light. Though he failed to send Maxwell a copy, Maxwell, who, it may be recalled, had learned to read Dutch to follow van der Waals' thesis, also read Lorentz's and commented favorably on it in a British Association report for 1876. In 1877 Lorentz was offered a chair of theoretical physics at Leiden, one which had originally been intended for van der Waals. It was the first chair of theoretical physics in Holland and one of the first in Europe. Lorentz accepted it, not yet twenty-five and even younger than J J Thomson when he succeeded Rayleigh at the Cavendish.

At Leiden, Lorentz was at first no more productive than his contemporaries abroad, concentrating on the molecular–kinetic interpretation of thermodynamics and publishing two widely used textbooks in Dutch. His fame dates from 1892 when, nearly 40, he returned to his earlier interests, optics and electrodynamics—fields which he would slowly but steadily elaborate on in a series of articles published between 1892 and 1904. For an authoritative discussion of all of this, the reader can do no better than turn to Russell McCormmack's article in the *Dictionary of Scientific Biography* [11-19]. In lieu of a full-fledged biography, his life is ably sketched in a charming memoir by his daughter, Gertruida Luberta de Haas-Lorentz (herself a physicist), in the commemorative volume she edited; it features contributions by Einstein, W J de Haas, A D Fokker, P Eherenfest, H B G Casimir, and others [11-20].

For our purposes it suffices to note Lorentz's first opus of 1892, the one introducing his atomistic interpretation of Maxwell's electrodynamics [11-21]. He took as his starting point Helmholtz's action-at-a-distance version of the subject, sidestepping certain aspects of a contrary version promulgated by Hertz in the wake of his own demonstration of electric waves in air—principally an æther dragged by ponderable bodies. Electric and optical processes, in Lorentz's theory, are ascribed to the behavior of positively or negatively 'charged particles' orbiting in atoms and molecules—particles which, in his second major exposition of 1895, he calls 'ions' [11-22]. (Only after 1899 does he refer to them as 'electrons'.) Ordinary matter and his imponderable, all-pervasive and stationary æther are distinct media, and the interaction between them occurs by means of electrons acting as intermediary agents. The interaction of light waves and matter results from the presence of electrons in matter; if set into motion these charged particles generate light waves in conformity with Maxwell's equations for the electromagnetic field. A major feature of Lorentz's 1895 memoir, accompanying the four Maxwell equations which are presented here for the first time in vector notation, is his own equation for the force on an ion acted on by an electromagnetic field. The *Lorentz force*, as it was to go down in history, constitutes Lorentz's 'own contribution to electrodynamics' [11-23] and has been described as 'the most important contribution to theoretical physics in the 1890s' [11-24].

Thus, Lorentz was by no means unprepared when Zeeman came calling on him in mid-autumn 1896. His ions, vibrating in simple harmonic motion within atoms moving in a magnetic field, were just what Zeeman needed. They would experience elastic restoring forces, like a pendulum displaced from its equilibrium position, and, in addition, a deflectional Lorentz force from the external magnetic field. The equations of motion for such an ion of charge e and mass m moving in the x, y-plane in a uniform magnetic field H perpendicular to this plane are

$$m\frac{\mathrm{d}^2x}{\mathrm{d}t^2} = -k^2x + eH\frac{\mathrm{d}y}{\mathrm{d}t}$$
$$m\frac{\mathrm{d}^2y}{\mathrm{d}t^2} = -k^2y - eH\frac{\mathrm{d}x}{\mathrm{d}t}.$$

Here the first term on the right is the restoring force, proportional to the displacement, and the second term the magnetic deflection force, proportional to the charge, field, and velocity; the latter force is directed perpendicular to both H and the velocity, and constrains the ions to move in a circular path in the x, y-plane. In the absence of a magnetic field, the solution is simple harmonic motion with an orbital frequency ν_0, and in the presence of a field the frequency becomes $\nu_0 \pm \Delta\nu$, where

$$\Delta\nu = \frac{eH}{4\pi m}.$$

The ion's original circular or elliptical motion becomes, in the presence of a magnetic field, a linear motion of constant frequency ν_0 parallel to the field, plus two circular components of motion in the plane perpendicular to the field, one of higher and one of lower frequency. The light emitted by such vibrating ions, if viewed in the direction of the field, is only emitted by the circular component of motion and will be seen as split into two lines, circularly polarized and of two different frequencies. If viewed at right angles to the field, the circular components are seen edge-on, so they yield two different frequencies of plane-polarized light in which the vibrations are perpendicular to the field direction; in addition, the linear motion emits light in the transverse direction, vibrating parallel to the field with the original frequency ν_0. Hence, viewed in this direction, a spectral line should appear as three plane-polarized components, a central line unshifted in frequency between two other lines symmetrically located and of frequency $\nu_0 \pm \Delta\nu$. Viewed in the field direction, the spectral line should appear as a pair of frequency-shifted lines, one right- and one left-handed circularly polarized.

To verify these predictions, the large electromagnet was equipped with pierced pole pieces and placed such that the axes of the holes could be oriented parallel to the field direction and aligned with the center of the grating. The sodium lines were observed with an eyepiece equipped with a vertical cross wire. Between the grating and the eyepiece were placed a quarter-wave plate and a Nicol prism—two optical components commonly used jointly in analyzing polarized light, and the same ones as used earlier by Zeeman in studying the Kerr effect. At this stage of the experiment, the apparatus was not yet powerful enough to split any spectral lines into doublets or triplets; just as well, in hindsight. In fact, Lorentz's predicted line splitting, known as the 'normal' Zeeman effect, does not apply to the D-lines of sodium—only the so-called 'anomalous' Zeeman effect, which is *not* explained by the classical theory. However, Zeeman did observe that the D-lines were broadened, something he already knew, *and* that the outside edges of the broadened lines were circularly polarized in the sense predicted by Lorentz.

Now, a property of a quarter-wave plate in conjunction with a Nicol prism is one of completely extinguishing circularly polarized light—something well known even to Faraday. Accordingly, while squinting through the eyepiece at the broad line centered on the cross wire, Zeeman adjusted the plate and prism so as to extinguish one border of the line. When the analyzer was turned through 90° or, alternatively, when the current through the magnet was reversed with a switch, the border disappeared from sight; in this way the visible border of the line could be made to jump back and forth. The electromagnet was now rotated through 90° in a horizontal plane until the field direction was perpendicular to the line of sight and to the line joining slit and grating. 'The edges of the widened line now appeared to be plane polarized, at least in so far as the present apparatus permitted to see' [11-25].

One additional aspect of the circular polarization involved here was amenable to testing with the quarter-wave plate and Nicol prism: the *polarity* of the emitting charge (since the direction of rotation of the polarization depends on whether the light emitters are positively or negatively charged). In the event, Zeeman got the sign wrong in his first attempt. Viewing the sodium light parallel to the field, 'the right-handed circularly-polarized rays appeared to have the smaller period [higher frequency]. Hence ... it follows that the positive ions revolve, or at least describe the greater orbit' [11-26]. It might be added that Zeeman ignored this slip, soon righted, in his subsequent Nobel address.

Finally, as Lorentz had also foretold, the line broadening provided a measure of the ratio of charge-to-mass of the ions; the ratio could be calculated simply from the separation between the outer lines or, to a good approximation, from the observed broadening. Since

$$\nu_2 - \nu_1 = 2\,\Delta\nu = \frac{eH}{2\pi m}$$
$$\frac{e}{m} = \frac{2\pi}{H}\,(\nu_2 - \nu_1).$$

Zeeman's value was 10^7 emu, in excellent agreement with Thomson's value just then; we will have more to say about this remarkable agreement presently.

These results were in hand in time for submission to the Amsterdam Academy at its meeting on November 28 1896 [11-27]. They were also communicated in the form of a short paper to the Deutsche physikalische Gesellschaft in Berlin, with thanks to Professor K Onnes for the interest he had shown in this work [11-28]. An English version of this paper, translated by Arthur Stanton (A Schuster's assistant), appeared in *Nature* on February 11 1897 [11-29]. Both reports communicated to the Amsterdam Academy, i.e. that communicated on October 31 and that communicated four weeks later, were subsequently translated into English and sent as a single manuscript to Oliver Lodge, editor of the *Philosophical Magazine*. Lodge, for reasons we shall see, communicated it posthaste to his journal for publication with a footnote to the effect that he himself had verified the author's results 'so far as related to emission spectra and their polarization' [11-30]. (Lodge caught Zeeman's slip about the sign of the effect.) This paper contains an appendix, not found in the original Dutch communication. The appendix reprints the entry cited earlier from Bence Jones, together with the excerpts from Faraday's *Diary* also cited; it also dwells on a paper by Peter Tait who, after Faraday, anticipated Zeeman's effect, and on measurements somewhat similar (albeit negative) to those of Zeeman by the Belgian astronomer Charles Fievez.

11.3. Mopping up

Zeeman's paper sent to the *Philosophical Magazine* did not catch Oliver Lodge entirely by surprise [11-31]. On February 11 1897, a note by him entitled 'The influence of a magnetic field on radiation frequency' was received and read at the Royal Society. It recounts his own repetition of Zeeman's measurements, and draws attention to Lorentz's role in the background of the Leiden experiments.

> To put myself in order [he reported], I will state that I have set up apparatus suitable for showing the effect, and have verified its primary features, viz., that both lines in the ordinary spectrum of sodium are broadened when a magnetic field is concentrated upon the flame emitting the light. [11-32]

The observation was made 'with a 1-inch flat reflecting grating containing 14,600 lines, and with an oxy-coal gas flame playing on pipe clay supporting carbonate of soda between pointed poles' [11-33]. The published note is accompanied by one under the same title but of more theoretical substance by Joseph Larmor. In it, the author emphasizes the direct lineage of Zeeman's discovery to Faraday's magneto-optic effect and points to a previous memoir by Helmholtz (1893), to the two by Lorentz of 1892 and 1895, and to a major paper of his own dated 1894—all relevant to Zeeman's effect [11-34].

Larmor had, in fact, been looking for just such an effect. At the time of Zeeman's experiment, he was lecturer at St John's College, Cambridge, where he would in time (1903) succeed Stokes as Lucasian Professor of Mathematics—an exalted chair he retained until 1932. In terms of his scientific work, Larmor bears close resemblance to H A Lorentz as one of the giants in the final phase of classical theoretical physics embracing electromagnetic theory, optics, mechanics, and geodynamics. Unlike Lorentz, however, he cannot be said to have actively guided the new generation of physicists in developing quantum theory and relativity; this must, in large measure, be due to his steadfast reluctance toward uncautious endorsement of loosely conceived ideas [11-35].

Larmor was born in Magheragall, County Antrim, Ireland, in 1857, the eldest son in a large family of a farmer who took up grocering in Belfast when the boy was six or seven [11-36]. After schooling in Belfast, he obtained his BA and MA degrees there from Queen's University. He entered St John's College, Cambridge, in 1877, where he became Senior Wrangler in the mathematical tripos in 1880 (with J J Thomson placing Second Wrangler that year). Having won a Smith Prize and with a fellowship, he served five years as professor of natural philosophy at Queen's College, Galway, before returning as a lecturer to St John's in 1885.

While still at Queen's College, Larmor had been developing an 'electron' theory, borrowing the term from his countryman G FitzGerald, whose uncle, G Johnstone Stoney, had coined it [11-37]. (Thus, with Stoney, FitzGerald, and Larmor, the Irish had a considerable stake in the electron.) His theory was expounded in two major papers entitled 'A dynamic theory of the electric and luminiferous medium', published in 1894 and 1895 [11-38]. The electron only replaced vortex rings in the æther in an appendix to the first of these, foreshadowing a bona fide electron theory. In the second paper, an all-pervasive æther of a type developed much earlier by another Irishman, James MacCullagh, proved surprisingly receptive to Stoney's electrons as mobile replacements for MacCullagh's physically unreal centers of rotational strain. The æther and ordinary matter interacted via discrete electrons much as in Lorentz's treatment. More pertinent to the magneto-optic problem at hand, Larmor showed that the effect of a magnetic field on a system of charged particles, all of the same e/m ratio and moving in a single central field, was equivalent to motion in the absence of a magnetic field except for a superposition of a common precession of angular frequency $\omega_L = eH/2mc$ (an expression we have already encountered and one named in honor of Larmor). The particles spiral around the magnetic field lines at Larmor's frequency, and the spectral lines of the emitted light are similarly shifted or broadened.

On the plausible assumption that the radiating electrons were not less massive than a hydrogen ion with $e/m \sim 10^4$ emu, Larmor concluded that the effect was far too small to be observable in practice. However, on receiving word of Zeeman's results, including his concomitant value of 10^7 for e/m, Larmor wrote to Lodge posthaste, doubtful as usual but urging him to check it out anyway [11-39]. Lodge, too, wasted no time; within a week he had confirmed Zeeman's results 'with such appliances as were to hand' [11-40].

In addition to his short note to the *Proceedings of the Royal Society* on his extemporized experiment, Lodge also dispatched a somewhat longer discussion of the subject to the *Electrician*. Under the heading 'The latest discovery in physics', it also appeared in the February issue of that journal [11-41]. In it, Lorentz's and Larmor's explanation is spelled out in some detail, and the state of polarization of the line edges, viewed along or across the magnetic lines of force with a quarter-wave plate and Nicol, is experimentally confirmed. (Observation of the light across the lines of force was comparatively easy, the results of which Lodge had already confirmed in his Royal Society note of February 11; observation *along* the lines of force was another matter, since the perforation of a pole piece interfered with the need for concentrating the field along the axis of vision.) Perhaps as a result of this paper, or more likely because he had written to Zeeman the moment he was tipped off by Larmor, Zeeman himself sent Lodge by post his aforesaid manuscript in English. As noted, Lodge passed on the manuscript to the *Philosophical Magazine*, where it appeared in the March 1897 issue [11-42].

In his article to the *Electrician*, Lodge had written rather casually about electrons and their motion, and 'the idea of radiation excited by the motion of *electrons* pure and simple' [11-43]. Looking back on all this, 25 years later, he felt compelled to set the record straight, explaining that

> ...the difficulty was that at that date we all—except perhaps Larmor and Lorentz—thought of an electron as of something attached to an atom, making it an ion, in accordance with Faraday's electrolytic ideas; and the notion of a free satellite electron, inside the boundary of an atom, was of later growth. In fact, it was a development largely brought about by Zeeman's discovery. [11-44]

In May, Lodge communicated yet another note on the subject to the Royal Society, read by the Society on June 3. He had, in the meanwhile, gained access to a concave Rowland grating ruled with 14 438 lines to the inch and measuring 10 ft in radius of curvature—the very one used by George Higgs in photographing the solar spectrum, and one just as powerful as Zeeman's. With the greater optical power,

> magnetization greatly widens the doubling [of each sodium line], pushing asunder the bright components very markedly; stronger magnetization reverses the middle of the widened dark band, giving a triple appearance; stronger magnetization still reverses the middle once more, giving a quadruple appearance to the line.

> In every case a nicol, suitably placed, cuts off all the magnetic effect and restores the original appearance of the line. [11-45]

Similar behavior was seen in the spark spectra of the salts of lithium and thallium, and particularly of the red cadmium line. The same month Lodge demonstrated the effect in an exhibit at a Royal Society soirée, writing in the Yearbook of the Society for 1897 that 'a nicol or other analyzer shows that the [sodium] light of changed refrangibility is polarized, as it would be if the source of radiation consisted of revolving electrified particles whose motion is accelerated or retarded by magnetic lines of force through the plane of motion' [11-46]. Even so, in hindsight Lodge disdains credit for the discovery of doublets or triplets.

> I did not apprehend it clearly as a pure precessional effect (akin to that which Dr. Johnstone Stoney had worked out long ago), and was inclined to suppose that the magnetic acceleration and retardation of frequency, acting on a random collection of molecules, would be likely to cause a confused broadening... I was still too much influenced by the idea of random atomic motions, instead of precise electronic orbits. [11-47]

Zeeman's effect had, in fact, been anticipated by others under different forms, long after Faraday's unsuccessful experiment in 1862. Among the first was Peter Tait of Edinburgh University. In 1875 he began repeated, albeit frustrated, attempts to detect the influence of a magnetic field on the selective absorption of light. In this, he was impelled by a theory of magneto-optic rotation of the luminiferous medium propounded by his esteemed colleague William Thomson (Kelvin)—a concept grounded in 'gyrostatic cells' and one also borrowed somewhat later by Larmor in his own dynamical theory noted earlier [11-48]. According to Thomson's theory, Faraday's rotation of the plane of polarization is explained by the fact that the plane-polarized ray, while traversing the luminiferous (magnetized) medium, is broken up into circularly polarized components rotating with or against the æther; the two components propagate with different velocities.

> Now [suggested Tait], suppose the medium to absorb one definite wave-length only, then—if the absorption is not interfered with by the magnetic action—the portion absorbed in one ray will be of a shorter, in the other of a longer, period than if there had been no magnetic force; and thus, what was originally a single dark absorption line might become a double line, the components being less dark than the single one. [11-49]

Unfortunately, no such effect was seen. Tait repeated the experiment 'again and again... Hitherto it has led to no result' [11-50].

Arthur Schuster, too had, at about the same time, or shortly after being appointed Demonstrator at Manchester in 1873, vainly attempted to repeat Faraday's experiment. Schuster, like Tait, was influenced by Balfour Stewart whose scientific intuition is said to have been 'greater than his command of experimental or theoretical technique, which prevented him from establishing his profound ideas' [11-51]. In Schuster's case, the immediate motivation seems to have been a mixture of the diamagnetism of Wilhelm Weber (in whose laboratory Schuster spent the summer of 1874) and his early impressions on reading Maxwell's *Treatise* which was to figure so prominently in his formative years [11-52]. If light is an electromagnetic disturbance, ought not Weber's circulating currents to have some influence on it? Taking a cue from Faraday, Schuster placed a sodium flame in the strongest magnetic field that Owens College could muster. To his delight, a spectroscope showed a widening of the lines. On closer inspection, however, this interesting observation proved to be nothing more than a result of the mechanical widening of the slit from the magnetic field drawing aside the steel spring controlling it.

> An experience such as this [admonishes Schuster] ought to shew the danger of taking too easily for granted, that an observation recorded in some old book is an anticipation of a later discovery, when in reality it is probably only the result of a careless experiment accidently simulating an effect which is real, but can only be detected by more refined methods. [11-53]

Closer to the mark was the Irishman G Johnstone Stoney, whom we have met and who spoke of a 'unit quantity of electricity' at Belfast even the year before Tait's first experimental attempt [10-34]. In 1881, when he at long last penned his ideas on the subject in print, Stoney had offered a numerical estimate for his quantum—an estimate fully two orders of magnitude too low, to be sure, but one offered long before anybody else's. The importance of his address before the Royal Dublin Society ten years after that was not so much that he first used the term 'electron' in public on that occasion, as that he linked the concept to the origin of the line spectra of the chemical elements. Stoney was, in fact, no stranger to spectra; he was a pioneer among those searching for order in the patterns of spectral lines. Thus, just as he seems to have anticipated Johann Jakob Balmer's 1885 classic publication on the hydrogen spectrum with his own paper on the subject in 1871 [11-54], his remarkable paper to the Dublin Society entitled 'On the cause of double lines and of equidistant satellites in the spectra of gases' [11-55] clearly anticipated Lorentz's explanation of Zeeman's effect.

Another contender to Zeeman's effect was closer to home. As Zeeman informs us in the appendix to his article in the *Philosophical Magazine*, during 1885–1886, Charles Fievez, an astronomer at the Royal Observatory of Brussels, brought before the Belgian Academy of Sciences two papers on the influence of magnetism on the character of spectral lines [11-56]. In them, he describes experiments of his own purporting to show not only widening and brightening of line spectra of a flame in a magnetic field, but also reversal and double reversal of the lines. However, from his description Zeeman was led (with Lodge and Kamerlingh Onnes) to conclude that 'the whole of the phenomena observed by Fievez can readily be attributed to a change of temperature by the well-known actions of the field upon the flame' [11-57]. 'If the true effect was seen at all [added Lodge], it was so mixed up with spurious effects as to be unrecognizable in its simplicity, and so remained at that time essentially undiscovered' [11-58].

Even in far-flung America, Zeeman's effect played a role late in the life of he who had indirectly contributed to its discovery in the first place, Henry Rowland; without his concave spectral grating, resolution of the line broadening and splitting would have been impossible. The time of Zeeman's discovery was a disappointing period in the otherwise busy life of this extraordinary American scientist [11-9]. During most of the 1890s, he had become distracted from basic research, involving himself instead in various commercial ventures, in part to provide a legacy for his newborn children and given but ten years to live as a result of then incurable diabetes [11-59]. Only at the end of the decade was he inspired to take up research once again, principally to repeat nagging, inconclusive charge-convection experiments of his own 25 years earlier during his *Wanderjahr* in Helmholtz's laboratory in Berlin [11-9]. Hard on the heels of even more inconclusive experiments by his students came Zeeman's announcement. Splitting of the sodium D-line appeared to Rowland as just what was needed: a magnetic effect explainable by his convection experiment. According to John Miller, who has devoted much time to Rowland,

> vibrating, electrified 'matter' within a molecule gripped the ether. This might produce a magnetic effect, which interacted with Zeeman's externally applied magnetism. Perhaps the rotating matter of the Earth likewise retained 'a feeble hold on the ether sufficient to produce the Earth's magnetism'. [11-60]

Alas, new charge-convection experiments undertaken in 1900 gave promising results only days before Rowland's untimely death in April 1901, 53 years old.

Though his effect was attracting increased attention at home and abroad, the last had not been heard from Zeeman himself. In January 1897, riding high on the strength of his sudden fame, he was appointed lecturer at the University of Amsterdam—an institution with which he would be associated until his retirement in 1935. He was promoted to Extraordinary Professor of Physics in 1900, and succeeded van der Waals on his mandatory retirement at 70 in 1908, both to his professorship and as Director of the Physical Institute. In 1923, Zeeman became Director of the new Laboratorium Physica erected in his honor, later renamed the Zeeman Laboratory. A prominent feature of the new Laboratory was a concrete block weighing a quarter of a million kilograms, mounted so as to be isolated from the floor and thus constituting an excellent, highly stable platform for experiments free of vibration. Zeeman could sorely have used this contrivance in 1897!

Installed in Amsterdam, Zeeman lost no time getting back to work. A suitable Rühmkorff electromagnet was conveniently at hand. Unfortunately, however, a Rowland grating on a par with the 10 ft one at Leiden was not available; Zeeman had to make do with one of 6 ft radius, though one still featuring 14 000 lines in^{-1}, or about as in the earlier grating. At any rate, the grating was good enough to confirm Lorentz's prediction that, in a strong enough field, the cadmium lines are not only broadened (as shown in Leiden), but also actually broken up into doublets or triplets, depending on whether the emitted light is viewed parallel or perpendicular to the field direction. Predictions concerning the sense of polarization from the different viewpoints also checked nicely [11-61]. With sodium, the matter was less clear-cut, as it proved 'difficult to see the phenomena pure and simple' [11-62]. Still, the sign of the circular polarization of the cadmium doublets agreed with what had formerly been seen with the sodium D-lines, with one important correction. The right-hand circularly polarized rays now exhibited greater power, and thus the charge of the vibrating particles emitting the radiation was evidently *negative*, not positive as first reported. 'Probably', allows Zeeman in a footnote, 'my mistake arose from a faulty indication of the axis of the $\lambda/4$-plate used' [11-63]. In an appendix, he also acknowledges a communication from Lodge to the effect that Lodge himself had demonstrated essentially the same doubling of lines before a Royal Society soirée that May, albeit somewhat incompletely, to which the Editor appended a disclaimer by Lodge insisting that he (Lodge) had observed pretty much the 'new effects, but without any intention of trespassing on the prerogative of the discoverer' [11-64].

On the whole, however, detailed observations at Amsterdam left something to be desired, particularly photographic measurements. Such measurements were more often than not spoiled by vibrations caused by human movement in the laboratory or Amsterdam street traffic—even at night! As luck would have it, Professor Hermann Haga of the University of Groningen offered a way out with a kind invitation for Zeeman to avail himself of his own grating in Groningen. Measuring 10 ft in radius, it had 10 000 lines in^{-1}, and, most importantly, was mounted 'in a very stable manner' [11-65]. With it, many further subtleties of the doublets and triplets were ascertained, as was a new value for e/m from the separation between the sodium D-lines, namely $\sim 1.6 \times 10^7$ emu (versus 1×10^7 emu earlier).

The main advantage of the new grating was in yielding good

photographic images of the line patterns. By late summer of 1897, the Zeeman effect had become a hot topic among physicists, and photography was clearly the right approach, at least in augmenting visual sightings. The spectral variations being small at best, photographic plates offered the possibility of obtaining permanent records of faint phenomena—records not subject to personal bias inevitably accompanying obscure visual observations. At first, results (by Alexander Anderson) were disappointing to say the least, however. There was simply 'no evidence of any effect of the magnetic field, though the definition of the lines was all that could be desired' [11-66].

> 'When it be remembered [adds the Editor of *Nature*] 'that a length of one centimetre in the photograph corresponds to a change of wave-length of 26 tenth-metres, and since an increase of breadth of a line of one-tenth of a millimetre (and probably much less than this) could easily be seen in the photographs, there could not have been a change in the period of oscillation of as much as one part in 20 000'. [11-67]

Only two reasons for this utterly negative result came to mind: either the magnetic field was not strong enough, or the exposure, though up to 30 minutes for the narrowest slit, was still not long enough.

Much more successful proved Thomas Preston of Dublin, an experienced spectroscopist [11-68]. Preston concurred with the need for photographic records of the effects in question, as urged specifically in the conclusion to the gloomy report in the September 2 issue of *Nature*, and in light of lingering doubts as to whether the effects amounted to a simple broadening of the spectral lines, to the production of bona fide doublets and triplets, or perhaps to a combination of all. At first, however, he too ran into the usual difficulties, despite having recourse to an excellent concave grating belonging to the Royal University of Ireland—an instrument measuring no less than 21.5 ft in radius and mounted according to Rowland's own specifications in the Physical Laboratory at Earlsfor Terrace, Dublin [11-69]. The instrument was fitted with a camera box accepting photographic plates fully 50 cm long, and thus capable of photographing a length of the spectrum equal to half a meter in a single exposure. With the ordinary U-shaped electromagnet at his disposal at the Royal College of Science he, too, nevertheless at first experienced difficulty in obtaining any effect whatsoever. However, by pushing the pole pieces very close together, using a short spark as the source of light, and using, in addition, an ordinary magnifying glass, he obtained images of doublets and triplets of the violet line of cadmium at 4678 Å—albeit images too faint for satisfactory reproduction in his preliminary report to *Nature* [11-70]. In a follow-up report to the Royal Dublin Society Meeting on December 22 1897, Preston had better results to present, aided by a more powerful electromagnet kindly lent by the

Rev Monsignor Molloy. The highlight of the new printed report was a plate showing the various line configurations with remarkable clarity [11-71]. The aforesaid pattern in cadmium proved considerably more complex than had originally appeared.

> While the line 4678 of cadmium and the adjacent line 4680 of zinc are both converted into triplets, the line 4722 of zinc and the celebrated blue cadmium line 4800 are not so resolved (in appearance), but show as a species of quartets in which most of the light is concentrated in the side lines. Finally, the line 4811 of zinc shows as nothing very definite, but may in this picture be described as a hazy doublet. It is to be remarked, however, that many lines show as perfectly distinct doublets when viewed across the lines of force, doublets which are scarcely rendered any clearer by the interposition of a nicol. [11-72]

The two D-lines of sodium were even more complex in appearance.

> It is interesting to remark that the two D-lines of sodium do not belong to the class which show as triplets in the magnetic field. Owing to the ease with which these lines reverse and to the constant variations going on in the vapour-density in the spark obtained from a solution of the salt, it is not easy to obtain the sodium lines in a steady state for any length of time. But when the sodium salt is contained in small quantity in the solution the lines may be obtained sharp and fairly steady. Under these circumstances one of the sodium lines D_2 shows as a sextet... of fine sharp equally-spaced lines of which the two border-lines are somewhat nebulous on their outside edge. The other line D_1 is of the double doublet type..., and shows as two pairs of sharp lines. It is thus seen that the sodium lines and the blue cadmium line 4800 do not show as normal triplets; and it is probably for this reason that different observers have differed in their accounts as to the effects observed in the case of these lines. [11-73]

Considering the 'unstable character' of the sodium lines, Preston was more inclined to be charitable with regard to the effects observed by Fievez but discounted by Zeeman as spurious. What Fievez saw was evidently real—effects 'true to nature when special precautions are not taken against reversals. The real point of importance [was] the broadening always observed' [11-74].

The French were not far behind. Already on October 18 1897, Marie Alfred Cornu of the École Polytechnique, spectroscopist and optical physicist perhaps best known for the optical spiral which bears his name, presented his first report on the Zeeman effect to the Paris Academy. Possessing a grating obtained from Rowland himself, he reported on triplets and their polarization which he found to agree with Lorentz's theory [11-75]. However, in a subsequent report to the Academy on January 17 1898 he corrected his earlier results, arguing, among other things, that the sodium D_1-line was actually split into a quadruplet, not a triplet as Lorentz's theory demanded [11-76]. Nor was Cornu alone in pursuing the matter in Paris. About the same time that he read his initial report to the Academy, Henri Becquerel, also at the École Polytechnique and hardly a novice when it came to polarized light in a magnetic field, took an interest in the new effect—perhaps as a respite from radioactivity. With Henri Alexandre Deslandres, he too observed quadruplets and other complex patterns in the spectra of iron. He devoted fully a year and a half to this matter before returning to radioactivity and a new line of inquiry—one which would have important consequences for the two ongoing, parallel investigations of the Zeeman effect and of cathode rays.

A heavyweight contender in exploiting Zeeman's discovery was the American Albert A Michelson, who had once studied under Cornu during his European *Wanderjahr* during 1880–1882. Michelson first applied his powerful interference technique to the Zeeman effect in 1897. His echelon spectroscope surpassed the resolution of the best gratings then available by a factor of five [11-77], though he was nevertheless thwarted by Preston who anticipated him in resolving the spectral lines into more than three components. Even more frustrated by Preston was Zeeman himself, although the Nobel Prize went to him, and to Lorentz, in 1902 [11-78]. Even though Zeeman qualitatively anticipated a fundamental relation between his line patterns and the laws of spectral series, credit for the quantitative elucidation of that relation belongs to Preston for his subsequent, systematic study of spectral lines which yielded the rule that bears his name—namely, that all lines of a given series of a substance exhibit the same pattern of components in a magnetic field. Moreover, analogous spectral lines of the same series, even if belonging to different elements, have the same Zeeman effect [11-79]. Still plagued by vibrations, Zeeman abandoned this fruitful line of investigation, absorbing himself instead in less exacting studies of his own effect and related matters. These researches occupied him some 15 years, researches for which the magneto-optics of Woldemar Voigt performed the same role that Lorentz's theory had for Zeeman's original investigation [11-80].

And so we conclude our digression on the Zeeman effect, an effect that was considered a triumph for Lorentz's theory of electrodynamics linking light and electricity, and sufficiently persuasive for the Royal Swedish Academy in 1902 to award Zeeman and Lorentz jointly the second Nobel Prize in physics (following Röntgen in 1901), this despite clear experimental evidence by then for the inadequacy of Lorentz's theory in dealing with ever more complicated spectral patterns as observations steadily improved. These patterns, constituting what Friedrich Paschen and Ernest Back a decade later named the 'anomalous

Zeeman effect', foreshadowed an aspect of the old quantum theory, ultimately explained by the spin of the electron.

The early history of the Zeeman effect is, indeed, replete with ironies. Larmor, as we know, was in position to calculate the effect to be expected by exposing sodium light to a magnetic field, and concluded it was too small to warrant an attempt to measure it, presuming the radiator was the size of an atom. Zeeman, according to Oliver Lodge who himself was at the center of these developments at the turn of the century, was 'undeterred by supertheory' [11-81], and found an effect after all. To be sure, he saw only a broadening, not even a simple splitting, but that was enough for our purpose. Merely a broadening, coupled with Lorentz's own theoretical insight, sufficed to yield a value for e/m for the radiators, or bound carriers of electric current in atoms. That value, $\sim 10^{-7}$, was in excellent agreement with Thomson's, Wiechert's, and Kaufmann's ratio from cathode rays, strongly hinting at the identity of the oscillating light particles in atoms with the free carriers of negative charge in cathode ray tubes.

We must share A Pais' astonishment at the lack of any expression of surprise in Zeeman's article in the *Philosophical Magazine* for March 1897 over such a large value of e/m or, rather, over such a small value of m, assuming e to be the order of magnitude value estimated much earlier by Stoney and well known to Lorentz [11-82]. J J's *Recollections* barely mentions Zeeman in quite another context, much less offers any clue to how he first learned of Zeeman's e/m result; however, in his Friday evening discourse at the Royal Institution on April 30 1897, Thomson noted without further ado that his own value for e/m agreed with Zeeman's value, probably quoting Zeeman's *Philosophical Magazine* article published the month before. In his article 'On the charge of electricity carried by the ions produced by Röntgen rays' in the *Philosophical Magazine* for 1898, Thomson noted that Lorentz's demonstration [11-83] that the charge on the ions whose motion gives rise to the Zeeman effect is of the same order as the charge on a hydrogen ion in electrolysis. Finally, in his Dover address of 1899, he refers to the e/m ratio and sign of the Zeeman effect as furnishing 'a reason for believing that there are many more corpuscles in the atom than the one or two that can be torn off' [11-84]. That is, if there are only one or two of J J's corpuscles in an atom, we should expect that only one or two lines in the spectrum would show the Zeeman effect; in fact, 'there are a considerable numbers of lines in the spectrum which show Zeeman effects comparable in intensity' [11-85].

Zeeman put the bottom line of his discovery most simply in his Nobel address. Namely, 'that which vibrates in the light source is the same as that which travels in cathode rays' [11-86]—a further hint at the universality of the charge-carrying corpuscles, whether inside or outside the atom.

Chapter 12

THE PHOTOELECTRIC EFFECT REVISITED

12.1. A blow to German science, and another blow to Lenard

On October 19 1899, Philipp Lenard presented to the Vienna Academy a report on 'The production of cathode rays by ultraviolet light', about a month after J J Thomson read the gist of his definitive paper 'On the masses of the ions in gases at low pressures' before the British Association at Dover [12-1]. In fact, the hapless Lenard had learned that Thomson shared his painfully acquired views on the true nature of the photoelectric discharge only a few weeks prior to his own presentation at Vienna; once again he was anticipated in one of his major discoveries by someone else [12-2].

The photoelectric effect, as it was to become known, had been studied most recently at Cambridge, following up old experiments by Elster and Geitel at Wolfenbütel—first by Thomson's 'research students' Zeleny and Rutherford, and then by the 'Prof' himself. Thomson had found, as we recall, that the effect was just what was needed at this critical juncture in his research program. It allowed simultaneous determination of the charge-to-mass ratio e/m and the absolute charge e for the carriers of negative electricity liberated by ultraviolet light. The value of the e/m ratio was found to be the same as for cathode rays. The charge, moreover, agreed with that found for ions liberated by x-rays. Knowing both e/m and e, Thomson had the mass in hand as well, seemingly identical for the otherwise disparate charge carriers showing up in the various discharge processes. More astounding, the mass was exceedingly small, only 10^{-3} times that of hydrogen ions in electrolysis—until then the smallest protyle mass known to man.

Lenard's paper, undoubtedly one of his definitive ones, is often dated one year later (in, among other sources, Thomson's subsequent treatise *Conduction of Electricity in Gases*, p 137), since a reprint issued in 1900 is more accessible and frequently cited [12-3]. In fact, either paper describes

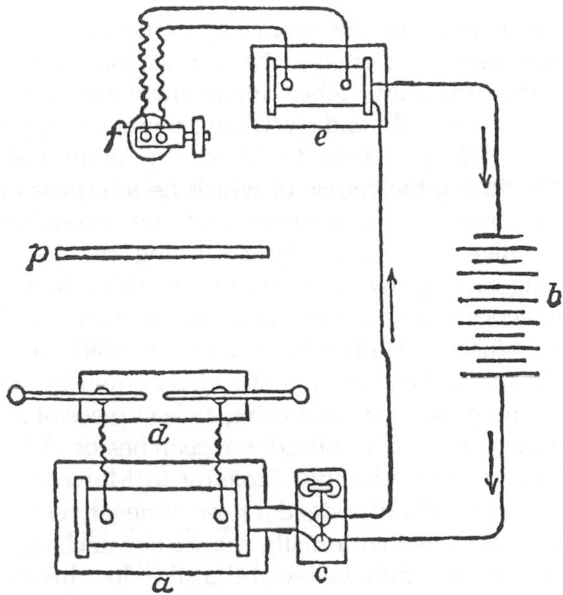

Figure 12.1. *Photoelectric experiment by Heinrich Hertz.*

work of his own on the photoelectric effect begun as early as 1890–1891—dates first revealed in the annotated second edition of his Nobel Lecture, as noted by B Wheaton [12-4]. The eight year delay in publication was deliberate, as we shall see.

The photoelectric effect was discovered quite accidentally at Karlsruhe in 1887 by Heinrich Hertz, Lenard's mentor, during the initial phase of his celebrated work on Maxwell's electromagnetic waves. We have touched on the circumstances surrounding the discovery earlier, but perhaps a somewhat fuller recapitulation does not hurt at this point. The experimental arrangement is depicted by Hertz himself in figure 12.1. Sparks were generated in the gaps *d* and *f* by Rühmkorff coils *a* and *e*, powered by the common Bunsen cells *b*. To observe faint sparks in the gap *f*, Hertz at first shielded against strong ambient light by enclosing the gap in a black box. In so doing, he soon found that the gap must be shortened to allow sparks to pass. At first annoyed by the effect, Hertz 'had no intention of allowing this phenomenon to distract my attention from the main object I had in view, but it occurred in such a definite and perplexing way that I could not altogether neglect it' [12-5]. Because sparks were known to emit ultraviolet light, he concluded that ultraviolet light from the primary spark gap facilitated the passage of sparks in the secondary gap. (The black box absorbed the radiation, hindering production of the secondary spark.) Hertz also found that the cathode surface of the spark gap was the sensitive spot.

Before long, Hertz was sufficiently captivated by the distracting phenomenon to interrupt his main experimental program and pursue it in greater detail. Perhaps, he wrote his father (a Senator in Hamburg), this discovery, albeit made inadvertently, of an apparent relationship between light and electricity, might one day yield important understanding [12-6]. True to form, he undertook a meticulous investigation, during the course of which he interposed a great number of substances between the primary and the secondary sparks. This effort was, indeed, rewarded by becoming the basis for important understanding accompanying a new effect which, before long, inspired near-simultaneous research programs in Germany, Italy, and Russia. As a start, Wilhelm Hallwachs found in 1888, while assistant to G H Wiedemann at Leipzig, that the effect could be observed equally well by discarding the spark and using a clean piece of zinc attached to a gold-leaf electroscope. The collective researches of Hallwachs, Augusto Righi in Bologna, and Alexandr Stoletov in Moscow soon established that the ultraviolet light dislodged a flow of negatively charged particles from a metal plate, whether initially charged or uncharged with negative electricity. It was generally believed that the flow involved dissociation of ambient gaseous molecules into charged atomic constituents, with the negative fraction subsequently accelerated away from the plate.

One who took up the problem was Philipp Lenard, three years after receiving his doctorate and while assistant to Georg Quincke in Heidelberg. Lenard viewed the dissociation explanation with skepticism. In 1887 Robert Nahrwold, a Gymnasium teacher, had concluded from experiments of his own that it was 'highly probable that a gas cannot be statically electrified' [12-7]. Nevertheless, that the charge is carried by negatively electrified particles was perhaps most conclusively demonstrated by Elster and Geitel, who had shown that a transverse magnetic field diminishes the photocurrent if the experiment is conducted in a well evacuated vessel. Furthermore, the effect persisted in the lowest gas pressures then attainable, again ruling out gas molecules as the charge carriers. To Lenard and his collaborator at the time, Max Wolf, who was then a *Privatdozent* at Heidelberg, it seemed clear that the explanation lay in charged 'dust' dislodged from the electrode. 'Dust can be electrified…, as a gas cannot', they declared [12-8]. Their experimental evidence in support of this statement, among other things that ultraviolet light roughens or pulverizes metal surfaces, soon became the subject of considerable criticism, however; to pin the matter down, Lenard devised several fresh experiments—experiments to which we now turn.

Lenard's definitive paper of 1899–1900, covering experiments begun during 1890–1891, opens with several general results. First, he admits that his original conjecture, that massive particles dislodged from the cathode carry the photoelectric discharge, was incorrect. The proof involved a clean platinum wire acting as anode and a polished surface of sodium amalgam as cathode, both maintained in an atmosphere of hydrogen. The photoelectric current was allowed to flow until about 3×10^{-6} C had passed through the circuit. If the carriers of charge were atoms of sodium, each atom could hardly be expected to carry a larger charge than its carrier in electrolysis. There should, therefore, have been deposited on the platinum wire at least 0.7×10^{-6} mg sodium. On removing the wire from the bulb, no trace of sodium could be detected spectroscopically. Second, in conformity with the findings of Righi, all known effects of ultraviolet light in air persist at the highest state of rarefaction, ruling out a dependence of the photoelectric discharge on the residual gas in the tube.

If neither gaseous molecules, nor molecules of the cathode itself, just what are the carriers of the photoelectric current? The answer came from a careful measurement of the charge-to-mass ratio of the charge carriers—the crux of Lenard's present investigation. His apparatus is shown in figure 12.2. Ultraviolet light from a spark between zinc electrodes entered the evacuated tube through a quartz window B and liberated negatively charged particles from the aluminum cathode U. The cathode could be biased with an external dc voltage supply. A metallic screen E, with a small hole in its center and connected to earth, served as anode; it also shielded the right-hand portion of the tube from the electrostatic action of the charged cathode. The small electrodes α and β were connected to electrometers.

When U was illuminated and charged to a negative potential of several volts, negative particles were liberated and accelerated toward the anode E. A few of them passed through the hole and continued at uniform velocity to the electrometer α, where their arrival was registered by the electrometer. If, by means of a pair of Helmholtz coils (indicated by the dashed circle in figure 12.2), a magnetic field directed toward the reader was produced in the region between E and α, the negative particles were deflected upward. For a particular value of the magnetic field strength they struck the collector β, where their charge was detected with the electrometer connected to that electrode.

The e/m determination went as follows. Assume first that the region to the left of E is unaffected by the magnetic field, and that the right-hand region is free from any electric field. Let a negative potential V (in electromagnetic units) be applied to the cathode U, the anode E presumably always being at zero potential. The particles, on reaching the anode, will have a kinetic energy given approximately by the usual expression

$$eV = \tfrac{1}{2}mv^2.$$

The velocity v is maintained after passing through the hole in E, and the only subsequent force acting on the particle is that due to the magnetic field H. Since this force, evH, is always perpendicular to the path,

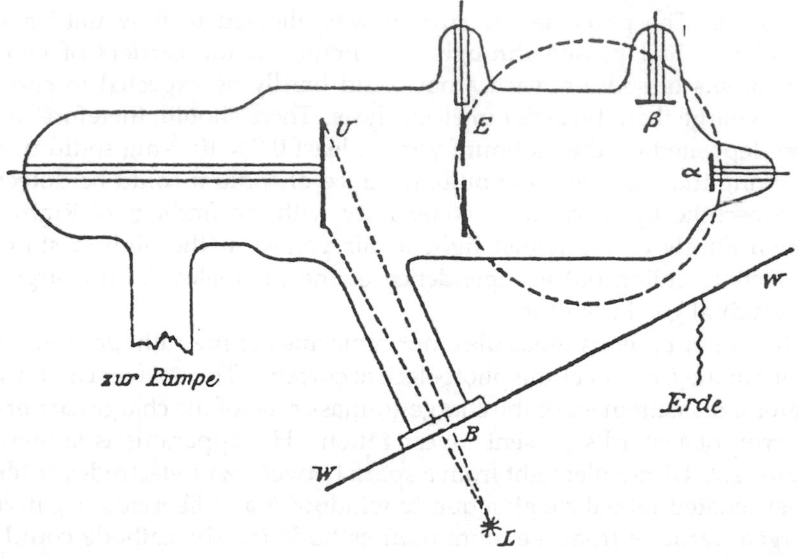

Figure 12.2. *Lenard's apparatus for measuring e/m of charge carriers from the photoelectric effect.*

the trajectory becomes a circle of radius r, determined by equating the centripetal and centrifugal forces on the particle:

$$evH = (mv^2)/r$$

where H is the magnetic induction in gauss just necessary to cause the particles to reach β, and r is the radius of the corresponding circular path, determined from the geometry of the apparatus [12-9]. From the two equations, we have

$$e/m = (2V)/(H^2r^2)$$

an expression also used by Lenard's countryman Wiechert in these years (pp 154–5).

Lenard's e/m ratio (or ε/μ in his notation) was approximately 1.15×10^7 emu g^{-1} [12-10].

> This value [he observed] deviates from that obtained earlier for cathode rays, 6.4×10^6. No compelling arguments suggest themselves for the deviation; the values could easily be off, since neither experiment was undertaken for the purpose of performing this particular measurement. All the more, a direct comparison between rays produced by the action of light and by a gaseous discharge is of interest. [12-11]

Indeed, the qualitative agreement for e/m not only held for a broad range of the accelerating potential between the electrodes U and E, but also for rays in a good vacuum *as well* as in a tube with sufficient air introduced to produce 'normal' cathode rays in the absence of ultraviolet light. All in all, it could scarcely be doubted that cathode rays are produced by ultraviolet light. In a good vacuum, the photocurrent is carried by cathode rays, pure and simple. In the presence of a gas,

> the cathode rays released from the illuminated surface [of the electrode] are absorbed in the gas and their charge deposited on it; from then on the gas transports the charge on [the original] path at comparatively low velocity. [12-12]

Why the extraordinarily long delay in publishing Lenard's photoelectric results? One reason must have been the nagging old sodium experiment, that first cast serious doubt, not only on Lenard's interpretation of the photocurrent, but also on the German view of the nature of cathode rays in the first place as some kind of ætheral, wavelike vibration in contradistinction to the particulate view of the English school. If the photocurrent was carried by cathode rays, and also consisted of a beam of charged particles, were not, then, the cathode rays a form of charged particles as well? While confronted with this very fundamental and troubling issue, Lenard was unexpectedly burdened with effectively having to take charge of the Institute in Bonn following the untimely death of Hertz in 1894. On top of that, the far from simple task also fell on him of having to edit and oversee publication of Hertz's *Gesammelte Werke*. (His successful completion of the latter task stood him in good stead when he applied for a theoretical post at Heidelberg two years later [12-13].) Late in 1894, moreover, he was offered an extraordinary professorship at Breslau, on the initiative of the Prussian Ministry of Education, that he could ill afford to turn down. In the event, since his duties at Breslau called for teaching physics by supplementing the teaching of the ordinary professor (Ernst Pringsheim), with little or no opportunity for experimentation, he lasted barely a year in his new post. He did better as assistant to Adolph Wüllner with a lectureship at the Technische Hochschule in Aachen. There he managed some experimentation supported by Wüllner, but was again interrupted when compelled to return, in 1896, to his old turf at Heidelberg, where he filled a newly created position as extraordinary professor. Two years later, he settled at Kiev as *Professor Ordinarius* and director of the physics laboratory—a position he would hold for nearly a decade.

Lenard had been enchanted by cathode rays since reading Cookes' celebrated lecture on 'Radiant matter' as a student in 1880. His preoccupation with them dated from 1892, the year he witnessed Hertz's discovery that thin metal leaf transmits the rays, as discussed in section 5.2. Using his own subsequent 'window' technique, Lenard was able to separate the ray production process, which requires a glow discharge in a low but finite gas pressure, from experiments on 'pure' cathode rays in a separate high-vacuum observation chamber; with this arrangement he embarked on a long series of experiments on the properties of cathode rays. In light of his additional familiarity with the photoeffect, he concluded, among other things, that the incident cathode rays are a form of light triggering the emission of charged particles in the thin foil, thereby accounting for the electrical conductivity accompanying any residual air outside the foil. Adding to the considerable confusion surrounding these experiments, some of the effects he attributed to cathode rays were undoubtedly due to x-rays—rays in whose discovery Lenard with some justification felt entitled to a share, as we know. While Röntgen maintained that x-rays and Lenard's 'pure' cathode rays were intrinsically different in nature, Lenard thought not; to him, both rays, in contrast to the photocurrent, were merely extreme manifestations of electromagnetic waves. This he thought despite increasing evidence for the English and French (Perrin's) particulate view of cathode rays, culminating in J J Thomson's isolation of his 'corpuscles' in 1897. Lenard had steadfastly maintained his position as late as the British Association meeting in 1896, but was soon forced to acquiesce to the British view, if only half-heartedly. Cathode rays, he now argued, were still 'ætheral', but 'latent motion of the æther' or 'individual pieces of the æther, hitherto unobserved, which move individually, possess mass (inertia), and seem to be identical to the carriers of electric charge' [12-14].

It was at this critical concurrence of events that Lenard, compelled to publicly modify his long-held stance on cathode rays, resumed his work on the photoeffect, newly ensconced at Kiev. His comprehensive report, beginning with the old sodium experiment at Heidelberg and culminating with the results of the e/m determination completed at Kiev, was submitted to the Vienna Academy in October, 1899. He then embarked on a further two-year program of systematic measurements destined to be among the underpinnings of Einstein's theory of the photoelectric effect in 1905. Dwelling on them would carry us too far astray in our chronicle, though touching on them briefly is scarcely avoidable. For an in-depth discussion of the measurements, their importance, and reception, the reader should consult Wheaton [12-15].

Lenard first took up the relationship between the photocurrent reaching the anode and the potential applied to the illuminated cathode. The measurements were initially performed with the apparatus in figure 12.2, and then with a tube of more flexible geometry. As always, in interpreting the results, the anode is assumed to be at zero potential. No current was observed when the potential was several volts positive. However, with a potential still of about 2 V positive, a small current was recorded, indicating that the charge carriers were not simply liberated from the cathode, but that some were ejected with high enough initial velocity to overcome the positive retarding potential. The current rose as the potential was reduced to zero, and became even greater when the potential was negative; however, it leveled off at a 'saturation' value at a potential of typically 10–15 V negative. This behavior, vaguely understandable in terms of an electrostatic layer of charge building up adjacent to the cathode (J J Thomson had observed similar effects), was nevertheless sufficiently puzzling to warrant a closer study of the transport of charge as a function of bias potential.

The new results, which we skip, were submitted to *Annalen der Physik* in March, 1902. The lasting fame of this very long paper stems from three particular conclusions [12-16]. First, by displacing the ultraviolet light source relative to the tube, Lenard found that the number of negatively charged particles emitted in unit time (the photocurrent) is strictly proportional to the light intensity [12-17]. Second, the null potential in a plot of accelerating potential against photocurrent, a fair measure of the maximum initial velocity of the charge carriers, was quite insensitive to variations in the light intensity [12-18]. Third and finally, for a given cathode material, there is a maximum particle velocity, or kinetic energy, proportional to the retarding potential required to stop the fastest charge carriers [12-19]; however, the velocity is independent of the light intensity. The *frequency* dependence of the particle velocity, it might be noted, was left for others to deduce, first by Erich Ladenburg in 1907 [12-20]. All Lenard could say on this point in 1902 was that the particle velocity depended on the particular light source used; that is, the photoelectric response depended on the spectral composition of the light [12-21].

Lenard's results would remain unexplained until Einstein published the first of his three celebrated, back-to-back journal articles in 1905, 'A heuristic point of view concerning the emission and transformation of light'. That was the article in which, among other things, he formulated the hypothesis of light quanta on the basis of Planck's quantum of action, and applied it in explaining the photoelectric effect [12-22]. In it, he acknowledged and made use of the 'pioneering' groundwork on the subject by Lenard who, as usual, would nourish a perpetual grudge against Einstein for appropriating the photoelectric effect. Einstein interpreted Lenard's results in terms of an interchange of energy between radiant light and matter, suggesting that the interchange takes place in energy quanta of magnitude $h\nu$, where ν is the frequency of the radiation absorbed or emitted, and h is a constant; he further assumed that the whole quantum $h\nu$ of radiant energy was absorbed by one electron, but that a part of it was expended by the electron in escaping from the emitter [12-23].

In fact, to Lenard the role of the light was rather incidental to the photoelectric process; the importance of the effect was as a tool for probing the atomic structure of matter [12-24]. In his view, that for some years was commonly accepted among most physicists, the photoelectric effect is a resonance phenomenon in which the light contributes no energy to the escaping electron (since the velocity at escape is independent of the light intensity or energy) but acts as a 'trigger' in releasing those electrons from inside an atom that are in tune with the frequency of the light and possessing pre-existing mechanical energy of vibration *à la* Zeeman. Thus, the velocity distribution of the emerging electrons somehow reflected the intrinsic, dynamic motion within the undisturbed atom. These considerations led Lenard to his dynamic atom of 1903 in which the atom is pictured as an assemblage of very small, widely spaced 'dynamids'; they had mass and consisted of electric dipoles containing equal charges of opposite sign. (The negative constituents were identified with the negative electrical constituents of cathode rays.) The number of dynamids in an atom was proportional to the atomic weight [12-25].

Lenard's triggering hypothesis held sway until about 1911, by which time it was reasonably clear to almost everybody that the photoelectric effect represents a *transformation* of energy, with the photoelectron energy derived directly from the energy of the incident light. Even so, the hypothesis was discarded, not so much by direct experimental repudiation as by having outlived its usefulness in overcoming various problems plaguing photoemission based on the classical electromagnetic theory of light [12-26]. It suffices to note that, in our present context, the importance of Lenard's photoelectric work is twofold: his independent verification of e/m for cathode rays, and the concomitant role of its derivation in undermining the German interpretation of those rays.

Chapter 13

THE β-PARTICLE

13.1. Are β-rays deflected in a magnetic field?

We last encountered radioactivity in the hands of Ernest Rutherford in 1898. That was the year that saw the abrupt ending of widespread apathy towards radioactivity in scientific circles, following its discovery by Henri Becquerel in early 1896. Unfortunately the discovery came hard on the heels of the announcement of x-rays by Wilhelm Röntgen, and Becquerel's rays were greeted as anticlimactic in the wake of the uproar surrounding those of Röntgen. At the British Association meeting of 1896 in Liverpool, J J Thomson devoted fully five pages in his address to x-rays, and scarcely five lines to Becquerel's rays. One reason radioactivity was dismissed as a largely obscure curiosity was the feebleness of its photographic shadow, in contrast to the startling images obtained with x-rays. Moreover, just about everybody considered Becquerel rays to be a mixture of primary and secondary x-rays of unknown origin [13-1]. This was most definitely the opinion of Rutherford, who announced in his seminal paper of 1899 that the rays of uranium and its salts consist of at least two components: α-rays (the 'secondary' rays), that are readily absorbed in matter (and ionize a gas rather well), and β- (primary) rays of much greater penetrating power (but which are less effective ionizers).

Rutherford had another announcement to make, correcting a serious error then making the rounds. Becquerel had unfortunately reached the spurious conclusion that his uranium rays could be reflected, refracted, and polarized—properties belonging to well-nigh all known ætheral waves—just when Röntgen announced that his rays were *neither* reflected *nor* refracted. This comforting conclusion for uranium rays threw everybody off the right track, save for Rutherford. In his 1899 paper, he reported finding no evidence for refraction of the rays. If anything, his refutation of Becquerel on this point made 'the similarity [with x-rays] complete' [13-2].

Before 1899, however, physicists were by and large preoccupied with searching for new radioactive substances, not with the rays themselves.

The upshot was a virtual deluge of radiators and their radiations. Real or imagined, they ranged from chlorides and sulfides to glow worms and fireflies. In the USA, A F McKissick of the Alabama Polytechnic Institute compiled an impressive, and largely dubious, array of 'active emitters', including sugar, chalk, and glucose. At Japan's Daisan Kotogakko, Kyoto, Hanichi Muraoka's evidence for the photographic effects of glow worms was in due course traced to vapors arising from the need to keep the worms moist! [13-3]. In a word, the field stagnated. Then suddenly, the search bore fruit in 1898 when Marie Curie, opting to work for her doctorate in Paris, set herself to the task of a systematic search for other possible radioactive elements besides uranium, fortified with a particularly sensitive electroscope which owed its success to her husband, Pierre Curie. (Its important feature was a compensating system utilizing a quartz crystal depending on the piezo-electric effect—one on which Curie was also an authority.) Indeed, subsequent progress owed much to the new-found practice of augmenting the tedious photographic (and spectroscopic) exposures with tests for electrical conductivity. Thorium, adjacent to uranium in the periodic table, seemed a good candidate. To her chagrin, Marie found that Gerhard Schmidt of Erlangen had anticipated her, if just barely, in verifying the radioactivity of thorium compounds. Schmidt then called it quits, but Marie, teaming with Pierre, had just begun her legendary quest.

It was already known that pitchblende, an oxide of uranium, was more 'active' than uranium. Might not the activity be 'attributable to some other very active substance included in small amounts' [13-4]? The rest is history, as the saying goes. Obtaining hundreds of kilograms of pitchblende ore from Saint Joachimsthal in Bohemia courtesy of the Austrian government, they isolated, by a Herculean effort, polonium, linked to bismuth (a component of pitchblende), and radium, tied to barium (another component of pitchblende) [13-5]. Substances many thousand times more active than the purest uranium or thorium, their availability sparked a resurgence of interest in radioactivity generally, and a renewed interest in the rays themselves.

An obvious starting point for investigating the rays was the by now time-honored test of their behavior in a magnetic field—magnetic deviability being the irrefutable signpost for a beam of charged particles. The first such experiments on Becquerel's uranium rays were reported to the German Physical Society at its meeting on May 5 1899 by the singular team of Julius Elster and Hans Geitel of Wolfenbüttel. We last met Elster and Geitel in connection with the photoelectric effect (section 10.2), and here is a good place to pause and take stock of this interesting pair. Their lasting friendship dated from when they were fellow pupils at the secondary school in Bad Blankenburg in Braunschweig [13-6]. They studied together from 1875 to 1877 in Heidelberg and then in Berlin, where Geitel took his examination for Gymnasium teaching and then was hired at the Gymnasium in Wolfenbüttel near Braunschweig. Elster received his doctorate under Quincke at Heidelberg in 1879, and, after he too passed his examination to become a teacher, he joined Geitel in Wolfenbüttel. There they shared the same household for much of their adult lives. Elster lived in the home of Geitel's parents until he married, following which Geitel joined the Elsters. Eventually, they built a large house in which they established their private, well-equipped research laboratory ministered by Frau Elster. Here, in the house still standing at Rosenwall 114 in Wolfenbüttel, they pursued their variegated researches until Elster died in 1920, all the while carrying on their teaching duties, beloved by their students. Their research earned them world fame, covering the conduction of electricity through gases and flames, the photoelectric effect, spectroscopy, and atmospheric electricity. The latter work culminated in their classic investigation of the terrestrial origin of radioactivity. Their papers were nearly always joint publications. Individually, each turned down the offer of more than one professorship; they even declined the offer of the Prussian Minister of Education for a prestigious dual university chair!

Elster and Geitel were initially drawn to Becquerel's rays in an attempt to determine the origin of their energy, in a circuitous extension of their main line of research: atmospheric electricity. Crookes had proposed that the air molecules with the greatest velocity stimulated the rays; thus energy was extracted from the surrounding air. Elster and Geitel mounted a piece of uranium in an evacuated glass vessel. Even in the most rarefied atmosphere attainable, the radiation remained constant. Blackening of a photographic plate was also independent of pressure. Thus, the radiation could not be stimulated by air [13-7]. In their earlier work, however, the two physicists had shown that the conductivity in gases produced by incandescent bodies, and by the photoelectric effect, decreased in a magnetic field; presumably the magnetic field deflected the gaseous ions, reducing their number reaching the collecting electrode in unit time [13-8]. Might the conductivity produced by Becquerel rays be similarly affected?

In the event, preliminary tests with uranium were inconclusive, so Elster and Geitel turned to their close friend and nearby colleague Friedrich Giesel for a more potent radioactive source. Giesel was an organic chemist employed at Buchler und Compagnie, a quinine factory in Braunschweig (Brunswick), a major center just north of Wolfenbüttel in Lower Saxony. A physician's son, born in Winzig, Silesia, in 1852, he had studied at the Köngliche Gewerbeakademie in Berlin, and received his doctorate from Göttingen in 1876 [13-9]. He then spent several years as assistant to Carl Liebermann at the Gewerbeakademie. At Buchler & Co., where he was appointed chief chemist in 1878, in addition to his main research in alkaloid and phytochemistry, Giesel became highly adept at concocting radium and radium preparations (though not necessarily

100% pure—a shortcoming that bears on our account) by fractional crystallization as a sort of hobby. His success in this side-line activity, he being then one of few commercial suppliers of radium, was such that he was given the rank of titular professor and of Honorary Doctor of Engineering [13-10].

Elster and Geitel obtained a sample of radium-containing barium chloride from Giesel, that they placed in a shallow aluminum container supported by a platinum wire fused into the bottom of a glass tube. The aluminum container, acting as an electrode, was maintained at a potential of 500 V, and the glass tube evacuated to a pressure of 1 mm of Hg. A second electrode, located above the first, was connected to a grounded Bohnenberger gold-leaf electroscope. A magnetic field could be generated by a horseshoe electromagnet. As in their earlier experiment, they found that in the presence of a field perpendicular to the axis joining the sample container and the collector plate (upper electrode), the rate of discharge of the electroscope was significantly diminished. The diminuition ceased when the field was extinguished. Still, though one more indication that the gaseous ions had been repelled from the plate by the magnetic field, they could not rule out that the Becquerel rays themselves were also deflected by the field; that is, if the Becquerel rays were charged particles, they too should experience magnetic deflection in the rarefied atmosphere.

To check out this possibility, Elster and Geitel mounted a phosphorescent screen of barium platinocyanide between the aluminum source container and the collector plate, and sealed the source off with a 1 mm thick airtight aluminum plate between the source and the screen. With this arrangement, the radium rays, well collimated into a pencil-thin beam by the orifice in the aluminum container, produced a bright spot on the screen. Behold, the spot was not noticeably affected by the magnetic field. If deflected at all, the rays were deflected much less than cathode rays, showing once more that Becquerel rays 'agree in this respect too—as in all other characteristics known to date—with the Röntgen rays' [13-11].

A few months later, Stefan Meyer, a *Privatdozent* at the University of Vienna, had his interest aroused in radioactivity while attending a demonstration of radium and polonium by Giesel at the Deutsche Naturforscherversammlung, held in Münich during September 1899 [13-12]. At the time, Meyer was studying the magnetic susceptibility properties of the chemical elements, and asked Giesel if he might borrow samples of the new materials for the same purpose. Giesel readily obliged, lending him 2 g radium–barium chloride. Meyer also obtained a small quantity of polonium preparation from Pierre Curie in Paris. Collaborating with his colleague Egon Ritter von Schweidler, then a fellow *Privatdozent* and lecturer in physical chemistry at Vienna, he lost no time in performing magnetization measurements as well as repeating the negative magnetic deflection experiment of Elster and Geitel. Their results were first presented by Ludwig Boltzmann, Meyer's mentor at the University, to the Royal Viennese Academy of Sciences in early November [13-13].

In repeating the Elster–Geitel experiment, Meyer and von Schweidler had the essential advantage of a much stronger electromagnet at their disposal, one capable of no less than 17 000 G. (Elster and Geitel did not specify the strength of their magnet, but retrospective scrutiny of their results indicates that it was good for, at most, 2 000 G—not strong enough, or at best marginal, for this experiment utilizing a very short (about 15 mm) spacing between source and screen [13-14].) They placed Giesel's radium sample, wrapped in paper, beneath an open, grounded brass tube connected to an electrometer. The apparatus was at atmospheric pressure. With the field applied, the rate of discharge of the electrometer fell, much as Elster and Geitel had observed at low pressure. Next, with the radium in the field but the rest of the apparatus outside the field, the rate of discharge fell even more, showing that the emission itself was affected by the field—either by retarding the emission or deflecting the rays [13-15]. The first possibility was eliminated by the observation that the luminescence produced in the wrapping paper by the radium (and observable in the dark) was about the same with the field on or off. The second possibility was shown to be the correct one by using a barium platinocyanide screen similar to that of Elster and Geitel. With the screen 1 cm in front of the wrapped radium specimen and the field applied, the fluorescence on the screen disappeared completely. With the (wrapped) radium placed directly on the screen, but without applying the field, a bright, sharply defined image of the sample was cast on the screen. With the field applied, two bright images appeared on either side of the original image, as if the rays had been emitted from both the top and the bottom of the sample and both rays bent back onto the screen—rays emitted from the top bent downward, and rays from the bottom bent upward. The direction of curvature of the bending in the field of known polarity indicated that 'the rays behave quite analogous to cathode rays' [13-16].

Unbeknownst to Meyer and von Schweidler, Giesel himself, the generous supplier of radioactive preparations to both the Wolfenbüttel and Vienna teams, also decided to check out the negative deflection results of Elster and Geitel upon reading their report. His experiment appears in Wiedemann's *Annalen* for December 1899. Like Meyer and von Schweidler, he did not use a vacuum, and he too had access to a stronger electromagnet in his industrial laboratory of Buchler and Co. Of particular interest, as we shall see, he used polonium as well as radium. (Polonium was also subsequently used by Meyer and von Schweidler.) With a freshly prepared polonium sample mounted in a small vessel between the poles of a vertically oriented U-shaped electromagnet and

~ 1 cm below a phosphorescent screen lying on the pole-tip surfaces, the luminous spot on the screen developed a comet-shaped tail on one side with the magnet energized. On reversing the poles, the extension of the spot was in the opposite direction. The rays were, obviously, deflected by the field. Similar, though less distinct, effects were observed with radium, even with the very sample borrowed by Elster and Geitel the previous spring. Unfortunately, however, Giesel failed to determine the sign of the charge of the electrified particles presumably involved, something his colleagues in Vienna had gone to some pains to do [13-17].

In spite of the latter shortcoming of his experiment, priority in detection of the magnetic deflection of Becquerel rays goes to Giesel, if only by a hairsbreadth, since his paper was received by *Annalen der Physik und Chemie* on November 2 1899, one day before the corresponding report was submitted by Meyer and von Schweidler to the Vienna Academy. This also in spite of the fact that the Austrian paper appeared in print before Giesel's (in *Physikalische Zeitschrift*, a brand new journal of the German Physical Society). In a letter to Meyer and von Schweidler, Giesel claims to have observed the magnetic deflection with both polonium and radium on October 21, while Meyer, in retrospect, claims that he and von Schweidler obtained their first results 'more than a week' prior to November 3, or on or about October 27 [13-18].

However, the story does not end there. As noted, Giesel's first radioactive source in this investigation was a polonium sample of his own. Now, pure polonium turns out to be an α-emitter only, and the α-particles are too energetic to have been sensibly deflected in Giesel's experiment (and too absorbable to have penetrated the paper wrapped around the sample). His polonium sample contained at least one β-emitting contaminant. A hint of the problem accompanying Giesel's anomalous discovery is apparent in his finding that the deviable activity of his freshly prepared polonium sample faded in a matter of days [13-19]. Moreover, in subsequent tests by Meyer and von Schweidler with polonium obtained from the Curies, *not* from Giesel, no hint of deviable rays was seen. The culprit in Giesel's polonium, Giesel himself found in 1906, was RaE (^{210}Bi), a β-emitter with a half-life of 5 d [13-20]. Pure polonium, Curie's speciality, has a half-life of 139 d and emits no β-particles.

13.2. A controversy brews

How did it happen that Giesel, a chemist, more or less beat Elster and Geitel, experienced physicists, to the magnetic deflection of β-rays, albeit somewhat fortuitously? Some light is shed on the actual circumstances by a contemporary exchange of letters between Giesel, Elster, and Stefan Meyer. The correspondence was reprinted by Meyer, interspersed with helpful commentary, on the 50th anniversary of the experiments, 1949

[13-21]. A few excerpts follow. (Bracketed material is from Meyer's commentary.)

> *Giesel to Meyer, 10 October 1899*: on Meyer's request, Giesel is dispatching 2 g of radium chloride on loan to him, wishing him good luck on his forthcoming elemental magnetization studies.

[The purity of the radioactive preparation so kindly supplied, Meyer cautions, was scarcely better than 1%. Nevertheless, following some further exchange of letters between the two, Meyer and his colleague von Schweidler obtained positive results on the magnetic deflection of Becquerel rays from one of Giesel's radium samples, in contradistinction to the negative experiment of Elster and Geitel with a similar preparation the previous spring. Their results were communicated to the Vienna Academy on November 3 and 9. In the meantime came yet another letter from Giesel—and a rude shock, it would seem.]

> *Giesel to Meyer, 22 October*: Giesel is turning out polonium samples in record amounts, yet has another piece of news to share with his Vienna colleagues 'in strict confidence'. To wit: he has observed a noticeable magnetic deviability in polonium as well as radium rays—a result perhaps borne out in the Viennese observations (?)

> *Meyer to Giesel, 24 October*: Using the radium preparation generously provided by Giesel, he and von Schweidler have found, among other things, that the rate of discharge of electrified bodies is strongly retarded by a magnetic field in air, something already shown by Elster and Geitel at low pressures. Moreover, by excluding all but the radium source from the field, they conclude that the emission itself is affected by the field, not only the gaseous ions; either the emission is diminished, or the rays are deflected. They have submitted a short note to the Vienna Academy, and are also writing to Elster and Geitel, trusting all three parties can come to an agreement on how to pursue their respective experiments without stepping on each other's territory.

> *Elster to Meyer and von Schweidler, 26 October*: Elster and Geitel assure Meyer that he need not worry about interfering with the Wolfenbüttel program, drawing at the same time attention to Giesel's 'most remarkable observation concerning Becquerel rays in a magnetic field'. They are encouraging Giesel to publish his observations without delay in the *Physikalische Zeitschrift*, thereby also leaving Meyer and von Schweidler a free hand vis-a-vis Giesel's researches.

> *Giesel to Meyer, 26 October*: Giesel is pursuing the matter with photographic plates, as well as the phosphorescent screen, and

asks that Meyer leave this aspect of the investigation to him [13-22]. He also notes that Meyer and Elster–Geitel are in accord on their respective lines of investigation. For his part, Giesel remains fully occupied with polonium, a substance notoriously difficult to prepare. Not only that, the material allows little time for experimentation, since 'its activity vanishes and cannot be recovered again'.

[It is clear from Giesel's letter, warns Meyer, that Giesel, who routinely wrapped his preparations in black paper, could not have observed any polonium α-rays. Instead, his source undoubtedly contained an admixture of RaE which has a half-life of about 5 d, compared to ca. 140 d for pure polonium. This also explains why unwrapped radium and polonium exhibited the same activity in a magnetic field.]

Giesel to Meyer, 4 November: Giesel notes with interest Meyer's communication to the Vienna Academy the day before, and is dispatching several more samples, including some freshly prepared polonium. He is quite anxious to pin down any difference in magnetic behavior between Curie's samples and those of his own—in particular, what effect, if any, Curie's polonium has on the phosphorescent screen. Thus, 'Elster and Geitel have found that my polonium preparation, which scarcely affects the screen any more, still exhibits some discharging ability'. By the way, he adds, it appears that Elster and Geitel no longer intend to pursue the behavior of the rays in a magnetic field. Nor does Giesel himself, due to the press of other matters. Consequently, Meyer's ongoing observations take on added importance. Giesel has already submitted his own results on the magnetic deflection to Wiedemann's *Annalen*, and will be glad to lend Meyer more active material if needed.

[From the foregoing, claims Meyer, it should also be evident that already then he and von Schweidler were aware of the difference in activity between unwrapped pure polonium, which emits α-rays, and the penetrating action of β-rays from radium.]

Elster to Meyer and von Schweidler, 3 November: Elster confirms that he and Geitel are giving up after one more unsuccessful attempt to observe deviable Becquerel rays, since their experimental arrangement has proved 'inadequate for the task'. For what it is worth, their experiment appears to show that Becquerel rays in a magnetic field, whether from (wrapped) polonium or radium, behave as a swarm of *positively* charged corpuscles, in contrast to cathode rays. Elster intends to report on the matter before a meeting of the Verein für Naturwissenschaften in Braunschweig on November 16.

[By now, however, adds Meyer, it was abundantly clear to him and von Schweidler that negative, not positive, particles were involved in these experiments with wrapped radium sources. On learning of Elster and Geitel's latest, once more spurious conclusion just then, he 'forestalled the erroneous announcement' intended for Braunschweig on November 16 with a telegram to Wolfenbüttel putting matters straight. He also dispatched a detailed account of their own results, including excerpts from their report to the Vienna Academy published in the *Physikalische Zeitschrift* for December [13-23]. From Wolfenbüttel came the following reply.]

> *Elster to Meyer and von Schweidler, 12 November*: After considerable effort involving observing the luminous patch on the phosphorescent screen as a function of the relative positioning of source, screen, and a lead block, and even substituting the cathode of a Röntgen tube for the radium source, they admit to agreeing with Meyer's finding; the Wolfenbüttel failure boiled down to an erroneous interpretation of the direction of propagation of the rays in the magnetic field (!)

> *Giesel to Meyer, 18 November*: Giesel is still awaiting details on the performance of Meyer's polonium source acquired from the Curies; in particular, does it retain its power of emitting deviable rays?

[The source in question, Meyer assures us, was sent to Giesel without delay. Unfortunately, no sooner done than a priority disagreement arose, which he attributes to the sensitivities of Giesel and his shortcomings as a physicist, coupled with the excessive modesty of Giesel's advisors Elster and Geitel, as evident from Elster's letter to Meyer and von Schweidler dated December 1. In it, Elster eschews credit in connection with Giesel's successful experiment, as might be inferred from the tone of Meyer's reference to the experiment in his latest communication to the Vienna Academy. In Meyer's opinion, he concludes his 1949 ruminations, Giesel owed his success in large measure to helpful interactions with his Wolfenbüttel and Vienna colleagues. To which we might add, the resources of a well-equipped industrial laboratory, including a powerful electromagnet (13-24)].

13.3. One more electron, but with a difference

Not long after Giesel and his Austrian competitors completed their experiments on the magnetic deflection of Becquerel rays, Henri Becquerel himself resumed his own work on radioactivity, after an unfortunate hiatus of a year and a half. During that lackluster interval, he had turned to the Zeeman effect in the company of H A Deslandres. What turned him back to radioactivity seems to have been his role as member of the Academy of Sciences. In that capacity, he became the intermediary through which the papers of Marie and Pierre Curie reached the Academy [13-25]. In the course of this interaction, the Curies would lend him radium and polonium samples from time to time.

With powerful new radioactive sources in hand, Becquerel wasted little time. For a start, he repeated his earlier unfortunate experiment on the reflection, refraction, and polarization of the rays. This time he was unable to confirm his previous, spurious results—a conclusion he reached in March 1899, only two months after the appearance of Rutherford's seminal paper in the *Philosophical Magazine* that had questioned Becquerel's original results in the first place. The existence of effects analogous to secondary radiation, Becquerel now decided, 'accentuated the close tie between the uranium and the x rays' [13-26]. Others were no more able to confirm his previous results, so that from that time it was generally taken for granted that Becquerel rays were incapable of reflection, refraction, and polarization. In effect, they behaved optically more like x-rays than light [13-27].

Next, Becquerel undertook a comparison of the behavior of various phosphorescent minerals exposed to radium rays and x-rays. Finding that the two types of radiation produced quite different effects, he at first attributed the difference to the rays having different wavelengths. Yet in some respects the two rays exhibited similar effects; thus both lengthened the phosphorescence period of certain substances, and both colored glasses (their greatest industrial asset). It seems likely that these studies led Becquerel to his tests of whether the rays from radium or polonium were deflected in a magnetic field, though he seems not to have harbored strong views on an 'x-ray theory' of radioactivity [13-28]. Moreover, Becquerel was aware that the Curies had tried to deflect uranium rays in a magnetic field [13-29], no doubt in keeping with Marie Curie's suggestion early in 1899 that the rays might be material in nature [13-30].

Unaware of the just-completed experiments in Braunschweig and Vienna, Becquerel began his own experiments on the behavior of radium rays in a magnetic field toward the end of 1899. He submitted his first report to the Paris Academy on December 11, and a second installment followed on December 26 [13-31]. He tested his preparations in fields of an electromagnet varying between 4 000 and 10 000 G, employing both the usual phosphorescent screen and photographic plates. (Dry photographic emulsion plates, in contrast to the earlier, cumbersome wet collodion plates, were commercially available by now.) His results were similar to those of Giesel in showing, among other things, the effect of a magnetic field in modifying the image of a radioactive source attached to a photographic plate placed horizontally between the poles of the magnet. In a non-uniform field, the radiation was concentrated at the poles, akin to the behavior of cathode rays in the experiments of the Norwegian physicist Kristian Birkeland. However, almost at once

Becquerel established unambiguously two additional experimental facts: that rays from polonium were *not* deviated, and that rays from radium or uranium contained one component that *was* deviable in a magnetic field, namely (it was soon established) Rutherford's β-rays, and another (the α-rays) that was not. This distinction, deviable or not, he emphasized, was much more important than the mere difference in *absorption* or penetrating power espoused by Rutherford. Magnetic deviability being an undisputable test for particulate behavior, identifying the deviable component of the rays with x-rays was ruled out once and for all [13-32].

Just how entrenched the skepticism was toward the notion that Becquerel rays might contain material particles is apparent from the reluctance of Rutherford himself, along with J J Thomson, to accept the notion, because of the great penetrating power of β-rays. In June 1899, now installed at McGill University in Montreal, Rutherford had written Elster and Geitel as follows.

> *12 June 1899*
>
> I was pleased to see that you had investigated the effect of a magnetic field at low pressure on the rate of discharge. Your results are exactly what one would expect on the theory of ions. [13-33]

By the end of the year, however, news of the latest experiments abroad, at least from Giesel, had reached Cambridge, and in December Thomson wrote Rutherford that

> *21 December 1899*
>
> I see Giesel makes out that the radiation from polonium is affected by a strong magnetic field. If this is so it might be worth trying whether your emanation were so affected. [13-34]

Shortly into the new year, upon hearing word of the additional magnetic deflection experiments in Paris, Rutherford replied to Thomson:

> *9 January 1900*
>
> The results of Giesel and Becquerel are very interesting and remarkable. I expect the 'emanation' in thorium is also true for polonium and radium when prepared in a special manner, and that the deflection due to the magnetic field is due to the action on a charged particle cast off from the active body. [13-35]

Following Rutherford's path-breaking experiments on radioactivity at the Cavendish, the subject had been all but dropped in England. The English were at a disadvantage, not having ready access to the potent sources of radium and polonium becoming available on the Continent. Rutherford had had to make do with much weaker preparations of uranium and thorium [13-36]. Rayleigh relates a visit by Becquerel to the Cavendish as late as March 1902 when he demonstrated a sample of radium to the thrill of the assembled laboratory staff [13-37]. Becquerel, he adds, developed a serious lesion on his stomach beneath the waistcoat pocket in which he usually carried his radium sample (as did Crookes, for that matter).

Once at McGill, however, there was no doubt in Rutherford's mind on what to take up, and who were the principal players back on the Continent. As he confided to his mother, albeit well after his new research program was in full swing at the Macdonald Laboratory,

> *5 January 1902*
>
> I have to keep going, as there are always people on my track. I have to publish my present work as rapidly as possible in order to keep in the race. The best sprinters in this road of investigation are Becquerel and the Curies in Paris, who have done a great deal of very important work in the subject of radioactive bodies during the last few years. [13-38]

The first order of business was to order from W H Hayles, chief lecture assistant at the Cavendish, some preparations of uranium and thorium. With them in hand, and with his electroscopes in working order, he set to work. The first fruit in Rutherford's renewed attack on radioactivity was thorium 'emanation' (a deliberately vague term of his own), evidence for which had, unbeknownst, bedeviled the earlier work in Cambridge. On looking back on those early experiments, J J Thomson recalled that 'those on uranium gave comparatively little trouble; the results got on one day could be repeated on the next' [13-39].

> The behavior of thorium, on the other hand, was most perplexing and capricious; changes in the surroundings which seemed quite trivial, such as opening a door in the workroom, produced a very large diminuition in the radiation, while, on the other hand, very large changes in the physical conditions produced no appreciable effect upon it. The radio-activity seemed to act like a contagious disease and infect solid bodies placed near the thorium. These, however, recovered in time if the thorium were taken away. These vagaries turned out, however, to be an illustration of the principle that difficulties in the experiments may be the seed of great discoveries, for Rutherford, when he went as Professor of Physics to Montreal, resumed these experiments and, in his attempts to unravel their intricacies, was led to a discovery of fundamental importance which was the origin of modern views about the processes going on in radio-active substances. [13-40]

The discovery in question was that thorium 'casts off' (as Rutherford put it), in addition to at least two kinds of radiation, radioactive gases—the skittish culprit at work, and something also seen by the Curies. This evanescent emanation wafting away with room air currents, Rutherford soon found, not only could make other bodies radioactive if they came in contact with it, causing 'excited' or 'induced' radioactivity, but also the activity decreased exponentially with time [13-41]. In effect, this marked the discovery that uranium and thorium atoms were not indivisible after all, but disintegrated (by the expulsion of radiation) down a chain of transformations; each radioactive fragment could be identified by its half-life, or the time it took to lose half its strength from a given level of activity.

Interesting indications of the immediate, world-wide impact of Rutherford's first two papers on the thorium emanation, published in January and February 1900, particularly in light of the leisurely pace of trans-oceanic mails ca. 1900, are found in letters from former fellow students and mentors alike. Thus, John Zeleny, then equally removed from Cambridge at the University of Minnesota, wrote that

> I read your papers with great interest. I am about ready to believe that most anything is possible. [13-42]

From Callendar, whom Rutherford succeeded at McGill and now securely installed at University College, London:

> I see you are still working at those fascinating rays which promise so much insight into the nature of things. [13-43]

From Trinity College, Dublin:

> We were very much interested in your thorium experiments [wrote FitzGerald]. There seems no doubt that there is some emanation from the thorium.... I have got some thorium but we are all too lazy here to do experiments and indeed between National Education Boards, Veterinary College Boards, Technical School Boards, etc., etc., one gets sick of doing anything. [13-44]

Not quite so lazy, but no more successful, was R J Strutt at Cambridge.

> I have been trying to repeat your experiments on 'induced radioactivity' without much success.... I think my thorium oxide must be at fault.... I am writing to ask if you could send me some thorium which you know to be efficient. [13-45]

(We will have more to say presently about Strutt, then a student at Cambridge, in connection with the non-deviable rays, the α-rays.)

Despite the exuberant letters from afar, the early days at McGill were not always easy for Rutherford, being 'the only worker in the field of excited radioactivity in the English-speaking world' [13-46]. None have put it better than Norman Shaw.

> When Rutherford was working on the detection and isolation of the numerous members of the radium family and developing the theory of the disintegration of matter, there were several occasions

when colleagues in other departments gravely expressed the fear that the radical ideas about the spontaneous transmutation of matter might bring discredit on McGill University! At one long-remembered open meeting of the McGill Physical Society he was criticised in this way and advised to delay publication and proceed more cautiously. This was said seriously to the man who had probably allowed fewer errors to creep into his writings and found it less necessary to modify what was once announced than any other contemporary writer. At the time, he was distinctly annoyed and his warm reply was not entirely adequate, for in his younger days he sometimes lost his powerful command of ready argument, when faced with unreasonable or uninformed criticism. [13-47]

While Rutherford was forging ahead (or, equally to the point, preparing for a well earned sojourn in New Zealand to marry Mary Newton in the midst of the experimental excitement at McGill), things were by no means quiet in Paris. Already in January 1900, Pierre Curie had shown, nearly simultaneously with Becquerel, that only the β-rays were sensibly deflected by a magnet [13-48]. Soon he and Marie also found that the β-rays carried a negative charge, a result reported to the Academy in March [13-49]. For the next development, we switch to Ernst Dorn in Halle. Dorn, born in Guttstadt, East Prussia (now Dobre Miasto, Poland), had been educated at the University of Königsberg, and then taught physics at the universities of Darmstadt and Halle. Shortly after the discovery of the thorium emanation, he showed that radium compounds also give off radioactive emanations. To his annoyance, Rutherford paid him skant attention, as we learn from J J Thomson.

Cambridge, 15 February 1901

I had a letter from Dorn lately, complaining that you had not sufficiently acknowledged his work in your paper, and wanting me to communicate with the Royal Society a paper asserting his claims. I wrote to him that all that had appeared was an abstract, and that I was sure he would find when he read the complete paper that you had suitably acknowledged his work. [13-50]

Of more immediate interest from our vantage point was Dorn's success in deflecting β-rays in an electric field [13-51]. In so doing, he set the stage for determining the charge-to-mass ratio as well as the velocity of the rays. The first such determination was reported by Becquerel to the Academy in January and March 1900, and again discussed by him before the International Congress of Physics held in Paris that spring on 'The new radioactive substances and the rays they emit' [13-52]. In view of its relevance for our chronicle, we pause and dwell briefly on this particular experiment.

The method employed by Becquerel was basically the same as J J Thomson's classic e/m determination for cathode rays in crossed electrostatic and magnetic fields, and Becquerel's procedure has been analyzed by Thomson himself in some detail in his equally classic monograph of 1903 [13-53]. The experiment was performed entirely at atmospheric pressure, allowable in this case because of the high penetrating power, and hence great range, of the rays in ponderable matter. The electrostatic deflection was found by measuring the displacement of a shadow cast on a photographic plate of the β-rays passing between a pair of metal plates charged to ca. 10 000 V; this yielded an expression for the relationship between the electric force on the particles, $F = Ee$, their velocity v, e/m-ratio, and their displacement parallel to the electric field lines, say y:

$$y = \tfrac{1}{2}\frac{eE}{mv^2}\, l(l + 2h)$$

where E is the electric field strength, l is the length of the path between the plates, and h is the vertical distance of the photographic plate above the upper edge of the metal plates. For the magnetic deflection, rays from a radium sample in a lead saucer placed on a photographic plate were bent by a strong magnetic field through a semicircle of radius R back onto the plate, producing an elliptical image in it. Measuring the coordinates of the loci where the rays struck the plate gave a second relation between e/m and v in terms of the magnetic force, or

$$x = \tfrac{1}{2}\frac{eB}{mv}\, l(l + 2d)$$

where x is the distance of the image from the radium, and B is the magnetic induction. (These expressions are discussed in some detail in section 15.1.) From these two results, Becquerel determined that the ions from the radium emanation moved with a velocity of (at least) 1.6×10^{10} cm s^{-1} with a ratio $e/m = 10^7$ emu g^{-1}. This ratio was in excellent agreement with that for the negative ions in the cathode rays and in Lenard's rays, as well as for those produced by ultraviolet light. Clearly, a constituent of the Becquerel rays was none other than J J Thomson's corpuscles, or electrons—albeit very fast electrons at that [13-54].

Electrons of the ordinary type they may have been, but the velocity of the β-particles, more than half the velocity of light, conferred on them a special status. One who appreciated this fact was Walter Kaufmann who, we recall, had, while in Berlin in 1897, performed an independent determination of e/m for cathode rays. Subsequently, during his Göttingen period, 1899–1902, he took the matter up again, supplied with several samples of radium chloride from the Curies, and intrigued with the notion that the 'apparent' mass of a moving charge should increase with its velocity—a concept usually ascribed initially to J J Thomson, as we also recall. Moreover, by this time the electron theory of H A Lorentz had yielded the theoretical dependence of mass on velocity usually associated with Einstein's third famous paper of 1905, even though in 1900 it was still not certain that the speed of light was the maximum attainable speed for any object [13-55]. Nor could the electron mass be measured with great accuracy; in fact, it could only be determined indirectly by the connivance of simultaneously measuring e/m and e. But why not let Kaufmann state the problem at hand, citing the opening paragraph of his paper of 1901 [13-56].

> 1.) The question as to whether the 'mass' of the electron calculated from the experiments on cathode rays or from the Zeeman effect is the 'true' or 'apparent' mass has recently been discussed quite extensively, although no direct experiments have yet been proposed in this direction. Now investigations into Becquerel rays have shown that these are deflected by electric and magnetic fields, and a rough measurement has given values for ϵ/μ (ϵ, charge; μ mass) as well as for the velocity v, which are of the same order of magnitude as for cathode rays. It must therefore be all the more striking that the Becquerel rays are quantitatively so different from cathode rays. The magnetic deflection of the former is much smaller and their ability to penetrate solids much larger than the latter. Since previous experiments on cathode rays have shown that with increasing speed the deflectability decreases and the penetrability increases, it was reasonable to conclude that the Becquerel rays have much higher speeds than the cathode rays. If the cathode rays have speeds anywhere from 1/3 to 1/5 the speed of light, we must assume that the Becquerel rays have speeds only slightly different from that of light. It is impossible for these rays to exceed the speed of light, at least in a path length large with respect to the size of the 'electron' (as these ray particles are now called) because during such motion energy is radiated until the speed is reduced to the speed of light.
>
> 2.) The purpose of the following experiments is to determine the speed as well as the ratio ϵ/μ as accurately as possible for Becquerel rays and also from the degree of dependence of ϵ/μ on v to determine the relation between 'actual' and 'apparent' mass.

Kaufmann, being a highly competent experimenter, used *simultaneous* electrostatic and magnetic deflections, a procedure even more in the manner of J J Thomson, but one we need not pursue in detail here. (It is essentially the field arrangement used by Thomson in his subsequent positive ray studies, as discussed in chapter 15.) Essentially, his method gave a curve as an image on a photographic plate, each point of which corresponded to a definite velocity and definite e/m value of β-rays emerging from a speck of radium bromide. He thus obtained in a

Table 13.1.

v (cm s^{-1})	e/m (emu g^{-1})
2.83×10^{10}	0.63×10^7
2.72×10^{10}	0.77×10^7
2.59×10^{10}	0.975×10^7
2.48×10^{10}	1.17×10^7
2.36×10^{10}	1.31×10^7

single exposure a series of observations from which the values of e/m for different velocities could be read off. Five pairs of values were singled out for analysis, and are here reproduced from table 34-1 of his paper.

Two conclusions were reached. The first is quite evident from table 13.1, namely, that, as the velocity of the electrons increases, e/m decreases. Since, moreover, the charge e is presumably constant, the mass of the electrons also increases with velocity. Kaufmann's second conclusion was reached with the aid of an expression for the 'apparent' mass (mass by virtue of the motion of the electron) as a function of $\beta = v/c$ (where c is the speed of light) based on a model of the electron by G F C Searle. (In Searle's model, the electron is distributed over an infinitely thin spherical shell.) 'The apparent mass', he declared, 'is of the same order of magnitude as the true mass [rest mass] and for the two fastest Becquerel rays the apparent mass is appreciably larger than the true mass'. Indeed, from a plot of the ratio of apparent to true mass (m/m_0) against v/c, which appears at the very end of his paper, it is dramatically borne out that, as v approaches the velocity of light, the apparent mass approaches infinity. The approach to infinity goes approximately as $mc^2/\sqrt{1-(v/c)^2}$, a conclusion that has withstood the test of time since 1901.

The supposedly non-deviable component of Becquerel's rays, the α-particles, took longer to subjugate. A hint of their peculiar nature came from absorption experiments by the Curies in 1900. First, Pierre Curie found that the detectable ionization produced by the rays from polonium or radium ceased abruptly after traversing a characteristic distance in air, quite at variance with Rutherford's exponential results for uranium [13-57]. On the very same day, Marie Curie found that, within the characteristic distance, the ionization current decreased more quickly with distance from the source than required by Rutherford's exponential law $I = I_0 \exp(-\lambda d)$, or that the absorption coefficient λ *increased* with distance. This singular behavior of the non-deviable rays, Marie Curie noted, differed from that of any other known radiation [13-58].

The hint was not lost on R J Strutt at the Cavendish. In his opinion, the α-rays were high-velocity, positively charged ions of atomic mass, particles which overcame the deflection in a magnetic field purely by virtue of their great momentum. Why positively charged? Perhaps, suggests Heilbron, simply because no positive ions of electronic mass had yet been uncovered [13-59]. One who was persuaded was Rutherford. It appears that he began his attempts to deflect the α-rays in a magnetic field some time in 1901, but was stymied for two years due to the exceedingly feeble radium preparations in hand—only 1000 times as active as uranium. However, in late 1902 he obtained from the Société Centrale de Purduits Chimiques, through the kindness of Marie Curie, some radium stronger than samples usually sold; namely, a preparation of radium 19 000 times as strong as uranium. In addition, he substituted for his usual ion collector and quadrant electrometer a simpler but highly sensitive gold-leaf electroscope cleverly incorporated into his collector, as we shall see. These instruments, we might note, electrometer as well as electroscope, were the mainstay of the physics laboratory at the turn of the century for measuring the rate of leakage of charge from ionized air and gases. Both were exploited by Rutherford through the first half of his career [13-60]. At any rate, these improvements did the trick, though the α-ray experiment was far from easy. In fact, confided Rutherford to Thomson, 'it was the most difficult piece of work I have tackled for some time' [13-61]. Just how difficult, Rutherford himself explains.

> The smallness of the magnetic deviation of the α rays, compared with that of the cathode rays in a vacuum-tube, may be judged from fact that the α rays, projected at right angles to a magnetic field of strength 10 000 c.g.s. units, describe the arc of a circle of radius about 39 cms., while under the same conditions the cathode rays would describe a circle of radius about 0.01 cm. [13-62]

In Rutherford's experiment, shown in figure 13.1, the α-rays from a thin layer of his radium preparation passed upward between a number of parallel metal plates, and then through a thin aluminum foil into the testing vessel housing the previously charged electroscope. The ionization produced by the rays in the vessel was measured by the rate of movement of the gold leaf as the electroscope was discharged; the rate of sink of the leaf was observed through a mica window in the vessel by means of a low-powered microscope fitted with a micrometer scale in the eyepiece. (The time taken for the edge of the gold leaf to pass over a fixed part of the scale had to be corrected for the natural leakage of the instrument, determined before the introduction of the active material.)

A magnetic field was then applied, causing the α-particles to swerve sideways into the plates, thus not reaching the electroscope. However, finding that the largest electromagnet in the physics laboratory was only able to deviate about 30% of the rays, Rutherford turned to Professor R B Owens of the Electrical Engineering Department for help. (Owens, on a Fellowship from Columbia University, and Rutherford began their McGill appointments at about the same time.) Together they partly

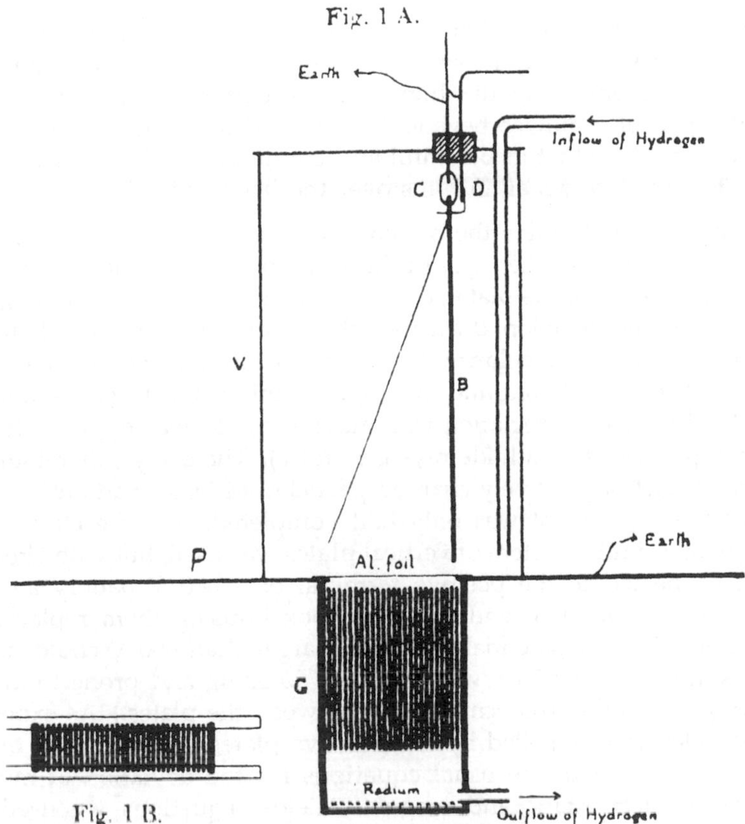

Figure 13.1. *Rutherford's α-ray detector. α-rays from the radium source at the bottom of the container G passed upward between a number of vertical metal plates, and through a thin aluminum foil, into the testing chamber V, made of brass and containing a gold-leaf electroscope. The gold leaf hangs at an angle with respect to a rod B, insulated by a sulfur or amber bead C inside the vessel. The rod is charged by a movable rod D. After charging, the upper, vertical rod is connected to the case of the instrument and to earth. Ionization by the α-particles discharged the gold leaf, causing it to sink. The rate of sink, observed with a microscope having a micrometer eyepiece, was a measure of the ionization. Hydrogen blown in the tube at the top swept back 'emanation' from the radium. P is a lead plate, shielding the electroscope from the radium rays.*

dismantled the largest dynamo in the electrical engineering laboratory, replacing its pole pieces with more suitable appendages. This makeshift contraption, generating a field of 8370 G, proved adequate to throw all the α-particles into the walls of the slotted box, signaled by a large reduction in the rate of discharge of the electroscope.

Eve quotes Frederick Soddy on the occasion of the first magnetic deviation of the α-rays:

I recall seeing [Rutherford] dancing like a dervish and emitting

> extraordinary imprecations, most probably in the Maori tongue, having inadvertently taken hold of the little deviation chamber before disconnecting the high voltage battery and, under the influence of a power beyond his own, dashing it violently to the ground, so that its beautiful and cunningly made canalization system was strewn all in ruins over the floor. [13-63]

In order to determine the direction of the deviation, and hence the sign of the charge of the rays, Rutherford covered half of the openings between the tops of the plates with little brass lids. When the magnetic field was in one direction, the α-particles were caught by these half-lids; on reversing the field polarity, they slipped through the openings, and the gold leaf would continue to drop. In this way, it was established that the direction of deviation in a magnetic field was opposite to that of the β-particles (or cathode rays generally). The α-rays, therefore, did indeed consist of positively charged particles, as Strutt had anticipated.

But the experiment was only half completed. For the electrostatic deflection, a similar system of vertical plates was used, but with alternate plates connected to the positive terminal of a 600 V battery and the other plates grounded, and the brass box housing them replaced by one of ebonite. (A potential difference larger than 600 V could not be used, since the α-particles were strongly ionizing and prone to initiate sparking across the 0.055 cm air gap between the plates.) As expected, the particles were repelled by the positive plates, and attracted by the negative ones. With the usual equations for the deviation of moving charges in electrostatic and magnetic fields, equations we need not repeat, Rutherford deduced that the velocity of the α-particles was about 2.5×10^9 cm s^{-1}, or about one-tenth of the speed of light, and that their e/m ratio was approximately 6×10^3 emu g^{-1}, indicating that the particles were atomic in size—either hydrogen or helium ions. The rays bore a canny resemblance to Goldstein's *Kanalstrahlen* (section 5.1), although they were considerably faster.

The above e/m apparatus is still on display at McGill University [13-64]. An improved determination several years later, using a homogeneous source of α-rays such as radium C (i.e. a source from which all the α-particles escape with about the same velocity) and a photographic method, showed that the e/m ratio was half that for hydrogen ions in electrolysis, or 5.1×10^3 emu g^{-1}, which strongly suggested identification of the particles with doubly ionized helium [13-65]. In a series of experiments reported two years later, Rutherford and Hans Geiger measured directly the number of α-particles emitted per gram of radium and the total charge carried by them. Dividing the charge by the total number of particles gave for the charge per α-particle 9.3×10^{-10} esu, which corresponds to two elementary charges. (Thomson's value for e, section 10.1, was 6.5×10^{-10} esu, compared with the modern value of 4.8×10^{-10} esu.)

> On the general view that the charge e carried by a hydrogen atom is the fundamental unit of electricity [wrote Rutherford and Geiger]... the evidence is strongly in favor of the view that [the charge of the α-particle] $= 2e$. [13-66]

The matter was put conclusively to rest in 1909, when Rutherford and Thomas Royds, a student, detected spectroscopically the helium gas formed when α-particles pass into a closed glass vessel.

As a postscript, we note that in *Comptes Rendus* for April 30 1900 Paul Villard at the École Normale in Paris reported finding very penetrating rays that were non-deviable in a magnetic field [13-67]. Though Villard is generally given credit for discovering γ-rays, even by Rutherford in his Nobel lecture of 1908, evidence points to Rutherford as a co-discoverer [13-68]. In his classic, first paper on radioactivity in 1899, we read in one place that 'these [absorption] experiments show that the uranium radiation is complex, and that there are present *at least* two distinct types of radiation...' (italics added) [13-69], and in another place

> While the experiments on the complex nature of uranium radiation were in progress, the discovery [by Schmidt] that thorium and its salts also emitted a radiation, which had general properties similar to uranium radiation, was announced. A few experiments were made to compare the types of radiation emitted by uranium and thorium. [13-70]

Subsequently it would become clear that the thorium rays, which Rutherford characterized simply as 'not homogeneous but... of a more penetrating kind [than the α-rays of uranium]' [13-71] were γ-rays. In 1902 Rutherford reported that both thorium and radium salts emitted such radiation, which he later the same year suggested might be exceptionally hard β-rays. However, in his first book *Radio-activity*, which was published in 1904 by mutual agreement with Frederick Soddy (or so Eve delicately puts it!) a few months before Soddy's own book on the same subject, he suggested that, since x-rays are pulses produced by the sudden stopping of electrons, γ-rays might be pulses produced by the sudden starting of β-rays—this because in 1904 it was generally believed that γ-rays always accompanied β-rays [13-72]. In fact, the nature of gamma-rays as similar to x-rays was only settled in 1914. That year Rutherford and Edward Andrade found that a beam of γ-rays striking mica crystals at specific angles was enhanced in certain directions and impoverished in other directions. From their deflection, they deduced that the wavelengths of the rays were comparable with those of x-rays [13-73].

Chapter 14

EVANESCENT RAYS: A FRENCH COTTAGE INDUSTRY

14.1. The N-ray episode

Alas, the last word on exotic rays had not been heard in French scientific circles. In early 1903, René Blondlot of Nancy, engaged in x-ray research, announced the discovery of radiation emitted from x-ray tubes that differed from x-rays in many important respects. He christened the rays N-rays in honor of his city and its university. Between 1903 and 1906 some 300 papers were published on the topic, among them 26 papers and a book by Blondlot alone, and no fewer than 38 by a close colleague. The rays were not a hoax, but simply a mistake; they are perhaps best described, as they were in due course by a committee of the French Academy, as the chimerical products of autosuggestion—the results of subjective viewing of feeble or alleged effects under difficult observational conditions at best. Remembered today chiefly for the insight the purported rays throw on the pathology or psychology of science [14-1], they are worthy of our attention, appearing as they did amidst the plethora of rays, cathode, x, Lenard, α, β, γ, infrared, ultraviolet, photoelectric, which by their multiplicity prepared scientists to be open to still more forms of radiation. They are also a reminder that the nature of x-rays was still troubling many physicists in the early years of the 20th century. One who was definitely open to new ideas, and far from satisfied with the prevailing view on x-rays, was Blondlot, a highly distinguished French physicist at the time.

René-Prosper Blondlot, born in Nancy in 1849, was the son of Nicolas Blondlot, a renowned physiologist and chemist [14-2]. Apart from obtaining his doctorate under Jules C Jamin at the Sorbonne in 1881, he spent virtually all his life in Nancy, where he had been recommended by Marcellin Berthelot for *maître de conférences* in physics at the newly unified University in 1882 [14-3]. Though never a member of the Paris Academy of Sciences, he was named *correspondant* of the Academy in 1894, taking the place of Helmholtz. He was the recipient of the Academy's Gaston Planté Prize in 1893, the LaCaze Prize in 1899, and, at the height of the N-ray controversy in 1904, the Le Conte Prize. Probably as a result of the ensuing scandal over the rays, Blondlot spent the rest of his career in relative obscurity, never really recanting in his views, and died without fanfare in 1930.

His scientific background led Blondlot naturally to research on x-rays, utilizing a laboratory technique that, while not unique, was his speciality from the very start. Already in his doctoral research under Jamin (on electric cells and their polarization), he employed a movable mirror in a study of the time lag between the onset of an electrical phenomenon and the appearance of the Kerr electro-optic effect in the oscillating discharge of a condenser. His next piece of research used basically the same technique of a rapidly rotating mirror. Inspired by Hertz's electromagnetic researches, like so many in his day, this investigation centered on measuring the velocity of propagation of electric waves in conducting wires. By photographing the light from electric sparks reflected in a rotating mirror, Blondlot found that the speed was about 3×10^{10} cm s^{-1}. This was very close to the value calculated for light, both from optical experiments and from the ratio of electrodynamic and electrostatic charges. The investigation was lauded by Henri Poincaré and J J Thomson for its importance for Maxwell's theory and as an exemplary experimental measurement, respectively [14-4].

Following Röntgen's discovery of x-rays in 1895, Blondlot threw himself into studies of the physical nature of the rays; in particular, were they particles or electromagnetic rays? In 1902, he claimed to have demonstrated that the rays are propagated with a velocity equal to that of light [14-5]. Ironically, this important experiment, though subsequently rejected as inadequate, is all but forgotten in the glare of the spotlight suffusing his imaginary rays (and this despite Blondlot's later conclusion that the experiment had actually measured the velocity of N-rays, not x-rays). In any case, if truly waves, x-rays could be polarized as they emerged from the discharge tube in which they were generated, despite proffered evidence to the contrary (including Röntgen's own judgment). Might not the experimental failure to polarize x-rays be due to their already being plane-polarized from the start of emission because of the unidirectionality of the parent cathode rays? [14-6].

In attacking the question, Blondlot again resorted to an electric spark as an analyzer, by orienting it for maximum brightness. This time, however, he did not photograph the spark, as he had in his earlier experiments on electric waves. Instead, he relied on *visually* detecting changes in luminosity—a notoriously unreliable technique far inferior to the photographic and ionization-measuring techniques of his

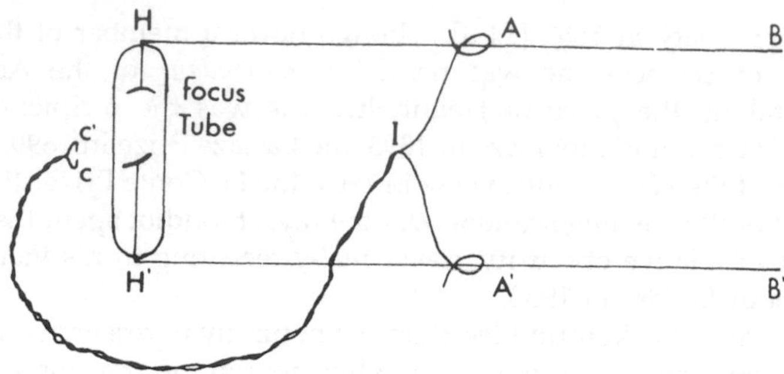

Figure 14.1. *Blondlot's apparatus for demonstrating the polarization of x-rays, alias N-rays.*

colleagues abroad, from Amsterdam to Montreal. The strong personal bias inevitably accompanying visual sightings of feeble variations of intensity would be a major source of criticism by physicists, and is somewhat reminiscent of the pitfalls accompanying the early studies of the Zeeman effect (section 11.3).

In Blondlot's apparatus (figure 14.1) a pair of insulated wires linked the terminals of an x-ray tube to the terminals of a Rühmkorff induction coil circuit equipped with a vibrating spring interrupter. Two other insulated wires, originating in rings around the primary wires, terminated in a spark gap; here, a small spark was produced from the inductive surge at each interruption of the coil current. The spark was shielded from the tube by an aluminum sheet, and the orientation of the spark gap could be readily charged, due to the flexible lead wires. If the gap were to be oriented such as to lie in the plane of polarization of the x-rays, the electric component of the electromagnetic wave should reinforce the energy of the spark, and thus increase its brightness, and that is just what happened. With the spark gap parallel to the direction of the cathode rays, the brightness of the spark visibly increased. Moreover, if the x-rays were collimated by a slotted lead plate, the orientation for maximum brightness could be changed by interposing quartz or sugar pieces between slit and spark, indicating rotation of the plane of polarization of the x-rays [14-7].

As emphasized, Blondlot observed changes in *brightness* of the spark, not in its length—the latter a more quantitative procedure that, as may be recalled, had been Hertz's procedure. Blondlot stressed brightness as a key property of the spark, insisting it must be short and weak to properly demonstrate polarization of x-rays by the variation in brightness with orientation. In any case, he soon found that the rays illuminating the spark could be *refracted* by a quartz prism, and hence concluded that they could not be x-rays; presumably, he reasoned, they constituted a hitherto unsuspected, new form of radiation that he first named n rays and later

N-rays. Alas, he did so lacking the intuition 'J J' Thomson had shown when confronted with the apparent absence of x-ray refraction some seven years earlier—a lack J J attributed simply to the high frequency of the Röntgen rays (p 117). To be sure, the new rays exhibited very different characteristics from x-rays. For one thing, reducing the x-ray tube voltage to a level where no fluorescence was discernible on the tube wall, presumably cutting off x-rays, did not inhibit changes in spark brightness. Nor did enclosing the spark gap in a cardboard box. In fairly short order, Blondlot undertook an exhaustive series of experiments uncovering a host of materials that proved transparent to N-rays but opaque to visible light, including wood, paper, mica, quartz, paraffin, iron, silver, and gold. Aluminum prisms and lenses bent and focused them; water and rock salt stopped them. Not unlike Hertz's systematic investigation of the effects of ultraviolet light on sparks, Blondlot's program nevertheless erred in not following up the possibility that many of the apparent results might, indeed, be due simply to photoelectric effects.

Blondlot soon found additional N-ray emitters besides x-ray tubes, initially guided by their index of refraction in quartz—about the same as that of infrared light [14-8]. They were emitted by a gas-burning Auer lamp (Welsbach mantle) as well as by the Nernst glower (a lamp in which zirconium oxide is heated to incandescence by a current) though, surprisingly, not by the Bunsen burner. They were also emitted by a sheet of iron or pieces of silver heated to a dull red, by a flame, and even by the Sun. The latter sources came to light in the discovery of new methods for N-ray detection; mainly, surfaces covered with spots of a properly prepared deposit of calcium sulfide. The luminous spots became still more luminous when exposed to light from an Auer lamp equipped with an aluminum window for transmitting the N-rays before being focused by a quartz lens. In his book on the subject (published in Paris in 1904 and in English translation two years later in London), Blondlot provided careful instructions for observing the rays, even including a sheet bearing 25 calcium sulfide deposits. Absolute silence and no smoking was a prerequisite, as was the need for averted vision in the manner of astronomers observing faint stellar images. In the event, quipped J J Thomson, 'no English, German or American physicist succeeded in finding them, while in France they seemed to be universal' [14-9].

Using his calcium sulfide detector, Blondlot found that N-rays could be stored and re-emitted by certain materials once exposed to the rays by an Auer lamp subsequently extinguished. Among these materials were quartz, feldspar, and Iceland spar; the strengthening action of the rays on the retina allowed faintly luminous objects to be viewed more clearly. Among the primary emitters were tempered steel, twisted rubber, and other solid materials in a state of strain—even a walking cane bent by the hands and held near the eye [14-10]. The length of storage was dramatically underscored in the discovery that weapons from the Graeco-Roman period were still emitting the rays [14-11]! To confuse the matter even further, certain rays, N_1-rays, *lessened* the luminosity of N-ray detectors and *decreased* visual acuity.

However, this was not all. The most sensational properties of N-rays were reported to the Paris Academy in the winter of 1903—not by Blondlot, but by another distinguished member of the Nancy faculty: Auguste Charpentier, professor of medical physics. He and several other colleagues at Nancy had taken up the subject of N-rays, and found them to be liberated by human and animal organs, as well as by plants and even seeds. Charpentier's spellbinding report was sponsored by none other than by the Sorbonne physicist, physiologist, and Academician Arséne d'Arsonval who himself expounded enthusiastically on the latest findings at Nancy when given the chance. Subsequently, Charpentier delivered a steady stream of reports to the Academy between December 1903 and July 1904, including seven papers in the month of May alone.

N-rays, and related 'physiological rays', according to Charpentier, were emitted from animal muscles and the human nervous system. Emanating while a person was talking from that region in the brain that governs speech, the 'Broca center', the rays were noticeably enhanced by human intellectual effort or psychic activity. They, in turn, augmented the sensitivity to vision, smell, taste, and hearing, as shown by the purported demonstration that elimination of nervous sensation by muscular relaxation, either by a dose of curare or by electrical shock, extinguished the rays. Both types of ray could be transmitted along wires. For instance, an experimenter, merely by placing himself behind a small copper plate fixed at one end of a copper wire which terminated in a calcium sulfide screen at the other end, caused an increase in brilliancy of the phosphorescent screen over its ambient value. An extensive list of medical and biological effects attributed to the physiological rays and the N-rays alike were chronicled in a contemporary treatise [14-12]. The rays appeared to be far superior to x-rays in depicting the human anatomy, and even promising in diagnosing diseases by the pattern of bodily emissions reflecting the selective sensitivity to levels of N-ray activity.

Several others soon challenged the findings of Blondlot, Charpentier, and their colleagues, claiming priority in discovering the rays and their effects. In the spring of 1903 Gustave Le Bon, a controversial psychologist and amateur physicist best known for his 'black light' and a perennial priority claimant, wrote Blondlot, insisting he had discovered a form of radiation that could penetrate metals with impunity. Before long, P Audollet petitioned the Academy directly, claiming that he, not Charpentier, had been the first to discover the emission of N-rays from living organisms. Similar claims were voiced by several spiritualists, among them Carl Huter. Perhaps they echoed, with

Audollet, the fervent interest in the unity of spiritualistic, pathological, and psychological phenomena that swept French circles in the last decades of the 19th century, preoccupied with mental pathologies [4-13]. Indeed, N-rays held great promise as a material explanation for spiritualistic and parapsychological phenomena—something Blondlot by no means disputed. However, the claims to the Academy were rebuffed in a report by d'Arsonval, maintaining that the priority belonged to Charpentier alone [14-14]. D'Arsonval himself, we might add, was keenly interested in psychics and spiritualism, and held that 'there exists around individuals a kind of atmosphere of an altogether special nature, a radiation for which one can determine physical constants absolutely, as for all other manifestations of energy' [14-15].

Which is not to say that only crackpots and extremists dabbled in N-rays, besides Blondlot and his followers in physics and medicine at Nancy. Blondlot got the movement going, but the French scientific community kept it going, abetted by various social, scientific, and institutional factors which contributed to the debate. Nye has identified four such factors: the renewed scientific interest in psychiatry and spiritualism around 1900, as noted; the perception that the international reputation of French science was slipping, particularly with respect to German physics, due to various embarrassing episodes in these years (including the N-ray affair, which was becoming a two-edged sword); the decentralization of the French national university system, which emphasized selected provincial universities, among them Nancy; and 'the inbred and hierarchical structure of the French scientific community' [14-16].

In addition to the Nancy faculty, two groups of scientists supported the N-ray investigation in particular. One consisted of influential contemporaries and close friends of Blondlot, among them Henri Poincaré, who had strong ties to Nancy, Marcel Brillouin, formerly of the Nancy faculty, Edmond Bouty of the Sorbonne, Éleuthère Mascart (Collège de France), and, of course, d'Arsonval. The other group of supporters included younger physicists and physiologists, usually associated with provincial faculties and eager to complete some piece of research to enhance their chance if and when some more prestigious professorship became available. The best known of these, if somewhat atypical, was Jean Becquerel; as son of Henri Becquerel and heir to the family dynasty, his academic future was hardly in doubt. At the time of the N-ray affair, he had just become assistant to his father at the Museum of Natural History in Paris, and paid a visit to Blondlot's laboratory; there he believed he had observed the effects of the rays on a calcium sulfide screen. For about a year, he remained an enthusiastic student of N-rays emanating from metals 'anesthetized' with chloroform and their deviability in a magnetic field. However, he then turned abruptly away from the whole business, setting the record straight by admitting that

> the author . . . had various reasons to believe in the existence of N-rays: as we have said, he had been engaged in this direction by a scientist whose previous work inspired only admiration; more, another physicist declared that he had seen the effects, and some of Blondlot's students (who since then have become illustrious) published notes on N-rays. Before such authoritative powers, a young man, just graduated from the Ecole des Ponts et Chaussées and never yet having done any research, can be excused somewhat for getting carried away. Finally, not one note of the author was published without having been submitted to Blondlot and without receiving his approval. [14-17]

We will, in fact, hear more from Jean Becquerel who, before long, would occupy center stage in another major controversy concerning new, purported rays. As to the N-rays, skepticism was indeed growing, even within the French scientific community. Jean Perrin, just starting his classic colloid studies at the Sorbonne, openly attacked the subjective method of detecting the rays from the outset. Paul Langevin, substituting for Mascart at Collège de France, failed to observe the rays while visiting Blondlot's laboratory, and had no better luck in Mascart's laboratory. Victor Henri voiced strong doubts, as did Aimé-Auguste Cotton, Paul Janet, and Pierre Curie. At Germany's Physikalisch-Technische Reichsanstalt, Otto Lummer and Heinrich Rubens found that they could reproduce Blondlot's results *without any source of illumination whatever*. Lummer concluded, and argued vehemently before the British Association, the German Physical Society, and other bodies, that their French colleague's observations were psycho-physiological, not physical, in nature, caused by the role of the retinal rods in peripheral vision. John Burke at the Cavendish concurred with Lummer's explanation, as did C C Schenk at McGill University. Rutherford, also at McGill, was no more successful in observing the rays, judging by W C Dampier Whetham's letter to him from Upwater Lodge, Cambridge:

> We were very much interested to hear you had been trying to detect N rays. Burke tried in the Cavendish Laboratory, also with negative results. Is it a physiological effect, or did Blondlot dream it? [14-18]

In countering the criticism of subjective visual observations of sparks or phosphorescent spots, Blondlot and others went to some pains to devise methods for obtaining photographic records of the rays, with limited success. In a typical experiment, a spark-gap detector was periodically moved manually back and forth over a photographic plate mounted horizontally under a source of N-rays. At one end of the side of the plate facing the rays the detector was shielded from exposure by a lead screen covered with moist paper, while at the other end it was unshielded; thus, the photographic plate registered the rays more brightly in this location. (Or, rather, the negative images were darker.) Nevertheless, Blondlot's critics argued that the manual back-and-forth movement (to the beat of a metronome) was open to the possibility of unconsciously exposing one end of the plate longer than the other end in the hands of Blondlot's overzealous laboratory assistant (who rigged up the apparatus and usually carried out the measurements). The American physicist Robert Wood went so far as calling the assistant, L Virtz, 'a sort of high-class laboratory janitor' [14-19].

It was, in fact, Wood who effectively administered the *coup de grâce* to the whole affair. Robert Williams Wood was professor of experimental physics at Johns Hopkins University, where he had succeeded Henry Rowland on his death in 1901 [14-20]. Educated mainly at Harvard and the University of Berlin, he never bothered completing his PhD, yet became an internationally known expert in physical optics and spectroscopy. He was a man of broad talents—an artist, musician, writer, and poet. He was also a great wit, ebullient, and highly unconventional. Between his serious books is a little gem, *How to tell the Birds from the Flowers*, which went through no less than 20 editions. Finally, he was not only a great prankster, but also a tireless pursuer and debunker of frauds, such as spiritualistic mediums.

It is, therefore, hardly surprising that Blondlot drew Wood's attention. He, too, had attempted to reproduce Blondlot's results at Baltimore, but 'failed to confirm them after wasting an entire morning' [14-21]. Apparently his celebrated visit to Blondlot's laboratory was made on the urging of Rubens (whom he had known well during his stint in Berlin) at a private discussion after the British Association meeting in Cambridge in September of 1904, where the Blondlot matter was thrashed out.

> [Rubens] felt particularly aggrieved because the Kaiser had commanded him to come to Potsdam and demonstrate the rays. After wasting two weeks in vain efforts to duplicate the Frenchman's experiments, he was greatly embarrassed by having to confess to the Kaiser his failure. Turning to me he said, 'Professor Wood, will you not go to Nancy immediately and test the experiments that are going on there?' 'Yes yes', said all of the Englishmen, 'that's the idea, go ahead'. [14-22]

Wood arrived in Nancy late the same month, received cordially by Blondlot. He spent some three hours in his laboratory in the company of Blondlot and Mr Virtz, treated to a series of experiments intended to demonstrate the many exciting features of the rays. He was shown cards with spots painted with luminous colors, but was unable to perceive any increase in luminosity when the N-rays were turned on. Nor did he trust feeble distinctions in brilliancy between photographic images obtained by exposure to the rays and exposures screened by wet cardboard. However, he held his tongue.

Then came a crucial test involving a spectrum of N-rays produced by an aluminum prism. Light from a Nernst lamp was first passed through screens of aluminum foil, black paper, and wood, interposed to mask ordinary electromagnetic radiation. It then traversed a vertical slit in a screen of wet cardboard, and finally fell on the aluminum prism. The refracted rays were located by means of a narrow strip of phosphorescent paint applied to a piece of dry cardboard which could be precisely moved through the spectrum by turning the screw of a ruling engine with a graduated wheel. According to Blondlot, at least four different locations, or N-ray wavelengths, could be identified by momentary changes in brightness when the deviated rays crossed the luminous strip.

> I asked him to repeat his measurements [recalled Wood], and reached over in the dark and lifted the aluminum prism from the spectroscope. He turned the wheel again, reading off the same numbers as before. I put the prism back before the lights were turned up, and Blondlot told his assistant that his eyes were tired. The assistant had evidently become suspicious, and asked Blondlot to let him repeat the reading for me. Before he turned down the light I had noticed that he placed the prism very exactly on its little round support, with two of its corners exactly on the rim of the metal disk. As soon as the light was lowered, I moved over towards the prism, with audible footsteps, but *I did not touch the prism*. The assistant commenced to turn the wheel, and suddenly said hurriedly to Blondlot in French, 'I see nothing; there is no spectrum. I think the American has made some *dérangement*'. Whereupon he immediately turned up the gas and went over and examined the prism carefully. He glared at me, but I gave no indication of my reactions. This ended the séance, and I caught the night train for Paris. [14-23]

As Wood wrote in a damning letter posted to *Nature* from Brussels the following morning, September 22, and translated as well in the semi-popular *Revue Scientifique*, 'the removal of the prism (we were in the dark room) did not seem to interfere in any way with the location of the maxima and minima in the deviated (!) ray bundle' [14-24].

Wood's account of his experience at Nancy effectively spelled the end of any lingering support for N-rays abroad, but not in France [14-25]. Though the editors of the *Revue Scientifique*, particularly sensitive to the French scientific standing, lost no time in launching an effort to resolve the matter once and for all by polling leading French scientists, most proponents at home remained unswayed; indeed, they rallied behind Blondlot by bestowing upon him the Academy's Prix Le Conte in December of 1904 [14-17]. Still, further seeds of doubt had been planted in the French community by the *enquête* in the *Revue Scientifique*. Though the Le Conte Prize was awarded to Blondlot, not to Pierre Curie as proposed by some members of the Prize committee, in the end it went to Blondlot 'for the whole of his work', with the N-rays barely mentioned. Very few new confirmations of the rays were published in the wake of the *Revue Scientifique* inquiry. Only one was presented to the Academy in 1905, and by 1906 the N-ray episode was essentially over, save for Blondlot's own opinion.

Blondlot himself had remained defiant amidst the mounting skepticism at home and abroad. Soon after Wood's visit, he wrote that

> ... several eminent physicists, who have been good enough to visit my laboratory, have witnessed [the photographic detection experiments]. Of ... forty experiments, one was unsuccessful.... I believe this failure, unique, be it noted, to be due to insufficient regulation of the spark, which undoubtedly was not sensitive. [14-26]

In addition to its poll of the French scientific community, the editors of the *Revue Scientifique* had attempted to persuade Blondlot to perform a particular 'control experiment' suggested by several physicists [14-27]. Blondlot did not bother to respond at the time, but in 1906 replied as follows:

> Permit me to decline totally your proposition to cooperate in this simplistic experiment; the phenomena are much too delicate for that. Let each one form his personal opinion about N-rays, either from his own experiments or from those of others in whom he has confidence. [14-28]

On that note we, too, let the N-ray matter rest. Blondlot remained a firm believer long after his retirement from Nancy in 1910, according to sealed records opened at the Academy upon his death in 1930 [14-29]. He retired as *Professor Honoraire;* in old age he maintained contact with associates at the university, and updated his two widely used texts on thermodynamics and static electricity, both of which went through several editions. He never married. His house and garden were willed to the city of Nancy in perpetuity as a place of rest and for young people to receive educational guidance. The park adjoining the estate is named Parc Blondlot.

14.2. Seeds for another controversy

We last heard of Jean Becquerel in 1904, after his unfortunate fling with Blondlot's N-rays. Two years later, comfortably ensconced as *Aide-Naturaliste* under his father, Henri Becquerel, at the Musée d'Histoire Naturelle, he turned to magneto-optics and related studies. These were topics in physical optics that had preoccupied his father too before his discovery of radioactivity. To Jean Becquerel's chagrin, his own preoccupation with the same subject would embroil him in a second, major scientific controversy destined to drag through the second half-decade of the 20th century. A controversy less familiar to most than the N-ray affair, it was nevertheless more important in the limited context of sorting out the multi-faceted evidence for or against positive carriers of electric charge ca. 1900.

To place the controversy in perspective, we must again backtrack and take particular note of several electron theories of matter that reached fruition in Jean Becquerel's formative years. One, involving carriers of electric charge vibrating but otherwise bound by harmonic forces in ponderable matter, we have already encountered in the electro-optics of Zeeman and Lorentz. The same carriers surfaced in another, closely related guise in the magneto-optics of Franz Neumann and Woldemar Voigt. Henri Becquerel's first extensive research, during 1875–1882, had dealt with the latter sub-topic, the rotation of plane-polarized light by magnetic fields (the Faraday effect). His next investigation of interest in the present context centered on the propagation of polarized light in colored crystals, chiefly didymium and its kindred substances. Together, these researches earned the elder Becquerel his doctorate in 1888 and election to the Academy the following year. Thereupon his research productivity ground to a halt, while he succeeded to his father's two chairs of physics (at the Conservatoire National des Arts et Métiers and at the Museum) and to Alfred Potier's professorship at the Polytechnique. Following the surprisingly short-lived excitement surrounding his discovery of radioactivity in 1896 (p 147), he had briefly returned to his earlier interests, primarily the Zeeman effect, before devoting himself to radioactivity and scholarly affairs.

The discovery by Zeeman of the effect which bears his name had confirmed what had been independently hypothesized by Lorentz and Larmor: the existence of bound carriers of charge with a negative charge and charge-to-mass ratio in excellent agreement with the free particle identified in cathode rays. In principle, the electro- and magneto-optic equations admitted positively charged bound carriers as well; in practice, agreement with experiment, both the Zeeman as well as the Faraday effect (the latter which we have not yet discussed), proved only possible on the assumption of negative charges at work alone. However, the possibility of a positive counterpart to the ubiquitous negative elementary quantum of charge—particles identical to electrons in every respect except for their charge and distinct from the murkier positive 'canal rays' not yet understood in 1900—has an earlier history in the Continental electron theory of metallic conduction. That theory, mainly the work of German physicists, arose to account for a long line of experiments on the electrical resistance of pure metals and alloys, experiments fully as old as the cathode ray investigations that have occupied our attention all along.

Experiments on the relative 'conducting power' of metals can be said to have had their serious start in the celebrated electrical researches by Henry Cavendish. (Tradition has it that Cavendish acted as his own galvanometer, estimating the strength of a current by the shock it gave him or a hapless visitor to his private laboratory in London's Great Marlborough Street.) Humphry Davy placed the measurements on a sounder footing, observing that the electrical conductivity of a wire falls with increasing temperature. Credit for the first quantitative determination of the fall in conductivity (or increase in resistance) goes to Emil Lenz, best known for his law linking induction to the ponderomotive interaction of currents and magnetic fields. His least-squares analysis of his own resistance data for silver, brass, iron, platinum, and other metals, collected over nearly three decades, was reported to the Saint Petersburg Academy of Sciences in 1833. The general trend in the variation of resistance with temperature, as closely proportional to the absolute temperature, belongs to the Norwegian physicist Adam Arndtsen, Adjunct and keeper of the physical cabinet at the University of Christiania, and to the analysis of his data in Poggendorff's *Annalen* for 1858 by Rudolf Clausius while serving his first professorship at the new Polytechnicum in Zürich [14-30]. As important as Clausius' opinion was the fact that Arndtsen's work, begun at Christiania, was completed during his stint under Wilhelm Weber at Göttingen during 1857–1858. Weber was a man of some importance.

Wilhelm Weber inherited the professorship at Göttingen in 1831. He was hardly a neophyte himself when it came to electrical experimentation [14-31]. In 1833 he had, in his long-standing collaboration with Karl Friedrich Gauss, personally installed one of the first practical long-range galvanic telegraphs between his laboratory (housed in a private residence) and the astronomical observatory (then in charge of Gauss), roughly a mile over the Göttingen rooftops. The telegraph was needed in order to facilitate simultaneous geomagnetic observations—observations predicated on magnetometers crafted by Weber himself. Later, he had collaborated with Rudolph Kohlrausch (father of Friedrich Kohlrausch) in determining the ratio between the electrodynamic and electrostatic units of charge. That investigation, and all the subsequent researches in Weber's long series of *Elektrodynamische Maassbestimmungen*, relied on his own electrodynamometer, but, for all his experimental flair, Weber was an equally gifted theorist. Though his fame rests primarily on his controversial law of force between electric current elements, of more lasting importance was his attempt to comprehend electrical resistance in terms of electric fluids or particles. His notions in this connection seem to have been inspired during his brief tenure at Leipzig from 1843 to 1846. Several professorships at Göttingen, his among them, had been revoked during the politically troubled period of Hannoverian rule under Ernst August, uncle of Queen Victoria. At Leipzig, a bastion of liberalism, he was able to assume the chair of physics vacated at the onset of the long neurotic illness of Gustav Theodor Fechner, a close friend and himself an ardent atomist.

Weber's theory of electrical conduction in metals occupied him off and on from his return to Göttingen in 1849 until his retirement in the 1870s. In one of its manifestations, it treats a metal as a lattice of fixed negative particles adhering to ponderable masses, surrounded by positive particles rotating in elliptical orbits. In the presence of an electromotive force, the positive particles move in ever-widening spirals until they come under the influence of a neighboring atom. Unlike electrolytic conduction, there is no net transport of mass in the process [14-32]. Weber's attempts to formulate these ideas quantitatively in terms of molecular properties were not very successful, however. They were bedeviled by his countrymen's insistence that rays in the ubiquitous discharge tubes are a form of ætheral wave propagation, by the frenzy of *Naturphilosophie* that shook German intellectual circles in these years, and by the prolonged feud between the Weber and Helmholtz camps over criteria for electrodynamic forces (whether velocity-dependent forces are conservative) [14-33].

Weber's model was extended by Wilhelm Giese of Berlin, who argued that metallic conduction must be associated, not with charged molecules, but with the *ions* of a metal that originate when molecules dissociate by collision [14-34]. In England, similar ideas were expressed by Arthur Schuster, who had spent some time both at Göttingen and at the old Berlin institute while still the residence of Gustav Magnus, and whom Giese accused of having stolen his thunder. Schuster declared that 'one molecule cannot communicate electricity in an encounter in which both molecules remain intact' [14-35]. Others sharing a viewpoint embracing two kinds of electricity included the Leipzig astrophysicist Karl-Friedrich Zöllner, Weber's last close collaborator and champion in the feud with Helmholtz, the mathematician Hermann Grassmann and his brother Robert (an amateur scientist), and abroad the Italian physicist F Ottaviano Mossotti. However, it was Weber's student, Eduard Riecke, who laid the groundwork for the classic electron theory of metals—a theory completed by Paul Drude, also of the Göttingen school.

Riecke, like his mentor, showed an early flair for both theory and experiment, studying mathematics under Carl Neumann and experimental physics under Eduard Reusch at the University of Tübingen [14-36]. On completion of his undergraduate studies in 1869, his graduate work under a state scholarship at Göttingen was interrupted almost at once by a year's conscription as an officer in the Franco-Prussian War. Returning to Göttingen in 1871, he completed his doctorate within the year on the magnetization of iron, and became Weber's assistant. The same year, now *Privatdozent*, he took over some of the experimental lectures from Weber, who was then unwell, and in 1873 he inherited all the lectures and was appointed extraordinary professor of physics. In

1881, he rose to the rank of ordinary professor, taking charge of Weber's laboratory and institute. In so doing, he inherited both the experimental and theoretical programs of his mentor. His own research, like Weber's, would be alternately theoretical and experimental during his 45-year tenure at Göttingen.

As an experimentalist, Riecke's work with Geissler tubes helped to establish the identity of negatively charged particles projected from the cathode—particles which, he argued, are none other than the free negative particles in a metal [14-37]. Subsequent researches of his on the motion of charged particles in electromagnetic fields would have a direct bearing on the theory of the aurora. However, there can be little doubt that his theoretical researches on the metallic state and metallic and electrolytic conduction would be of greater enduring importance. His major contribution to the subject was published in 1898 [14-38]. In this work, he envisaged the interstices between the ponderable molecules of a metal swarming with both negative and positive particles of the same mass—particles or ions dislodged from each molecule by collisional excitation. The freed ions drifted randomly through the metal, like a gas, Riecke assuming that the average velocity of the random motion was nearly proportional to the square root of the absolute temperature. Unfortunately, he failed to derive the constant of proportionality; for this and other reasons (chiefly too many arbitrary physical parameters generally) his rather elaborate formulæ for both the thermal and electrical conductivity of a metal produced no quantitative predictions [14-39].

Whereas Riecke was a disciple of Weber and his electrodynamics, Paul Drude was a pupil of the magneto- and electro-optic school of Franz Neumann, father of Carl Neumann, and Woldemar Voigt [14-40]. He set out to be a mathematician, studying at Göttingen, Freiburg, and Berlin, then returned to Göttingen where he switched to theoretical physics under Voigt who became co-director of physics with Riecke in 1883. (Riecke was in charge of the experimental institute, and Voigt inherited the theoretical institute on the death of J B Listing, who left no assistant.) Drude completed his dissertation in physical optics under Voigt in 1887, stayed on as Voigt's assistant, and became *Privatdozent* at Göttingen in 1890. He moved to Leipzig in 1894, where he pursued both theoretical and practical researches on the propagation of Hertzian waves, and continued his work on physical optics, only now from the electromagnetic standpoint (as opposed to the older mechanical view of light). His *Physik des Æthers*, published in 1894, which truly launched his academic career, was the first German text to treat electricity and optics from the standpoint of Maxwell's theory.

Drude's growing interest in correlating optical, electrical, thermal, and chemical properties of matter culminated in his classic papers on the electron theory, mostly dated 1900 and all written at Leipzig [14-41]. Drude, too, like Riecke, at first found it necessary to assume the simultaneous presence of positive and negative free particles in a metal, partly by analogy with the coexistence of cathode rays and canal rays in Geissler tubes. Drude purposely used Stoney's term *electron* for his particles, reserving the terms *corpuscles* or *ions* 'for the aggregate electrical particles and ponderable masses encountered in electrolytes'. He thus left the question of whether or not an electron carries a very small ponderable mass 'undetermined for now' and somewhat irrelevant [14-42]. However, Drude's major advance over Riecke lay in explicitly adopting the viewpoint of the kinetic theory of gases. That is, Drude assumed the electrons of the 'gas' move freely and are in thermal equilibrium with their surroundings through the collisional exchange of energy with the atoms; thus he ignored mutual collisions between the electrons. This view implies a kinetic energy of the electrons equal to that of a gas molecule at the same temperature, or proportional to the absolute temperature. In Drude's notation,

$$\tfrac{1}{2}mv^2 = \alpha T$$

where v is the average electron velocity of thermal motion, m its mass, T the temperature, and α is Loschmidt's number (the constant Riecke failed to specify), also given by $\frac{3}{2}k$ where k is Boltzmann's constant. Drude's celebrated expression for electrical conductivity is, in conventional notation [14-43]

$$\sigma = \frac{1}{6}\frac{e^3 n\lambda v}{kT} \tag{14.1}$$

where n is the number of free electrons in unit volume, and λ the mean free path. A temperature gradient in a metal will also cause an electron current to flow, and Drude's analysis applies equally well to thermal conductivity. His expression for the coefficient of thermal conductivity is

$$K = \tfrac{1}{2}nv\lambda k. \tag{14.2}$$

According to these two expressions, the ratio between the thermal and electric conductivities is the same for all metals at the same temperature, in good agreement with the old (1853) rule of Wiedemann and Franz, or

$$\frac{K}{\sigma} = 3\left(\frac{k}{e}\right)^2 T. \tag{14.3}$$

The agreement with Wiedemann and Franz, Drude now decided, suggests that 'in metals only a single type of [free] particle is present', with 'double particles... only encountered in electrolytes as a bivalent ion or bound particle' [14-44].

Drude's conclusion with regard to a single species of free (negative) electrons is echoed in the opening paragraph of H A Lorentz's tome of 1905 presenting his own, more rigorous electron theory of metals [14-45]. (Lorentz's model treated the electron velocity as a statistical quantity, in accordance with Maxwell's velocity distribution.) 'I shall begin [announced Lorentz] by assuming that the metal contains but one kind of free electrons, having all the same charge *e* and the same mass *m*' [14-46]. The *negative* sign of the charge (and the charge-to-mass ratio) had been surmised by Zeeman and himself nearly a decade earlier, and independently established in many quarters, as we know—by J J Thomson and others for J J's corpuscles, by Jean Perrin for the same, by Giesel and company for β-rays, and by still others. Yet, despite the authority of Lorentz, troubling aspects of the Hall effect did not quite rule out positive electrons as well [14-47]. Much other speculation on positive electrons at this time remained simply that, pure speculation [14-48].

14.3. The positive electron: another French coup?

On the whole, then, the mood ca. 1905 was one of favoring negative electrons only, rejecting positive electrons in the restricted sense of positive mirror images of cathode ray electrons. J J Thomson, for one, had long been of this opinion (p 188). By 1907, leading physicists all agreed that 'we ought surely to give up all attempts to explain phenomena by the assumption of two kinds of movable electrons' [14-49]. Thus it came as quite a surprise when, in March 1906, Jean Becquerel announced experimental evidence for the existence of positive electrons, albeit not necessarily *free* electrons.

Recovered from his bout with Blondlot's N-rays, Becquerel had turned to a more tractable subject, magneto-optics. Taking up where his father had left the subject, the young Becquerel set out to determine whether crystalline bodies exhibit the same kind of electro- and magneto-optic effects as do gases [14-50]. It had been known for over half a century that a beam of linearly polarized light sent through a vapor in the direction of an applied magnetic field has its plane of polarization rotated—an effect in due course known as the Faraday effect after its discovery by Michael Faraday in 1845 [14-51]. The magnetic rotation may also be complicated by the fact that a particular spectral absorption line may be split into two or more lines of different frequency and polarization by a process known as the 'inverse Zeeman effect' (the Zeeman effect obtained in absorption). The line with increased frequency corresponds to circularly polarized light with a positive ('left-handed') direction of rotation, and the low-frequency line corresponds to a negative ('right-handed') direction of rotation. To complicate matters further, circularly polarized light incident on the vapor will be shifted toward higher frequency, irrespective of its sense of rotation [14-52]. All these effects were nicely accounted for by Lorentz, Larmor, Woldemar Voigt, and others in terms of charged particles orbiting in atoms whose motions are modified by an external magnetic field, along the lines we have already touched on in the case of the normal Zeeman effect (pp 198–9). Only one essential stipulation proved necessary for securing agreement between experiment and theory—namely, that the electrons in the absorbing vapor carry *negative* charges.

These were the salient facts when Jean Becquerel began his magneto-optic experiments on salts of certain rare-earth elements, initially studying their absorption band spectra in a magnetic field and subsequently as a function of temperature. His samples, principally didymium-containing mineral crystals, were conveniently on hand at the Musée d'Histoire Naturelle from his father's old program. Before long he made a puzzling discovery. In some cases, the circularly polarized light was shifted both toward higher *and* lower frequency. Measured low-frequency shifts, in conjunction with the Lorentz–Voigt model, indicated positive vibrating charges with an *e/m* ratio in order-of-magnitude agreement with that for the ubiquitous negative electron [14-53]. Then, in 1907, exactly a year later, came reports of a truly *free* positive electron detected in cathode ray experiments. On March 22 of that year Julius Edgar Lilienfeld, a young Polish physicist working under Otto Wiener in the institute of physics at Leipzig University, submitted a paper at the meeting of the German Physical Society entitled 'On novel phenomena in the positive light column in a glow discharge' [14-54]. The 'novel phenomena' appeared to be positive particles comprising what he described as a positive column of red fluorescent light, moving co-jointly with the usual negative cathode ray particles but deflected in the opposite direction in a magnetic field. Using both electrostatic and magnetic deflection, he deduced an e/m ratio for the positive particles in the range 5×10^7–8×10^7 emu g^{-1}, implying positively charged electrons with a mass somewhat less than that of the negative electron. Presumably, he speculated, the particles were liberated in the presence of the strong electric fields in the discharge tube from gaseous atoms which, in common with the view of many, he pictured as some kind of conglomeration of negative and positive charges. (Note, however, that about equally many contemporary scientists were of the opinion that the positive charge attributed to atoms results from a deficiency of negative electrons, *not* from a constituent of opposite charge.)

Lilienfeld's claim was quickly refuted by the leading cadre of German cathode ray physicists, including the oldest and foremost authority on positive rays, Eugen Goldstein. Repeating his experiments, none of them saw anything more than ordinary cathode rays and Goldstein's canal rays (positive ions, not electrons). Evidently Lilienfeld's red light was caused by low-velocity electrons of the usual negative variety [14-55]. Nevertheless, Lilienfeld seems not to have been swayed in his opinion; as late as 1910 he insisted that new experiments of his own upheld his earlier claim [14-56].

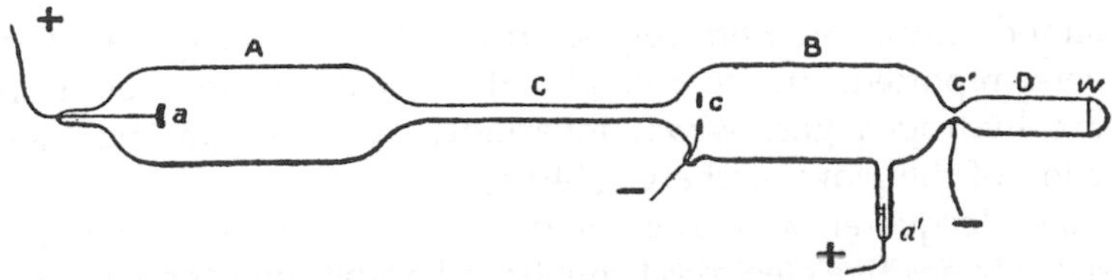

Figure 14.2. *Becquerel's discharge tube of 1908, purported source of free positive electrons.*

Meanwhile, in Paris Becquerel's own investigation continued. On the chance that the puzzling magneto-optic effect was caused, not by positive electrons, but by internal, inhomogeneous magnetic fields in the crystals, he extended the measurements down to the temperature of liquid air (80 K) with cryogenic apparatus in place at the Museum. Any internal field due to 'Amperian molecular currents' ought to change with temperature, and so should the effect in question. In the event, no significant change was seen, underscoring his belief in the positive electron [14-57]. By 1907, he had further spectral evidence for positive electrons with e/m ratios four to nine times larger than the ratio for cathode ray electrons, in fortuitously (as it would turn out) good agreement with Lilienfield's claim [14-58]. Despite the poor reception of the latter's results, Becquerel now decided that an independent test for *free* positive electrons in gaseous discharge tubes was in order, since 'the question has remained in suspense, Lilienfeld having, so far as I know, never since confirmed or disowned his first experiments' [14-59].

His new experiment was reported to the French Academy in June 1908. The discharge tube, shown in figure 14.2, was a rather elaborate contraption, blown by his assistant, M Matout. The cathodes c and c' were of aluminum, as were the anodes a and a'. The cathode c was perforated, allowing canal rays to penetrate into section B, as was the cathode c' at the entrance to the smaller section D. An auxiliary cathode was connected to a phosphorescent (willemite) screen W. In addition to ordinary cathode and canal rays, positive rays were indeed observed; they were very sensitive to magnetic deflection, and comparable to cathode rays in their deviability. From the sense of deflection, their charge was evidently positive. Depending on the magnetic field strength and residual gas pressure, they could be coaxed into forming a colored patch on the wall of tube B or D, or on the screen W.

> The only interpretation [declared Becquerel] which seems to me at all likely is that the beam, in the region where it can be deflected, is not formed of ions, but of *positive electrons*, comparable to the negative electrons, or at least possessing a ratio of charge to mass of the same order of magnitude. [14-60]

The particles only appeared in the simultaneous presence of both canal and cathode rays, and were very short lived. They disappeared rapidly, Becquerel reported, due to recombination with negative electrons and forming hydrogen gas, which invariably appeared in the tube after extinction of the glow discharge [14-61].

While Becquerel was thus preoccupied with his discharge tube, evidence for positive electrons came from another quarter, namely, Johns Hopkins University in Baltimore. Robert Wood, of all people, having decisively debunked Blondlot's N-rays four years earlier, announced independent observation of magnetic rotation of some of sodium's absorption D-lines in one direction and others in the opposite direction. In light of Zeeman's demonstration that the 'positive rotation' of the D-lines of sodium was accounted for by negative electrons, the reverse 'negative rotation' of the sodium D-lines presumably indicated the existence of vibrating positive electrons [14-62]. 'It is perhaps unwise to speak of a positive electron', Wood nevertheless cautioned, 'since electron has come to mean the disembodied negative charge, after it has been expelled from the atom'.

> Whether the two types of magnetic rotation proves the existence within the atom of both positively and negatively charged discrete particles is for the theoretical physicist to answer. The observations recorded in this paper merely prove that some of the absorption-lines give a rotation opposite to that given by the D lines. [14-62]

At about the same time as Wood's announcement, Becquerel shifted his magnetic rotation apparatus to Kamerlingh Onnes' laboratory at Leiden—a laboratory in which his father, Henri Becquerel, had been a frequent guest collaborator. Jean Becquerel's own collaboration with senior Leiden staff on magnetization experiments in rare-earth compounds would continue off and on for many years after Kamerlingh Onnes passed away in 1926—particularly with Onnes' successor and laboratory co-director Wander Johannes de Haas. Before long Becquerel was a familiar figure in the Leiden corridors. Hendrik G B Casimir, who himself began experiments in low-temperature physics at Leiden in the 1930s, and no doubt knew Becquerel then, recalls that

> Becquerel was a lively little fellow, quick of speech and quick in his movements. On helium days he would always get one or two refills and while waiting for the cryostat to come back from the liquefier he would walk up and down the corridor explaining with eloquent gestures and in rapid French—not withstanding his frequent visits, he never learned a word of Dutch—to anyone who would listen how things were going. [14-63]

We know from Kamerlingh Onnes' New Year letter to Woldemar Voigt at Göttingen in January 1908 that Becquerel initially arrived for a longer stay at Leiden early that month [14-64]. The cryogenic apparatus involved—the handiwork of *Glasbläsermeister* O Kesselring whom Onnes had imported from Jena—was far from simple, as is obvious from figure 14.3. The crystals were fixed with wax on a platinum foil attached to a rod which extended down into two concentric, double-walled Dewar vessels. The inner Dewar contained liquid hydrogen and the outer liquid air. Since the cryostat had to operate in a magnetic field between the poles of a Weiss-type electromagnet with tapered pole pieces only 15 mm apart, the lower end of the cryostat had to be very slender: the inner diameter of the hydrogen vessel was 4 mm, and the outer diameter of the liquid air vessel was 8 mm.

Casimir testifies to the vulnerability of the apparatus by relating a somewhat later incident involving a young Dutch physicist who had been assigned the task of assisting Becquerel in his experiments. Intending to check the liquid level with a flashlight, the young man forgot that the magnet was still energized. The flashlight was jerked from his hand, with disastrous consequences for the glass vacuum system [14-65].

Numerous magneto-optic measurements suggested themselves in the new temperature domain in liquid hydrogen, 20 K. High priority was placed on verifying the constancy of the magneto-optic frequency shift observed earlier in liquid air. Indeed, reported Becquerel and Kamerlingh Onnes before long to the Amsterdam Academy,

> ... the difference of frequency of vibration of the two components had then proved to be independent of the temperature. It follows now in a still more convincing way from the comparison of the divergence of the two bands at the temperature of liquid hydrogen with the divergence at the temperature of liquid air, that within the limits of errors of observation, the difference of frequency of vibration is entirely independent of the temperature. According to the theory of Lorentz this constancy of the divergence of the bands, which is observed both for those which behave in the sense of the Zeeman-effect as for those which behave in opposite sense, must be considered as proceeding from the invariability of the relation e/m. Accordingly the observations in liquid hydrogen seem to furnish a strong support to the argument in favor of the existence of positive electrons derived from the constancy of this quotient. [14-66]

Moreover, the band spectrum of tysonite in particular (a fluoride of a mixture of cerium, lanthanum, neodymium, and praseodymium), 'behaves as if it were owing both to positive and to negative electrons with the same period of vibration, and the same ratio e/m ...' [14-67]. The measurements would subsequently be extended and confirmed down to the temperature of solid hydrogen (14 K). The preliminary measurements in liquid hydrogen are also discussed in Kamerlingh

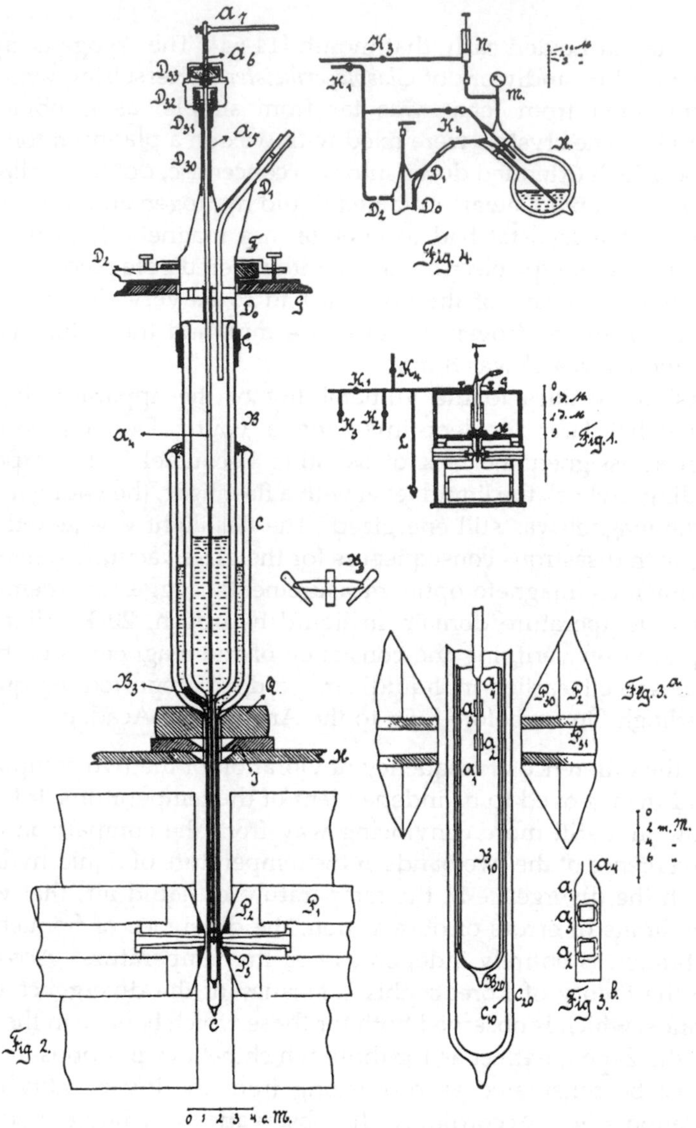

Figure 14.3. *Becquerel's low-temperature apparatus of 1908. The apparatus was the same with and without the magnetic field. The crystals were fixed with wax on a small piece of platinum foil which was carried by a rod protruding down into the hydrogen Dewar (figure 3(a)). B is the double-walled hydrogen Dewar of non-silvered vacuum glass, surrounded by another double-walled glass Dewar containing liquid air (c). At its narrow end, the outer Dewar is only 8 mm in diameter, and the wall clearance between the two Dewars is $\frac{1}{2}$ mm. The pole pieces (figure 3(a)) are perforated with a hole 3 mm in diameter. The rod a_4 can be turned and moved up and down. D_1 is a tube for siphoning liquid hydrogen from a supply bottle; the siphoning operation is indicated diagrammatically in figure 4. Stringent precautions were taken to prevent mixing of hydrogen and air. Once filled with cryogens, observations with the crystals could be made uninterruptedly for several hours.*

Onnes' letter to Voigt on November 5, in which he (perhaps uneasy with the primary conclusion) notes that the report was written by Becquerel alone [14-68].

Amidst Becquerel's success at Leiden came good news from Paris as well. Alexandre Dufour had reached the same conclusion regarding the need to invoke positive electrons in order to explain the inverse Zeeman effect in the band spectra of gases [14-69]. Confident, Becquerel wrote Lorentz that his latest experiments rendered the positive electron hypothesis all but a certainty [14-70]. Alas, his optimism proved premature. The very next year, 1909, Dufour had second thoughts on the matter, now attributing the anomalous magneto-optic effects to intermolecular magnetic fields [14-71]. He went as far as repeating Becquerel's discharge tube experiment, concluding that the purported positive electrons were nothing more than secondary canal rays produced when the cathode rays struck the glass wall [14-72]. Others pounced on Becquerel's cathode ray experiment as well, including the Italian Augusto Righi, A Bestelmeyer in Göttingen, and Marcel Moulin at Paris' l'École de Physique et de Chimie [14-73]. In Baltimore, Wood too expressed reservations about his earlier sodium observations, which he now decided could probably be explained by negative electrons alone [14-74].

Woldemar Voigt, who had in some sense, with Lorentz, precipitated the controversy on theoretical grounds in the first place, had followed the Leiden experiments with mixed feelings. He now re-examined the results in light of the two contending hypotheses: positive electrons versus intermolecular magnetic fields. Castigating 'the irresponsible introduction of electrons which have... masses and signs different from each other', he warned that in doing so 'we should lose what has been up to the present one of the chief advantages which characterise the electron theory, the simplicity of the fundamental conception, and thus make the whole hypothesis of less value' [14-75].

> The chief objection of J. Becquerel to the hypothesis of inner-molecular magnetic fields [Voigt continued] lies in the fact that the Zeeman-effect is notably independent of temperature. But as we do not know anything definite about the cause of the inner field, it appears to me, that one cannot assert anything about the sensibility to temperature with certainty. [14-76]

Lorentz, too, was privately becoming weary of vibrating positive electrons by 1909 [14-77], and at the Winnipeg meeting of the British Association that August J J Thomson, whom we have not heard of much on the matter, voiced similar skepticism in his inaugural address as President of the Association. In cases when the positive particles in discharge tubes are more readily deflected than canal rays, in his experience,

> . . . the ease with which they are deflected is due to the smallness of the velocity and not to that of the mass. It should, however, be noted that M. Jean Becquerel has observed in the absorption spectra of some minerals, and Prof. Wood in the rotation of the plane of polarization by sodium vapour, effects which could be explained by the presence in the substances of positive units comparable in mass with corpuscles. This, however, is not the only explanation which can be given of these effects, and at present the smallest positive electrified particles of which we have direct experimental evidence have masses comparable with that of an atom of hydrogen. [14-78]

Rutherford, too, in his opening address as President of the Section on Mathematics and Physics at Winnipeg, agreed that 'there is yet no decisive proof of the existence of a . . . positive electron' [14-79].

Becquerel himself never really abandoned his stance on the positive electron, though he seems to have soured on the concept after 1910, the last year he made explicit reference to it. He would continue his magnetization studies at Leiden into the 1930s, explaining, with de Haas, the Faraday effect in rare-earth crystals within the context of the new quantum mechanics and in terms of diamagnetic and paramagnetic effects [14-80]. As for his earlier particulate views, his retrospective musings on the positive electron hypothesis lays the matter to rest in the commemorative volume issued in 1922 on the occasion of the 40th anniversary of Onnes' professorship.

> Allow me to remark that, in the current state of science, we are no longer justified in claiming the existence of positively charged particles possessing a ratio of charge to mass of the same order of magnitude as for cathode ray particles. [14-81]

With that, enough said about a conceptually important controversy which, however, played a minor role in the history of the electron. Though the positive electron faded from view in favor of a single negative subcarrier of electricity, *positive rays*—doubtless a factor in the aforesaid protracted dispute—did not, as we now shall see.

Chapter 15

POSITIVE RAYS

15.1. Positive rays and Thomson's crisis

While the positive electron came and went during 1906–1909, the same period ushered in definitive studies of another species of positive particles in glow discharges. For the latter developments we must return to J J Thomson who had, on the whole, stayed out of the controversy surrounding Becquerel's ongoing researches in Paris and Leiden. The four years culminating in 1899 with J J's 'corpuscles' would prove to have been the most productive experimental period in his career. There followed a quiet period from about 1900 to 1906, during which Thomson took up the theoretical problem of atomic structure, prepared his monumental treatise on *Conduction of Electricity through Gases* [15-1], and otherwise occupied himself with administrative, teaching, and other matters claiming his day-to-day attention as the Cavendish Professor. Thomson's second important experimental period began partly in response to a crisis in his theoretical ruminations, culminated ca. 1912, and now begs our own attention. First, however, we must make our acquaintance with Willy Wien, who laid the groundwork for Thomson's new experiments.

Eugen Goldstein, we may recall, had found in 1886 that, by piercing the cathode in a discharge tube with fine holes, self-luminous rays could be observed streaming 'backwards' into the space behind the cathode away from the discharge. Their color was strongly dependent on the residual gas in the tube; they also appeared insensitive to the deflection by magnetic fields available to Goldstein with permanent magnets—fields strong enough to readily deflect the non-luminous cathode rays. Having coined the term *Kathodenstrahlen* for the latter rays earlier, Goldstein now tentatively dubbed the luminous rays *Kanalstrahlen*—a term which also would stick along with the subsequent 'positive rays'. The magnetic (and electric) deflection of Goldstein's stubborn *Kanalstrahlen*, showing that they consisted of positively charged particles, was first accomplished in 1898 by his sometime student, the brilliant and versatile Wilhelm Wien.

Wilhelm ('Willy') Wien was born the only child to a farmer and landowner on January 13 1864 in Gaffken near Fischhausen, East Prussia [15-2]. Torn between becoming a gentleman farmer and studying mathematics and physics, he attended the Gymnasium in Rastenburg for a period. He then marked time with a private tutor, finally graduating from the Gymnasium in Königsberg in 1882. Still undecided, he briefly studied mathematics and physics at Göttingen, and wandered about for a period. He resumed his studies at, successively, the University of Berlin (under Helmholtz), Heidelberg (G H Quincke), and again in Berlin. He received his doctorate in 1886 under Helmholtz, albeit on the basis of a poorly judged final examination, with a dissertation on the diffraction of light upon striking sharp metallic edges.

With his doctorate in hand, he returned to the family agricultural estate, by now relocated to Drachenstein, again inclined to take up farming. In this he was actually encouraged by Helmholtz, who in 1888 was appointed the first president of the newly founded Physikalisch-Technische Reichsanstalt (PTR) in Charlottenburg, and by Helmholtz's successor in Berlin, August Kundt. However, when drought forced his parents to sell the farm in 1890, Wien accepted a position as assistant to Helmholtz at the PTR. There he launched his career as a physicist. Despite a shaky start, he would soon make his mark, equally adept in experiment and theory—a distinction rare but not unheard of, witness Eduard Riecke. Wien's major research was to center on the theory of black-body radiation, including his famous displacement law and attempts to obtain a formula for the distribution of energy in the normal spectrum—work that brought him the Nobel Prize in 1911 [15-3]. However, for our purpose his cathode ray experimentation is the thing of immediate interest.

Wien remained at the PTR until 1896, when Friedrich Kohlrausch succeeded Helmholtz as head of the Reichsanstalt. Dissatisfied with the inflexible research plan drawn up by Kohlrausch, Wien gladly accepted an offer to succeed Philipp Lenard as extraordinary professor (*Extraordinarius)* at the Technische Hochschule of Aachen. In 1899, he was appointed full professor (*Ordinarius)* at the University of Giessen, but left after only six months to succeed W C Röntgen at Wurzburg. There he remained for the next 20 years, until he assumed his last appointment at Munich in 1920, again succeeding Röntgen.

When Wien succeeded Lenard at Aix-la-Chapelle, he also inherited the laboratory left by his predecessor—a laboratory well equipped for cathode ray studies, as Lenard was quick to point out [15-4]. Wien himself took up cathode ray studies in early 1897. In so doing, he joined his countrymen Wiechert and Kaufmann (and, to some extent, Lenard, we might add) in establishing the particulate nature of cathode rays, simultaneously with J J Thomson abroad. Using one of Lenard's vacuum tubes which, equipped with an aluminum window, allowed the rays to pass into a tube extension with a highly rarefied vacuum, he first confirmed Jean Perrin's discovery two years earlier that the rays consist of high-velocity negatively charged particles [15-5]. By magnetic and electrostatic deflection he also found that the particles have a velocity about equal to one-third that of light, and a mass-to-charge ratio of 5×10^{-8} g emu^{-1}, in qualitative agreement with e/m values for electrons deduced by his colleagues at home and abroad. The following year, having pursued the cathode rays to his satisfaction, he turned to a more timely question, the nature of Goldstein's *Kanalstrahlen*.

> ...the thought came to me that the canal rays observed by Goldstein, which cannot be deflected appreciably by ordinary magnets, and which proceed backwards through a pierced cathode, might carry the positive charge. In this case it was not possible to shut out the field of observation completely from the discharge tube, because in spite of many attempts I could find no substance through which the canal rays will pass. [15-6]

The apparatus that finally did the trick is depicted in figure 15.1. Here, the 'canal' is simply a 2 mm hole bored in an iron plate K (the cathode), to which glass tubes are cemented on each side. The anode is sealed into the far end of the left-hand tube, as shown. In the right-hand tube a pair of electrodes are mounted, spaced 1.7 cm apart and connected to a high-voltage accumulator. NS are the poles of a Rühmkorff electromagnet; the rest of the shaded area represents a shield of soft iron to prevent stray magnetic field from upsetting the discharge in the left-hand tube. With this arrangement it can be seen (figure 15.2) that Wien's method of simultaneously deflecting the canal rays by magnetic and electric fields differed from J J Thomson's method for cathode rays in one important respect: the deflections, in Wien's scheme, instead of being *opposed*, as in Thomson's, were applied at *right angles*, or, what amounts to the same thing, the magnetic fields between the two poles were parallel to the electric field lines joining the two electrodes.

In practice, when the discharge tube was sufficiently exhausted, there emerged from the hole in the cathode a beam of canal rays, terminating in a spot of yellow–green fluorescence on the glass wall 9 cm to the right of the iron plate. With a potential difference of 2000 V applied to the electrodes, the beam was deflected 6 mm in the direction of the negative electrode, but not for very long; the canal rays soon increased the conductivity of the residual gas between the plates to the extent of inducing a spark discharge. As Goldstein knew full well, the magnetic deflection was more problematic. Thus, a powerful ordinary horseshoe magnet barely produced a discernible deflection—at any rate one too small for measurement. It took the Rühmkorff electromagnet, charged

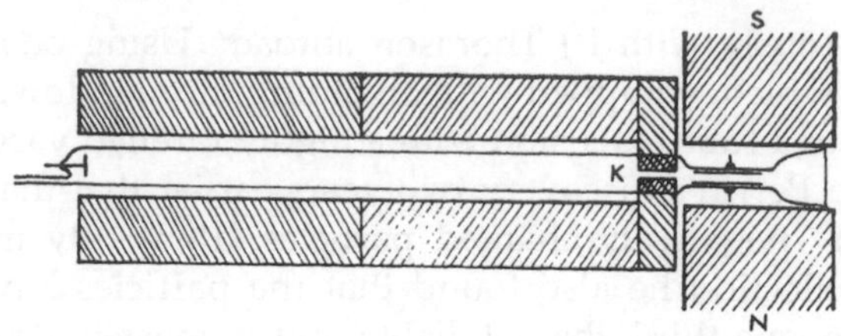

Figure 15.1. *Wien's positive ray apparatus of 1898. K is the cathode pierced with a single 'canal'. NS are the poles of an electromagnet, between which is the deflection tube extension containing a pair of electrostatic deflection plates and a flat surface coated with a Willemite screen.*

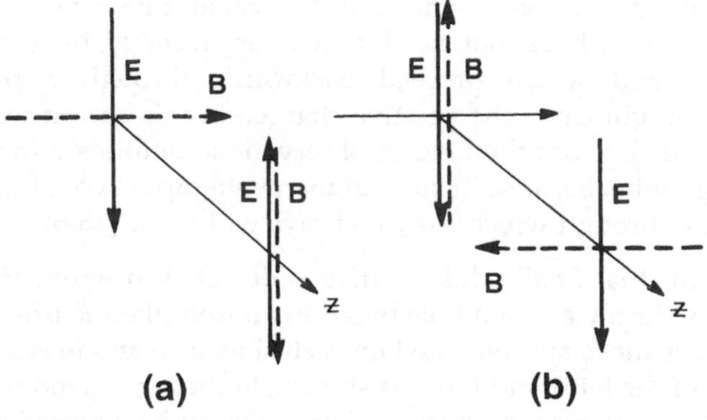

Figure 15.2. *A comparison of the deflection schemes for cathode rays (Thomson, a) and positive rays (Wien and Thomson, b). The electric and magnetic field vectors* **E** *and* **B** *are shown in the rear, and the resulting deflections in the front. (Z is the direction of propagation of the negative and positive particles, respectively.)*

to a poletip field of 3250 cgs units (or 3250 G, as measured with a Stenger bifilar galvanometer) to produce a deflection of 6 mm. The electric deflection is inversely proportional to the kinetic energy of the particles (as we will show somewhat later),

$$y = \frac{1}{2}\frac{e}{mv^2}El(l + 2d)$$

where E is the electric field, v is the velocity, l is the length of the electrodes, and d is the distance from the end of the plates to the far wall of the deflection tube. The magnetic deflection goes inversely with their momentum, or

$$x = \frac{1}{2}\frac{e}{mv}Bl(l + 2d)$$

where B is the magnetic field (or, strictly speaking, the induction). Both deflections are thus directly proportional to the charge of the particles [15-7].

If the particles are acted on simultaneously by both fields, no two particles of unequal velocity or charge-to-mass ratio will strike a suitably placed fluorescent screen in the same spot. However, it is not difficult to show that all particles of the same e/m-ratio should fall on a parabola whose vortex is the point where the undeflected beam meets the screen. In addition, particles with the same velocity but different e/m-values will lie on straight lines passing through the undeflected point. (We will have more to say about these relationships presently.) Wien did, indeed, observe streaks passing through this point, which he interpreted in terms of particles with a roughly constant velocity of about 3.6×10^7 cm s^{-1} and minimum m/e-value of about 3.2×10^{-3} (hence $e/m \sim 300$). Comparison with table 9.1 or 9.2, above, shows that this velocity is two orders of magnitude lower than that typical of cathode ray particles, while the charge-to-mass ratio is about 1/30 000 of the cathode ray value, or about as for a hydrogen ion in electrolysis.

Thomson was certainly familiar with Wien's positive ray researches. He referred to Wien's *Annalen* article of 1898 in his own definitive paper the following year, 'On the masses of the ions in gases at low pressures'. In it, he noted that, in accordance with Wien's e/m-ratio, 'the carriers of positive electricity at low pressures seem to be ordinary molecules, while the carriers of negative electricity are very much smaller' [15-8]. He also remarked on Wien's e/m-result in an oral presentation before the Cavendish Physical Society at about the same time. However, when, afterwards, Rayleigh pressed him for details of Wien's measurements, Thomson claimed not to remember. 'He had concentrated on the electron, and had evidently given only very superficial attention to Wien's work at that time' [15-9].

Despite his preoccupation, Thomson himself started work on the positive rays in 1906. He now regarded the rays 'as the most promising subjects for investigating the nature of positive electricity'. Accordingly, he undertook a more careful 'series of determinations of the value of e/m for positive rays under different conditions' than hitherto possible [15-10]. It was, in fact, not his first experimental encounter with the rays; his initial attempt at their magnetic deflection had been made the year before, in connection with his study of the emission of ions from incandescent metals. The method then employed was the same as described earlier (p 186) for determining e/m for negative ions produced under the action of ultraviolet light. The positive rays were produced by raising an iron wire to a red heat in an atmosphere of oxygen under reduced pressure; the wire ran parallel to a metal plate connected to an electrometer. However, the results were highly inconsistent, the ions more often than not refusing to be deflected in the strongest fields then available at the Cavendish, 12 400 cgs units (G). Thomson was at first inclined to blame the erratic behavior on irregular heating of the wire, which relied on a transformer connected to the Cambridge public supply grid fed by an ancient power station on the bank of the Cam near Magdalene College. (It supplied alternating current at the unconventional rate of 93 cycles s^{-1} and 200 V!) Rayleigh recalls that Thomson 'frequently and vigorously' demanded over the telephone that the supply voltage be kept constant.

> The then manager of the Supply Company was not the man to take this lying down and he retorted by sending in a bill: 'To man's time listening to complaints on the telephone, 1 s'. This lay about in the entrance lobby of the laboratory for a long time. [15-11]

In the end, Thomson did manage to deduce a rough value for e/m of about 400 emu g^{-1}, which 'is about the value for e/m for the ion of iron in electrolysis' [15-12].

> [However, he added,] it does not... prove that the carriers of positive electricity are the atoms of iron, for if m were the mass of a molecule of oxygen and e the charge on a hydrogen ion in the electrolysis of solutions e/m would be about 310, and the difficulties of the experiment are so great that we cannot say that this result differs from that actually found by more than the possible errors of the experiment.
> We see however that for the positive ions e/m is of the same order as in ordinary electrolysis of solutions, and is variable, while for negative ions it is of an entirely different order, and is constant. [15-13]

Thomson's positive ray program marked a critical juncture in his theoretical notions on electricity and matter. Keeping in mind his above admonition, by 1905 he fully expected to observe positive protyle units of electricity analogous to his negative corpuscles. At the same time, as we shall see, he took the hydrogen ion, H^+, as the positive subunit, since it invariably showed up in gaseous discharges among the many other ionic species prone to appear, depending on the residual pressure. Only in 1910, after frustrating experiments, prolonged discussions with Wien, and after listening attentively to his new research assistant, F W Aston, was he persuaded otherwise. However, that is leapfrogging the sequence of events by several years.

To appreciate the intellectual crisis behind Thomson's new experimental program, we must dwell briefly (at the very least) on certain aspects of his views on the structure of the atom ca. 1906. As may be recalled (section 10.2), in his paper of 1899 already cited in the above, he proposed to make do with negative corpuscles alone in accounting for the structure of neutral atoms, neglecting how one might account for an atom's mass simply with corpuscles (except vaguely in terms of 'æthereal mass'). The neutralizing positive ingredient was provided by a massless mortar, much as in a cohesive liquid, that held the corpuscles together, and positively or negatively electrified atoms were explained by corpuscles moving from or toward these neutral atoms.

Alas, a single-fluid theory proved unsustainable by 1906, as Thomson himself would show [15-14]. To be sure, Becquerel's short-lived positive electron surfaced in France about then, but the majority of physicists, among them Thomson, saw no need for it. What undermined Thomson's unitary viewpoint was n, the number of corpuscles in the atom. The very small mass of the electron, with a single-fluid theory of electricity, at first suggested 1000 electrons in a hydrogen atom, and correspondingly more in heavier atoms—an estimate echoed in the closing pages of Rutherford's book of 1906, *Radioactive Transformations*, and borne out in the richness of spectra emitted from even the lightest of atoms [15-15]. Indeed, Oliver Lodge suggested 20 000–30 000 electrons in the sodium atom and no less than 100 000 in the mercury atom. However, the same year, 1906, Thomson himself, in an important paper, marshalled evidence via three different lines of reasoning that the number of corpuscles per atom was actually far less than that—about equal to the atomic weight [15-16]. Herein lay part of the 'crisis'; reducing n so drastically reopened what had earlier been seen as a stability problem, but brought under control by a certain scheme of Thomson's—a scheme we discuss in chapter 17.

As Thomson knew full well, accelerating charged corpuscles radiate energy, as demanded by classical electromagnetic theory; they slow down, and an atom containing several thousand such corpuscles becomes unstable. An atomic model of this type, consisting of a sphere of uniformly distributed positive electricity in which discrete electrons, or 'electrions', were embedded, was first proposed by Kelvin in his *Æpinus Atomized* of 1901–1902 [15-17]. In Kelvin's version, stability was ensured with the charges at rest, in accordance with a fanciful scheme predicated on untold numbers of constituent particles interacting via Boscovichian forces (chapter 17). Thomson had reexamined Kelvin's static atom in 1904, showing that for a rapidly rotating ring of electrons within the positive, diffuse sphere, stability grew with the number of electrons in the ring [15-18]. Alas, the depleted electronic population of 1906 brought the radiation problem to the fore again. Moreover, presuming with Thomson all matter to be electrically charged, it followed that the positive electricity carried most of the atomic mass [15-19]. That was not all. Comparison of theory with experiment in the case of the dispersion of light showed that the mass of the carrier of unit positive charge was large compared with that of the negative unit [15-20]. Herein lurked an even greater crisis: positive electricity ceased to be a nearly weightless substance, something largely an artifice in Thomson's atomic scheme. The unavoidable fact of the existence of a fundamental, positive particle—a massive positive counterpart to the corpuscle—had to be reckoned with, and so Thomson did:

> The most obvious interpretation of this result [that the mass of the particles of positive rays was never less than that of a hydrogen ion and that it might be variable] is that... all gases give off positive particles which resemble corpuscles, in so far as they are independent of the nature of the gas from which they are derived, but which differ from the corpuscles in having masses comparable with the mass of an atom of hydrogen. [15-21]

And thus it came about that Thomson embarked on an entirely new experimental program, one focusing his efforts almost exclusively on the canal rays hitherto the province of his German colleagues, and one to occupy him for the better part of a decade.

15.2. An experimental dead end

Most of Thomson's positive ray results were published in a series of eight papers in the *Philosophical Magazine* between 1907 and 1912. The first appears in the *Philosophical Magazine* for May 1907. Simply titled 'On rays of positive electricity', it reports on experiments largely completed late the year before, as we know from Thomson's extant laboratory notebooks, and judging by the tone of Thomson's letter to Rutherford in mid-December 1906 [15-22]. As usual, Everett assisted Thomson [15-23]. The rays, deflected by magnetic and electric fields, were observed by the phosphorescence they produced on a screen at the flattened end of a discharge tube. A number of phosphorescent substances were tested for maximum brightness, as were methods for adhering them to glass backings. (Most adhesives tended to 'outgas' under bombardment in vacuum.) In the end, the method chosen was dusting finely powdered Willemite (a mineral) over a glass plate smeared with 'water-glass' (sodium silicate).

The apparatus, as sketched in Thomson's notebook for October–December 1906, is shown in figure 15.3. The date of entry of the sketch appears to be late November [15-24]. The same apparatus, as depicted in the published paper, is also shown in figure 15.4; letters in the description that follow correspond to those in the latter figure. Most of the features of Wien's earlier arrangement (figure 15.1) are apparent. A hole was bored through the cathode C, into which was firmly fixed the fine tube F. Here we have a decided improvement over Wien's 2 mm wide 'canal'. In Thomson's apparatus, the bore of the tube F was made as narrow (and as long) as possible to collimate the pencil of rays accurately and obtain a well defined fluorescent spot on the screen. A long, narrow tube was equally important for maintaining a high differential pressure across the cathode; that is, maintaining a suitably high pressure in the discharge tube for keeping the glow discharge going (say, 0.01 mm Hg) without requiring excessive tube voltage, while at the same time maintaining

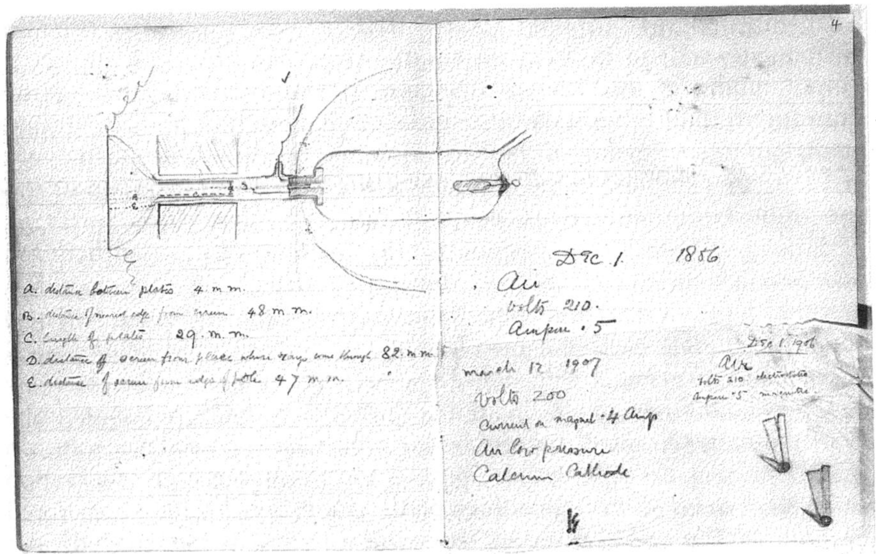

Figure 15.3. *Postive ray tube sketched in Thomson's notebook of 1906. Also shown are traces of positive ray figures, drawn on pieces of tissue paper fastened to the notebook page with straight pins.*

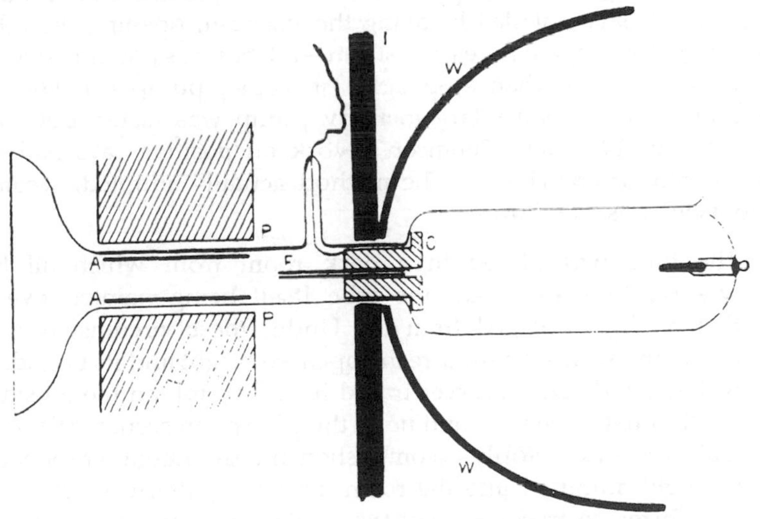

Figure 15.4. *Thomson's first positive ray apparatus.*

the pressure in the observation chamber as low as possible (say 0.001 mm Hg) to minimize collisions with residual air molecules. To this end, Everett did as well as could be expected with his glass blowpipe; hypodermic needles were also used, which Thomson found to 'answer excellently for this purpose' [15-25]. (Later, even finer tubes, 0.1 mm in diameter and up to 7 cm in length, were made by drawing down copper tubing to the desired diameter and straightening the twisted strand by rolling it between two pieces of steel with accurately ground plane surfaces.) Several of the early positive ray tubes, if not the very first one, are exhibited in the museum of the Cavendish Laboratory, and one of the latter ones may be seen in London's Science Museum [15-26].

Emerging from the tiny aperture, the positive rays passed between two parallel aluminum or brass deflection plates AA, mounted in a portion of the observation tube squeezed between the poles PP of a powerful Du Bois electromagnet. The electric and magnetic fields caused vertical and horizontal deflections, respectively, of the fluorescent patch on the Willemite screen. To shield the discharge in the right-handed tube from the magnetic field, the discharge tube was encased in a soft-iron vessel W, and the tube section joining the discharge and observation chambers extended through a thick plate of soft iron I with a hole bored through it. The soft iron served the same function as the (shaded) iron shield in figure 15.1 of Wien's apparatus.

When undeflected, the rays produced a bright spot in the center of the Willemite screen. When deflected, the fluorescent spot was drawn into bands on the screen. Their detailed shape could have been recorded by substituting a photographic plate for the Willemite screen. However, this would have entailed breaking the vacuum, opening the tube, and pumping down again for each test run—a laborious and time-consuming process at a time when a Spengler or Töpler pump was still worked by hand. The Gaede rotary mercury pump was faster, but was only invented in 1905, and Thomson's work on positive rays had initially to do without one [15-27]. The method actually adopted, according to Thomson, was as follows:

> The tube was placed in a dark room from which all light was carefully excluded, the tube itself being painted over so that no light escaped from it. Under these circumstances the phosphorescence on the screen appeared bright and its boundaries well defined. The observer traced in Indian ink on the outside of the thin flat screen the outline of the phosphorescence. When this had been satisfactorily accomplished the discharge was stopped, the light admitted into the room, and the pattern on the screen transferred to tracing-paper; the deviations were then measured on these tracings. [15-28]

The fragile tissue paper scraps with their traced patterns and gas type, magnet current, and deflecting voltage added are meticulously preserved, still attached to the pages of Thomson's laboratory notebooks with straight pins. An example, apparently the very first, is shown in figure 15.3 [15-29].

The electric deflection of a particle of charge e and mass m in passing through a region of potential difference V between a pair of plates separated by a distance s is readily derived, if we neglect end effects. If l is the distance of uniform field (the length of the plates, in our approximation), and v is the initial velocity of the particle, it experiences a constant transverse acceleration given by $a = (e/m)V/s$, and in the time l/v acquires a transverse velocity $v_x = a(l/v) = (e/m)(V/s)(l/v)$. Upon leaving the region between the plates, the particle will have undergone a vertical deflection

$$x_1 = \tfrac{1}{2}at^2 = \tfrac{1}{2}(e/m)(V/s)(l/v)^2.$$

It takes a time d/v to travel the remaining field-free distance d to the screen, so the additional deflection is

$$x_2 = v_x(d/v) = (e/m)(V/s)(l/v)(d/v),$$

and the total deflection is the sum of the two, or

$$x = \frac{1}{2}\frac{e}{mv^2}\frac{V}{s}l(l + 2d). \qquad (15.1)$$

(This is the same expression for the electric deflection as given in section 15.1, since the field strength $E = V/s$, and where, however, the coordinate conventions for x and y are reversed.)

The magnetic deflection is not so simple analytically (or experimentally!), even for a uniform field. To start with, by equating the magnetic force acting on the particle, Bev, to the centripetal force mv^2/r, where r is the radius of the circular path, we know (e.g. p 159) that $1/r = Be/mv$. If y_1 is the transverse displacement of the particle traveling initially in the direction z, the equation for the circular path is readily shown to be

$$r^2 = (r - y_1)^2 + z^2$$

which can also be written as

$$2ry_1 - y_1^2 = z^2$$

(a relationship also discussed on p 165). If, as is usually the case, $y_1 \ll r$, we may neglect y^2, and obtain

$$y_1 = \tfrac{1}{2}(z^2/r).$$

Substituting in our earlier expression for $1/r$,

$$y_1 = \frac{1}{2}\frac{e}{mv}Bz^2.$$

In the field-free region the particles move in straight lines, and the deflection y_2 is given by

$$y_2 = d \tan\phi = d\left(\frac{\mathrm{d}y}{\mathrm{d}z}\right)_{z=l} = d\left(\frac{eBz}{mv}\right)_{z=l} = d\frac{eBl}{mv}.$$

(A similar treatment, $x_2 = d\,(\mathrm{d}x/\mathrm{d}z)_{z=l}$, would have worked in the field-free electric case.) The total magnetic deflection is

$$y = \frac{1}{2}\frac{e}{mv}Bl(l + 2d). \tag{15.2}$$

We may express equation (15.1) and equation (15.2) as

$$x = C_1\frac{e}{mv^2}$$

$$y = C_2\frac{e}{mv}$$

where C_1 and C_2 are constants depending only on the magnitude of the electric and magnetic fields and the dimensions of the apparatus. We may equally well write x and y as

$$y^2 = \frac{C_2^2}{C_1}\left(\frac{e}{m}\right)x \tag{15.3}$$

and

$$y = \frac{C_2}{C_1}vx. \tag{15.4}$$

These two expressions, relating y and x for constant values of e/m and v, are the equations of a parabola and straight line, respectively. Thus, according to equation (15.3), for a constant value of e/m, particles of all velocities will trace a parabola upon reaching the screen, the length of the parabola depending on the range of velocities present among the ions in the beam. Taking the position of the undeflected beam as the coordinate origin, half of the curve (that with positive values for both coordinates) will lie wholly in one quadrant with its vortex at the origin; by reversing the magnetic field, the other half of the curve (with negative coordinates) is obtained. A separate parabola is described for each value of e/m present. From the parameters of each parabola, Thomson could, in principle, calculate the corresponding e/m-values.

According to equation (15.4), particles of a given velocity will trace a straight line on the screen, a separate line for each velocity present in the beam, with the length of each line depending on the range of e/m-values for that velocity.

In practice, the magnetic field was far from uniform. An exact calculation of the magnetic deflection would have required mapping the

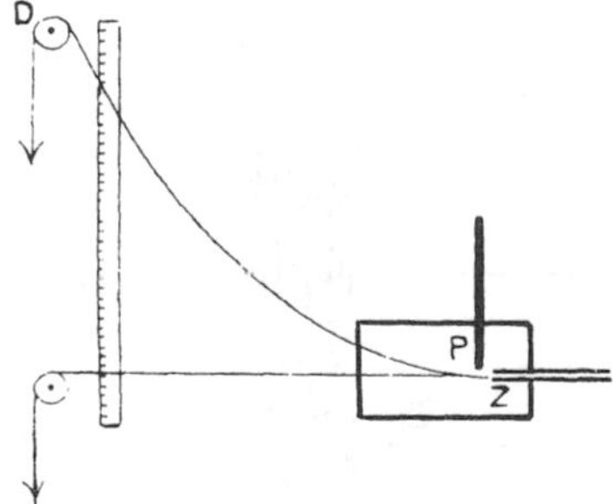

Figure 15.5. *Thomson's method for determining the magnetic field distribution by measuring the deflection of a current-carrying wire in a magnetic field.*

field along the beam path in considerable detail—a laborious process. Instead, Thomson at first resorted to a scheme which was ingenious, if almost as cumbersome. It depended on the fact that a flexible, current-carrying wire under tension in a magnetic field will assume the shape of a curve which is a simple function of the current, field, and tension in the wire [15-30]. A fine phosphor bronze wire soldered at one end to the end of the narrow tube F was passed over a pulley, and carried a weight in the free end, as indicated in figure 15.5. The pulley was first placed so that the shape of the stretched wire, when undeflected by a magnetic field, coincided with the path of the undeflected rays. When the field was turned on, the wire was deflected. The amount of deflection was measured by observing with a microscope a vertical scale placed at the same distance from the tube F as the phosphorescent screen in the actual ray experiment. Observing the vertical height of the wire with and without the field, y_1 and y, in both cases taking care that the soldered wire end was horizontal as it left the opening of the tube, gave a relationship between the two heights of the form

$$\frac{y}{y_1} = \frac{e/mv}{i/T}$$

a relationship involving the current i and the tension T, but one from which the functional dependence of the magnetic field has been eliminated [15-31]. Since y, y_1, i, and T were readily measured, the expression gave e/mv without further ado. (Deflection by the electric field gave e/mv^2.)

It appears from Thomson's notebooks that he and G W C Kaye, who assisted him in these stretched wire measurements, devoted rather less time to them than might be expected, considering their sophisticated and time-consuming nature [15-32]. Thomson's appointment as Professor of Natural Philosophy at the Royal Institution in 1905 (while retaining his Cambridge chair), on the resignation of Rayleigh, had entitled him to an

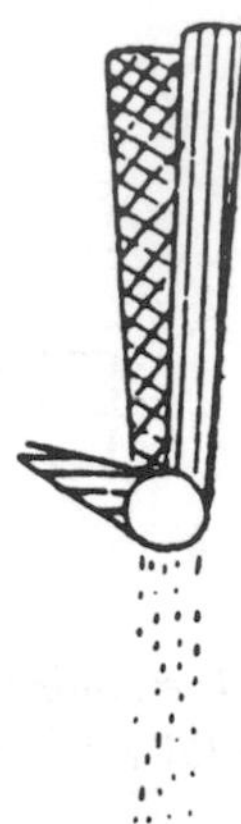

Figure 15.6. *The 'high-pressure' fluorescent pattern. 'The deflexion under magnetic forces alone is indicated by vertical shading, under electric force alone by horizontal shading, and under the two combined by cross shading'. The faint vertical traces are from negatively charged rays.*

assistant; the managers kindly agreed that Thomson might avail himself of his services, not only at the Institution, but wherever he desired [15-33]. George W C Kaye was selected—an excellent choice. He had studied physics for a short time under Oliver Lodge at Liverpool University, from where he migrated to the Royal College of Science [15-34]. There, after taking his degree with first-class honors in physics, he remained as a demonstrator and was appointed a teaching associate in 1904. From Kensington he went to Trinity College, Cambridge, where he enrolled as a research student in 1905. With his thorough grounding in science, he also proved a good complement to the faithful Everett and his purely mechanical talents, who could not be entrusted with the responsibility of making observations. (Note, however, that on occasion Everett could make subtle contributions, such as in preparing special, electropositive cathodes, as Thomson made a point of informing Rutherford [15-35].) In time Kaye moved on to a distinguished career at the National Physical Laboratory; in the short run Francis W Aston would succeed him as Thomson's research assistant, thereby opening up another chapter in the positive ray saga, but that is a story we defer for now.

The 1907 results were very sensitive to the tube pressure. At 'high pressures'—ca. $\frac{1}{50}$ mm Hg—the fluorescent spot was drawn into approximately straight bands, as in figures 15.6 and 15.7: a bright band attributed to positively charged rays, and a band of much fainter traces from negatively charged rays. The bright band indicated rays of nearly constant velocity, but e/m-values ranging from zero to 10^4 (in the usual units). The maximum value corresponded to the singly charged hydrogen ion (H^+)—ions invariably present regardless of the

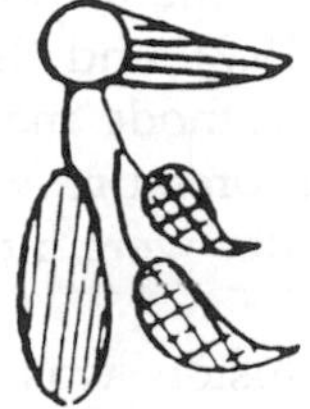

Figure 15.7. *A typical 'low-pressure' fluorescent pattern.*

gas in the tube and despite deliberate efforts to exclude hydrogen gas. (Doubly charged hydrogen *molecules*, H_2^{2+}, were another, if less likely, possibility) [15-36]. The faint traces of negative charges were of little interest; they could be explained by multiple charge exchange or as the result of hydrogen ions acquiring more than one residual corpuscle along their flight path. More importantly, the fluorescent pattern was distinctly curved in a residual hydrogen atmosphere, suggesting a bulge from an additional positive ion species, besides the usual band (as in air) with e/m corresponding to hydrogen, 1.2×10^4. The maximum e/m-value in the second band was 5×10^3, and could either be attributed to singly charged molecular hydrogen or doubly charged helium atoms. Moreover, in helium gas a third band showed up, indicating an e/m-value almost exactly one-quarter the hydrogen value in air, or 2.9×10^3, again suggesting atoms of helium (now with a single charge).

> The case of helium is an interesting one [wrote Thomson]; for the class of positive rays, known as the α rays, which are given off by radioactive substances, would *a priori* seem to consist most probably of helium, since helium is one of the products of disintegration of these substances. [15-37]

With that, Thomson let the enticing matter of the α-particle rest for the time being.

More difficulty was encountered in working at substantially lower, but unspecified pressure. (Thomson's characterization of the tube pressure leaves much to be desired. His notebooks are peppered with statements such as: 'brilliant green phosphorescence'; '1 cm dark space'; 'good phosphorescence'; 'little lower pressure' [15-38].) The discharge could only be coaxed through the tube with such high voltage that sparking punctured the the glass bulb almost at once, destroying the vacuum before Thomson and Kaye could trace the phosphorescent image on the tube wall. The problem was overcome by the use of special, electropositive, cathode materials, such as the liquid alloy of sodium and potassium smeared over the cathode. Later in the investigation, an easier remedy was found in simply enlarging the volume in the discharge tube (thereby lengthening the extent of the 'dark space' between the cathode and the glass wall). At the lower pressure, the bands broke up into isolated patches (figure 15.7), and the negative spots disappeared, irrespective of the gas or the cathode material. One patch gave 10^4 for e/m, one 5×10^3, and (if the pressure was not too low) another gave $\sim 2.5 \times 10^3$—that is, the results were not qualitatively different from those at higher pressure.

Thomson's primary conclusion was twofold: the range of e/m-values, whether from bands or patches, represented variations in *electric charges*, not in *masses*, and that H^+ was the fundamental unit of positive electricity—the positive, protylean counterpart to his corpuscle. The failure to observe 'characteristic rays' (rays for which $e/m = 10^4/M$, where M is the atomic weight of the gas in question), except for helium, was, Thomson felt, purely a technical problem: one caused by the inability to produce positive rays with much smaller potential differences than hitherto employed. High potentials rendered asunder the heavier gases which undoubtedly took part in the discharge, as attested to by the nitrogen bands clearly revealed in a spectroscope trained on the discharge.

Unbeknown to Thomson, the trouble was the fluorescent screen. All fluorescent screens, and photographic plates for that matter, are very insensitive to heavier ions [15-39]. The proportion of hydrogen gas was normally not particularly high in these experiments, except when hydrogen-containing gas was deliberately admitted. The insensitivity of the screen was due to the fact that, for equal energy (or 'cathode fall of potential') the lighter ions are the fastest and most effective in exciting fluorescence in, say, Willemite.

The second major paper in Thomson's positive ray program appeared in the *Philosophical Magazine* for October 1908 [15-40]. The new round of experiments was intended to account for the predominance of H^+, and to bolster Thomson's notion that, like the corpuscles, the H^+ ions could be produced from well-nigh any gaseous substance by ionizing collisions. Perhaps, he now speculated, a first stage of ionization might be the formation, not of a single corpuscle or positive ion, but an electrically neutral 'doublet' containing a corpuscle revolving around a much more massive positive particle. Secondary cathode and positive rays, in this scenario, were due to the subsequent breakup of the doublet, their nearly constant velocity reflecting the definite attractive force between the positive and negative constituents. Attempts to detect such doublets, the details of which need not concern us here, proved highly inconclusive, however [15-41], though Thomson was not yet shaken in his faith in positive as well as negative protylean subunits of matter.

If nothing else, the latest experiments, using a tube incorporating auxiliary electrodes capable of producing counteracting potentials, dramatically underscored the complexity of the discharge tube phenomena at hand: an indiscriminate potpourri of positive, negative, and neutral rays, some primary, others secondary, and possessing varying amounts of charge along their paths, including 'retrograde' rays—positive *or* negative rays of mass similar to that of the normal positive (canal) rays, but moving in the opposite direction. In the end, Thomson was compelled to characterize his results as 'only preliminary', with 'more experiments... necessary' [15-42].

Despite the desultory results, Thomson's papers continued to appear with uncommon regularity. The next one, published in the December 1909 issue of the *Philosophical Magazine* [15-43], covers material presented in a paper opening with a discussion on positive rays in Section A (Mathematics and Physics) at the Winnipeg meeting of the British Association that August and September, where Thomson assumed presidency of the Association as a whole, and Rutherford of Section A [15-44]. The most pressing order of business now was pinning down the elusive neutral doublets. Thomson had not yet given up on them, though he was growing more doubtful whether they did, in fact, represent the first stage in ionization. If, as he put it, they were 'shot out from the molecule by explosions' [15-45] in the vicinity of the cathode, they should fly off in all directions. In actuality, they seemed to be projected preferentially in directions normal to the surface of the cathode; 'in short [in directions] approximately [those] of the lines of force close to the cathode' [15-46]. If so, the doublets must be formed outside the confines of the atom or molecule, and constitute an intermediate, not first, stage between atom and ion.

> Is the doublet the thing set free from the atom when it is ionized? and are the ions produced by the splitting up of the doublets, or are the ions first produced and the doublets produced by a combination of the positive and negative ions? [15-47]

If secondary, Thomson felt, the neutral doublets might account for the hitherto unalterable fact of the nearly constant velocity of the positive rays, independent of the fall in potential across the discharge, which he now explained in terms of a narrow limit in binding energy of the doublets. If the relative velocity of the positive and negative constituents before combining were too high, 'the charges would fly past each other without entering into combination' [15-48]; on the other hand, 'as the kinetic energy of the doublet decreases it gets more and more stable and less likely to dissociate' [15-49].

However, if the neutral doublets did in fact represent a second stage of ionization, positive particles moving too fast to combine with corpuscles should have registered as positive rays. Where were they? In one more paper the following March, Thomson argued vaguely that

> ... it seems probable that they would be so deflected by the intense electric forces in the neighbourhood of the cathode, that they could not travel down the long narrow tube interposed between them

and the phosphorescent screen without striking against the sides and so getting absorbed. [15-50]

Indeed, Thomson seems not to have been convinced on the point; the constant velocity of the positive rays, quite independent of the potential difference between tube terminals, remained a vexing problem. Perhaps the problem had somehow to do with unsteady output of the induction coil employed. To guard against this possibility, he went to the considerable trouble of replacing his Wimshurst machine with a Wehrsen electrostatic induction machine—a Mercedes, as it was known [15-51]. It allowed experimentation with stable tube potentials up to 40 000 V.

> [The Mercedes] is far ahead of any other Wimshurst Machine I have ever come across [he wrote Rutherford] and with two moving plates I get as much as from an ordinary Wimshurst with twenty moving plates. The only draw-back to them is the price, they cost about £40. [15-52]

Alas, he continued, 'the results with this machine are just the same as those with the coil' [15-53]. The positive ray velocity remained unchanged and independent of tube potential; there was nothing wrong with the original apparatus on this score.

Equally preplexing was the predominant emission of the positive rays, including their retrograde counterparts, normal to the cathode, or along the electric field lines. Why then did their velocity not depend on the potential difference, which could be varied by over a factor of ten, or between 3000 V and 40 000 V? In the end, Thomson returned to his original view of the doublets constituting the primary form of emission, with the additional dubious suggestion that 'the electric field produces some kind of polarization in the molecules that makes them eject the uncharged doublets along the line of motion of the cathode rays in their neighbourhood' [15-54].

On that shaky note, Thomson's preoccupation with the positive ray velocity and neutral doublet emission came to an end. The last measurements in the current series of experiments, recorded in the notebook for October–November 1909, are dated November 12 [15-55]. The paper of March 1910 was the last in which these topics played a significant role [15-56]. The same year marked an abrupt change in emphasis in Thomson's positive ray program. The reason was largely the arrival on the scene of Francis Aston, a man whose unique talents were sorely needed to resuscitate the flagging experiments.

15.3. A new tool in hand

Francis William Aston suited Thomson's purpose to a tee, as Thomson had good reasons for suspecting when, in 1909, he offered Aston a joint assistantship at the Cavendish and Royal Institution, financed by the

Royal Institution and replacing Kaye. (Kaye was leaving for the National Physical Laboratory, reports Thomson to Rutherford in his holiday letter that December [15-57].) From an early age, Aston had displayed a strong affinity for mechanical contrivances and the stuff of chemistry; by 1909 his gift for devising apparatus and physical experimentation was the talk of the community. To him, 'experimental methods... [were] a joy in themselves, approaching that to be gained from the results they gave' [15-58].

Aston was born in 1877, the second and only surviving son of William Aston, a metal merchant, in a suburb of Birmingham. After primary education in a vicarage school, he began the study of science and mathematics at Malvern College. Finishing with high honors in 1893, he entered Mason College, where he studied for the London intermediate science examination with the chemists William A Tilden and P F Frankland and the physicist John Henry Poynting [15-59]. Here he also developed his skill as a glass blower, a talent that would serve him in good stead throughout his career. In 1898, he earned a Forster Scholarship to work with Frankland in organic chemistry. As a practical matter, he also took a course in fermentation chemistry, on the basis of which he earned a living, between 1900 and 1903, as a brewery chemist with a company in Wolverhampton. In 1903, he returned on a scholarship to the University of Birmingham (as Mason College became known from 1900) as a physics research student under Poynting.

Aston's enthusiasm for physics, and its instrumentation in particular, was never far from the surface in his early years—especially in the wake of Röntgen's discovery of x-rays in 1895. The following year he outfitted a disused loft in his father's house as a private laboratory and workshop. There he began his notable improvements to the Sprengel and Töpler pumps, and with them evacuated small discharge tubes contrived from chemical glassware; he also wound an induction coil capable of throwing a 3 in spark. His return to the University of Birmingham under Poynting coincided with the publication of Thomson's *Conduction of Electricity through Gases*, and he now threw himself into work on the gaseous discharge. His studies on variations of the Crookes dark space with current and pressure, using specially crafted Geissler tubes of his own, proved of lasting value. In 1908, he discerned a new 'primary cathode dark space' immediately adjoining the cathode—nowdays known as the 'Aston dark space'. The same year, as a result of a legacy on the death of his father, he took time off for a trip around the world. Returning in 1909, he was accepted as lecturer at Birmingham. However, after but one term, he eagerly accepted Thomson's invitation (on Poynting's recommendation) to join him as his private assistant [15-60].

Aston's timing could not have been better. The first breakout from the positive ray quagmire was communicated in a paper read by Thomson at the meeting of the British Association on September 1 1910 [15-61]. The latest experiments were, in a sense, the first fruits of Aston's stepping in, as even the laboratory notebook shows [15-62]. For one thing, much larger vessels for the discharge portion of the tube, introduced on Aston's suggestion, were used for the first time, allowing lower pressure without destructive sparking [15-63]. More uniform Willemite screens, thanks to Everett, were also employed.

Behold! As the pressure sank below a certain value, the straight, luminous bands, by now attributed to secondary positive rays (rays charge exchanging along their path) moving with constant velocity independent of the tube potential, gave way to the first convincing traces of separate parabolas—each for a range of velocities but constant e/m-ratio, quite in accordance with equation (15.3). These parabolas, yielding e/m-values inversely proportional to the atomic weight of a given gaseous constituent, were ascribed to primary positive rays moving, as they should, with a kinetic energy determined by the potential difference between the electrodes. They signaled, at long last, the appearance of the 'characteristic rays' that had eluded Thomson since 1907. There were parabolas (or at least 'spots along parabolic paths') in abundance, corresponding to the hydrogen atom, hydrogen molecule, helium atom, and atoms of carbon, oxygen (or possibly nitrogen), and mercury, depending on the gas under examination [15-64].

More improvements in experimental technique were not long in coming. The very next paper (the seventh in the series), in the *Philosophical Magazine* for February 1911 [15-65], features the use of liquid-air-cooled charcoal (derived from coconuts) in conjunction with a motor-driven Gaede pump. The enormous absorptive power of charcoal chilled to $-185\,°C$, soaking up well-nigh all gases except hydrogen, helium, and neon, was discovered by James Dewar in 1905, the year Thomson, too, began his professorship at the Royal Institution [15-66]. The liquid air plant used for the purpose at the Cavendish was the gift of Thomas Cecil Fitzpatrick, a former lecture demonstrator who later rose to be President of Queens' College, Cambridge. It was actually installed in 1904, along with a 10 h.p. gas engine to work it, as we learn from Thomson's letter to Rutherford that April.

> We have been busy at the Laboratory installing a liquid air plant which is not running yet. We have however just got the new gas engine running, and I expect we shall have the whole thing running by the time you come. [15-67]

(Rutherford, still at McGill, was scheduled to give the Bakerian Lecture to the Royal Society in May, as well as a lecture at the Royal Institution, and was wined and dined in great style at Cambridge on the occasion.) Brief reference to charcoal pumping is also made in the previous paper, and the technique is featured throughout Thomson and Aston's notebook covering the summer experiments of 1910, so it was clearly in routine use by June 1910 [15-68]. Indeed, it appears to have been a sporadic practice earlier at Cambridge, before being temporarily discontinued for one reason or another [15-69]. Still, it proved a boon to Thomson's current line of investigation; Rayleigh recalls Thomson saying 'I never saw a parabola until I put on the charcoal and liquid air exhaustion' [15-70]. However, one more improvement in vacuum technique was needed to truly do the trick. With his assistants, Thomson managed to lower the pressure in the deflection chamber substantially below that in the discharge vessel, by sealing one from the other and pumping directly on the deflection chamber; the only vacuum connection to the discharge vessel was by way of the narrow cathode channel.

By now, Thomson also had concluded that the photographic method was superior to the awkward method of tracing the evanescent phosphorescent pattern on the Willemite screen. At first, he tried photographing the luminosity on the screen with the aid of a very large portrait lens; that proved impractical, requiring exposures lasting several hours! Not only was it excruciatingly tedious, but also it was quite impossible to keep conditions in the tube steady for such long periods. But why not dispense with the screen altogether? Indeed,

> ... a much more sensitive and expeditious method is to put a photographic plate inside the tube itself and let the positive rays fall directly on the plate instead of on to the willemite screen. The photographic plate is very sensitive to these rays, and the places where they fall are recorded when the plate is developed. The plate is much more sensitive than the willemite screen, and an exposure of three minutes shows curves on the plate which cannot be detected on the screen. [15-71]

Several methods were explored. Aston, true to form, won the day with the ingenious arrangement depicted in figure 15.8.

> In this method the plate is suspended by a silk thread wound round a tap which works in a ground-glass joint; by turning the tap the silk can be rolled or unrolled and the plate lifted up or down. The plate slides in a vertical box of thin metal, light-tight except for the opening *A*, which comes at that part of the tube through which the positive rays pass .... When the silk is wound up, the strip of photographic plate in the box is above the opening, so that there is a free way for the positive rays to pass through the opening and fall on a willemite screen placed behind it, so that the state of the tube with respect to the production of positive rays can easily be ascertained. The box is large enough to hold a film long enough for [taking] three photographs [in succession]... without opening the tube. [15-72]

Aston, by the way, was a lone worker, often away on his travels abroad for long stretches of time. Mark Oliphant remembers his

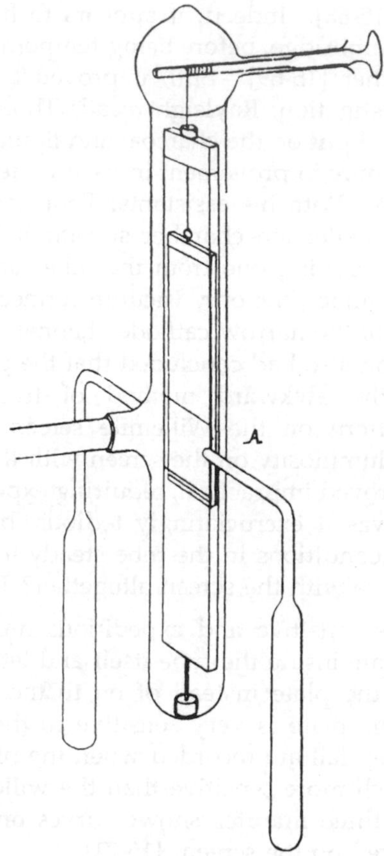

Figure 15.8. *Aston's photographic plate holder, mounted inside a positive ray tube.*

laboratory as a dingy, darkened room in a corner of the ground floor of the Cavendish building. For all the junk and discarded apparatus, Aston was, as we have gathered, an excellent glass blower and top-rate experimentalist. The same cannot be said for Aston's lectures, which were invariably dull. The only worse lecturer at the Cavendish, by unanimous consent, was C T R Wilson, whose audience invariably dwindled to one or two unfortunates who did not have the courage to sneak out [15-73].

The first occasion in which Thomson proudly demonstrated his photographic plates in public, according to Rayleigh, was at a meeting commemorating the 25th year of Thomson's tenure as the Cavendish Professor, held at the Laboratory on November 12 1910 [15-74]. (The anniversary actually fell on December 12 1909.) At that meeting, Richard Glazebrook presented a bound copy of the History of the Laboratory to the Professor [15-75]; afterwards, there was a *conversazione*

Figure 15.9. *Thomson measuring positive ray parabolas with calipers, ca. 1912.*

with experiments on view. 'I can well remember [adds Rayleigh] his characteristic grin of pleasure as he showed them to a group of whom I was one' [15-76].

With these improvements in his apparatus, Thomson was almost ready to get on with his earlier stated goal of using the rays as a tool for chemical analysis. First, however, several improvements in the positive ray tube itself were made, introduced specifically to increase the brightness of the rays. The eighth, and last paper of the *Philosophical Magazine* series, dwells particularly on the optimum shape of the business end of the cathode, and on its position in the spherical flask comprising the discharge tube [15-77]. Even then, and despite the major innovation of introducing a photographic plate inside the vacuum to catch the rays, the fine channel through the cathode (varying from 0.1 to 0.01 mm in diameter) still necessitated very long exposures, up to 3 h. This, in turn, necessitated 'special arrangements' for keeping the magnetic and electric fields constant over such a protracted interval; Thomson does not explain how, and his notebooks offer no clue. As to the photographic plates, Thomson adds that

> ... inasmuch as the rays do not penetrate right through the film, the most suitable plates for these experiments might be expected to be those which had very thin films containing as much of the silver salts as possible. Guided by this consideration, I tried the old Daguerrotype silver plates treated with iodine, but did not

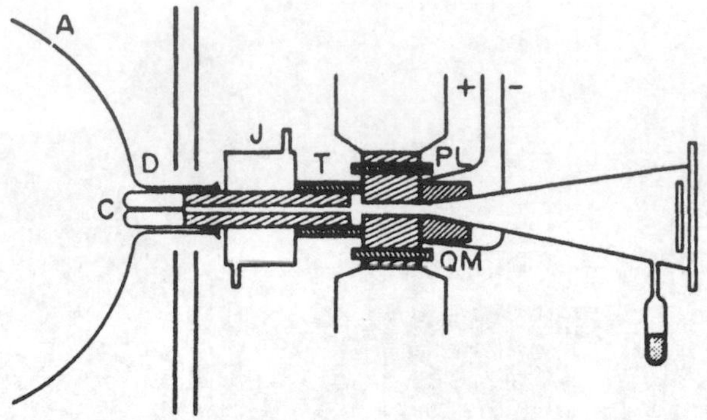

Figure 15.10. *Final version of Thomson's positive ray apparatus.*

> obtain such good results as with the [Imperial] Sovereign plates. [15-78]

Finally, for some experiments it was deemed important that the magnetic and electric fields be as coterminous as possible. As usual, the task fell on Aston. In his arrangement, this was accomplished by having the flat ends of the soft-iron pole-tips, suitably insulated, also serve as the plates producing the electric deflection fields. The final version of the apparatus, ca. 1913, is shown in figure 15.10. Here

> *A* is a large bulb . . . in which the discharge passes, *C* the cathode placed in the neck [*D*] of the bulb. . . . The cathode is fixed into the glass vessel by a little wax. . . . The wax joint is surrounded by a water jacket *J* to prevent the wax being heated by the discharge. . . . An ebonite tube is turned so as to have the shape shown [at *T*], *L* and *M* are two pieces of soft iron with carefully worked plane faces, placed so as to be parallel to each other, these are connected with a battery of storage cells and furnish the electric field. *P* and *Q* are the poles of an electromagnet separated from *L* and *M* by the thin walls of the ebonite box; when the electromagnet is in action there is a strong magnetic field between *L* and *M*; the lines of magnetic force and electric force are by this arrangement parallel to each other. [15-79]

All these improvements in experimental technique were highly opportune. By 1910, Thomson's faith in the evidence for a fundamental unit of positive electricity was dwindling fast [15-80], giving way instead to increasing optimism that the positive ray analysis might 'furnish a valuable means of analysing the gases in the tube and determining their atomic weights' (the latter something spectrum analysis was not capable of) [15-81]. The change in direction of his positive ray program was a direct outcome of the refinements introduced, or at least put into practice, by Aston. The refurbished apparatus revealed a complex mixture of ionic species, atoms and molecules, characteristic of the gases admitted into the tube, or otherwise present from a variety of causes.

The advent of the new tool for chemical analysis coincided with the availability of two samples of gases from the residues of liquid air furnished by James Dewar at the Royal Institution. Dewar had exploited the different condensabilities of gases over charcoal to separate out the constituents of a gas mixture, collecting the liberated gas in fractions as the temperature rose; he also separated the rare gases from air by this method [15-82]. One of the two first gas samples supplied to Thomson contained the heavier constituents, the other the lighter ones. The first presented no surprises, reported Thomson in his Bakerian Lecture to the Royal Society on May 22 1913. 'The positive ray photograph... gave the lines of xenon, crypton, argon, and a faint line due to neon; there were no lines on the photograph unaccounted for' [15-83]. The second sample, containing the lighter gases, had a definite surprise in store, in that the photograph revealed, 'in addition to helium and neon,... another gas with an atomic weight about 22' [15-84].

> There can, therefore [decided Thomson]... be little doubt that what has been called neon is not a simple gas but a mixture of two gases, one which has an atomic weight about 20 and the other about 22. [15-85]

The concept of *isotopes*, elements of different atomic weight, but chemically indistinguishable, was in fact fairly widespread by this time, if not the term itself—albeit a concept associated with radioactivity, and neon is manifestly not radioactive. The concept seems to have been floated as early as 1910 by Frederick Soddy, who took it up in earnest in early 1913 and introduced the term late that year. Soddy, whose greatest claim to fame rested on his discovery, with Rutherford a decade earlier, of the transmutation of the elements through radioactive decay, had an unorthodox and obstinate personality worthy of a slight digression at this point.

Born in Eastbourne, Sussex, in 1877, Soddy was the youngest son of a London grain merchant [15-86]. His lifelong attitude, aloof yet shy, critical but kind and generous, reflected the Calvinistic traditions in the Soddy household. From Eastbourne College and University College of Wales, Aberystwyth, he went to Merston College, Oxford, where he graduated in 1898 with first-class honors in chemistry. After two years of research at Oxford, he was accepted as demonstrator in chemistry at McGill University in Montreal, where he took up problems of radioactivity in collaboration with Rutherford. This teamwork proved surprisingly successful, considering the strong opinions of both partners. Their disintegration theory was promulgated during 1902–1903. Returning from Canada in 1903, he briefly joined Sir William Ramsay at University College, London, where he continued his studies of radium emanation. He and Ramsay were able to demonstrate spectroscopically that helium was produced in the radioactive decay of radium. From 1904 to 1914, he held an appointment as lecturer in physical chemistry and radioactivity at the University of Glasgow—the decade that proved his most productive scientific years. In 1914, he became professor of chemistry at the University of Aberdeen, and, in 1919, settled as Dr Lees Professor of Chemistry at Oxford until his retirement. However, the intervening war years, and his own involvement in the war effort, altered his outlook and sense of values more and more, and he did virtually no more work on radioactivity after 1914. He increasingly occupied himself with the social and economic ills of mankind. Despite his Nobel Prize in Chemistry in 1921 and numerous honors in his scientific heyday, he gradually fell out of tune with his scientific colleagues, and retired prematurely from academic life in 1937.

Soddy had spoken on the existence of isotopes before the Royal Society on February 27 1913 [15-87]. Unfortunately, he used the phrases 'atoms of the same *chemical properties*', non-separable by any known process'. Arthur Schuster promptly and correctly took him to task in a letter to *Nature*:

> Mr. Soddy believes in the existence of a number of bodies which differ in molecular weight but 'are non-separable by any known process'; these are also supposed to have identical spectra. Among 'known processes' I count gravitation, diffusion, and mechanical processes, such as separation by centrifugal forces, among which diffusion, perhaps, is the only available one. Is there any reason to suppose that molecules which, ex hypothesi, differ in mass, cannot be separated by diffusion? [15-88]

In his reply, Soddy sheepishly admitted that by 'non-separable by any known process' he had in mind chemical, not physical processes.

> I agree with Prof. Schuster that it should ultimately be possible partially to separate by purely physical methods certain members of these chemically identical groups by virtue of the slight differences in their molecular masses. [15-89]

Soddy expounded further on the problem of isotopes (still without using the term) before the British Association in Birmingham in September 1913 [15-90]. On that occasion, 'the discussion on radio-active elements and the periodic law attracted a very large audience', according to a résumé of the BA discussion in *Nature*. 'Unfortunately', continued *Nature*, 'the counterattractions of Sir. J. J. Thomson's new gas limited it to an hour and a half, but Mr. Soddy, who opened it, was properly very brief' [15-91]. Indeed, Soddy managed to cover the group displacement law, according to which emission of an α-particle causes a shift of two places to the left in the periodic table, and the loss of a β-particle causes a shift of one place to the right [15-92]. The discussion concluded with an overview of experiments by Alexander Fleck, a student of Soddy, listing a number of radioactive substances that are chemically identical.

The notion of isotopes was, therefore, very much in the air in the spring of 1913, even if only linked to radioactive, not stable, elements. Opinions differ whether, alternatively, Thomson subscribed to the possibility that the heavier constituent might be the compound NeH_2, a hydride which fit the observed atomic weight within the limits of error. The atomic weight, 22, is also half the molecular weight of carbon dioxide (CO_2), a substance invariably present from stopcock grease [15-93]. Then again, did it perhaps signal an entirely new element, *meta*-neon? For reasons still debated, Thomson was reluctant to draw a firm conclusion. When Georg von Hevesy, a member of the audience of his Bakerian lecture in May, wrote him that perhaps the two kinds of neon were analogous to radium D and lead, Thomson replied simply as follows:

> Many thanks for your letter. I do not think myself that it is impossible to separate the gas with atomic weight 22 from Neon (20) by fractionation; in fact F. W. Aston is at present engaged on this work. [15-94]

The task of separating the neon mixtures thus fell on Aston. His first line of attack was fractional distillation over charcoal cooled with liquid air—much the same procedure as that used by Dewar in obtaining the sample in the first place. The density of the fractions was determined with a special quartz micro-balance devised by Aston for the purpose; it was sensitive to 10^{-9} g. The attempt failed utterly, despite thousands of operations. The second method, fractional diffusion through pipeclay, was a qualified success, indicating a difference of about 0.7% between the densities of the heaviest and lightest fractions. Although too small, the change in density was far too great to be ascribed to experimental error.

In the event, the onset of the war interrupted further work on the subject, and Aston joined the Royal Aircraft Establishment at Farnborough. There he contributed his technical skills to research on aeroplane dope and fabrics, in the company, among others, of G P Thomson and Frederick A Lindemann, later Lord Cherwell and a recent protégé of Hermann W Nernst. G P Thomson recalls long nocturnal discussions of the neon results between Aston and Lindemann—the latter prone to favor NeH_2 or CO_2 as the culprit [15-95]. When the war ended, Aston returned to the Cavendish and the diffusion problem. Curiously, his new results were even poorer; this time he attained only half of the previous yield. However, he more than saved the day with his famous mass spectrograph, the principle of which he had contrived while still at Farnborough [15-96]. No sooner was it in working order than it verified that neon is, indeed, isotopic with *three* masses: one 20.0, a second 22.0, and a third 21.0 (too rare to have registered in Thomson's apparatus).

As for Thomson, Rayleigh adds that J J seems to have been perpetually haunted by his suspicion about hydrogen compounds, and consequently he only accepted Aston's isotopic results with considerable reluctance [15-97].

Chapter 16

THE ELECTRONIC CHARGE REVISITED, AND ONE MORE CONTROVERSY

16.1. Robert Millikan: from clouds to drops. First round of experiments

The years 1906–1907, when J J Thomson took up his positive ray studies in earnest, also marked a turning point in the experimental career of Robert Millikan in America. In the near future, his dogged research and singleness of purpose, born of a sense of academic urgency, would be devoted to obtaining a definitive value for the charge of Thomson's electron.

To be sure, while the concept of a fundamental subunit of positive electricity proved short lived, a corresponding role for the ubiquitous electron *vis-à-vis* negative electricity was hardly in doubt by 1900. This was the case despite a lingering epistemological dispute as late as 1910 in certain Continental circles, both over the reality of atoms generally and the discrete nature of electricity in particular. The evidence for a unique value of the charge-to-mass ratio e/m for the electron from multifarious glow discharge studies was persuasive, thanks largely to Thomson. Augmented by separate measurements of e by Townsend, by C T R and H A Wilson, and by the 'Prof' himself, they indicated a subatomic carrier of negative electricity in remarkably good agreement with the bound charges implicated in the magneto-optic studies of Zeeman and Lorentz.

All the same, the electronic charge measurements were far from definitive. They were either indirect, or based on measurements on an aggregate of charges with some unknown statistical distribution. Their accuracy was exceedingly poor. Thus H A Wilson's most recent determinations in 1903 ranged from $e = 2.0 \times 10^{-10}$ esu to 4.4×10^{-10} esu, a spread of more than 100% in what was then arguably a fundamental constant of the greatest importance [16-1]. A direct determination of *e*, based on measuring individual charges, had not been attempted, much less had systematic studies on whether measured charges were integral multiples of a single elementary subunit.

Millikan now decided the time was ripe for such an attempt, building on the shortcomings of the concerted British efforts. In so doing, his push for a renewed determination of the electronic charge would propel him, 38 years old and largely (or so he felt) with ten years of teaching and textbook writing under his belt, into the scientific limelight in a very short time. The push would also precipitate yet another controversy engendered by a highly experienced European research physicist, in an affair painfully reminiscent of the Becquerel dispute over positive electrons laid to rest just as the new dispute erupted. By way of introducing the final chapter in the saga of the electron, let us turn to he who played the leading role in these unfolding events, Robert Andrews Millikan.

Robert Millikan was born in 1868, in Morrison, IL [16-2]. The second of six children of a Congregational Minister, his boyhood was spent mostly in Maquoketa, IA, where he grew of age in an atmosphere stressing simple virtues and in a setting not unlike that of Mark Twain's Tom Sawyer. In 1886, he entered Oberlin College, OH, where he majored in the classics but found a course in physics 'a complete loss' [16-3]. Nevertheless, at the time of his graduation in 1891, the college lost one of its tutors in physics and, of all things, mainly on the strength of his performance in Greek, he was asked to fill the tutorship. This he did, and the positive experience thereby gained launched him on a career in physics second to none.

Receiving a master's degree at Oberlin in 1893 on the basis of his physics teaching while simultaneously studying Silvanus P Thompson's *Dynamic Electric Machinery* on his own, he then enrolled on a fellowship as the sole graduate student in physics at Columbia University. There he came under the tutelage of Michael I Pupin who, though trained at Cambridge and to his dismay, deplored the atomic theory of matter, yet left his mark on Millikan for his deftness in mathematical physics. Equally important was Millikan's exposure to the experimental dexterity of Albert A Michelson at the University of Chicago during the summer of 1894. His actual doctoral research, on the polarization of incandescent light, was carried out under Ogden N Rood who left yet another mark on Millikan in his predilection for making precision measurements. He received his doctorate in 1895, whereupon, on Pupin's insistence, he journeyed to Europe for the obligatory postdoctoral stint—financed by a loan from Pupin.

What better year than 1895 for a neophyte physicist's *Wanderjahr* in Europe! Within months of his arrival in Walther Nernst's laboratory in Göttingen, Röntgen's discovery took everybody by storm; that of Henri

Becquerel was not far behind. While pursuing his research under Nernst on the dielectric constant, he sat in on a lecture by Poincaré in Paris, and attended a syllabus by Planck in Berlin. In 1896, offered an assistantship under Michelson in the physics department of the University of Chicago, he nevertheless cut short his stay in Göttingen and hastened homeward.

The University of Chicago had only opened in 1892, but was already a distinguished institution, generously endowed by John D Rockefeller and Marshall Field. The first faculty included eight former university and college presidents. Indicative of the intellectual influx in the Chicago suburb, Millikan's boarding house counted Thorstein Veblen and Harold Ickes among his fellow lodgers. The physics facility, the 'Ryerson Physical Laboratory', was named by Martin A Ryerson, son of a wealthy lumber tycoon who had largely underwritten the financing of the facility; the laboratory was completed on New Year's Day 1894. Having thus arrived on the scene at a timely hour, Millikan immediately found himself effective in establishing departmental policies with respect to teaching and research (as he was to be much later at Pasadena's newly renamed California Institute of Technology). For the next decade, he devoted much of his energy to pedagogical matters: organizing courses, teaching, and writing textbooks. He coauthored a number of texts and laboratory manuals in physics which spread his name far beyond the Chicago campus; these ranged from a highly elementary *First Course in Physics* (1906) to a translation (with Charles R Mann) of P Drude's *Theory of Optics* (1902). All the while, initially as a result of his teaching, he rose steadily on the academic ladder, from Instructor in 1899 to full Professor in 1910.

Despite his pedagogical load, Millikan was only too aware of the importance of research in an institution such as the University of Chicago. For his doctoral thesis, he had reinvestigated the old problem of the polarization of light emitted from incandescent solid and liquid surfaces, a phenomenon first observed by François Arago in 1824. At Göttingen, he began research under Nernst on an experimental test of the Clausius–Mossotti formula for the dependence of the dielectric constant ϵ on the density of liquids; the reduction of his data and writing of his paper was completed upon his return to America. At Chicago, owing to the enlightened academic atmosphere, some of the day-to-day teaching activities naturally bordered on research in their own right.

> I soon found myself responsible for the weekly seminar in physics, which Professor Michelson asked me to take off his hands.... Furthermore, I soon began to give advanced courses on the electron theory, on the kinetic theory, and on thermodynamics... [After 1900 Michelson...] asked me to assign research problems to three of the prospective candidates for the doctor's degree... so that by 1902 and 1903 I had quite a group of problems going in addition to my own... [16-4]

The first research problem 'on his own' at Chicago, undertaken jointly with George Winchester, involved the photoelectric effect which we last left in the hands of Philipp Lenard (section 12.1). Lenard's e/m determinations in connection with the effect, published ca. 1900, had revealed the essential identity of the photocurrent emitted from a metallic surface exposed to ultraviolet light with Thomson's cathode ray corpuscles. His results of 1902 on the dependence of the photocurrent on the intensity of the incident light, and on the velocity of the emitted charge carriers, would remain unexplicable until the publication of Einstein's classic paper in 1905, explaining them in terms of an interchange of energy between light quanta and photoelectrons.

It appears that Millikan was not yet aware of Einstein's paper when he and Winchester took up the photoelectric effect. They had set out to determine whether the photocurrent, and the concomitant saturation potential, exhibited a dependence on temperature of the emitting metal; if the photoelectrons derived their energy from within the metallic atoms, heating the cathode should increase their speed of emission. In the event, no effect was seen [16-5]. Though disappointed in this apparent setback, Millikan nevertheless resolved to forego his heretofore strong card, churning out textbooks, and henceforth stick to research.

> I knew by this time from the way the textbooks were beginning to succeed, and the way the teaching interested both me and my pupils, that I could have a satisfying career in the field of scientific education; and at this period I sometimes raised the question with myself as to whether it might not be more rewarding to devote myself to writing and teaching, and to abandon the endeavor to make good in a research career. Also, I well knew from my experiences with the Nernst episode, as well as from my research failures at Chicago, how much prospecting one could do in physics without striking a vein that had any gold in it.... At any rate, I resisted the temper, decided to stick to the gamble that had originally brought me to Chicago, and in 1908... kissed textbook writing good bye... [16-6]

But what problem to take up? There really was not much of a choice for someone of Millikan's penchant for precision measurements. No sooner had he seen their manuscript on the lack of dependence of the photoelectric emission on temperature safely off in the mail to the *Philosophical Magazine*, than he set out on what he would later call 'my oil-drop venture (e)' [16-7]. 'Being quite certain that the problem of the value of the electronic charge... was of fundamental importance, I started into it' [16-8]. Nor was there any question of how to get started. The first thing to do, obviously, was to repeat an experiment conducted by H A Wilson in 1903—the latest in the long line of Cambridge determinations of the charge of the electron.

The experiment of Harold A Wilson, lately a student of J J Thomson [16-9], was but one of several modifications of the falling cloud technique of C T R Wilson (unrelated to Harold) introduced since 'C T R' first devised it. It utilized essentially a copy of J J Thomson's apparatus of 1898 (figure 10.1), but with a pair of horizontal brass plates mounted in chamber *A*. As usual, a cloud was produced in the chamber by expansion; the droplets of the cloud condensed on ions produced by x-ray irradiation of the saturated air. In H A Wilson's version, the top of the cloud was timed as it descended between the two plates, first under action of gravity alone, and then under the combined action of gravity and an imposed vertical electrical field acting on the ions [16-10]. If v_1 and v_2 are the velocities of descent without and with the electric field, *m* the average mass of a drop with a single charge *e* in the top surface of the cloud, and *E* the electric field, then the ratio of the two velocities is given by

$$\frac{v_1}{v_2} = \frac{mg}{mg + Ee}. \tag{16.1}$$

Combining the equation with Stokes' law for the velocity of free fall,

$$v_1 = \tfrac{2}{9}(ga^2\sigma)/\eta \tag{16.2}$$

(section 10.1) and eliminating *m* by using $m = \frac{4}{3}\pi a^3\sigma$ (*a* is the radius, σ is the density of the drop, and η is the viscocity of air), gave the following expression for *e*:

$$e = 3.1 \times 10^{-9} g/E(v_2 - v_1)v_1^{1/2}. \tag{16.3}$$

Wilson's method, by focusing only on the *top layer* of the cloud, eliminated the earlier need for the *ad hoc* assumption that each drop contained only a single ion, since the more heavily charged drops were forced downward more rapidly by the electric field. All the other objections to Thomson's experiment still held, however: e.g. the assumed validity of Stokes' law for small drops, assumed lack of evaporation while timing the rate of fall, and ignoring the effects of convection currents. In addition, Wilson's method entailed a fresh and rather serious assumption; namely, that successive clouds, with and without the applied field, were identical with respect to the size distribution of the drops. Hence, it is perhaps not surprising that the method gave a rather broad range of *e*-values with a mean of $e = 3.1 \times 10^{-10}$ esu—about 35% too small, in hindsight.

It appears that Millikan started on his new 'venture' early in 1907, assigning the leg-work to his graduate student Louis Begeman [16-11]. From the outset, two aspects of H A Wilson's experiment seemed especially open to criticism. One was the use of x-rays as the ionizing agent, which were subject to large fluctuations in intensity. There was also the possibility of the sinking of the cloud being affected by evaporation. The first objection was countered by introducing radium as the ionizer, and the second by reducing the distance through which the cloud sank—hence, the period of observation. Millikan and Begeman reported their results at a meeting of the American Physical Society in Chicago during December 30 1907–January 2 1908, and again in an abstract to the *Physical Review* for February 1908 [16-12]. (The *Physical Review* began publication while Millikan was in graduate school.) Their mean value from ten different sets of observations, ranging from 3.66×10^{-10} esu to 4.37×10^{-10} esu, amounted to 4.03×10^{-10} esu, compared with H A Wilson's value of 3.1×10^{-10} esu. As to whether it was better or worse than Wilson's result, Millikan wrote in hindsight that

> ... we stated at the time that although we had not eliminated altogether the error due to evaporation, we thought that we had rendered it relatively harmless, and that our final result, although considerably larger than either Wilson's or Thomson's (3.1 and 3.4, respectively), must be considered an approach at least toward the correct value. [16-13]

A better assessment of the validity of the result was not long in coming, furnished by none other than Ernest Rutherford, newly relocated from McGill to the University of Manchester. (We cover the circumstances of his relocation in the next chapter.) One of the finest pieces of research to be carried out by Rutherford during his tenure at Manchester, in collaboration with the spectroscopist Thomas Royds, was proving that α-particles are ionized helium atoms—something long suspected. Strong, if not conclusive, evidence was obtained some months before in a preliminary experiment in which the α-particles expelled per second from 1 g radium could be counted individually by electrical means. The counter was a 'proto-Geiger counter', an alternative to the widespread scintillation counting technique, and one devised with the aid of Hans Geiger, Rutherford's new assistant.

Hans Geiger was spending a post-doctoral stint in England, reversing the usual practice of graduates in the English-speaking world heading for the Continent. On obtaining his doctorate at Erlangen in 1906, he had joined Arthur Schuster as his assistant at Manchester. When Rutherford succeeded Schuster in 1907, he persuaded Geiger to remain in Manchester and pursue jointly research in radioactivity [16-14]. Their ionization chamber was a low-pressure cylinder equipped with a thin wire stretched coaxially. The wall of the cylinder had an entrance window at one end, and was maintained at a high negative potential with respect to the wire. An α-particle entering parallel to the wire generated a secondary avalanche, multiplying the primary ionization 1000-fold. The resulting voltage step on the central wire was read by an electrometer [16-15]. The total charge carried by α-particles from a radium source, measured in a separate experiment, divided by the total number of α-particles counted, then gave the charge per α-particle, namely, 9.3×10^{-10} esu [16-16]. This value, noted Rutherford in his report read to the Royal Society on June 18 [16-17], was conspicuously close to twice the charge e carried by a hydrogen atom—a value by then nearly universally accepted as the fundamental unit of electricity. Hence, concluded Rutherford and Geiger, the charge carried by an α-particle is given by $2e$; and $e = 4.65 \times 10^{-10}$ esu.

The value for e, thus determined, added Rutherford, was somewhat higher than found by Millikan and Begeman, unfortunately misquoting the Chicago value as 4.06×10^{-10} instead of 4.03. This erroneous value would stick in the literature on account of Rutherford's authority, even quoted by Millikan himself in his subsequent papers and in *The Electron*, his retrospective tome [16-18]. More to the point, the Rutherford–Geiger value was higher yet than the values deduced still earlier by H A Wilson and J J Thomson. Adding delicately that 'it is far from our intention to criticise in any way the accuracy of the measurements made by such careful experimenters' [16-19], the authors correctly pointed out that evaporation tends to overestimate the number of ions (droplets) in falling cloud experiments, and hence underestimate e. 'If the error due to this effect were about 30 per cent in the experiments of J. J. Thomson and H. A. Wilson, and 15 per cent in the experiments of Millikan, the corrected value of e would agree with the value 4.65×10^{-10} . . .' [16-20]. Rutherford's confidence in his own value was reinforced by the excellent agreement with that deduced by Max Planck on theoretical grounds nearly a decade earlier, $e = 4.69 \times 10^{-10}$ [16-21]. In his recollections of the Manchester period, Eve underscores the point by adding that 'when I protested to Rutherford that [his] result must be wrong, he pointed out that eight years previously Planck had obtained a similar figure to his own' [16-22].

Suspecting that the evaporation problem was, indeed, still not under control, Millikan now decided the next step should be deployment of an electric field strong enough to hold the top of the cloud stationary by counter-balancing the force of gravity acting on the cloud. By so doing, the rate of evaporation could be observed directly and allowed for in the calculations. This called for a 10 000 V storage cell battery—in itself no small undertaking when built in 1906. Everything was at long last ready in the spring of 1909, however. The chamber was duly expanded, the cloud formed, and Millikan threw the knife-edge switch.

> What I saw happen was the instantaneous and complete dissipation of the cloud—in other words, there was no 'top surface' of the cloud left to set cross hairs upon as Wilson had done and as I had expected to do. [16-23]

Obviously the long-held assumption that each drop contained a single charge was invalid, and the whole idea of relying on the timing of the rate of fall of an ionized cloud was in serious jeopardy. Disappointed, Millikan nevertheless repeated the test to be quite sure. Behold!

> I saw at once that I had something before me of much more importance than the top surface of a cloud upon which to set the cross hairs of my observing telescope. For repeated tests showed that whenever the cloud was thus dispersed by my powerful field *a few individual droplets would remain in view.* [16-24]

These droplets were the ones that by chance had a ratio of charge to mass such that the upward force of the electric field just balanced the downward pull of gravity. Fortuitously, the problem at hand was suddenly transformed from one of experimenting on a cloud of drops to one on individual, charged drops. Each drop, carrying an integral number of electronic charges, could be suspended almost motionless for some time in the field of view of the short-focus telescope by carefully switching the field on or off; there they swam as bright specks against a black background, slowly rising or sinking between the cross hairs of the eyepiece. During this quiescent period, the rate of evaporation could be closely monitored; if the drop remained stationary, it was no longer evaporating, and the subsequent rate of fall would be correspondingly unaffected. Next, by switching off the field and short-circuiting the plates by means of a double-throw switch, the freely falling drop was timed with a stopwatch; Stokes' law then gave the weight of the drop. Finally, knowing the strength of the electric field, the amount of charge necessary to balance this weight was readily determined.

> These balancing charges actually always came out, easily within the limits of error of my stop-watch measurements, 1, 2, 3, 4, or some other exact multiple of the smallest charge on a droplet that I ever obtained. Here, then, was the first definite, sharp, unambiguous proof that electricity was definitely unitary in structure, and that it only appeared in exact multiples of that unit, no fraction of that unit ever appearing, at least in these experiments. To this day no case of fractional electronic charge has ever been found by anyone of the many who have repeated my work while taking such precautions as I took to avoid the errors arising both from droplet impurities and from Brownian movements (random motions due to molecular impacts). [16-25]

The value of e deduced from these first measurements on balanced water droplets was $e = 4.65 \times 10^{-10}$ esu—a significant increase over the earlier value, and indeed a value coinciding with Rutherford's value from α-particles. This remarkable agreement would play some role in the subsequent train of events, as we shall see. At any rate, these balanced-drop measurements were completed in the late summer of 1909, just in time for the start of the meeting of the British Association in Winnipeg during late August, but too late for inclusion in the printed program. It was at Winnipeg, it may be recalled, that J J Thomson was inaugurated as president of the Association and Rutherford as president of Section A (Mathematics and Physics). Millikan showed up, unsure whether he would be given the opportunity to present his results [16-26].

Rutherford's opening presidential address, delivered on August 26, dealt with the present status of atomic theory, particularly in light of evidence from Brownian motion of suspended particles, and from α-particles from radioactive sources, since

> ...at the time when, in the vision of the physicist, the atmosphere is dim with flying fragments of atoms, it may not be out of place to see how it has fared with the atoms themselves, and to look carefully at the atomic foundations on which the great superstructure of modern science has been raised. [16-27]

(The Winnipeg agenda also featured the statesman of chemistry, H E Armstrong, as president of Section B on chemistry. He was no friend of the new physics, as we recall from his remarks at Dover, and is abundantly clear from his own presidential address at Winnipeg.) At any rate, the pillars of the 'atomic foundations' referred to by Rutherford were the various methods for determining the fundamental atomic constants. Prominent among the constants was the unit quantity of electricity, from which sprang the absolute number of molecules in unit volume of any gas, and a host of other physical relationships. Rutherford cited the recent measurements converging on the electronic charge: those by Townsend (3.0×10^{-10} esu); J J Thomson, using C T R Wilson's cloud method (3.4×10^{-10}); H A Wilson (3.1×10^{-10}); and most recently Millikan and Begeman (4.06×10^{-10}, Rutherford being unaware of Millikan's new value). From other quarters came additional values for e, namely, by Felix Ehrenhaft (4.6×10^{-10}) from Brownian motion in air of ultra-microscopic dust of silver, and by Moreau (4.3×10^{-10}) from negative ions produced in flames, not to speak of Rutherford and Geiger's own value from α-particles (4.65×10^{-10}). Last but not least, there was the old value by Planck (4.69×10^{-10}) to be reckoned with; 'for reasons we cannot enter into here [stressed Rutherford], this theoretical deduction must be given great value' [16-28].

Returning to his own work, Rutherford allowed that

> ...it is difficult to fix on one determination as more deserving of confidence than another; but I may be pardoned if I place some reliance on the radio-active method..., which depends on the charge carried by the α particle. [16-29]

Acknowledging that

> ... the proof of the existence of corpuscles or electrons with an apparent mass very small compared with that of the hydrogen atom... we owe mainly to the genius of the president of this Association..., [he noted that] the existence of the electron as a distinct entity is established by similar methods and with almost the same certainty as the existence of individual α particles. [16-30]

Alas, unaware of the gist of Millikan's forthcoming post-deadline presentation, he added that

> ... while it has not been found possible to detect a single electron by its electrical or optical effect, and thus count the number directly as in the case of the α particle, there seems to be no reason why this should not be accomplished by the electric method. [16-31]

Millikan got his chance to address the Winnipeg audience on the latest charge determinations from Chicago's Ryerson Laboratory on August 31. The record is silent on Rutherford's reaction, but, according to Millikan's autobiography, Sir Joseph Larmor 'was intensely interested in my paper and raised questions as to whether these experiments were not precise enough to warrant looking into the accuracy of Stokes' Law of Fall for particles of the size used. He promised, himself, to look into the theoretical side of this question, and actually did so when he returned to England' [16-32].

Nor do we have a record of Millikan's oral remarks at Winnipeg. However, he read an abstract of his forthcoming paper proper at a meeting of the American Physical Society at Princeton on October 23, published in the *Physical Review* for December 1909 [16-33]. The full paper, submitted for publication to the *Philosophical Magazine* on October 9, appeared in *Philosophical Magazine* for February 1910 [16-34]. This time Millikan was the sole author, but the tone of the narrative leaves no doubt that the new experiments had been performed by 'Mr. Begeman and myself'. Indeed, individual observations of rates of fall in the tabulated results are carefully credited to 'Millikan' or 'Begeman', as appropriate. Furthermore, 'none of the observations were worked out until after all of the observations had been taken, so that neither Mr. Begeman nor myself had any predispositions as to what should be the correct time of fall in the case of any drop' [16-35].

The weighted mean of a number of 'very concordant' observations came to, as noted earlier, $e = 4.65 \times 10^{-10}$ esu, or the very result of Rutherford and Geiger, with an error not exceeding 2%. In arriving at this value, Millikan exercised judicious selection among the raw data, discarding observations for one reason or another. An underlying article of faith, it would appear, was the implicit assumption of integral multiples of a single unit of charge. Sub-multiple charges were ruled out on the grounds that 'the observed charges would then represent only even multiples of this elementary charge, and there is no reason why odd as well as even multiples of the elementary charge should not have been observed' [16-36]. The possibility of fractional charges was ignored outright, his autobiographical reminiscences notwithstanding. As to the validity of Stokes' law—'the only theoretical assumption involved in this determination'—Millikan was quite emphatic. Since 'the spheres under observations in these experiments have diameters which... are from 35 to 50 times the mean free path of the air molecules... it is scarcely conceivable that Stokes' law fails to hold for them' [16-37]. Moreover, in light of the agreement between Millikan's and Rutherford's value, the Manchester α-particle experiments could be viewed as independent verification of Stokes' law, at least for water drops of the size in question. For good measure, Begeman had by now obtained, independently, a value of $e = 4.66 \times 10^{-10}$ by the 'regular Wilson method', but confining himself to multiply charged cloud layers [16-38].

In addition to the independent, as yet unpublished Begeman value, however, three other e-determinations were judged 'deserving...of much weight', namely, those of Planck, Rutherford and Geiger, and a value by Erich Regener obtained by counting α-particles from polonium (yielding $e = 4.79 \times 10^{-10}$). Assigning equal weight to each of these five determinations (Planck, Rutherford and Geiger, Begeman, Regener, Millikan) 'which seem least open to question', Millikan obtains a grand most probable value for e, namely, $e = 4.69 \times 10^{-10}$.

Complicating matters, four other 'important and interesting pieces of work' leading to estimates of e, were left out in the final weighing, on the ground that they involved 'larger elements of uncertainty than are found in any of the foregoing methods' [16-39]. They included three determinations in France (by Perrin, Maurice de Broglie, and G Moreau), and one by Felix Ehrenhaft in Vienna. Perrin's work involved Brownian movement in liquid emulsions and 'so many assumptions of questionable rigor that it can scarcely be regarded as furnishing a determination of e with a certainty which is at all comparable with that found in any of the five methods first considered' [16-40]. De Broglie's and Moreau's results relied on Perrin's assumptions and results, respectively, and were thus ruled out as well. However, Ehrenhaft's method was actually quite similar to Millikan's, at least in principle. Measuring the velocity of charged particles thrown off by an electric arc, first under gravity and then under an electric field, gave $e = 4.6 \times 10^{-10}$—a result in good agreement with Millikan's own latest value. Yet Millikan rejected it on several grounds [16-41]. In so doing, whether plausibly or guided by some feeling of vulnerability in his own research [16-42], he flung in effect before Ehrenhaft an intellectual challenge the Viennese scientist could ill afford to overlook. The stage was set for the last major confrontation in the saga of the electron.

16.2. Subelectrons?

The paper of Ehrenhaft downplayed by Millikan had been submitted for publication to the *Physikalische Zeitschrift* on April 10 1909, fully six months before Millikan submitted his own paper to the *Philosophical Magazine* [16-43]. Unbeknownst to Millikan, it was the last of three reports on the electronic charge issued by Ehrenhaft during the brief span of March 4–April 10 1909. Moreover, these papers were the first to report charge determinations based on observations of the motion of *individual* charged particles. The mean value deduced by Ehrenhaft for e, 4.6×10^{-10} esu, was well above the extant value of Millikan and Begeman from 1908, but in excellent agreement with that of Planck and of Rutherford and Geiger. (Rutherford had emphasized the agreement in his Winnipeg address.) As if to add insult to injury, by the time Millikan's revised value for e appeared in print, Ehrenhaft had much more startling conclusions to report, casting doubt altogether on all e-determinations, from Planck onward. Who, then, was Ehrenhaft, and what was going on in Vienna?

Felix Ehrenhaft was born into a prosperous Viennese family in April 1879 [16-44]. His father was a physician, his mother the daughter of a prominent Hungarian industrialist. (She was also the niece of a student of J B L Foucault.) He studied in Vienna, first at the Technische Hochschule and then at the University of Vienna where he obtained his doctorate in 1903. The physics department of the university was going through a period of unrest just then, prompted in part by the departure of Ludwig Boltzmann, director of the physics institute since 1894, for Leipzig in 1900 [16-45]. On Boltzmann's return to Vienna in 1902, Leipzig not having met his expectations, the 'physical cabinet' under Viktor von Lang was reorganized into what was called the 'first physics institute'. Franz Exner's 'physical chemistry institute' became the 'second physics institute', and Boltzmann, succeeding himself, took on the 'institute for theoretical physics'. Ehrenhaft became *Assistent* to Lang the year he finished his doctorate, and in 1905 he advanced to the rank of *Privatdozent*.

A fellow student of Ehrenhaft in these years was Paul Ehrenfest, the theorist who received his doctorate under Boltzmann in 1904. Apart from the similarity in their names, the two had little in common, notes Martin Klein. Because of their differences, the unavoidable confusion in names, and most particularly because Ehrenfest resented receiving credit for Ehrenhaft's somewhat dubious achievements in years to come, he worked tirelessly to dispel the confusion [16-46].

Ehrenhaft's earliest research, ca. 1900, was in the mainstream of current research interests, however, dealing with metallic colloids—ultra-light particles much larger than molecules but small enough to show Brownian motion when immersed in a fluid [16-47]. Ehrenhaft authored several papers on the topic during the period 1902–1905. The field received a substantial boost during 1905 and 1906, when Albert Einstein and Marian von Smoluchowski, respectively, took up the problem of Brownian motion and its interpretation in terms of molecular kinetics; their authority helped lend credence to the 'reality' of atoms, as opposed to simply a heuristic artifice. (Atomism was still a subject of some debate in certain Continental and New-World circles; we have, for instance, touched on the anti-atomic stance of Michael Pupin, Millikan's teacher [16-48].) In 1907 Ehrenhaft published an investigation of his own on Brownian motion of silver, zinc, and other particles in air, building on the ideas of Einstein and Smoluchowski. This piece of work earned him the prestigious Lieben Prize for 1910 from the Vienna Academy of Sciences.

It was thus altogether natural for Ehrenhaft to turn to measurements of the electric charge carried by his 'ultramicroscopic' particles, and hence, to a determination of the electronic charge e. In his first publication on the topic, a one-page summary in the *Anzeiger* of the Vienna Academy, he explained that metallic colloidal particles emitted from an arc lamp now and then exhibited an electric charge, revealed by their motion in a horizontal electric field [16-49]. Measuring the vertical particle motion under gravity and its horizontal component with an electric field superimposed, and using Stokes' law to obtain the particulate mass, allowed him to determine their charge, much as had Millikan, except for the important fact that Millikan used a *vertical* electric field. Two weeks later, in a longer paper to the Academy dated 18 March, and again in the paper earlier referred to in the *Physikalische Zeitschrift*, he presented a detailed account of his actual results [16-50]. His value for the 'elektrischen Elementarquantum', as he called it, was $e = 4.6 \times 10^{-10}$ esu; Ehrenhaft drew conspicuous attention to the fact that his value was in good agreement with Planck's and Rutherford's values, but not with Millikan's (1908) value [16-51].

Unfortunately, as Millikan would justifiably object, Ehrenhaft's determination was flawed in several respects. This was so despite his sophisticated instrumentation which, among other things, included a Zeiss version of the slit ultra-microscope developed by Henry Siedentopf and Richard Zsigmondy in 1903 and newly modified for his purposes by Ehrenhaft himself, in contrast to Millikan's simple, short-focus telescope. (In the ultra-microscope, the particles are illuminated with a cone of light at right angles to the optical axis of the instrument, which, in turn, is diffracted by the particles into a compound microscope, rendering polarized diffraction disks as images. Those particles to be made visible are selectively illuminated, by focusing the light through a slit and system of lenses and collimators.) The most serious shortcoming in Ehrenhaft's procedure was the fact that his velocity measurements with and without an electric field, albeit a horizontal field, could not be made on the same set of particles. Indeed, assured Ehrenhaft in his second report to the Academy, measurements with a vertical electric field, a prerequisite for making the two velocity determinations on the same particles, was intended in follow-up measurements soon to come [16-52]. Alas, a year was to pass before further word was forthcoming from Vienna.

The silence was broken in the third week of April 1910, fully one year later, when we hear again from the two protagonists literally within days of each other. Neither Ehrenhaft nor Millikan had been idle in the intervening period, with a new round of charge determinations largely completed both at Vienna's Physikalische Institut and at Chicago's Ryerson Laboratory. We shall hear from Millikan presently. First to report was Ehrenhaft, in a two-page note to the Vienna Academy on April 21 1910 [16-53]. By now he too, as promised, had deployed a horizontal condenser producing a vertical electric field—one strong enough to cause metallic particles to rise against gravity, and allowing measurements of the rate of rise and fall with and without the superimposed field acting on the same particles. However, the improvement in technique was not the only thing he had to report. His new results were quite another matter. His sparse note bore the intriguing title 'On the *smallest* measurable quantity of electricity' (italics added). Twenty-two measurements culled from over 300 observations on platinum and silver particles from electric arcs indicated a *continuous* range of charge values from 7.63×10^{-10} esu down to 1.38×10^{-10} esu—the lowest value representing a 70% drop in the near-sacrosanct value for the electronic charge.

We hear again from Ehrenhaft and his supporters [16-54] in Institute No 1 three weeks later when, on May 12, he presented a slightly expanded *Anzeiger* report to the Academy; a full paper appeared in the issue of the *Sitzungsberichte* for the same date [16-55]. Much the same material appeared in a paper submitted to the *Physikalische Zeitschrift* on May 23 [16-56]. Coining the term 'Subelektronen', and in so doing repudiating his own 'unteilbare Elektrizitätsatom' and the doctrine of the electron altogether, he reports on May 12 that 'the smallest measurable value [of the electrical charge] for gold particles appears to be of the order of 5×10^{-11} esu', and that 'the value for gold ranges continuously from 5×10^{-11} esu to a maximum of about 1.75×10^{-10} esu, or [to] a third of the electronic charge' [16-57]. With 'all reserve' he suggests that 'if indivisible atoms of electricity exist in nature, they must be less than 1×10^{-10} esu' [16-58]. He throws caution to the wind in his two major reports for May, declaring flat out that 'these experiments do not allow the author to maintain the fundamental hypothesis of the electron theory', namely indivisible electrons [16-59].

Ehrenhaft took the occasion to subject Millikan's February 1910 *Philosophical Magazine* article to an in-depth and lashing critique, re-analyzing the Chicago data in terms of his own methodology. Retaining observations rejected by Millikan as unreliable, and recalculating the charge for each separate observation, instead of following Millikan's procedure of lumping runs of a given series to form weighted subaverages as input to the final averaging, Ehrenhaft obtained a large spread in *e*, ranging from 8.6×10^{-10} esu to 29.82×10^{-10} esu. If nothing else, notes Holton, the re-evaluation showed that 'the same observational record could be used to demonstrate the plausibility of two diametrically opposite theories' [16-60].

What saved the day for Millikan against Ehrenhaft's onslaught was the timely announcement of his own new results. They were read by him at the Washington meeting of the American Physical Society on April 23 1910—two days after Ehrenhaft's opening salvo at the Vienna Academy. And this time Millikan, too, had fresh ammunition in store.

16.3. Oil drops; isolation of the electron

As Millikan recalls it, the crucial idea for his 'oil drop venture' came to him on returning from the Winnipeg meeting in the first days of September 1909.

> Riding back to Chicago from this meeting I looked out the window of the day coach at the Manotoba plains and suddenly said to myself, 'What a fool I have been to try in this crude way to eliminate the evaporation of water droplets when mankind has spent the last three hundred years in improving clock oils for the very purpose of obtaining a lubricant that will scarcely evaporate at all.' [16-61]

A somewhat different version of events has been handed down to us by Harvey Fletcher, who would be Millikan's protégé in the ensuing experiments from late fall 1909 into spring the following year. The closest to Fletcher's counterpart in Ehrenhaft's camp might have been Karl Przibram, Fletcher's senior by six years. Both Fletcher and Przibram received possibly less credit than they deserved in the shadow of their respective mentors—a not uncommon experience in academia, then as now. Coauthorship with the professor was reserved for more senior staff members, or for student or junior colleagues only when the time was ripe. In Przibram's case, moreover, matters came to a head in a disagreement with Ehrenhaft over experimental results [16-62]. Such was not the case with Fletcher, though his account of his role *vis-à-vis* Millikan warrants retelling, among other things for the gap it fills in Millikan's own telling of the final phase of the ongoing experiments in the Ryerson Laboratory.

Harvey Fletcher was born into a family of nine children of a contractor and builder in Provo, UT, in 1884. He was schooled in Provo, including Brigham Young University where he received his BS degree in 1907. He entered the graduate school of the University of Chicago in the fall of 1908, albeit conditionally, on account of the lesser preparatory standards at Brigham Young. In the beginning of his second year, having proved himself to the satisfaction of the University authorities, he approached Millikan for advice on a suitable thesis topic for the PhD [16-63]. Millikan, just back from Winnipeg, was a busy man. Finally Fletcher caught up with him, and was invited down into the Ryerson for a chat and to view Millikan and Begeman's electronic charge apparatus.

The well-used equipment needs concern us mainly with respect to the modifications soon to come. A tube led from an expansion chamber to a small box attached to a microscope and containing a pair of horizontal conducting plates connected to a battery; a switch in the circuit enabled the electric field between the plates to be thrown on or off at will. A sudden expansion of the air in the box caused a cloud of vapor to form by condensation. Viewed through the microscope, the mist was seen to consist of a large number of water droplets, too small to be observed visually. The droplets would sink under gravity, but their fall could be arrested by switching on the electric field; with just the right field strength, selected drops could be suspended between the plates. From the rate of fall, and the retarding field strength, the charge on the droplets could be determined. Unfortunately, the water forming the droplets evaporated too fast for the droplets to remain in view longer than about 2 s, placing an unacceptable limit on the accuracy of the charge determination. What to do? A less volatile cloud was obviously called for.

> We discussed ways and means of getting around the difficulty, and I think we all agreed that we should have a droplet that did not evaporate if we could get it small enough and could control it. Mercury, oil, and two or three other substances were suggested. In a discussion of that kind, it is rather difficult to be sure who suggested what. I left with the impression that I had suggested oil for it was easy to get and handle. However, in his memoirs Millikan said he had been thinking of this before this conference. Of course, I cannot say yes or no to that, but I do know what happened after this conference.
>
> Professor Millikan said to me, 'There is your thesis; go try one of these substances which will not evaporate'. [16-64]

Fletcher lost no time rigging up a simple demonstration device incorporating an atomizer (perfume sprayer) procured from the corner drug store, some watch oil, an arc light improvised from a projection lantern (Fletcher earned pocket money taking care of the departmental stock of projectors), and a cathetometer telescope—a standard laboratory instrument of the day. (He need not have resorted to a drug store atomizer. The atomizer method of producing minute spherical drops for investigating Brownian motion had been developed and honed to a fine art at the Ryerson in early 1908, and had been in continuous use ever since.) A crude condenser with a pinhole in one plate was fashioned from sheet metal in the student shop; it could be charged by one of the cart-mounted, 1000 V storage batteries cluttering the laboratory hallways.

Behold, the apparatus worked perfectly well on the first attempt. A spray of oil squirted from the atomizer caused a fine mist of oil particles to drift through the hole in the condenser plate into the region below where the droplets twinkled like starlets against the backdrop of the dark-field illumination. By switching the electric field on or off with the proper timing, an individual oil drop, charged by friction as it left the atomizer, could be kept in view for a considerable length of time. Before the day was over, Fletcher had a fair value for *e*. Cornered within a day or two, Millikan needed no convincing. The shop foreman was called in, and Millikan and his student sketched improvements to the apparatus—mainly a more accurate condenser, enclosed in a vessel to prevent air drafts, and an x-ray source to enhance the ambient ionization of the air.

Updating the apparatus took about a week. There followed an intense 5–6 weeks while Millikan and Fletcher immersed themselves in the experiment—Millikan as time allowed. Sometime in December 1909, results were good enough to warrant an open house for the public and the press. The local papers waxed enthusiastically, and VIPs from near and far dropped in to have a look. Among the prominent visitors, recalls Fletcher, was Charles Steinmetz, the electrical wizard from the General Electric Company. Despite his towering contributions to electrical engineering, Steinmetz was still dubious about electrons. (To him, all electrical phenomena boiled down to strains in the æther.) Still, he spent the better part of an afternoon bent to the apparatus, watching the dance of the little droplets. 'I never would have believed it, I never would have believed it', he muttered as he bade his farewell [16-65].

The publicity was all good and well, but a formal, scientific unveiling it was not. About four months elapsed, at some point during which Millikan and Fletcher started writing up their results in earnest. Much of the writing was left to Fletcher, according to his account, with Millikan smoothing out rough spots in the manuscript. 'All the while', recalls Fletcher, 'I thought we were to be joint authors' [16-66], though as time went by he became increasingly concerned whether this effort was indeed shaping up to become his thesis. A preliminary report on the new experiments on single charged drops of oil and mercury was read by Millikan at the Washington meeting of the American Physical Society on April 23. The results were discussed again by him at a Sigma Xi meeting in Chicago about a month later [16-67]. A one-page summary of the Washington presentation, entitled 'The isolation of an ion and a precision measurement of its charge' was at long last published in July, albeit under the authorship of Millikan alone [16-68].

Meanwhile, continues Fletcher in his autobiographical account, sometime in mid-June Millikan himself stopped by in Fletcher's apartment on the west side of the Chicago campus, ostensibly to chat but mainly to discuss who was to be the author of the forthcoming paper, by then about ready for release. Four other manuscripts covering various subtopics of the same investigation, e.g. Brownian motion and gaseous ionization, were also in different stages of preparation by then.

> It was obvious that he wanted to be the sole author of the first paper. I did not like this, but I could see no other way out, so I agreed to use the fifth paper... [on Brownian movements] as my thesis. [16-69]

In the event, the paper in question was published, at the request of the editor, as a 'special article' in the September issue of *Science* [16-70]. Millikan was the sole author, but Fletcher's collaboration is duly acknowledged. He got right to the point.

> There is presented herewith [Millikan began] a new method of studying gaseous ionization, with the aid of which it has been found possible:
> To catch upon a minute droplet of oil and to hold under observation for an indefinite length of time one single atmospheric ion or any desired number of such ions between 1 and 150.
> To present direct and tangible demonstration, through the study of the behavior of electrical and gravitational fields of this oil drop carrying its captured ions, of the correctness of the view advanced many years ago and supported by evidence from many sources that all electrical charges, however produced, are exact multiples of one definite, elementary charge.
> To make an exact determination of the value of the elementary electrical charge, which is free from all questionable theoretical assumptions and is limited in accuracy only by the accuracy which is attainable in the measurement of the coefficient of viscosity of air. [16-71]

Curiously, the paper is devoid of any figure depicting the apparatus (a shortcoming of the earlier *Philosophical Magazine* paper as well), though one is provided in a longer follow-up paper under the same title subsequently published in the *Physical Review* for April 1911 [16-72]. We reproduce that figure here. Once again the basic experimental procedure, which differs in principle from the Millikan–Begeman procedure mainly in the method of producing and admitting the charged droplets, merits retelling.

Referring to figure 16.1, a mist of oil was blown from the atomizer A into the dust-free chamber C. One or more of the droplets fell through a pinhole P into the space between the plates MN held apart by ebonite posts, and the pinhole was then closed by an electromagnetically operated cover (not shown) in order to shut out air currents. The droplet, once inside the condenser, was illuminated through a small window by

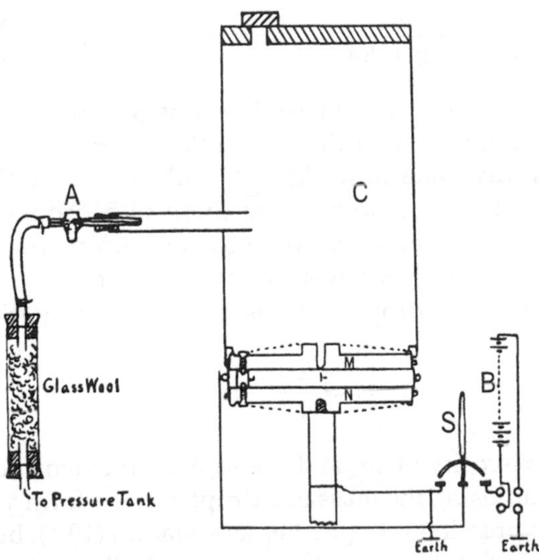

Figure 16.1. *Millikan's oil drop apparatus of 1910. Oil drops are blown from atomizer A into chamber C, where they fall under gravity through pinhole P into the space between condenser plates MN supported by ebonite posts a. A thin strip of ebonite encloses the space between the plates, pierced by three glass windows. A beam of light enters one window and emerges through the second; the third window is for observing the illuminated drops floating between the plates. B is a battery, and S is a switch for short-circuiting the charged plates.*

a beam from an arc light, and was observed through another window, appearing as a brilliant star against a black background in the field of view of a cathetometer telescope. It fell at constant velocity under gravity (owing to the viscosity of air for such a small drop). However, before it reached the lower plate an electrical field of between 3000 and 8000 V cm^{-1}, created by the battery B, was switched on. Provided the drop had received a frictional charge of the proper sign at it was blown out through the atomizer (or had picked up charges subsequently from the natural atmospheric ionization augmented by a nearby radium or x-ray source), the drop was pulled back up towards the upper plate. Before it struck, the plates were short-circuited by the switch S, and the time required by the drop to descend under gravity between the cross hairs of the telescope was noted. The rate of rise of the drop under the influence of the field was also recorded (again a constant velocity, because of the air resistance).

The relationship between the mass m of a drop, the charge e_n which it carries, its speed v_1 under gravity and v_2 under an electric field E was given as equation (16.1) in section 16.1, but we rewrite it here in a slightly altered form:

$$\frac{v_1}{v_2} = \frac{mg}{mg + Ee_n} \quad \text{or} \quad e_n = \frac{mg}{E}\left(1 + \frac{v_2}{v_1}\right) \tag{16.4}$$

where the subscript n now denotes the (integral) number of elementary charges (ions) adhering to a drop; e_n, rather than the absolute value e, constituted the raw data in Millikan's analysis. (Note that the mass of an ion is negligible compared to the mass of a drop. Moreover, m is actually the difference between the mass of a drop and the mass of the displaced air.) We may eliminate m by means of the usual expression for the rate of fall of a drop of radius a under gravity (Stokes' law), or

$$v_1 = \frac{2}{9}\frac{ga^2}{\eta}(\sigma - \rho) \tag{16.5}$$

where η is the viscosity of air, and σ and ρ are the densities of oil and air, respectively, and using the mass of a droplet in the form $m = \frac{4}{3}\pi a^3(\sigma - \rho)$. (We gave the expression for v_1 earlier as equation (10.1), but repeat it with some changes in notation, to conform with Millikan's usage.) We then obtain the expression for the charge in the final form used by Millikan:

$$e_n = \frac{4}{3}\pi\left(\frac{9\eta}{2g(\sigma - \rho)}\right)^{3/2}\frac{\sigma g}{E}(v_1 + v_2)v_1^{1/2}. \tag{16.6}$$

We return to the matter of Stokes' law in a moment; it would play a larger role than foreseen in the final data reduction.

> Mr. Harvey Fletcher and myself, who have worked together on these experiments since December, 1909, studied in this way between December and May from one to two hundred drops which had initial charges varying between the limits 1 and 150, and which were upon as diverse substances as oil, mercury and glycerine, and found in every case *the original charge on the drop an exact multiple of the smallest charge which we found that the drop caught from the air.* The total number of charges which we have observed would be between one and two thousand, and *in not one single instance has there been any change which did not represent the advent upon the drop of one definite invariable quantity of electricity, or a very small exact multiple of that quantity.* (Italics in original.) [16-73]

By way of illustrating the reliability of his data, Millikan leads off the discussion of his results with a tabular record of the observations of a drop which was kept within the field of view for no less than four and a half hours, as it was patiently coaxed up and down between the cross hairs of the telescope [16-74].

> How completely the error arising from evaporation, convection currents, or any sort of disturbances in the air, are eliminated, is

> shown by the constancy during all this time in the value of the velocity under gravity. This constancy was not attained without a considerable amount of experimenting.... [16-75]

The near-constancy of the successive times of descent is, indeed, ample confirmation of the author's point; during the four and a half hours, the drop evaporated ever so slowly, if at all. The times of *ascent* under the electric field are quite another matter, changing discontinuously up or down by a factor of five, showing that the droplet, originally with a negative charge, caught or released one or more ions in the air, and that the positive or negative ions caught carried exactly the same charge. It did not take many of these jumps in speed to recognize the minimum jump corresponding to the capturing or discarding of single, isolated charges [16-76].

We have referred repeatedly to the crucial role of Stokes' law in the analysis under discussion. In its original form, this law, which relates the force X acting on a spherical particle to its velocity, radius, and viscosity of the medium encountered, $X = 6\pi va\eta$, soon proved insufficiently accurate for the purposes of the oil drop experiment. Its failure was revealed by a slight dependence of the apparent charge e_1 (e_n of equation (16.6) with $n = 1$) on velocity, traced to a breakdown of Stokes' law for very small drop sizes. This was corrected for to first order by introducing, quite *ad hoc*, a functional dependence of the molecular mean free path in air, l, of the form

$$X = 6\pi va\eta\left(1 + A\frac{l}{a}\right). \qquad (16.7)$$

Hence follows

$$v_1 = \frac{2}{9}\frac{ga^2}{\eta}(\sigma - \rho)\left(1 + A\frac{l}{a}\right) \qquad (16.8)$$

where A is a numerical coefficient to be determined empirically. This turned out to agree surprisingly well with the Cunningham correction, introduced on the basis of kinetic theory considerations, and something we have referred to on several occasions.

This correction was largely the handiwork of Fletcher, who wrote up that portion of the paper in *Science*.

> We'd been plotting 'e' versus the radius; the radius of the little droplet, or a over l, the mean free path. And we'd get a curve. Then I said to Millikan one day, 'what fools we are. Why don't we take two-thirds of that and there'll be a straight line?' [16-77]

That is to say, a plot of observed values of $e_1^{2/3}$ as ordinates and l/a as abscissæ should yield a straight line intercepting the axis of ordinates at a value $e^{2/3}$ (and with a slope determined by the constant A), and hence, give the absolute value of e without further ado. This is fairly evident by combining equation (16.6) and equation (16.8) in the following form:

$$e\left(1+A\frac{l}{a}\right)^{3/2}=e_1. \tag{16.9}$$

(The 3/2 power arises from the fact that e_n varies as $v_1^{3/2}$ in equation (16.6).) To obtain the indicated plot, the pressure in the observation chamber C, hence the mean free path l, was systematically varied while keeping the droplet in the field of observation. The end result was not only a valuable, independent check of the so-called Stokes–Cunningham formula, but also provided a mean value for e given by 4.9016×10^{-10} esu, with a probable error of about 0.1% [16-78].

Subsequent to the publication of this result in *Science*, the Ryerson acquired a more reliable Weston Laboratory Standard Voltmeter. It enabled a new calibration curve to be constructed, and with it a new, reduced value for e, obtained in time for the expanded paper submitted to the *Physical Review* on November 28 [16-79]. In addition, 'the m of equation (1) [our equation (16.4)] was through oversight treated as the real mass instead of the apparent mass' (difference between actual mass and buoyancy of the air) [16-80]. This minor blunder further reduced e to a final value $e = 4.891 \times 10^{-10}$ esu with a probable error of 0.04%.

With Millikan's latest paper in the April issue of the *Physical Review*, the ball between him and Ehrenhaft bounced back into Millikan's court. By the previous fall he had been familiar with Ehrenhaft's latest publication on subelectrons, namely, the paper mailed from Vienna on 23 May and published in the *Physikalische Zeitschrift* for July 15 [16-81]. That journal was apparently received on a timely basis at the University of Chicago; Millikan may not have had as easy access to Ehrenhaft's preliminary *Anzeiger* and *Sitzungsberichte* reports to the Vienna Academy (both dated May 12), in part due to frequent delays in their publication [16-82]. Though Ehrenhaft had taken Millikan soundly to task in the *Physikalische Zeitschrift*, Millikan replied in kind in the closing section of this, his own major article of 1911 on the electron, using a simple argument to make his point.

Millikan acknowledged Ehrenhaft's newly adopted procedure—to be sure, one used all along at Chicago—of deploying a vertical electric field, allowing observations to be made on a single charged particle. There the similarity ended. Ehrenhaft's use of the ultra-microscope was a handicap in disguise; it restricted his determination of rates of fall or rise to exceedingly short distances (0.01 cm, versus 1.3 cm in Millikan's simpler apparatus). In Vienna, a particle was followed in its motion up and down but once, and kept under observation for at most a minute, instead of 4–5 h, as was the practice at Chicago. The clincher, however, was Ehrenhaft's use of metallic dust fragments, ranging in size from the very smallest of

Millikan's droplets down to one-tenth as large; therein lay the principal source of Ehrenhaft's problem. Millikan and Fletcher had found from experience that consistency in readings of successive rates of fall under gravity could no longer be expected when the particles were smaller than some minimum size. The reason was simple: 'the displacements of such particles due to their Brownian movement became compatible with the displacements produced by gravity' [16-83]. Millikan illustrated his argument with a table of rates of fall and rise for one such particle, adding in a footnote that 'a more complete analysis will be given in a separate paper by Mr. Fletcher'. Though the drop carried a single elementary charge, its apparent value calculated from individual observations varied by over a factor of four. 'The Brownian movement theory', he insisted, 'can be made to account quantitatively as well as qualitatively for all the irregularities of [the table] as well as for those in Ehrenhaft's tables' [16-84]. Indeed, he added, Ehrenhaft's colleague Przibram had obtained more consistent results than had his mentor, precisely because he used larger particles.

Thus it came about, by a curious circumstance, that the phenomenon of Brownian motion conspired to undermine the radical findings of Ehrenhaft, himself an authority on the very subject of *Brownschen Bewegung*. However, so was Harvey Fletcher by the fall of 1910. The forthcoming paper by Fletcher referred to by Millikan was in the manuscript stage when the two had their 'friendly talk' in Fletcher's apartment in June of that year. On Millikan's rightful insistence, for a published paper to constitute a doctoral thesis, the candidate must be its sole author. One of the papers in the pipeline seemed a good choice in Fletcher's case. As an outgrowth of the oil drop experiments, there had begun at the Ryerson Laboratory in early 1910 a program of investigation in its own right of Brownian motion in gases—one lauded by Theodor Svedberg in his 1913 review of what was by then a burgeoning subject of research [16-85]. In particular, by combining the cornerstone equation of the oil drop method, equation (16-4), with Einstein's equation for Brownian movements relating the mean particle displacement during random motion to the time τ and viscosity η [16-86],

$$\Delta\bar{x}^2 = \frac{RT}{N}\left(\frac{1}{3\pi a\eta}\right)\tau$$

it proved possible to obtain the product Ne, where N is Avogadro's number, without reference to the size of the particles or resistance of the medium. Moreover, comparison of Ne with the same product found in electrolysis led to a demonstration of Einstein's equation with greater precision than attained elsewhere in Brownian movement studies until then. For the experimental phase of the investigation, Fletcher used the oil drop apparatus, suitably modified, in which he measured the Brownian displacements Δx in terms of the fluctuations in the time required for a particle to fall a given distance. The upshot was a preliminary paper by him completed on February 3 1911, 'Some contributions to the theory of Brownian motion with experimental applications' [16-87], and a full paper submitted to the *Physical Review* on April 25: 'A verification of the theory of Brownian movements and a direct determination of the value of Ne for gaseous ionization' [16-88]. This work earned Fletcher the PhD *summa cum laude* in 1911—the first such honor awarded in physics at Chicago.

Millikan claims in his *Autobiography* that he never regarded the numerical value of e 'as either the major objective or the major result of the oil-drop method, which I took to be instead the final settling of the controversy which was violently raging at that time as to the unitary nature of electricity and of matter itself'.

> Indeed, I think that by far the most direct, simple and accurate verification, not only of the unitary theory of electricity but also of the Brownian movements equation, came out of the oil-drop technique, as initiated by Harvey Fletcher and myself and as carried out with much skill and more elaboration by Fletcher in his thesis. [16-89]

From Ehrenhaft's perspective in Vienna, the 'final settling' was not to come, and it was slow in coming in the contemporary view of many scientists whose reputation was beyond reproach. As seen from Chicago, the final settling came as early as 1913, and was at any rate a *fait accompli* by 1917 in the retrospective judgment of most observers of the controversy. In the remainder of this section, we touch on some of these milestones and points of contention.

A steady stream of papers flowed from the Vienna institute in the aftermath of Ehrenhaft's barrage during April–May 1910, some authored by Ehrenhaft and some by Karl Przibram, his student. For a starter, in the same volume of the *Physikalische Zeitschrift* that carried Ehrenhaft's major paper of May 23 (and in which he took Millikan to task), he published what amounted to a revision of that paper, now dealing with measurements by Przibram, at Ehrenhaft's behest, on droplets produced by blowing moist air over white phosphorus [16-90]. The paper was presented in a lecture demonstration at the Versammlung of the Deutscher Naturforscher und Ärzte (Meeting of German Scientists and Physicians) in Königsberg during September 18–24 1910. The session, on September 19, chaired by Walter Kaufmann, included a discussion from the audience, among it, Sommerfeld, Planck, Born, Siedentopf, and Kaufmann [16-91]. In this paper, while virtually ignoring Millikan's latest results, Ehrenhaft reiterated his argument that if indivisible atoms of electricity exist, they are smaller than 10^{-10} esu; if Przibram's data, encompassing some 1000 charge measurements, showed any consistent grouping of charges at all, it occurred in a spurious fashion or at values much too low 'for the author to... explain these experiments by adhering to the fundamental hypothesis of the electron theory' [16-92]. In 1914, by then promoted to associate professor at Vienna's Physikalische Institut No 1, Ehrenhaft had reduced his upper limit on the quantum by a factor of ten. If there are quanta of electricity to be found, they 'should be sought at most on the order of 10^{-11} e.s.u.' [16-93].

Always buttressing his argument with fresh experimental results, if not necessarily of his own, Ehrenhaft invoked smaller and smaller particles as time went by, steadfastly ignoring the warning of critics of the breakdown of Stokes' unmodified law with the onset of Brownian movements. The smaller the particles, the smaller became his limit on the quantum of charge [16-94]. All the while, he vigorously defended his findings. His critics, he countercharged, were biasing their arguments with trumped-up hypotheses containing indivisible electrons as built-in elements, and their larger particles provided stable bases for clustering of 'subelectrons'; he, himself, on the other hand, relied on a purely empirical approach relying solely on 'direct facts of nature' without the need for *ad hoc* assumptions or selection of data. His credibility as an experimenter received a boost with his demonstration in 1917–1918 of what he coined photophoresis (the effect of light on the motion of aerosol particles that scatter as well as absorb light) and various allied phenomena, such as 'negative photophoresis' (particles moving against the direction of light propagation) and magnetophoresis [16-95]. These are, to be sure, complicated effects far from being fully understood to this day.

Meanwhile, Millikan too had new results to report as early as 1912. He had worked them up during a six month sojourn in Europe, mostly in Berlin, since uninterrupted time for such analysis proved impossible to find at Chicago [16-96]. For that reason his new results, ergo his 'final absolute value' for e, were first read before the Deutsche physikalische Gesellschaft in June 1912 [16-97]. They were presented again before the British Association at Dundee, Scotland, that September, shortly before the Millikan family sailed from Liverpool for the USA [16-98]. The final paper—the culmination of Millikan's program on the electronic charge in the years up to World War I—was completed on June 2 1913, and published in the *Physical Review* for August of that year [16-99]. 'Inaccuracies in readings and irregularities in the behavior of the drops' were no longer of paramount concern. Rather, the main focus was now on systematic errors in the various parameters entering the final formula for e_n, equation (16.6), such as the viscosity of air, the density of oil or air, the electric field, and the correction factor A in Stokes' law. The updated, final value for e was 4.774×10^{-10} esu with an uncertainty of two parts in 1000. Since the Faraday constant, the number N of molecules per gram molecule multiplied by e, was by then fixed by international agreement, the redetermination of e also furnished a new value for N, 6.062×10^{23}, with the same uncertainty.

> The difference between these numbers and those originally found by the oil-drop method, viz., $e = 4.891$ and $N = 5.992$ is due to the fact that this much more elaborate and prolonged study has had the effect of changing every one of the three factors η, A, and d (=cross-hair distance) in such a way as to lower e and to raise N. The chief change however has been duc to the elimination of the faults of the original optical system. [16-100]

In a tit for tat, Millikan had nothing to say about Ehrenhaft's subelectrons in this, his grand finale, though the paper contains a section devoted to 'Comparison with other measurements' [16-101]. One who did speak up was Fletcher, who began a paper of his own while still at Chicago, in which he recalculated the latest Chicago results according to the prescription of Ehrenhaft—that is, virtually without data discrimination; the exercise led, to nobody's surprise, to no obvious value for e [16-102].

A renewed determination of e was completed by Millikan in August 1916, with brand new apparatus featuring condenser plates flat to within two wavelengths of sodium light and all auxiliary constants once again painstakingly re-evaluated. It gave precisely the same result for e to four significant figures, save for a slight improvement in the uncertainty [16-103]. At this point, the oil drop experiments came to an end at Chicago (as did the work on verifying Einstein's photoelectric equation, Millikan's second, major experimental program and one under way at the Ryerson since 1912). Millikan was leaving the University to devote his energies to organizing scientific research for wartime purposes.

The papers of 1913 and 1917 have been said to have 'definitely settled the question of the uniqueness of the electronic charge' [16-104]. Their value for the charge would remain the standard value for over a generation, and the accuracy in Millikan's relative results from those years would remain virtually unequalled [16-105]. The *absolute* value for e would undergo a slight upward revision in time, mainly due to re-evaluation in the 1920s of the coefficient of viscosity of air. Suitable corrections, undertaken at the Norman Bridge Laboratory of Physics at the California Institute of Technology and elsewhere, applied to Millikan's determination of 1913 and 1917, give a value $e = 4.799 \times 10^{-10}$ esu 'which must be considered the most accurate directly obtainable by the oil-drop method' [16-106].

With that, we have disposed of Ehrenhaft's subelectron too cavalierly. His results were by no means discounted out of hand in European circles ca. 1920, particularly in light of his success with photophoresis. In the 1916 edition of his *Theory of the Electron*, H A Lorentz concluded that 'the question cannot be said to be wholly elucidated' [16-107]. The Nobel Prize archives for 1920 show that Svante Arrhenius argued that year that, while most scientists sided with Millikan, the case was not closed and that, consequently, Millikan should not be recommended for the Prize [16-108]. Two years later, R Bär, in his review of the dispute, reiterated that while 'most physicists discounted the reality of the subelectron, ... the experiment left, at the very least, an uncomfortable feeling' [16-109]. Moreover, while Millikan's data has been extensively scrutinized, Ehrenhaft's laboratory records were lost when he fled Austria at the onset of World War II. Ehrenhaft himself never wavered in his faith in subelectrons; his last major paper on the subject is dated 1941 [16-110].

The lingering question remains: why did Ehrenhaft repudiate the electron, and with it atomism, so suddenly and utterly within the short span of April–May 1910? Holton offers several possible reasons [16-111]. Millikan's rebuff in February of that year was perhaps one, but could hardly have been a major factor. A clue may be found in a letter from Anton Lampa, professor of experimental physics at Prague, to Ernst Mach, his former teacher and senior spokesman for the school of anti-atomists (Wilhelm Ostwald having rejected his own anti-atomic views some time before) [16-112]. It is dated May 1 1910, or between Ehrenhaft's first two subelectron communications to the Vienna Academy that spring. Dwelling first on Planck's recent attack on Mach's (and Lampa's) philosophy, he also recounts his recent visit to Vienna. Should Ehrenhaft's preliminary measurements bear out, 'then the electron would be divisible'. Indeed, 'it would be just too beautiful if the electron were now to undergo the same fate as the atom did as a result of cathode rays ...' [16-113].

Another clue, emphasizing the importance of Mach's anti-atomic stance for Ehrenhaft in these years, is found in a speech Ehrenhaft gave on a public occasion in Vienna in 1926, commemorating the tenth anniversary of Mach's death. Contrasting Mach's views with those of most physicists, he lauded Mach

> ... as an advocate of the much more modest, phenomenological point of view which finds satisfaction merely with the description of the phenomena and despairs of other possibilities—and the others are advocates of views which, through statistical methods and speculative discussions concerning the constitution of matter, are reflected in atomism, and who believe themselves able to get down to the true Being of things. [16-114]

With that we leave the controversy, save for one more tantalizing question: might some of the fractional charges observed by Ehrenhaft (or Millikan) have been 'quarks' in the parlance of modern elementary particle physics? Among the underpinnings of the standard model—the theoretical edifice encompassing the fundamental constituents of matter and the forces between them—is the quark structure of nucleons; that is, of protons and neutrons [16-115]. Quarks, along with leptons, are believed to account for the scant dozen or so basic building blocks of matter; the quarks are subatomic particles possessing, among other esoteric properties, either $+\frac{2}{3}$ the electronic charge ('up' quarks), or $-\frac{1}{3}e$ ('down' quarks). The most familiar form of lepton is the electron. According to the quark model, protons are quasi-elementary, composite structures with two 'up' quarks and one 'down' quark irretrievably confined within them (giving a net charge of +1, in units of the electronic charge); neutrons consist of one 'up' and two 'down' quarks (hence, have zero charge). Around the nucleons orbit one or more electrons, making up atoms, molecules, and everyday matter.

Anomalous charges that could have been construed as quark candidates are found in both the Vienna and Chicago data. Thus, a diagram in Ehrenhaft's last paper of 1941, plotting number of observed charges against their measured values, shows, in addition to a broad peak centered on the nominal value for e, a second peak at approximately $\frac{2}{3}e$ [16-116]. Moreover, in his first major paper of 1910 (reporting, to be sure, on water drops, not oil drops), Millikan refers to having 'discarded one uncertain and unduplicated observation, apparently upon a singly-charged drop, which gave a value of the charge on the drop some 30 per cent. lower than the final value for e' [16-117]. Again, then, we have $\frac{2}{3}e$.

These anomalous charges were invariably associated with the *smaller* particles, which were *a priori* suspect according to Millikan, who was a very cautious experimenter indeed. One who has studied the matter is P A M Dirac, among the architects of modern physics, and one whose opinion counts. The fact that both Ehrenhaft and Millikan found anomalous charges for their smaller particles, in his view,

> ... does not constitute evidence for quarks. It merely shows there was some experimental error, perhaps the same for both of them, affecting their smaller particles. It must be counted as an unexplained coincidence that their anomalous charges tended to be close to two-third of the electronic charge. [16-118]

Chapter 17

DAWNING OF THE ATOMIC AGE

17.1. Background for Rutherford's atom

At the very time, spring 1909, that Millikan's young graduate student Begeman was squinting through his low-power eyepiece at Chicago, so was an assistant and a student of Rutherford in Manchester, England. Just as the newly launched Chicago experiment would lead to the definitive isolation of the electron as the first subatomic particle in nature, the experiments under way at Manchester would usher in the last triumph of classical atomism in the nuclear model of Ernest Rutherford.

At this point in our chronicle, it is easy to overlook that the electron was not Millikan's progeny, but J J Thomson's. Moreover, the standard history has Rutherford's planetary atom springing full-blown from an inspired refutation of Thomson's 'plum pudding' atomic model through analysis of the famous scattering experiment of Geiger and Marsden. In fact, Thomson's atom, or more properly his research program generally, played a much greater role in fostering Rutherford's triumph of 1911 than is commonly supposed. The α-particle scattering experiment, too, had its harbinger in J J Thomson's researches, leading Thomson, as Heilbron puts it, 'close to, but still some distance from the central principle of atomic theory, the doctrine of atomic number' [17-1]. Indeed, in the ultimate of ironies, one of the main arguments for Thomson's plum pudding was its support of α-particle scattering experiments—experiments which sealed the fate of his model [17-2]!

In recounting the immediate background for Rutherford's atom in the context of the discovery and elucidation of the electron, we must also keep in mind that serious atomic model-making preceeded Rutherford's model by some years. Not only were there explicit precursors for his nuclear atom, but also Thomson's pudding, in turn, was actually appropriated from, and amounted to an elaboration on, Kelvin's static atom of 1902. We have touched on these matters here and there in previous chapters, but some of it bears recapitulation at this point.

The first model maker to concern us, then, is Lord Kelvin. His atom first appears in the jubilee volume to Professor Johannes Bosscha of the Polytechnic School of Delft, Holland, under whom Kamerlingh Onnes served his apprenticeship. Kelvin named it in honor of Æpinus, whose one-fluid doctrine of attraction and repulsion we encountered early on. Kelvin's adoption featured discrete 'quanta' or 'electrions' embedded in a sphere of uniformly distributed positive electricity. Stability was maintained by upwards of a million electrions at rest, positioned at certain geometrical centers of equilibrium, and interacting via forces promulgated by one Father Boscovich, a Croatian Jesuit who is said to have been the first exponent of Newtonian physics in Italy [17-3]. Scarcely a year after its publication, J J Thomson adopted and modified it for his own purposes. Meanwhile, Kelvin himself went on to still more complicated atomic schemes, culminating in his Boscovichian atom of 1905. In it, 'vitreous and resinous' electricity occupied alternating shells, with 'the total vitreous greater than the resinous'. The atom thus lacked in stability, and electrions could be 'shot out' from it with varying speeds, in the manner of β-particles issuing from radioactive atoms, such as radium.

The omnipotent Kelvin had, one gathers, very definite views on radioactivity. Soon after the annual meeting of the British Association at York in 1906, he provoked a famous controversy with a letter to *The Times* in August of that year. Taking to task the notion of evolution of the elements, he argued that radium was simply a molecular compound. In the ensuing, protracted exchange in *The Times* and *Nature*, he had the sole (and embarrassing) support of the equally opinionated H E Armstrong, no friend of physicists generally, who, Armstrong declared, 'are strangely innocent workers under the all-potent influence of formula and fashions' [17-4].

Whereas Kelvin's atom of 1902 was a forerunner of Thomson's, a more direct precursor to Rutherford's subsequent nuclear atom was Jean Perrin's less well known solar atom of 1901—albeit devised purely on speculative grounds, launched in Perrin's campaign of expounding the atomic–molecular hypothesis among hostile French scientific circles, not founded on any direct experiment.

> Each atom [he wrote] will be constituted, on the one hand, by one or several masses very strongly charged with positive electricity, in the manner of positive suns whose charge will be very superior to that of a corpuscle, and, on the other hand, by a multitude of corpuscles, in the manner of small negative planets, the ensemble of their masses gravitating under the action of electrical forces, and the total negative charge exactly equivalent to the total positive charge, in such a way that the atom is electrically neutral. [17-5]

Perrin, too, had an explanation for spontaneous radioactivity, and one more rational than Kelvin's. In heavier atoms, he suggested, some of the corpuscles are well removed from the center of the atom and hence not held as firmly; they are easily severed from the atom [17-6].

A notable atomic candidate of the planetary variety was the Saturnian model of 1904 by the respected Japanese physicist Hantaro Nagaoka, professor of physics at the University of Tokyo. Nagaoka had entered the graduate school at Tokyo in 1887, where he began experimental research in magnetostriction under the British physicist C G Knott, a disciple of P G Tait who resided in Japan between 1883 and 1896 [17-7]. After receiving his doctorate, Nagaoka spent the years 1893–1896 in Europe, dividing his time between the Cavendish and the universities of Berlin, Munich, and Vienna. He was a keen observer of the western style of conducting science, urging as early as 1888 his countrymen to adopt European intellectual habits [17-8]. In 1911, the banner year under Rutherford at Manchester, he wrote Rutherford that 'however poor the laboratory may be, it will flourish if it has earnest investigators and an able director' [17-9]. Indeed, the modern Japanese tradition of experimental and theoretical physics is largely the legacy of Nagaoka and his successors, among them Kotaro Honda, Yosho Nishina, and his protégé Hideki Yukawa.

Nagaoka was inspired to study atomic structure and radioactivity by the Curies in their report to the first international congress of physics in Paris in 1900, and by Marie Curie's demonstration of radium at the congress. (He, too, had been invited to deliver a paper, on magnetostriction, on that occasion.) Starting from Maxwell's famous paper on the stability of motion of Saturn's rings, Nagaoka devised an atom in which electrons revolve with nearly the same velocity in rings around a positively charged body; small oscillations of the electrons about their equilibrium orbits accounted for certain spectral regularities [17-10]. Though Rutherford deemed the model worthy of note in his paper of 1911 in which he proposed his own nuclear atom, the stability of Nagaoka's atom was soon called into question [17-11]. The crux of the instability lay in the fact that the forces between Maxwell's gravitating particles are attractive, while those between Nagaoka's electrons are repulsive, leading to radial instabilities in the Saturnian atom [17-12].

Which brings us to J J Thomson, who proved a more pertinacious modeler. His views on atomic structure were first expounded in the Silliman Lectures which he gave at Yale University during his second visit to America in May 1903. They were promptly published in a small volume entitled *Electricity and Matter* [17-13]. Much the same material is treated in his article of 1904, 'On the structure of the atom' [17-14].

On that visit to America, Thomson traveled alone, his wife having given birth to a daughter some weeks earlier. Among his impressions this time were a nasty dispute in New Haven between the University community and striking tramcar drivers, even leading to shots being fired; an outbreak of typhoid fever in the city; the rivalry between Yale and Harvard, 'Quite as keen as that between Cambridge and Oxford'; and the relative obscurity, even in Yale University circles (but not to Maxwell!), of the great Josiah Willard Gibbs, who died the very year of Thomson's visit, 1903 [17-15].

In adopting a dynamic version of Kelvin's diffuse sphere, purely as a mathematical artifice, Thomson neglected for the time being the awkward problem of accounting for an atom's mass (as well as its neutrality) by negative corpuscles alone. No positive corpuscle was known; nor, for that matter, by Earnshaw's theorem, could a mixed assembly of negative and positive corpuscles form a stable configuration. Nor, moreover, did Thomson have much need for the positive electrification. As he wrote Lodge in 1904,

> I have... always tried to keep the physical conception of the positive electricity in the background because I have always had hopes (not yet realized) of being able to do without positive electrification as a separate entity, and to replace it with some property of the corpuscles. When one considers that all the positive electricity does... is to provide an attractive force to keep the corpuscles together, while all the observable properties of the atom are determined by the corpuscles, one feels, I think, that the positive electrification will ultimately prove superfluous and it will be possible to get the effects we now attribute to it, from some property of the corpuscles. [17-16]

Thomson's model, then, consisted of a uniformly charged sphere of positive electricity (the pudding), with discrete corpuscles (the plums) rotating about the center in circular orbits, whose total charge was equal and opposite to the positive charge. They were subject to their mutual repulsion as well as the central field of the positive 'fluid' and, if displaced from their equilibrium positions by a small amount, were subject to a restoring force proportional to the displacement. At first, the number of corpuscles was judged to be very large, both on the basis of the size of the corpuscle compared to the hydrogen ion, and from the abundance of lines in atomic emission spectra. At least, the number of corpuscles was thought adequate to avoid significant energy loss by electromagnetic radiation (as opposed to the *mechanical* instability that undermined Nagaoka's Saturnian atom [17-17]). *Some* radiation loss was admissible; Thomson made a virtue of the energy loss by viewing radioactivity as the end point of ever so gradual radiation drain in ancient atoms. As he wrote Rutherford in February of 1904,

> ... There are some interesting cases in which the stability of the atom depends on the rotation of the corpuscles. Just as the stability of a spinning top depends on its rotation, the equilibrium

ceasing to be stable when the velocity of the corpuscles around their orbits falls below a certain value, now a moving corpuscle is radiating energy so that its velocity is continually falling though very slowly. A time must come, though it will be long in coming, when the velocity falls below the critical value the previous arrangement becomes unstable. There is an explosion and some of the parts have then kinetic energy very much increased. In fact, they would behave somewhat like radium. I think a spinning top is a good illustration of the radium atom. [17-18]

For the actual distribution of his corpuscles in the diffuse sphere, Thomson appealed to certain floating magnet experiments carried out ca. 1878 by Alfred Marshall Mayer, a purely self-taught physicist at Stevens Institute of Technology in Hoboken, NJ. Despite his lack of an earned degree, Mayer had a distinguished research and academic career, was a member of the National Academy of Sciences, and had founded the physics department at Stevens Institute. These experiments echoed calculations by Thomson himself to a remarkable degree [17-19]. Mayer's experiment is shown in figure 17.1. He had suspended a long bar-shaped electromagnet vertically over a dish of water [17-20]. In the water floated a number of magnetized needles, each stuck in a cork, its upper end flush with the top of the cork, and all needles with the same pole pointing up. Both the suspended magnet and the floating magnets were long enough that only the poles nearer to the water surface were effective. The active pole of the suspended magnet was assumed to be positive; the poles of the floating needles pointing up were all negative. (Reversal of sign convention would not have affected the experiment.) Under the influence of the large suspended magnet, the floating magnets, acted on by a horizontal component of force proportional to the distance, distributed themselves in concentric planar configurations, which could be recorded by pressing a piece of paper against the inked tops of the corks.

Mayer found that floating magnets numbering between three and five arranged themselves at the corners of a single polygon: five at the corners of a pentagon; four at the corners of a square; three at the corners of a triangle. (Two, naturally, assumed equilibrium positions on opposite sides of the center; a single magnet resided in the center.) When the number exceeded five, the magnets formed concentric polygons or rings, some with a magnet in the center. Thus, 15 magnets had one in the center, surrounded by a polygon of five and a circle of nine. The essential feature of the experiment, for Thomson's purpose, was that successive rings of magnets formed models imitating many of the known properties of atoms—particularly the periodic changes with increasing atomic weight according to the table of Mendeléev. For example, there were certain sudden changes in the configuration of magnets with the addition of a single magnet; five magnets formed a single group, but

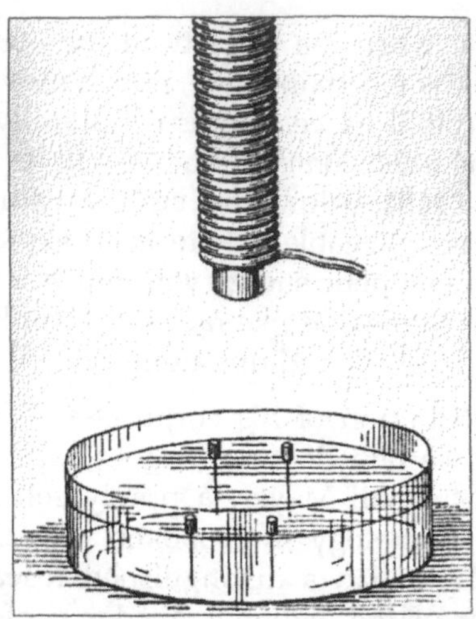

Figure 17.1. *Mayer's floating magnet experiment of 1878.*

throwing a sixth magnet into the water caused the group to break up into two. Analogously, among the chemical elements, in order of increasing weight, there occur certain sudden differences, as between fluorine and sodium; beginning with sodium there follows a mostly gradual variation in properties up to chlorine, where a second discontinuity sets in, and so forth, on up the periodic table [17-21].

For its crudity, it was a remarkably effective demonstration. Rayleigh notes that Thomson used to perform these experiments in his elementary lectures, even before the electron was nailed down with certainty. 'It was rather too strong meat for some of his students and I remember a fellow-student remarking to me that he thought it altogether too fanciful' [17-22].

A few years later, Thomson's former student Harold A Wilson, newly settled at the Rice Institute (now Rice University) in Houston, TX, brought an interesting version of Mayer's experiment to his attention. Devised by one of Wilson's own students, it furnished 'exceedingly interesting confirmation of [Thomson's] views on the constitution of atoms' [17-23]. Instead of the large suspended bar magnet, it employed a circular current-carrying coil, in conjunction with the usual dish of floating magnetized needles. The plane of the coil was horizontal, and its height was equal to the length of one of the small needles. It was mounted concentric with the dish, such that the tops and bottoms of the needles were coplanar with the top and bottom of the coil. The strength of the current was adjusted so that the average distance between the floating magnets was equal to their length, and Mayer's experiment was repeated. Needles were thrown in, one by one, and formed pretty much the same patterns as in Mayer's original experiment.

> He [Wilson's student, E R Lyon] finds that if he takes a number of magnets equal to an atomic weight then they arrange themselves in rings and the number in the innermost or the two innermost rings is equal to the maximum valency of the element. [17-24]

We have noted earlier the reasons for Thomson's subsequent drastic reduction in n, the number of electrons per atom. By computing the contributions of his atomic model to the scattering of x-rays, β-rays, and light (dispersion), he concluded in 1906 that n, far from numbering in the ten thousands, ranged between 0.2 and two times the atomic weight; in short, was of the same order as the atomic weight. In particular, for hydrogen this number 'cannot differ much from unity' [17-25]. This re-evaluation, albeit on rather insecure grounds, had two important consequences. First, it implied that the mass associated with the positive electricity—the diffuse sphere—was *not* small compared to nm, the aggregate electronic mass in the atom. It hinted of a massive, positive counterpart to the electron, however configured, and set in motion Thomson's positive ray program of 1907–1912. Second, it reopened the atomic stability problem associated with radiation drain, hitherto cured by the 'social pressure' of many electrons [17-26]. Thomson's response to this latter problem was characteristic; he essentially ignored it.

What led to the atomic *number*, not weight, as the quantity of importance in atomic structure, however, was not the re-evaluation of n *per se*, but Thomson's return to the problem of β-ray scattering in 1910. Thomson had taken the lead in using β-ray scattering, or 'absorption' as the process was then more commonly called, as an experimental tool for extracting atomic information. In this he was anticipated, we recall, by Lenard, who had used photoelectrons for the same purpose, and, in so doing, came up with his own 'dynemide' (dipole) model of the atom [17-27]. The simplest interpretation of the relative 'transparency' of matter to the rays, in Lenard's opinion, was that atoms are largely vacuous. Thomson, however, ventured well beyond the mere transparency of atoms [17-28]. Treating the scattering in terms of multiple close encounters between β-particles and bound atomic electrons, each deflection barely significant, and on the erroneous assumption that β-rays obey an exponential law of absorption [17-29], he derived an expression for the coefficient of absorption of the rays. (Comparison of this formula with β-ray absorption data by Becquerel and Rutherford had provided Thomson with one of the pieces of data underlying his re-evaluation of n.) More to the point, Thomson's theory sparked an experimental program on x-ray, β-ray, and γ-ray scattering at the Cavendish. Only one component of this effort concerns us here [17-30].

In his refined treatment of multiple scattering, Thomson considered the net deflection as the composition of many independent deflections, compounded in a statistical manner as the square root of the mean value (root mean square, rms) of the deflection resulting from a 'random walk' through the target material [17-31]. The final deflection was a statistical average of the scattering by individual atomic electrons, by the positive 'fluid', and by the number of atoms per unit volume in the target. For our purpose, the thing of importance is that the rms deflection was proportional to n, the electrons per atom. Soon after Thomson had completed his new theory of multiple scattering, it was put to an experimental test at the Cavendish by James A Crowther, then a Fellow of nearby St John's College [17-32]. A pencil of nearly homogeneous β-rays, produced by sorting β-rays from radium bromide in a magnetic field, entered a pair of exhausted ionization chambers; the second chamber measured the velocity of the rays transmitted through absorbing screens of varying thickness interposed in the first. The quantity measured was the ratio $mv^2/\sqrt{t_m}$, where $mv^2/2$ was the kinetic energy of the incident particles and t_m the thickness of material required to diminish the radiation intensity to one-half its original value; according to the scattering theory, this quantity should remain a constant for a given target material. Indeed, despite the primitive experimental conditions, and various doubtful features of the theory, Crowther's results, submitted to the Royal Society in June 1910, appeared to bear it out, and, with it, Thomson's atomic model [17-33]. In particular, the experimental value for the rms deflection yielded the important result $n \sim 3A$ where A is the atomic weight. Crowther thereupon repeated earlier x-ray scattering experiments by Charles Barkla, which had indicated that n was more nearly proportional to $A/2$, except for hydrogen. Crowther's confirmation of this result was conveniently overlooked.

With this nagging possible exception, it appeared by mid-1910 that Thomson's theory had been confirmed. If so, the multiple-scattering theory should hold for α-particle scattering as well. The α-particle was by then known to be a doubly ionized helium atom, and should therefore harbor approximately ten electrons. In short, it could be regarded as a stucture of atomic dimensions, as had, for example, been maintained for some time by R J Strutt. That was also the early conclusion by Ernest Rutherford on the basis of his singular success during 1902 in 'deviating' α-rays in magnetic fields, and from early scattering experiments of his own, discussed below. Only gradually does it seem to have dawned on Rutherford, or perhaps it was his habit of 'wait and see' [17-34], that, at least from the point of view of scattering, α-particles behaved more like point charges. This profound realization was first impressed on Rutherford in 1908, when experiments with his student Thomas Royds showed that α-rays too, like β-rays, had a far greater penetrating ability than previously supposed. Something was amiss with the many-electron view of the α-particle, and, with it, the helium atom. Might other atoms need revision as well?

17.2. Geiger's clue

The age of the nuclear atom can, with some stretch of the imagination, be considered to have been born on a Sunday afternoon during February 1911, in a private residence in Withington, a suburb of Manchester, England. The large and comfortable villa, on 17 Winslow Road, was the home of Ernest Rutherford, about two miles from his laboratory at the Victoria University of Manchester. Among those present for an afternoon supper at the Rutherfords was Charles G Darwin. A Trinity man, Darwin was a postgraduate student of Rutherford, son of the mathematician and astronomer George H Darwin, and grandson of Charles R Darwin himself [17-35]. Before supper, Rutherford held forth on his novel ideas on the constitution of the atom. In looking back on the occasion nearly three decades later, Darwin reflected that 'I count it as one of the great occurrences of my life that I was actually present half an hour after the nucleus was born' [17-36].

The experimental program leading to Rutherford's memorable afternoon gathering had its infancy late in his professorship at McGill University in Montreal, Canada. Irked by recent doubts cast on the prevailing view, shared by Rutherford, that α-particles are not scattered in crossing matter, but plunge straight through without any measurable diversion (and in so doing, exhibit a characteristic range), he had himself taken up the problem of the retardation of α-particles early in 1906. A preliminary paper on the subject, issued in January, was followed by a fuller experimental report in July [17-37]. In that investigation, a ray of homogeneous (monoenergetic) α-particles from RaC passed through a slit partially covered with mica. Those particles passing through the mica formed a diffuse band in a photographic plate, indicating that they had been deflected from their course by as much as 2°. 'Such a result', concluded Rutherford, 'brings out clearly the fact that the atoms of matter must be the seat of very intense electrical forces—a deduction in harmony with the electronic theory of matter' [17-38]. The significance of this conclusion is not so much the diffuse scattering *per se*, but the second half of Rutherford's observation. It suggests, notes Heilbron, that at that time, in regard to the α-particle scattering, Rutherford did not see the need for any modifications in the generally accepted 'plum pudding' picture of the atom [17-39].

Soon afterwards, Rutherford also began an experiment designed to determine the number of α-particles released in unit time from a gram of radium. This experiment, like the earlier one, would only be completed some years later in another setting, with the able assistance of Hans Geiger. In the meanwhile, Rutherford considered his position at McGill sufficiently secure to purchase a piece of land on the hilly outskirts of Montreal. Alas, a home in the West Mountains was not to be. September brought a letter from Arthur Schuster at the University of Manchester, England. Just as a communication from Röntgen had spelled an abrupt change in Schuster's experimental researches, so did Schuster's letter profoundly alter Rutherford's routine, and possibly presage a transition in the intellectual affairs of the western world. Unfortunately, Schuster's letter is no longer extant.

Arthur Schuster, who held the Langworthy Chair of Professor of Physics at Manchester, had been feeling the strain after 25 years in academia, and had eyed Rutherford as a worthy successor for some time. In July 1906, he had tentatively broached the subject to Rutherford. Explaining in confidence his entertaining the idea of retiring, and feeling Rutherford out, he added that 'I really do not know of anyone else to whom I would care to hand over the office' [17-40]. Though Schuster himself was no longer active in experimental work, the physics department boasted a cadre of research students of the first rank, and occupied a well stocked new laboratory building erected under Schuster and opened by Rayleigh in 1900 [17-41]. Much of the equipment was quite on par with that at McGill. What the department lacked for work in radioactivity—McGill's strong card—it made up for with spectroscopic equipment second to none. Moreover, improvization in technique was one of Rutherford's great strengths. Along with the proffered chair went several additional inducements, among them Schuster's head laboratory steward and lecture assistant, Mr William Kay, and, of paramount significance, Schuster's research assistant, Hans Geiger. To sweeten the deal, Schuster had persuaded the University Council to institute a Readership in Mathematical Physics, funded for some years out of his own pocket. It was first filled by Harry Bateman, then by C G Darwin, and after him Niels Bohr.

Schuster's formal invitation, as noted, no longer exists. Rutherford, for his part, needed little convincing.

> I very much appreciate your very kind and cordial letter and am inclined to consider very favorably the suggestion of becoming a candidate for the position you propose. The fine laboratory you have built up is a great attraction to me as well as the opportunity of more scientific intercourse than occurs here.
>
> I need hardly tell you how much I appreciate the suggestion coming in the way it has: I fully recognize the spirit of self abnegation displayed by you in your letter. Nothing could give me greater pleasure than to have you a member of the department to add your strength in the branch—Mathematical Physics—in which I should most value assistance... [17-42]

From the letter it is also clear that Rutherford had considered applying for Callendar's vacated chair at King's College in London, thus nearly replacing Callendar a second time. (Rutherford had succeeded Callendar in the chair at McGill.) In the event, he was dissuaded because of the poor laboratory facilities at King's College, among other things.

J J Thomson concurred in the Manchester offer, but for one small reservation.

> ...I was very glad to hear from Schuster that there was a chance of your coming to England. I hope it is true. I think you would like the Manchester people and the climate is not nearly so bad as its reputation. I ought to know as I was born there and lived for the first twenty years of my life. The laboratory is a very good one. The only problem I see is the possibility of a dual management of the laboratory by Schuster and yourself proving a little inconvenient. I think it would be conducive to a smooth working in the future to have the respective parts to be played by Schuster and yourself defined as clearly as possible at the outset. It will be very delightful to have you back again. [17-43]

Rutherford stepped down with few trepidations, explaining to Principal Peterson at McGill that 'the determining factor in deciding to go to Manchester was my feeling that it is necessary to be in closer contact with European science than is possible on this side of the Atlantic' [17-44]. The family sailed for Manchester on the *Empress of Ireland* in May 1907, and Rutherford took up his new post that October.

Manchester was then England's second largest city—a prototypical city of the industrial revolution, at the zenith of its golden age economically and technologically. It was, as well, not without its intellectual and cultural virtues. Its strength lay in music, in Manchester Royal College of Music and the world famous Hallé Orchestra and Choir, which dated from 1858. The city also boasted several world-class libraries, among them the new John Ryland Library of old and rare books, and the *Manchester Guardian*, the leading liberal and provincial newspaper in its day. The Manchester Literary and Philosophical Society is England's oldest continuing scientific society, apart from the Royal Society of London. It began in 1781 as an informal meeting of friends, first in each other's homes, then in a tavern, and eventually a house was built. John Dalton, the great chemist who discovered the law of multiple proportions in chemical combinations, joined the Society in 1794, became its Secretary in 1800, and served as its President from 1817 until his death in 1844. The Society was home to his laboratory and retained his records until they were destroyed in World War II. As President, he was not a fluent speaker; he is said, when introducing a paper being read, to typically comment that 'this paper will no doubt be found interesting by those who take an interest in it' [17-45]. Dalton was succeeded as President by James Prescott Joule, who had been his pupil, was then only 26, and had published his famous paper on the mechanical equivalent of heat the year before. Twin statues by Alfred Gilbert of Joule and Dalton stand side by side in the Town Hall of Manchester. We will encounter the Literary and Philosophical Society again before long; both the Society and the *Guardian* would figure in the upcoming events ushering in the atomic nucleus.

As for the University of Manchester, we have dwelt elsewhere on its embryotic origins as Owens College with its extraordinarily brilliant faculty, where J J Thomson (and Schuster) got started. The Victoria University of Manchester became the first and largest of the great English civic universities. Among Rutherford's faculty acquaintances during his tenure at Manchester were Chaim Weizmann, then a lecturer and reader in biochemistry, Thomas F Tout, the historian, Samuel Alexander, the philosopher, and Nathan Laski, father of the political scientist Harold J Laski.

We covered Rutherford's first experimental success at Manchester in a previous section. His and Geiger's two back-to-back papers on the counting of α-particles were characterized by Rutherford himself as 'of great solidity' and 'of the first magnitude' [17-46]. They were read before the Royal Society on June 18 1908—one year after the Rutherfords had arrived in Manchester looking for a house and getting acquainted. In addition to pinning down the charge of the α-particle, the experiment showed that 'the α-particle, after it has lost its positive charge, is a helium atom' [17-47]. That α-particles are indeed ionized atoms of helium was conclusively demonstrated some months later in an elegant experiment by Rutherford and the spectroscopist Thomas D Royds, capitalizing on the excellent spectroscopic facilities at Manchester. Royds was a former recipient, like Rutherford, of an 1851 Exhibition Scholarship to Cambridge.

The essential feature of that famous experiment—surely the centerpiece of Rutherford's Nobel Lecture late the same fall—was a tiny glass tube blown so thin (about $\frac{1}{100}$ mm) that it allowed α-particles emitted from radium emanation encapsulated within it to penetrate with negligible change in velocity, yet was impervious to helium gas [17-48]. The tube, still preserved in Cambridge, was the handiwork of Otto Baumbach, imported from Germany by Schuster; his role in Rutherford's laboratory is reminiscent of Kamerlingh Onnes' *Glasbläsermeister* O Kesselring in Leiden and his craftsmanship—witness figure 14.2 in an earlier chapter. (The experiment with Royds is sometimes referred to as the Rutherford–Royds–Baumbach experiment.) Unfortunately, Baumbach was German to the core, and would prove the source of some annoyance to the rest of the Manchester staff at the outbreak of World War I. (His interment by the British authorities was to be a major factor in bringing to a halt what meager experimentation was still going on at Manchester during the war.) In any case, his tube was surrounded by a second, highly evacuated glass vessel with thicker walls, topped by a spark discharge extension and certified free of helium gas at the start of the experiment. After several days, spectroscopic tests revealed the yellow helium lines, and then gradually the complete helium

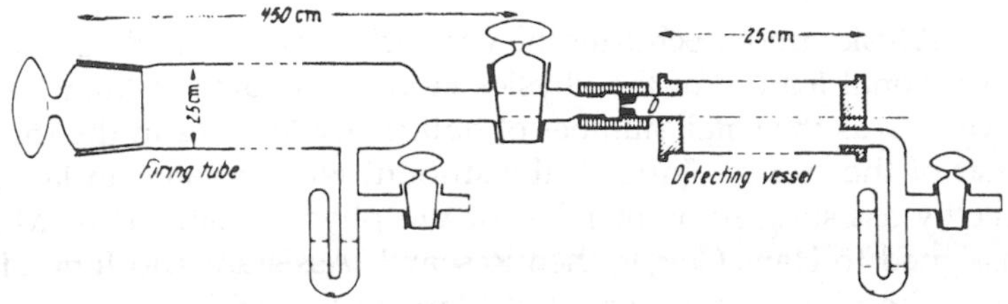

Figure 17.2. *Rutherford and Geiger's apparatus for counting α-particles electrically. α-particles from a radioactive source fixed at one end of an evacuated tube entered the counting chamber through a thin mica window fixed in a glass tube at the other end (at D). The detecting vessel was an exhausted brass cylinder, 25 cm long, with a central insulated wire passing through ebonite corks at either end. The wire was connected to a quadrant electrometer, and the outside tube to the negative terminal of a battery, the other pole of which was grounded. In practice, the intensity of the active matter in the 'firing tube' and its distance from the opening was arranged so that only a few α-particles entered the detecting vessel per minute; ionization by collision caused a magnification of the single particle's charge sufficient to give the electrometer a measurable 'kick.'*

spectrum, showing that helium accumulated in the outer vessel with the passage of time. Elaborate controls ensured that helium could not have diffused through the glass walls, but that it could only have come from the α-particles entering the exhausted space between the emanation tube and the outer vessel and, in so doing, capturing electrons.

For our purpose, however, the importance of both the counting and spectroscopic experiments was not so much the identification of the α-particles with ionized helium, but the ease with which they went through the glass of the latter experiment, and were knocked off their course through thin metallic foils in the former. In the first place, it seemed highly unlikely that particles of atomic dimensions could penetrate glass. Secondly, accurate α-counting proved impossible, traced in due course to scattering in the mica or aluminum window separating the 'firing tube' and the 'detecting vessel', alias prototype Geiger counter (figure 17.2). As Rutherford put it, 'the scattering is the devil' [17-49]. It was downright surprising, in the opinion of William H. Bragg, who followed the Manchester experiments with keen interest and had himself recently been engaged in highly pertinent α-particle experimentation at the University of Adelaide. 'The scattering of the α particles is certainly remarkable, more than I ever thought', he declared in a letter to Rutherford [17-50].

In light of the problematic scattering, Geiger began an experimental investigation of his own on α-particle scattering *per se*. A preliminary report was read before the Royal Society on June 18, at the same meeting at which the two aforesaid papers co-authored with Rutherford were also reported [17-51]. Serious work on the scattering problem would require an assistant, however. Rutherford agreed that a good choice might be Ernest Marsden, a young undergraduate who had begun a course of

Honours Physics under Schuster the year Schuster resigned his chair. An unconventional feature of the Physics Honours course at Manchester in those years was the small number of obligatory lectures in the third and last year of the course. Instead, the student was initiated in laboratory practice by assisting some member of the physics staff. Thus Marsden was assigned to Hans Geiger, then Research Assistant and John Harling Fellow in the physics department at Manchester [17-52].

The year 1908 had its distractions, to be sure. They began on an upbeat note in March with Rutherford's receipt of the prestigious Bessa Prize from the Turin Academy of Science. September brought two honorary degrees, from Trinity College, Dublin, and from the University of Giessen in Germany. By then Rutherford had 'plenty of these insignia [actually four] but they are gratifying to the extent that they are meant to be an appreciation of one's scientific work' [17-53]. Public lectures demanded his time as well, at the Royal Dublin Society (in February), Liverpool University Chemical Society and Friday Evening Discourse at the Royal Institution (both in March), and the Cavendish Laboratory (June). The culminating distinction that year was the Nobel Prize 'for his researches concerning the disintegration of elements and the chemistry of radioactive substances', two years after J J Thomson's prize and, no doubt to Rutherford's amusement, for chemistry, not physics.

> I much appreciate your kind congratulations [he wrote Hahn] and wishes on the award. It is of course quite unofficial but between ourselves I have no reason to doubt its correctness. I must confess it was very unexpected and I am very startled at my metamorphosis into a chemist. [17-54]

He shared the prize on December 11 1908 with Gabriel Lippmann (physics), Paul Ehrlich, and Élie Metchnikoff (both for physiology and medicine). In his banquet speech at Stockholm's Royal Palace that evening, he took a lighthearted shot at the crux of his upcoming Nobel Lecture the following day, noting that he had 'dealt with many different transformations with various periods of time, but that the quickest he had met was his own transformation in one moment from a physicist to a chemist' [17-55].

To get him going, Geiger set Marsden to counting α-particles fired down glass tubes up to 4 m long. In his preliminary experiment the previous year, in Room B5 of the ground floor of the Schuster Physics Building, Geiger had employed the tube shown in figure 17.3. It was 1.7 m long, interrupted about mid-way by a slit and closed at its far end with a zinc sulfide screen. With the tube exhausted, the scintillations in the screen, observed in a microscope, remained within the geometrical image of the slit; however, when air was admitted, or the slit was covered by foil, the α-particles exhibited scattering through 'quite an appreciable angle'—some 1–2° [17-56]. Note that these

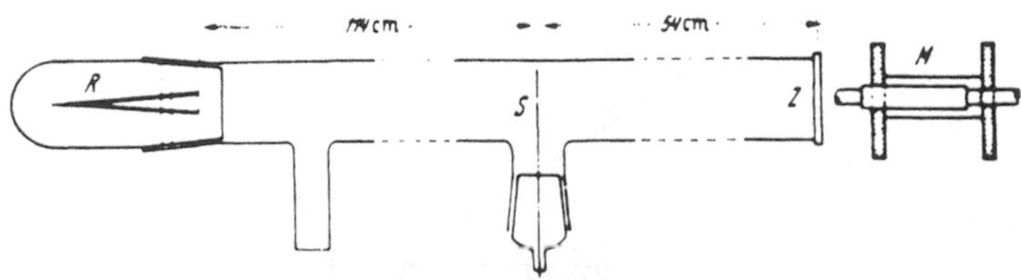

Figure 17.3. *Geiger's apparatus for measuring small-angle scattering of α-particles by the scintillation method. R is the source, S the scatterer, and Z a zinc sulfide screen viewed by the microscope M.*

measurements, counting feeble scintillation flashes, were performed in the dark, the observer squinting for agonizing stretches of time through a low-powered microscope eyepiece, first spending half an hour getting 'dark-adjusted'. (Rutherford himself had no patience for it.) Marsden's new arrangement was rigged up in the dank and gloomy basement of the laboratory, amidst a hot steam pipe at head level and water pipes under foot. The troublesome scattering, or what appeared to be particles diffusively reflected from the walls of the tube (by molecular-size protuberances, it was thought), was initially obviated by interposing a series of diaphragms in the tube, thereby yielding reproducible particle counts [17-57].

A different experimental tactic was soon suggested by Rutherford. 'One day [recalls Marsden] Rutherford came into the room where we were counting the α-particles at the end of the 4 1/2 metre firing tube'. He turned to Marsden and said, 'see if you can get some effect of alpha-particles directly reflected from a metal surface' [17-58]. Such scattering, or reflection, Rutherford fully realized, was interpreted in the case of β-particles as the accumulated effect of many small scatterings, but not expected for the fast and massive α-particles. Perhaps, suggests Badash, it was simply an example of Rutherford's willingness to try 'any damn fool experiment' on the chance it might work, and Marsden agrees [17-59].

The arrangement tried by Marsden is depicted in figure 17.4. A lead plate was mounted in such a position that no α-particles from the source AB could strike directly the zinc sulfide screen S. The source was similar to what Geiger had developed for his own firing tube, namely a glass tube drawn conically, and filled with radium emanation and sealed tight at end B with a mica window transparent to α-particles. Behold, with a reflecting foil of gold or platinum placed at R, the screen, viewed from below with the microscope, lit up at once. (Mostly from β-particles, to be sure, but also from a fair number of α-particles, it was soon established.) Various comparative measurements were made on reflecting materials ranging from aluminum to gold and lead, with corrections applied for particles reflected from the air and from glass backings for the fragile

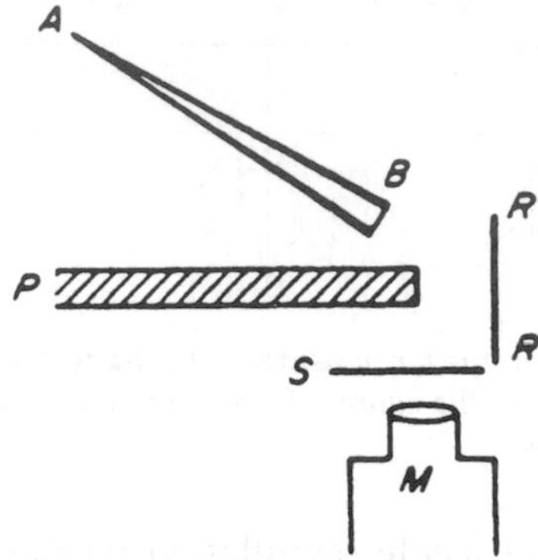

Figure 17.4. *Geiger and Marsden's arrangement for studying the reflective powers of different materials RR struck by α-particles fired from the source AB. A lead plate P prevented α-particles from striking directly the zinc sulfide screen S, viewed from below with the microscope M.*

foils, and allowing for the thickness and atomic weight of the reflectors. (The number of reflected β-particles was known to decrease with the atomic weight.) On average, one in every 8000 of the incident α-particles was reflected. Adds Marsden in his retrospective account:

> I remember well reporting the results to Rutherford a week after, when I met him on the steps leading to his private room... A few weeks later Rutherford instructed that I should round off the experiment with Geiger in a form suitable for publication. [17-60]

The paper of Geiger and Marsden was submitted to the Royal Society on May 19 1909, and read before the Society on June 17, one day short of a year since the two back-to-back papers by Rutherford and Geiger, and Geiger's own scattering paper, were read before the same august body [17-61].

Rutherford's incredulous reaction to the unexpected positive result from a long-shot experiment counts among a handful of rare pronouncements enlivening the infancy of early 20th century physics. Yet, curiously, his words cannot be authenticated. 'It was quite the most incredible event that ever happened to me in my life', he is purported to have exclaimed. 'It was almost as incredible as if you had fired a 15-inch shell at a piece of tissue paper and it came back and hit you' [17-62]. In Heilbron's opinion, his supposed reaction is overstated, and may be retrospective fiction, particularly in light of a lecture Rutherford delivered at Clark University in September 1909. From it, it seems clear that, at that time, Rutherford still adhered to the concept of multiple scattering. It goes without saying that he also subscribed to Thomson's plum pudding atom which still held sway and was made to order for multiple scattering by β-particles [17-63]. In his lecture at Worcester, he speaks of one in

8000 α-particles emerging on the side of incidence 'by its *encounters* with the molecules' (italics added) [17-64]. Furthermore, only with the advent of the nuclear atom could deflections through 90° or more in a *single* collision be contemplated in the first place. That multiple scattering also held for α-particles in this time frame appears from close study of a paper Geiger read to the Royal Society on February 17 1910, on the average scattering of α-particles by thin foils [17-65]. Nevertheless, that paper served as well to reinforce the anomalous scattering behavior brought to light in the Geiger–Marsden results. The probability of multiple scattering of α-particles by more than 90° was negligibly small, as cursory examination of the underlying physics bears out [17-66].

In the end, Geiger let the matter drop, observing merely that 'it does not appear profitable at present to discuss the assumption that might be made to account for [the large scattering angles]' [17-67]. It is owing to Rutherford's dogged genius that he himself was not similarly dissuaded.

17.3. The atomic nucleus

The precise circumstances which led Rutherford to abandon Thomson's multiple-scattering theory in favor of one based on single, large-angle collisions for α-particles have been subject to much scrutiny, yet remain somewhat murky to this day. Rutherford's correspondence prior to December 1910 offers no clue. We do know from his letter to a friend, the radiochemist Bertram B Boltwood at Yale, that, as of December 14, he had been doing 'a good deal of calculation on scattering'.

> I think I can devise an atom much superior to J. J.'s, for the explanation of and stoppage of α- and β-particles [continued Rutherford], and at the same time I think it will fit extraordinary well with the experimental numbers. It will account for the reflected α particles observed by Geiger, and generally, I think, will make a fine working hypothesis. [17-68]

Alas, the detailed genesis of Rutherford's inspiration, in as far as its exact timing is concerned, is obscure in the recollections of the most important onlooker—nay, participant, Hans Geiger. Referring vaguely to some time in the winter of 1910–1911, Geiger tells us as follows:

> One day Rutherford, obviously in the best of spirits, came into my room and told me that he now knew what the atom looked like and how to explain the large deflections of the α-particles. On the very same day I began an experiment to test the relation expected by Rutherford between the number of scattered particles and the angle of scattering.... [17-69]

As to the actual calculations, the bulk of them have by good luck been preserved at Cambridge University, in the form of rough notes and calculations in Rutherford's hand [17-70]. Though undated, it is generally

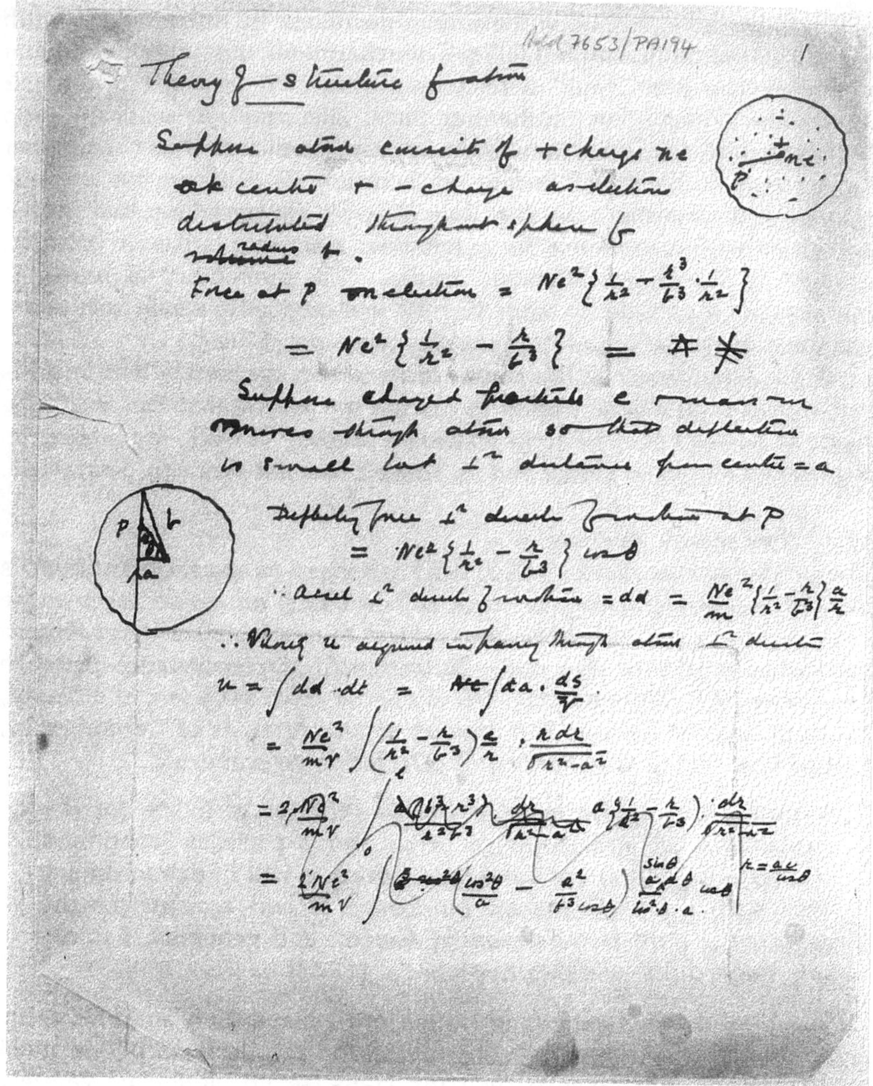
Add 7653/PA194

Theory of structure of atom

Suppose atom consists of + charge ne at centre & − charge as electrons distributed throughout sphere of radius b.

Force at P on electron = $Ne^2\{\frac{1}{r^2} - \frac{r^3}{b^3}\cdot\frac{1}{r^2}\}$

$= Ne^2\{\frac{1}{r^2} - \frac{r}{b^3}\}$

Suppose charged particle e mass m moves through atom so that deflection is small but ⊥ distance from centre = a

Figure 17.5. *A sheet from Rutherford's first rough notes and calculations on his nuclear structure of the atom. Thomson's theory is still very evident.*

agreed that these manuscript sheets, written in different batches with different pens and in pencil, date from the winter of 1910–1911. Two of the sheets, most likely among the earliest, are reproduced here as figures 17.5 and 17.6.

The best clue to the timing of Rutherford's atom seems to be a letter from Bragg, in reply to one from Rutherford apparently no longer extant. Rutherford must have written almost immediately after he settled on

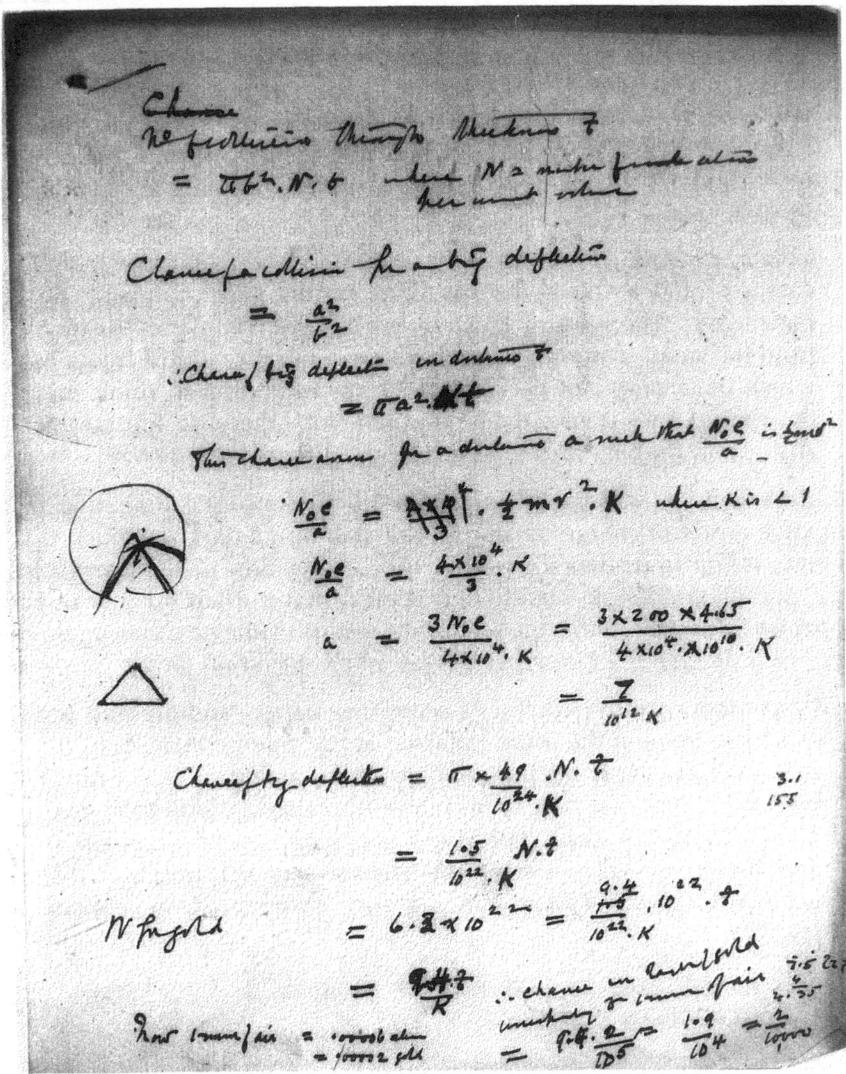

Figure 17.6. *Another sheet from Rutherford's notes and calculations, in which he estimated the probability of large-angle deflection of α-particles by gold.*

his own atomic construct. Bragg had assumed his professorship at the University of Leeds in 1909—a relocation that initially proved trying, one reason being the industrial grime of Leeds in the wake of sunny Adelaide. The upshot was that he relied strongly on Rutherford's company, when the occasion cropped up. In his return letter, dated December 21 1910, in which he invites the Rutherford family to join them at Leeds for the

New-Year holidays, he adds 'I would have liked to have seen you before this to hear about the new atom... The atom sounds fine. My boy [W L Bragg] wants to know if he may hear about it too, as he has been going to J. J.'s lectures...' [17-71].

We hear again from Rutherford in his letter to Bragg on February 8 of the new year. In it, he confirms that Geiger, despite the latter's professed pessimism the year before, is indeed tackling the scattering problem anew.

> Geiger is working out the question of large scattering [of α-particles] and as far as he has gone results look promising for the theory. The laws of large scattering are completely distinct from the small scattering, and there are a number of directions in which the theory can be verified. I am beginning to think that the central core is negatively charged, for otherwise the law of absorption for beta rays would be very different.... [17-72]

Readers will not fail to be intrigued by Rutherford's reference to a negative center of charge—a short-lived stab in the dark we touch upon below. Aside from that interesting morsel, relating to the energy loss of β-rays on sudden deceleration, it is clear that Rutherford was mainly occupied with the Cambridge scattering results. Thus he wrote again the very next day, taking Crowther to task in no uncertain terms.

> I have looked into Crowther's scattering paper carefully and the more I examine it the more I marvel at the way he made it fit (or thought he made it fit) J. J.'s theory. As a matter of fact, I find I can explain the first part of his curve of scattering with thickness, in *large* scattering alone. I believe it is only the use of imagination, and failure to grasp where the theory was inapplicable, that led him to give numbers showing such an apparent agreement. [17-73]

Boltwood, too, was kept abreast of progress in Manchester. As Rutherford wrote on February 1,

> Geiger is examining the distribution of the α particles which suffer a large deflection. As far as the experiments have gone, the results agree well with the distribution deduced from my special atom. I think that this type of work will ultimately throw a great deal of light on the intensity and distribution of the electric field within the atom. [17-74]

For the details of the electric field distribution, we can turn to C G Darwin's recollections of Rutherford's remarks during the memorable Sunday afternoon supper gathering at the Rutherfords that same month, February 1911. Explaining that he had been preoccupied with the large-angle scattering of α-particles and its implied revelation that there must be enormous forces at work in the atom, Rutherford had ventured that 'it would require a charge of something like a hundred electrons in the case of the gold atom'. However, adds Darwin, 'that may have been added a few days later' [17-75]. In contrast to Thomson's diffuse sphere, Rutherford's charge was concentrated at the center, repelling the α-particles according to the inverse square law demanded by classical theory, and causing them to describe a hyperbolic path.

> Rutherford was not a profound mathematician [observed Darwin], and I would doubt if he had thought about the properties of the hyperbola since he had been at school, but he had remembered just exactly enough of them to serve his purpose, and he told us the trigonometrical formula that should give the number of alpha-particles deflected through any angle. Indeed, as far as I can remember, on that very evening he asked me to check over his work, which was of course perfectly correct. [17-76]

Fortunately, Darwin's supporting calculations are also extant, in the form of handwritten manuscript sheets, in the Rutherford archives, Cambridge University Library [17-77].

Rutherford gave a public preview of his new scattering theory and atomic model in a talk to the Manchester Literary and Philosophical Society on Tuesday, March 7 1911, well over a century after Dalton had presented his chemical atom before the 'Lit and Phil' in 1803 [17-78]. Rutherford's talk was reported two days later in the *Manchester Guardian* under the matter-of-fact heading 'The structure of the atom'. The *Guardian* thus had the distinction of breaking the scientific news in print, though the short paragraph afforded it hardly rated as featured coverage. Curiously, the previous day's *Guardian* contained an even briefer report of another talk, in a different vein, namely one by J J Thomson on the potential of solar power [17-79].

The justifiably famous paper that followed, 'The scattering of alpha and beta particles by matter and the structure of the atom', was published in the *Philosophical Magazine* for May 1911. It, and Rutherford's succinct arguments in support of his new scattering theory, are too well known for us to dwell on here, except cursorily. It is, however, worthy of note that the original manuscript, with numerous corrections and editing, all in Rutherford's hand, is among the draft manuscripts of Rutherford's published papers extant at Cambridge University Library [17-80]. It is perhaps of equal interest to note that Rutherford also summarized his case that March in a reply letter to the Japanese physicist Hantaro Nagaoka, whom we met earlier with his Saturnian atom, and who he must have felt rated an in-depth personal preview of his paper. Nagaoka, the tireless traveller, had most recently completed a Grand Tour of the European laboratories during the last quarter of 1910. His itinerary included Rutherford's laboratory that September, or perhaps just when Rutherford was becoming absorbed in the scattering problem. On February 22 1911, Nagaoka penned a long thank you letter to his Manchester host, in which he described in insightful detail his impressions of Kamerlingh Onnes' Leiden laboratory, Boltzmann and Exner's old laboratory in Vienna (Stefan Meyer's radium institute was not yet completed), Righi's in Bologna, Weiss' in Zürich, Lenard's radiological institute in Heidelberg, and the institutes in Frankfurt, Leipzig, and Breslau [17-81]. As was his unfailing habit, Rutherford duly replied in a long letter dated March 20. We recall that he had learned of Nagaoka's atom from Bragg on the very day he presented his own planetary model at the Manchester Society [17-82]. Explaining the α-particle dilemma for the Thomson–Crowther scattering mechanism, he continued

> I have devised an atom which consists of a central charge '*ne*' surrounded by a uniform spherical distribution of opposite electricity, which may be supposed, if necessary, to extend over a region comparable with the radius of the atom as ordinarily understood. [17-83]

Describing his model in some detail, he went on:

> You will note that the structure assumed in my atom is somewhat similar to that suggested by you in a paper some years ago. I have not yet looked up your paper; but I remember that you did write on that subject. I gave a preliminary account of my results to the Manchester Literary and Philosophical Society recently, and hope soon to be published in the Philosophical Magazine. [17-84]

In the *Philosophical Magazine* paper that resulted, Rutherford gets to the point without further ado: the importance of charged particle scattering as a tool for probing the structure of an atom. J J Thomson's theory of compound scattering assumed that the average angle of deviation was the cumulative effect of a large number of atomic collisions, any one of which was inappreciable. In such a random scattering situation, the total deviation is the average deviation in a single collision not multiplied by the total number of collisions, but multiplied by the square root of this number. This, in fact, was not what Geiger and Marsden found. In their experiments on α-particles traversing a thin gold foil, they observed that about one in 20 000 particles suffered a deflection of 90° or more.

> A simple calculation based on the theory of probability shows that the chance of an α particle being deflected through 90° is vanishingly small. In addition, it will be seen later that the distribution of the α particles for various angles of large deflexion does not follow the probability law to be expected if such large deflexions are made up of a large number of small deviations. It

> seems reasonable to suppose that the deflexion through a large angle is due to a single atomic encounter, for the chance of a second encounter of a kind to produce a large deflexion must in most cases be exceedingly small. A simple calculation shows that the atom must be a seat of an intense electric field in order to produce such a large deflexion at a single encounter. [17-85]

Consider then, with Rutherford, an atom consisting of a central, positive charge in the midst of electrons of equal and opposite polarity, distributed uniformly throughout the remainder of the atomic volume. A positively charged α-particle directed with a velocity vector close to the atomic center is acted on solely by the field of the central charge, if we neglect the field of the electrons. (Individual electrons, because of their small mass, would not appreciably affect the motion of the swiftly moving α-particles.) It is subject to a repulsive electrostatic force varying as the inverse square of the distance between the particle and the center of force; according to classical Newtonian physics, the particle's orbit must be a conic section, specifically, a hyperbola with the center of force at one of its focal points. (The hyperbolic orbit applies, whether the central force is repulsive *or*, as in the planetary case, attractive.) Using a theorem of geometrical optics he must have retained from his undergraduate syllabus, namely one relating the eccentricity of the orbit to the angle between its asymptodes, Rutherford derived an expression for the angle of deviation ϕ of the α-particle in terms of the so-called impact parameter (perpendicular distance between the center of force and incident velocity vector). The impact parameter (Rutherford's p) cannot be invoked directly by experiment, however, since it is not possible to track an individual particle. Rather, it is necessary to transform the expression for ϕ, the experimentally relevant parameter, into a form independent of p. Rutherford next converted an expression for the fraction of particles having a narrow range of impact parameters to one giving the fraction of particles scattered within an increment of angle between ϕ and $\phi + d\phi$. Assuming a scattering foil of silver or gold sufficiently thin that the chance of a second encounter involving a large deflection was vanishingly small, he finally derived his famous scattering formula (equation (5) in his paper), showing that the number of particles striking unit area of zinc sulfide screen per unit time, in a single scattering event, was proportional to

(i) the inverse $\sin^4 \phi/2$, where ϕ is the angle of deflection,
(ii) the thickness of the scattering foil,
(iii) the square of the nuclear charge, and
(iv) the inverse fourth power of the velocity (inverse square of the kinetic energy) of the bombarding α-particles [17-86].

Following a discussion of the not inappreciable loss of velocity for encounters with lighter atoms (say air or aluminum), he proceeded to a comparison of single and multiple scattering, and of theory with experiment. With regard to multiple versus single scattering, according to Rutherford's theory, the fraction of α-particles scattered through any given angle in passing through a foil thickness t should be proportional to nA^2t, where n is the number of atoms in unit volume of scatterer and A its atomic weight. Now, Bragg had shown that the stopping power for α-particles was proportional to $\sqrt{A}$. Hence, nt should vary as $1/\sqrt{A}$, and the fraction of α-particles back-scattered, divided by $A^{3/2}$, should be a constant. 'Considering the difficulty of the experiments', Rutherford judged the agreement with experiment to be 'reasonably good'. As to the most probable angle of scattering in thin foils, Geiger had shown that it was nearly proportional to A. Admitting that multiple (small-angle) scattering could not be ignored, Rutherford continued:

> Since the atomic weight of platinum is nearly equal to that of gold, it follows from these considerations that the magnitude of the diffuse reflexion of α particles through more than 90° from gold and the magnitude of the average small angle scattering of a pencil of rays in passing through gold-foil are both explained on the hypothesis of single scattering by supposing the atom of gold has a central charge of about $100e$. [17-87]

Though this estimate is much too high, Rutherford correctly concluded that 'the central charge in an atom is approximately proportional to its atomic weight'.

The paper ends with some 'general considerations', among them the sign of the central charge, which remained an open question in Rutherford's approach. His scattering theory is independent of whether it is positive or negative, though in his treatment he assumed it to be positive 'for convenience'. It appears, as we saw earlier, that Rutherford first inclined toward a negative central charge—a matter disputed by Darwin, in contradistinction to the claim of Eve in his official biography of Rutherford [17-88]. This hunch of a negative core, if indeed so, must have been partly from β-ray absorption evidence (having to do with the mass dependence of accelerated charges), and possibly in analogy with comets orbiting the Sun under gravitational attraction; in the present paper, Rutherford takes note of Nagaoka's 'Saturnian atom' in his concluding section. He also reserved judgment on the detailed nature of the compensating electronic charge, other than being distributed throughout a sphere of atomic dimensions; in any case, the electronic cloud played no role in his scattering mechanism. Finally, he deferred addressing the stability of his atom for the time being. 'The question of the stability of the atom proposed', he declared at the outset of the paper, 'need not be considered at this stage, for this will obviously depend upon the minute structure of the atom, and on the motion of the constituent charged parts' [17-89].

Rutherford's paper stirred hardly a ripple. No mention of it appears in *Nature* for 1911, or in the British Association report for that year; nor is it found in the *Physikalische Zeitschrift*, in *Comptes Rendus*, or in *Science*. E N da C Andrade, then studying under Lenard at Heidelberg, recalls considerable local excitement caused by C T R Wilson's first publication on the cloud chamber, but no talk about Rutherford's paper [17-90]. It is missing as well from the reports of the Solvay Congress of 1911, which Rutherford attended though did not address, nor is it even mentioned in the series of lectures J J Thomson gave on the 'Structure of the atom' at the Royal Institution as late as 1913 [17-91]. It is, however, noted in passing in Thomson's report under the same title at the second Solvay Congress in 1913, devoted to 'The structure of matter'. In his report to the Congress, Thomson mentioned Rutherford's ideas and α-particle scattering results, but disagreed with Rutherford's interpretation, suggesting instead that the pronounced angular deviations were due to a particular interaction between the incident α-particle and α-particles within the atoms [17-92]. In the discussion following Thomson's report, to be sure, Rutherford himself did draw attention to the excellent experimental evidence for his nuclear atom [17-93].

The correspondence with Rutherford is equally silent on his new results. Thus Boltwood, his regular sounding board, had nothing to say on the subject. What little attention it drew, was in polite reply to Rutherford's letters, not in spontaneous response to the *Philosophical Magazine* paper [17-94].

The lackluster interest comes as no surprise. A plethora of atomic models had recently made the rounds, by J J Thomson, James Jeans, Nagaoka, Lenard, Johannes Stark, Walther Ritz, and Jean Perrin; each model was tailored to account for certain particular physical properties, say spectral regularities, atomic periodicities, radioactive emissions, or scattering behavior. Rutherford's reticence did not help matters, hedging on specifics, such as the electronic arrangement or even sign of the central charge, much less addressing the paramount question of stability. It is generally agreed, however, that at this stage Rutherford viewed his atom mainly as a construct for deriving the laws of scattering for α- (and β-) particles. He was, at any rate, biding his time with respect to its broader significance, weary of prematurely extrapolating from meager evidence, and with other eggs on his plate as well. (Among other things, he was writing another book just then: *Radioactive Substances and their Transformations*.) As he wrote Boltwood on April 11: 'I hope in another year or two to say something fairly definitive of the constitution of the atoms' [17-95].

Fortunately, Rutherford's discovery was not lost on everybody. One person, in particular, took a lively interest in the new developments surrounding it. Wrote Rutherford to Boltwood in March 1912, 'Bohr, a Dane, has pulled out of Cambridge and turned up here to get some experience in radioactive work' [17-96]. The young Dane thus arrived, by fortuitous timing, on the scene in Manchester just a year after Rutherford read his preliminary account before the Literary and Philosophical Society. He would seize upon Rutherford's concept precisely because of the challenging problems it presented and Rutherford wisely chose to overlook. Within two years Rutherford's nuclear atom, joined with Planck's quantum of action, would, in the brash newcomer's capable hands, be brought to its quantum-theoretical maturity in a masterly tour de force, thereby also signalling the decisive break with classical physics.

Niels Bohr had arrived in Cambridge in early fall 1911. Born in 1885, he was the second of three children of Christian Bohr, a professor of physiology at the University of Copenhagen, and Ellen Adler, who came from a prominent Jewish family [17-97]. (Niels' younger brother, Harald, would become an eminent mathematician.) Niels made his early mark in 1906 at the University of Copenhagen, with an experimental research project involving precision measurements on the surface tension of water; it won him a gold medal from the Academy of Sciences in Copenhagen. His doctoral dissertation (under Professor C Christiansen), on the other hand, was a purely theoretical but equally competent work on the electron theory of metals [17-98]. Upon receiving his doctor's degree in 1911, he received a Carlsberg Foundation fellowship for a year's study in England. He arrived in Cambridge at the end of September, and enrolled at Trinity College.

Bohr lost no time paying his respects to J J Thomson, and they had a little chat, during which J J suggested to Bohr an experiment on positive rays. The upshot was a frustrating start for Bohr, who wrote his mother that 'you have no idea of the confusion reigning in the Cavendish Laboratory, and a poor foreigner, who does not even know what the different things that he cannot find are called, is in a very awkward position...' [17-99]. In the event, nothing came of the experiment, but then, of greater concern to Bohr was getting his thesis published in English, and the 'Prof's' reaction to it. (J J himself had also been working intermittently on the electron theory of metals, and his opinion was of paramount importance for Bohr in any case.) During their initial chat, J J had expressed his interest in seeing it when it was finished. 'You can imagine how happy I was when I left', Bohr wrote his fiancée, 'and I looked forward to getting the formulae written in quickly. I am so anxious to learn what he will think of the whole thing and also of all the criticism' [17-100].

Bohr had hastily prepared an English translation of his dissertation before leaving Denmark, assisted by a friend, Carl Christian Latrup, whose physics proved as shaky as Bohr's English. ('Electric charge' in Danish became 'Electric loading'.) Once in England, however, with formulae added at the last moment and the English straightened out with the help of a young physicist named Owen, Bohr arranged with Thomson to have the dissertation submitted to the Cambridge Philosophical Society for publication. Alas, not until next May was Bohr informed that the document was too long, and hence expensive, to publish as submitted; if it could be cut in half, its publication could still be considered. Bohr weighed the alternatives, including publishing the whole thing privately out of his own pocket, then reluctantly shelved the matter. Equally disappointing, Thomson never got around to reading his thesis, or discussing it properly with Bohr; probably in equal measure due to pressure of work, failure to appreciate Bohr's contributions, and flagging interest in the subject [17-101].

In October things took a turn for the better, when, at the annual Cavendish Dinner, Bohr saw Rutherford for the first time. Though the two did not personally meet on that occasion, Bohr 'received a deep impression of the charm and power of [Rutherford's] personality by which he had been able to achieve almost the impossible wherever he worked' [17-102]. In November he had the opportunity of talking with Rutherford in Manchester, while visiting one of his father's colleagues who happened to be a close friend of Rutherford as well. It was subsequently arranged that Bohr would spend four months in Manchester, starting in May 1912 when the Cambridge term was up.

Bohr arrived in Manchester 'in a hopeful, but sobered mood' [17-103], still smarting from his devastating experience in Cambridge, but buoyed by Rutherford's encouragement. His stay in Manchester was ostensibly to further his experimental skills under Geiger's instruction, coinciding with an opportune period in the ongoing researches under Rutherford. Geiger and Marsden were already busy checking out Rutherford's predictions one by one; many other investigations were in the works. However, it was not the experimental possibilities opened up by Rutherford's discovery that would hold Bohr's interest, but its theoretical implications and the problems it posed. Even so, he did not jump into a full-fledged theoretical study of these ramifications without further ado; far from it, he intended to serve out the standard introductory course in experimental techniques in radioactivity offered by the Manchester physics department, dividing his time between the laboratory and library. As it happened, his research problem was interrupted for want of radium emanation, and he used the occasion to go over a paper on the absorption of α-particles recently published by Darwin, the only mathematical physicist around besides himself. In a letter to his brother Harald, just then, he also mentions Rutherford's atom for the first time, and hints that he has given it some thought.

> I am not getting along badly at the moment; a couple of days ago I had a little idea with regard to understanding the absorption of α-rays (it happened in this way: a young mathematician here, C. G. Darwin (grandson of the real Darwin), has just published a theory about this problem, and I feel that it not only wasn't

> quite right mathematically (however only slightly wrong) but very unsatisfactory in the basic conception), and I have worked out a little theory about it, which, even if it isn't much, perhaps may throw some light on certain things connected with the structure of atoms...
> In the last few years [Rutherford] has worked out a theory of atomic structure which seems to have a much more solid basis than anything that we had formerly. Not that my theory is of the same kind of significance... [17-104]

Bohr's theoretical turn had the blessing of Rutherford, who shrewdly sensed his particular bent [17-105]. Bohr saw perhaps better than his host the broader promise of the new atomic model, not simply useful for explaining scattering experiments, but as a potential foundation for elucidating the properties of atoms and molecules generally. In his next letter to Harald, he explains that

> ... perhaps I have found out a little about the structure of atoms. Don't talk about it to anybody, for otherwise I couldn't write to you about it so soon. If it should be right it wouldn't be a suggestion of the nature of a possibility (i.e. an impossibility as J. J. Thomson's theory) but perhaps a little bit of reality. It has grown out of a little information I got from the absorption of α-rays (the little theory I wrote about last time). You understand that I may be wrong; for, it hasn't been worked out fully yet (but I don't think so); also, I do not believe that Rutherford thinks that it is completely wild; he is a man of the right sort, and he would never say that he was convinced of something that was not fully worked out. Believe me, I am eager to finish it in a hurry, and to do that I have taken off a couple of days from the laboratory (this is also a secret). [17-106]

For a synopsis of the brilliant insight Bohr brought to bear on the problem at hand, none have put it better than Léon Rosenfeld, one of Bohr's closest colleagues in later years.

> Bohr already realized, from his critical study for his thesis, that a radical departure from classical conceptions was needed for the account of atomic phenomena... He greeted Rutherford's nuclear atom just because its radiative instability, inescapably following from classical electrodynamics, created such an acute contradiction to chemical and physical evidence about the stability of atomic and molecular structures. In fact, the realization that the stability of the Rutherford atom was beyond the scope of classical theories meant a considerable simplification, inasmuch as he could, so to speak, take it for granted without further analysis and proceed to unfold the further consequences of the model. [17-107]

Pursual of the 'consequences unfolded' would lead us too far off our path, particularly in light of the vast literature surrounding the genesis of the Rutherford–Bohr atom; however, Bohr's activities in Manchester bear dwelling on, if only briefly. From the outset, he saw that the new atomic model offered a sharp distinction between, on the one hand, the ordinary physical and chemical properties of the elements, which are determined by the peripheral electrons loosely bound at radii large compared with nuclear dimensions, and, on the other hand, radioactivity, which must depend on the structure of the nucleus itself. Moreover, he surmised that, owing to the large mass of the nucleus and its small atomic extent, the electronic configuration and with it the element's place in the periodic system, must depend primarily on the *nuclear charge* or atomic number, not the atomic weight, as hitherto generally supposed [17-108]. As a direct consequence of these concepts, Bohr soon came upon the notion of isotopes, and the so-called radioactive displacement law [17-109].

In developing these concepts, Bohr relied particularly on discussions with the young Hungarian aristocrat George Hevesy for his chemical knowledge; it was Hevesy, too, who drew the rather shy and introverted Dane into the social and intellectual life of the laboratory. Rutherford himself was less open to the logical cogency of Bohr's arguments, in part because of his cautious attitude, and thus was unprepared to attribute the broad significance to his model that Bohr did.

As we gather from his letters to Harald, Bohr spent his last month in Manchester largely engaged in a more concrete task, and one of particular relevance for the ongoing experimental research, namely a study of the energy loss of α-rays in passing through matter. The energy loss mechanism was complementary to α-particle scattering, in that the former problem involved collisions with the atomic electrons, the latter with the atomic nucleus. Bohr analyzed the ionizing process in terms of the Rutherford model; his resulting formula for the stopping power in terms of the relative velocity of incident and encountered particle, atomic number of the medium, and binding energies of the atomic electrons (which Darwin had ignored in his own analysis) has largely withstood the test of time, even in the face of modern quantum mechanics.

All the while, Bohr did not lose track of the inherent instability plaguing Rutherford's atom in the first place. As J J Thomson had known only too well, an electron can be in stable equilibrium with a central positive nucleus only if it revolves around it in a closed orbit, with the central attractive force balanced by the outwardly directed centrifugal force. However, the electron would quickly lose energy by radiation, spiralling into the nucleus, and be lost. Bohr realized that a fundamental revision was needed, and was conveniently offered in Planck's quantum of action, first put to practical use by Einstein in accounting for specific heats. Quite independent of the new experimental evidence for the structure of the atom, Planck's quantum was 'in the air', having already been the focal concept for the first Solvay Congress the previous fall.

In effect, what was needed was a new stabilizing quantity involving the dimension of length to define the stable configuration of the orbiting electrons in the atom. In Thomson's plum pudding, there occurs naturally such a dimension defining the size of the atom, namely the diameter of the sphere of positive electricity. An atom of this kind cannot collapse to a dimension smaller than this length. Alas, such a length was not an inherent feature among the parameters characterizing Rutherford's atom. However, Bohr saw, it was provided by Planck's universal constant h. Its value squared, divided by the product e^2m, where e and m are the charge and mass of the electron, respectively, and with $4\pi^2$ thrown in, is equal to 0.5×10^{-8} cm, which is the classic radius of the atom. The quantum of action, then, had the required dimension and magnitude to serve as the stabilizing quantity fixing the distance and velocity at which the electrons would have to move around the nucleus in stable orbits which, for the sake of simplicity, Bohr took to be circular [17-110].

Making use of such a stabilizing condition based on the quantum hypothesis, Bohr refined his ideas about the structure of atoms and molecules in their normal or 'ground' state during his final weeks in Manchester. Many of the concepts would be elaborated in the second and third parts of the famous 'trilogy' of 1913, 'On the constitution of atoms and molecules'. Missing at this time, however, was the centerpiece for the first part of the trilogy, the link with optical spectra (Balmer's formula) and the concomitant frequency condition which would allow extending the concept of a ground state to a series of 'stationary' states with quantum transitions between them. Shortly before leaving Manchester, Bohr jotted down his ideas taking form in a memorandum as backup for discussions with Rutherford—a famous document still extant, except for an important missing piece, and known as the 'Manchester Memorandum'. The memorandum, in Danish, consists of six numbered manuscript sheets glued together in strips of different lengths, enclosed in an envelope inscribed in Bohr's hand:

> First draft of the considerations contained in the paper 'On the constitution of atoms and molecules' (written up to show these considerations to Prof. Rutherford)
> (June and July 1912) [17-111]

Bohr returned to Copenhagen at the end of July 1912, becoming assistant to Professor Martin Knudsen at the University of Copenhagen that fall. (Knudsen had, in the meanwhile, succeeded Bohr's mentor, Christiansen, in the chair of physics.) First, however, Bohr married his fiancée, Margrethe Nørlund, in August. The couple had considered spending their honeymoon in Norway. Unfortunately, however, Bohr had not yet finished his paper on α-particle stopping; therefore, they compromised and went to England instead. Bohr finished the paper during a short stopover at Cambridge, then personally delivered it to Rutherford in Manchester, in a warm reunion accompanied by Margrethe. (Not only would Margrethe play the role of charming hostess to the physics community in years to come, but also many of the drafts of Bohr's papers are in her handwriting.)

Back in Copenhagen, Bohr made steady progress on his atomic theory, keeping in touch with Rutherford who admonished him to take his time, 'as there does not seem to be any one else working on the same subject' [17-112]. Only in February 1913 did he come up with the crucial missing piece of the puzzle, Balmer's radiation formula and the spectral regularities it implied, first drawn to his attention by a friend and doctoral candidate at the Polytechnic School in Copenhagen, H M Hansen. His model now incorporated the familiar feature of electrons confined to a set of discrete, stationary orbits, unable to radiate energy except by jumping from one orbit to another, and then only in discontinuous quantum packets. He dispatched what he called 'the first chapter' (on line spectra) of his great treatise to Rutherford on March 6 1913, requesting that he communicate it to the *Philosophical Magazine*. He added,

> I hope that you will find that I have taken a reasonable point of view as to the delicate question of the simultaneous use of the old mechanics and of the new assumption introduced by Planck's theory of radiation. I am very anxious to know what you may think of it all. [17-113]

In his reply, Rutherford admitted that he had, indeed, difficulties with assimilating the 'mixture of Planck's ideas with the old mechanics', offering one difficulty in particular: 'how does an electron decide what frequency it is going to vibrate at when it passes from one stationary state to the other?' Mainly, though, Rutherford thought the paper was too long, since 'long papers have a way of frightening readers'. 'P.S.', he added, 'I assume you have no objection to my using my judgement to cut out any matter I may consider unnecessary in your paper?' [17-114]. Horror-stricken at the idea of any tampering with the manuscript, Bohr departed for Manchester at once to 'fight it out with Rutherford', as he put it long afterwards. In the event, several long evenings of discussions and cajoling Rutherford saw to it that only linguistic changes were made to the paper. Bohr finished part II (on electronic ring distributions) and part III (on hydrogen molecules) of the trilogy during the late spring and summer 1913. Mailed from Copenhagen, they were forwarded for publication by Rutherford without further comments, and all three papers appeared successively in the July, September, and November issues of the *Philosophical Magazine* [17-115].

The immediate reaction to Bohr's theory was, understandably, mixed. It was given its first public airing in a special session on 'radiation, radio-active emanations, and the structure of the atom' at the meeting of the British Association in Birmingham in mid-September 1913 [17-116]. There, James Jeans hailed Bohr's work as 'most ingenious and suggestive', to Bohr's everlasting appreciation. Mme Curie skillfully parried questions by praising Rutherford. Lord Rayleigh let pass with a witty remark about the inadvisability for people over sixty to pass judgment on modern ideas. Lorentz confined himself to Thomson's atom. ('It was highly ingenious—as it could not otherwise be—but the point was, did it represent the truth?' [17-117].) Hevesy, even before the British Association meeting, responded in typical fashion to Bohr, upon receiving preprints from Copenhagen, with a moving letter of appreciation on August 6 [17-118]. Arnold Sommerfeld, too, sent a complementary card from Munich in early September, though it is unclear from it to what extent he really appreciated Bohr's break with classical concepts [17-119]. In Göttingen, seat of a rather conservative physics department, the reception of Bohr's work was rather cool; however, they had great respect for Harald Bohr, the mathematician, who often dropped in to collaborate with Edmund Landau. If Harald assured them that Niels' work was important, that was good enough with them.

What, finally, about J J Thomson's reaction? At Birmingham, his interaction with Bohr was limited to a a short scrap with him late in the meeting about the controversial gas X_3, during which he dismissed Bohr's suggestion on the nature of the gas as 'useless' [17-120]. In the session on radiation, he merely picked up on Larmor's preceding talk on the equipartition of energy, voicing suspicion 'if a calculation in probability required more than half a sheet of notepaper'. However, in the opening session ('not so interesting as it would have been a year ago, when the supporters of the corpuscular theory would have been in force', according to the editors of *Nature*), he spoke on the structure of the atom; in effect, he gave the gist of a paper under the same title he published in the same issue of *Philosophical Magazine* in which Bohr's trilogy appeared [17-121]. In that unfortunate paper, he treated an atom which has electrons subject to the usual attractive force varying inversely as the square of the distance in certain separated pie-shaped regions, as well as to a repulsive inverse cube force operating everywhere. By suitable adjustments of constants, he was able to reproduce many of Bohr's results, as well as Einstein's law for the photoelectric effect, or $mv^2/2 = h\nu$ (Thomson omits the work function; he uses n for the frequency ν.) Neither Bohr's nor Einstein's names appear in this paper. 'It will be long before his illustration of the quantum theory by pint-pots is forgotten', opined *Nature* in reference to Thomson's showmanship [17-122].

At the second Solvay Congress, in late October, Thomson referred to Bohr, but not to his theory, which was, strictly speaking, not on the Solvay agenda. In his *Recollections and Reflections*, Bohr rates only a short paragraph, in reference to his 'series of researches on spectra, which... have in some departments of spectroscopy changed chaos into order' [17-123].

If Thomson blew hot and cool on Bohr's ideas, perhaps, mused G P Thomson, his son, it was rather like Picasso. 'You could either say it was nonsense or you could say this was the greatest thing that could be, or you could say something a bit in between' [17-124].

On that lighthearted note we end our chronology. J J Thomson could not yet admit that scientific events were passing him by. Nor was he the only doubter. Wrote Rutherford to Geiger in March 1914:

> ...I go down tomorrow to open a discussion at the Royal Society on the structure of the atom. It is rather an innovation for the R. S. and I expect there will be some fun, expecially if J. J. T. turns up, for he sticks closely to his old love of the well developed positive sphere. [17-125]

Events would, in any case, soon be overtaken by the onset of the Great War. The nuclear atom of 1913 had been engendered by Thomson's 'discovery' of the electron in 1897, by his own atomic model, and by the experimental program the model set in motion in Cambridge and Manchester. By 1913, the reality of atoms was no longer in serious doubt in the scientific community; for that matter, the Bohr–Rutherford atom embodied even stranger concepts. The electron, the conspicuous ingredient of that atom, and the first elementary particle in nature so identified (if we overlook the photon), was firmly entrenched in the lexicon of physics, though the physical nature of the positive counterpart would remain obscure for a few more years.

Though Thomson is rightly considered the father of the electron, and 1897 the year of its birth, we have in this volume identified numerous progenitors in a lineage extending back at least to Faraday. The immediate chronicle began, arguably, in 1858 with Plücker and his greenish phosphorescent batch caught in a magnetic field; he was followed by his student Hittorf with his shadow-casting rays (1869), and by their countryman Goldstein, who coined the rays *Kathodenstrahlen* in 1876. Next on the scene came the Victorian amateurs, among them Varley and his 'little talc' rattling in the glow discharge, and above all Crookes, who glimpsed his 'radiant matter' in 1879. The same year brought Lenard's telltale metallic 'window' and all it stood for. Schuster, J J Thomson's dogged competitor, nearly beat him to his corpuscle in 1890, and ended up throwing his considerable support behind Rutherford two decades later. Wiechert and Kaufmann, too, came within a hair's breadth of anticipating Thomson's electron in the very year 1897, two years after Perrin caught the rays in his Faraday cage. About the same time, Zeeman, along with Lorentz and the followers of their magneto-optic school, offered a bound variety of the same particle. The century ended with a spirited race for the subatomic entity in another guise, β-rays, hotly contested by Elster and Geitel, Meyer, Giesel, Becquerel, and finally Rutherford—the heavyweight among particle hunters. Millikan's oil drops signaled the decisive end of the chronicle a decade later.

Though the electron was a *fait accompli* by 1913, the wondrous technology it spawned would mature but slowly. The atomic nucleus required another two and a half decades to fully yield its secrets, ushering in an altogether new age.

Chapter 18

EPILOGUE: THE NEXT TWENTY YEARS

18.1. The war intervenes; some later experiments

One last piece of work on the structure of the atom begs our attention, completed as the clouds of war were gathering over the Continent. By 1913 it was clear, from the scattering of x-rays by Charles G Barkla, from Rutherford's analysis of α-particle scattering, and from Bohr's insight into it all, that (except for hydrogen!) the charge of the atomic nucleus was roughly half the atomic weight. However, the constant of proportionality was only known to within a few percent. The situation was analogous to the state of knowledge with regard to the electronic charge a decade earlier. In contrast to Millikan's frontal assault on the latter problem, however, the resolution of the former uncertainty came indirectly, from the systematic study of x-rays by Henry G J Moseley.

In setting the stage for Moseley's epochal contribution, we are mindful of the discovery by Barkla [18-1] and Charles A Sadler in 1907 that x-ray targets give off discrete lines of secondary radiation—radiation as distinctly characteristic of the target element as are lines in its optical spectrum. The 'hardness' of the characteristic radiation increased regularly with ascending placement of the elements in the periodic table—the factor exploited so successfully by Moseley.

Henry Moseley came from a long line of distinguished scientists [18-2]. Among other things, his father had been the chief naturalist on the Challenger expedition, and was a protégé of Charles Darwin. It is thus fitting that the younger Moseley would strike up, if only temporarily, with Darwin's grandson. He won a scholarship to Eton, and then to Trinity College, Oxford, where he studied physics and graduated in 1910. The same year he applied to Rutherford, who set him up as lecturer and demonstrator in the physics department at Manchester. While thus engaged in teaching undergraduate students of marginal aptitude, he applied himself enthusiastically to a small research problem Rutherford had assigned him to develop his skills in laboratory techniques. Two years later he won a fellowship which left him free to devote himself full time to research. As luck would have it, news of the discovery of x-ray diffraction by Max von Laue reached Manchester just then, communicated at one of the regular Friday afternoon physics colloquia in May 1912. Moseley suggested to C G Darwin, also then with time on his hands, that they take up x-ray research together, what with Moseley's newly acquired experimental skills and Darwin's theoretical bent for wave phenomena. Rutherford, at first 'distinctly discouraging' [18-3], gave his go-ahead. Initial results were not long in coming, Moseley being a born workaholic and night owl [18-4]. Though the two amicably split up in about a year, Darwin turning to a mathematical analysis of the diffraction phenomena at hand, Moseley continued the experiments on his own.

The *experimentum crucis* lay in the technique honed to perfection by the two Braggs, reflecting a beam of x-rays from a cleavage plane in a crystal. After passing through a slit, the rays, generated in a Coolidge tube 'of rather irregular habits' [18-5], fell on the plane of the crystal mounted on the prism table of a spectrometer, and the reflected rays were detected by a photographic plate. According to the Laue–Bragg law of x-ray diffraction, the angle of incidence on the crystal face, selected by rotating the prism table, gave the wavelength of x-radiation striking the photographic plate without further ado [18-6]. The upshot, not without considerable initial confusion, was the characteristic line spectra of a sequence of elements from aluminum to gold. The regularity of the progression in characteristic wavelengths from one element to the next was striking.

> We have here a proof [reported Moseley] that there is in the atom a fundamental quantity, which increases by regular steps as we pass from one element to the next... Now atomic weights increase on the average by about 2 units at a time, and this strongly suggests that *N* [the atomic number, or nuclear charge, nowdays denoted by *Z*] increases from atom to atom always by a single electronic unit. We are therefore led by experiment to the view that *N* is the same as the number of the place occupied by the element in the periodic system. [18-7]

Moseley had intended to take part in the British Association meeting, held in Melbourne, Australia, during September 1914. While *en route* to Sidney in the company of his mother, the war broke out in Europe [18-8]. After reporting on his results at the meeting, he returned home posthaste, enlisting as Signal Officer in the Royal Engineers. In June 1915 he sailed for the Dardanelles, where he participated in the infamous Suvla Bay landings in a last desperate effort to gain Sari Bair ridge. He fell from a Turkish bullet through the head on August 10.

The outbreak of the First World War in August 1914 brought the normal routine to a halt at Manchester, Cambridge, and almost everywhere else. The younger members of Rutherford's team were soon scattered to the four winds. Darwin ended up in France, engaged in acoustic gun ranging (locating enemy guns by recording at three or more stations the time of arrival of the sound of the discharge)—a project in charge of the younger Bragg, operating out of Allied GHQ on the western front [18-9]. Fortunately Niels Bohr replaced Darwin as Schuster Reader, returning to Manchester in early fall 1914 for a two-year stay. Marsden, newly appointed to a relatively insignificant professorship at Victoria University College in Wellington, New Zealand, was in due course seconded as Lieutenant to the Royal Engineers; he, too, ended up with the Sound-Ranging Section in France, steadily advancing in grade and eventually wounded in action [18-10]. While at the front, he is said to have received from Geiger, serving on the opposite front, a letter congratulating him on his Wellington appointment, forwarded through Bohr, then back in Denmark.

Rutherford himself devoted much of his time to the war effort. In this capacity, he journeyed to America where he caught up with Lieutenant Colonel Robert Millikan, then head of the US Signal Corps' science and research division. The basement laboratory at Manchester housed a large tank for testing submarine detection devices. What little scientific experimentation still going on at Manchester early in the war, with Bohr joining in, came to a sudden end when delicate quartz apparatus, intended for following up on the experiments of Franck and Hertz on the excitation of atoms by electron impact, was ruined by an accident. Unfortunately, the volatile German glass blower, Baumbach, had been interned by the British authorities shortly before, and the experiments were called off [18-11].

At Cambridge, those who remained patrolled the streets to see that lights were out, thus keeping Zeppelins away [18-12]. The cloisters of Trinity College became a hospital for wounded soldiers. Regiments were billeted in the courts, and officer candidates received intensive instruction in hope of a commission. The Cavendish was relinquished for war work entirely, and the workshop employed in fabricating gauges. J J Thomson served on the Board of Inventions and Research, instituted by Lord Fisher to render expert technical assistance to the Admiralty by screening inventions and sponsoring research.

In Paris, on the other hand, life went on with a modicum of normalcy, though Mme Curie grew haggard with her radiological war effort: equipping, and even chaffeuring, ambulances with x-ray equipment, and organizing accelerated courses in radiology. Though the Netherlands remained precariously neutral during the conflict, low-temperature research at Leiden coasted to a halt for want of helium gas. In Belgium it was quite another matter; thus Louvain, the ancient seat of learning, was sacked by the advancing German armies, with the famous university library burned to the ground.

In Berlin, physicists rallied to the cause of the nation, by signing the Manifesto of 1914, an 'appeal to the cultured people of the world', declaring 92 spokesmen for German art and science to be solidly behind the German army and repudiating charges of the Allied Powers that German forces had committed atrocities in Belgium [18-13]. At first scientists corresponded with colleagues in countries at war with Germany, and generally kept abreast of what scientific developments did occur abroad through journals. Planck brought Lorentz to Berlin, and Lorentz responded in kind, inviting Planck to Leiden, to see things—attitudes, charges, and manifestos—from the other point of view. Still, attendance in Planck's institute dwindled away [18-14], and a mood of pessimism soon permeated German scientific circles. Voigt anticipated that German physicists would not recover from loss of their invaluable relations with their counterparts abroad for a long time to come [18-15]. Einstein, ordinary professor in the Prussian Academy in Berlin, told Lorentz that he was 'constantly very depressed over the immeasurably sad things which burden our lives', and that it no longer helped 'as it used to, to escape into ones's work in physics' [18-16].

If things were bad enough in the depleted German physics institutes, they were as nothing compared to the front. Gustav Rümelin fell in the first months of the war. Maximilian Reinganum died from a shell splinter near Le Menil on the Vosges. Hasenöhrl fell during the battle for Caporetto on the River Isonzo. Heinrich Willy Schmidt also fell, and Gustav Hertz, nephew of Henrich Hertz, was seriously wounded. Andrade and Erwin Schrödinger served as officers in the artillery, Ladenburg in the cavalry. Nernst, at 50, initially volunteered his services to the driver's corps of the army in its dash for France, then turned to advising the German High Command on chemical warfare in the program spearheaded by Fritz Haber, as did James Franck, Otto Hahn, and Gustav Hertz.

Among the researches carried out in wartime Germany despite the hardships, we note those of Arnold Sommerfeld, professor of theoretical physics at the University of Munich [18-17]. Following the appearance of Bohr's atomic theory in 1913, and though 'somewhat skeptical in principle toward atomic models', he hailed Bohr's contribution and courteously offered to have a try at the Zeeman effect with Bohr's theory [18-18]. The outcome was his celebrated extension of Bohr's atom, incorporating relativistic corrections, and featuring Keplerian elliptic orbits with a precessing perhelion. Despite the war, Sommerfeld was able to send his papers to Bohr in Manchester. 'I do not think', Bohr wrote back, 'that I ever have enjoyed the reading of anything more than I enjoyed the study of... your most interesting and beautiful papers' [18-19].

Figure 18.1. *Rutherford, ca. 1930.*

Amidst the wartime disruptions, Rutherford, too, undertook, during these very years, what became arguably his crowning experimental achievement. It was carried out pretty much by him alone, save for his laboratory steward, William Kay, and then only when he could find 'an odd half-day' for it. Bohr, for one, had returned to Denmark to assume a newly created professorship in theoretical physics at the University of Copenhagen. Fortunately, Rutherford's laboratory notebooks for this hectic and noteworthy period, covering his researches in Manchester during 1917–1919, and continuing at Cambridge from 1919 onward, are all extant [18-20].

For the significance of this, his last scientific work at Manchester, we must backtrack to the α-particle scattering experiments of Geiger and Marsden. The salient feature of those experiments was the utilization of target nuclei sufficiently massive that recoil from close encounters with the much lighter projectile could be ignored. What, however, if an α-particle encountered a much lighter atomic nucleus, say hydrogen with only one quarter of the mass of the projectile? Darwin had worked out the scattering theory for just such a case, finding that, in a head-on collision between an α-particle and the nucleus of a hydrogen atom, the latter would acquire a velocity 1.6 times that of the incident α-particle and, assuming Bragg's law of absorption held, would have a range up of to four times that of the incident particle. Marsden, on the eve of leaving Manchester, had observed long-range 'H-particles' (hydrogen ions) in fair agreement with Darwin's prediction. However, under certain conditions, including absorption in air, he obtained anomalous results in the form of H-particles with exceptionally long range. Much time and effort was spent excluding possible hydrogen contamination of the α-particle source. The 11th-hour investigation, with W C Lantsberry, came to a premature end when Marsden left for his new post at Wellington, just before war broke out. The results were hurriedly written up for a paper to the *Philosophical Magazine*, which ended on an intriguing note: 'There seems a strong suspicion that the H particles are emitted from the radioactive atoms themselves' [18-21].

> Rutherford was excited at the possibility of his beloved radioactive atoms giving off a new kind of particle [relates Marsden] and on his return from Australia, where he had attended a British Association meeting, he wrote to me in New Zealand asking me if I 'minded' if he went on with the experiment as I had not the facilities... However, I went to France in the Army to join all my other colleagues from the laboratory and during the next four years... Rutherford repeated and extended my observations. [18-22]

In Rutherford's experimental arrangement, an RaC source of α-particles was mounted inside a rectangular brass vessel which could be exhausted or filled with the gas to be examined. One end of the vessel was closed with a brass plate with a small rectangular opening covered with a thin plate of silver. The silver was located far enough from the radium source to be beyond the range of α-particles in the gas under study. A zinc sulfide screen was fixed close to the opening and just outside the vessel. Scintillations from long-range particles striking the screen were observed through a microscope. The whole apparatus was placed in a strong magnetic field to keep β-particles out of the way [18-23].

With dry oxygen or carbon dioxide admitted, the number of scintillations, from what appeared to be hydrogen-like particles, decreased with increasing gas concentration, as expected. However, admitting dry air instead brought a definite surprise: the number of scintillations actually *increased*. Having shown that oxygen alone decreased the count, the increase could only be due to the nitrogen in the dry air. This, in turn, could only mean that the α-particles had collided with nitrogen atoms and knocked hydrogen ions out of them.

The earliest record of the experiments on the disintegration of nitrogen by α-particles is found on pp. 39–40, the two last pages of the Manchester notebook for January–March 1919 [18-24]. The experiments as a whole were completed by the end of that year, and published in four back-to-back papers in the *Philosophical Magazine* for 1919, under the common title 'Collision of α particles with light atoms'. Their significance was twofold. First, they provided the first example of artificially induced transformation of the chemical elements, namely nitrogen into oxygen [18-25]. Second, they established that among the constituents of atomic nuclei are 'hydrogen atoms' or *protons*, a term subsequently coined by Rutherford in memory of Prout and on the suggestion of Darwin and F Fowler [18-26]. As Rutherford put it in the fourth of the seminal papers of 1919:

> We must conclude that the nitrogen atom is disintegrated under the intense forces developed in a close collision with a swift α particle, and that the hydrogen atom which is liberated formed a constituent part of the nitrogen nucleus... The results as a whole suggest that, if α particles—or similar projectiles—of still greater energy were available for experiments, we might expect to break down the nucleus structure of many of the lighter atoms. [18-27]

The 'two-particle orthodoxy' [18-28], that matter consists solely of electrons and protons, held sway for well over a decade. On that supposition, the helium nucleus consisted of four protons and two electrons; in general, a neutral atom of mass number A and atomic number Z was supposed to contain A protons, all in the nucleus, and A negative electrons, $A - Z$ in the nucleus and the rest making up the external electron shells. To be sure, increasing theoretical difficulties with this nuclear scheme gradually surfaced with the advent of quantum mechanics [18-29]. Moreover, Rutherford himself had speculated, in his Bakerian Lecture to the Royal Society in June 1920, on the possible existence of an additional, neutral particle in the form of a bound state of a proton and electron. Such a particle would have to posess novel properties. 'It should be able to move freely through matter..., and it may be impossible to contain it in a sealed vessel' [18-30]. Indeed, the concept of a neutral constituent hung in the air over the ensuing decade, actively searched for off and on in γ-rays spontaneously produced in the act of forming the electron–proton union or in highly penetrating material particles. On the whole, however, the intimate association between the two known elementary charged particles and the two basic units of electricity—the electronic charge and the numerically equal and opposite nuclear charge—prevented the notion of a third quasi-fundamental particle from seriously taking hold [18-31].

The complacency ended in 1932, that *annus mirabilis* of nuclear physics. January of that bountiful year brought Harold Urey's announcement of a heavy isotope of hydrogen (something also anticipated in Rutherford's prescient Bakerian Lecture), which he named 'deuterium'. In February came news of the long-suspected neutron, the experimental triumph of James Chadwick. In April, John Cockroft and E T S Walton achieved the first disintegration of light nuclei with artificially accelerated protons—Rutherford's long-sought more energetic projectiles. Cockroft and Walton were outgunned later in the year by the cyclotron in the American hands of Ernest O Lawrence, M Stanley Livingston, and Milton White. In August, Carl D Anderson's photographs of cosmic ray tracks revealed, at long last, the positively charged electrons, soon named positrons. Alas, the positron bore scant relationship to Becquerel's discredited positive electron of 1906–1909, and need not concern us further here. By way of ending our chronicle, we recount instead the circumstances surrounding Chadwick's momentous discovery.

James Chadwick, born in Bollington, England, in 1891, attended Manchester Municipal Secondary School, then won a scholarship to the University of Manchester [18-32]. He went to work under Rutherford, receiving his MSc degree in 1913. Simultaneously, he received the 1851 Exhibition Scholarship and joined Geiger, by then at the Reichsanstalt in Berlin. There he was stranded when World War I broke out in 1914, being interned under rather decent conditions in the Ruhleben camp for civilian prisoners, along with C D Ellis of Cambridge, for the duration of the war [18-33]. Released in 1919, he rejoined Rutherford, and moved to Cambridge with him when his mentor was appointed Cavendish Professor in 1919. He became assistant director of research under Rutherford at the Cavendish Laboratory, and took day-to-day charge of all in-house atomic and nuclear investigations. The first half of the 1920s was mainly devoted to transmutation experiments under α-ray bombardment, involving a range of elements besides nitrogen, as in Rutherford's original experiments. During the second half of the decade, Rutherford and Chadwick turned jointly to problems of nuclear structure raised by α-particle scattering experiments. All the while, the idea of a neutral particle inside the nucleus remained a nagging possibility for Chadwick, and he made several unsuccessful attempts to detect it.

Chadwick's luck turned for the better in the summer of 1931, when he was following up on the discovery of Frédéric and Irène Joliot-Curie that radiation from beryllium excited by α-particles could eject protons from paraffin wax or any other hydrogen-containing material—in itself a burgeoning research topic in Continental circles. As Chadwick recalls it,

> One morning I read the communication of the Curie-Joliots in the *Comptes Rendus*, in which they reported a still more surprising property of the radiation from beryllium, a most startling property. Not many minutes afterwards [Norman] Feather came to my room to tell me about this report, as astonished as I was. A little later that morning I told Rutherford. It was a custom of long standing that I should visit him about 11 a.m. to tell him any news of interest and to discuss the work in progress in the laboratory. As I told him about the Curie-Joliot observation and their views on it, I saw his growing amazement; and finally he

> burst out 'I don't believe it'. Such an impatient remark was utterly out of character, and in all my long association with him I recall no similar occasion. I mention it to emphasize the electrifying effect of the Curie-Joliot report. Of course, Rutherford agreed that one must believe the observations; the explanation was quite another matter.
>
> It so happened that I was just ready to begin experiment, for I had prepared a beautiful source of polonium from the Baltimore material [old radon tubes from the Kelly Hospital in Baltimore]. I started with an open mind, though naturally my thoughts were on the neutron.... A few days of strenuous work were sufficient to show that these strange effects were due to a neutral particle and to enable me to measure its mass: the neutron postulated by Rutherford in 1920 had at last revealed itself. [18-34]

Chadwick's experiment, in short, consisted of a source and a detector. The source was a silver disk coated with radium, placed close to a disk of pure beryllium; both disks were enclosed in a small vessel which could be evacuated [18-35]. The detector consisted of an ionization chamber connected to a linear amplifier—an arrangement featuring a counting technique newly developed at Cambridge and elsewhere. When a sheet of paraffin wax, or some other light element, was interposed in the path of the mysterious radiation emerging from the beryllium, the number of deflections on the recording oscillograph rose markedly, due to an influx of secondary particles ejected from the absorber; the particles were readily identified as recoil protons. The 'beryllium radiation' could not be explained on the assumption that it consisted of high-energy photons, say due to the Compton effect [18-36]. However, it made immediate sense if it consisted of particles of mass nearly equal to that of a proton and with no net charge.

Chadwick's letter to *Nature* announcing the discovery of the neutron appeared on February 27 1932 [18-37]. Three days earlier, Chadwick had mailed a copy to Bohr in Copenhagen [18-38]. The discovery figured prominently in the discussions of the informal conference hosted during the second week of April at Bohr's Institute of Theoretical Physics—an annual affair at Blegdamsvejen 17, and this one by chance marking the tenth anniversary of the Institute's founding. The neutron also figured in the script of a play, a parody on Goethe's *Faust*, written and performed by students of Bohr during the meeting. The Finale, 'Apotheosii of the True Neutron', delivered by Wagner, alias Chadwick (himself, in fact, unable to attend the meeting), goes, in part, as follows:

> The *Neutron* has come to be.
> Loaded with Mass is he.
> Of Charge, forever free.
> Pauli, do you agree? [18-39]

The excerpt can serve as our finale, too—one more poignant allusion to the controversies which have enlivened the quest for the fundamental particles of nature, starting with the electron.

18.2. The personalities

The Great War effectively ended academic physics for just about everybody, J J Thomson included. As noted, J J's wartime energies were mainly expended in connection with the Admiralty's Board of Inventions and Research. In 1918, additionally, he became Master of his old college, Trinity—a Crown appointment involving much pomp and circumstance in the investiture [18-40]. In 1919, feeling the strain of directing both a College and a Laboratory, and sensing the new physics to be passing him by, he resigned his Cavendish Chair in favor of Rutherford [18-41]. (Thomson continued as professor of physics without stipend.) Rutherford also succeeded Thomson as Professor of Natural Philosophy at the Royal Institution in 1921—a largely honorific appointment, however, that entailed a few lectures each year. He retained tight control over the Cavendish until the end from a strangulated umbilical hernia in 1937. Thomson, his senior by a decade and a half, outlived Rutherford by fourteen years. Both are interred in Science Corner in Westminister Abbey, along with Newton, Herschel, Darwin, and Kelvin.

As for Rutherford's Lieutenant, James Chadwick, he had an unfortunate falling out with his boss in the early 1930s [18-42]. Chadwick pushed for the installation of a cyclotron in the Cavendish Laboratory, but Rutherford refused. Chadwick left in 1935 for Liverpool University, where he built up the physics department virtually from scratch, with a cyclotron the centerpiece of its George Holt Laboratory. Among others who left their individual stamp on the Cavendish in the post-WWI years, C T R Wilson and F Aston rank high on the list. The indefatigable C T R, who had begun his career on the summit of Ben Nevis, retired among the lofty heights of Arran in the Scottish highlands, still pondering atmospheric electricity. At 86 he learned to fly, observing from above the clouds he had studied from below, and which had inspired his famous chamber. Francis Aston, the tireless world traveler, was equally at home in high country, climbing the rocks of North Wales and Skye. He led a life of uninterrupted success. Between his mass spectrographs, he photographed eclipses, went skiing in the Alps and surf-riding in Honolulu, and managed financial investments with equal aplomb. In consequence of his monetary skills, he left a considerable legacy with a host of scientific institutions, among them the Cavendish Laboratory and the Fitzwilliam Museum in Cambridge.

Geiger and Marsden, serving on opposite sides of the western front during the War, went their separate ways in later years. Marsden's academic tenure at Victoria University College in Wellington proved short lived, as did his personal research, when he entered government service in 1922: first with the New Zealand Education Department, and then as Permanent Secretary of the newly created State Department of Scientific and Industrial Research. During World War II he took charge of radar production in New Zealand, turning out state-of-the-art sets for Allied forces in the Pacific. He then served a stint as New Zealand's Government Scientific Adviser in London, topping off his career as senior scientific statesman until his retirement in 1954.

After Manchester, Geiger assumed the directorship of the laboratory for radium research at the Physikalisch-Technische Reichsanstalt in Berlin-Charlottenburg, interrupted by his service as artillery officer during the war [18-43]. In 1925 he took up his first teaching position as professor of physics at the University of Kiel; it was at Kiel that he put the final touches, with Walther Müller, to the famous counter begun under Rutherford in 1908. From Kiel he went to Tübingen, and thence back to Berlin, accepting the chair of physics at the Technische Hochschule. During the second world war he came down with a recurrence of the rheumatic condition he first contracted during WWI. Nevertheless, his affliction did not keep him from speaking out on behalf of unpoliticized physics [18-44]. In 1945 he fled from his home in Babelsberg to nearby Potsdam, where he died shortly afterward.

In the 1930s Jean Perrin, with Frédéric Joliot, Irène Curie, and others, played an important role in convincing the government to establish the Caisse Nationale des Researches Scientifique, an agency serving to further French science and technology independent of the powerful university lobby in French intellectual affairs. In still later years, Perrin became actively involved in French politics as a militant leftist, supporting the de Gaulle movement in World War II and escaping France for New York, where he died in 1942.

We have touched sufficiently on Lenard's Aryan politics, and let it go at that. After World War II, he was ordered out of Heidelberg, and died in the small village of Messelhausen.

The fates of two more German physicists, whose names are identified with the penultimate e/m experimentation on the electron, deserve mention. During the first world war, Emil Wiechert, physicist turned geophysicist, applied his expertise to determining atmospheric features by sound waves, and to geophysical prospecting by seismic methods. In later years he suffered from extreme deafness and serious illness. He died in 1928. His younger compatriot Walter Kaufmann survived him by nearly two decades, though after 1906 he abandoned his variegated electromagnetic researches, advancing up the academic ladder and pursuing a potpourri of investigations until his retirement.

World War I occupied Robert Millikan as director of science and research for the Signal Corps, US Army, and of the Special Board of Antisubmarine Devices, US Navy. In 1921 he assumed directorship of the Norman Bridge Laboratory of the newly renamed California Institute of Technology (formerly Throop College of Technology) in Pasadena, and became, in effect, president of the school. Amid escalating public responsibilities and fame, his research agenda never slackened. He worked on cold emission of metals, and extreme ultraviolet spectroscopy. Above all, he pursued the burgeoning field of cosmic ray research with untiring energy, sometimes chasing blind alleys, but amassing a valuable fund of experimental data. He was interred, at 85, under solemn circumstances in the Court of Honor in Forest Lawn Memorial Park, Los Angeles, to the strains of the Roger Wagner Chorale.

Felix Ehrenhaft's later career was, alas, a sad story. Forced to leave Austria after the Nazi takeover in 1938, he settled briefly in England, and then in the United States. All the while his professional reputation steadily sank, as he stubbornly clung to his subelectrons against all evidence, and, among other will-o'-the-wisps, to magnetic monopoles, for which he claimed experimental evidence. Thus Einstein, originally on the best of terms with him, came to regard him as a once capable experimenter who had 'gradually developed into a kind of swindler' [18-45]. Things got to the point where Ehrenhaft was not allowed to speak at meetings of the American Physical Society, and his research support dried up. In 1946, though a US citizen, he returned to Vienna as Guest Professor and director of the combined First and Third Physical Institutes. He held these offices until his death in 1952.

ABBREVIATIONS

The following abbreviations are used extensively in the notes.

ACP	*Annales de Chimie et de Physique*
AHES	Archive for History of Exact Sciences
AHQP	Archive for the History of Quantum Physics
AJP	*American Journal of Physics*
AP	*Annalen der Physik*
APC	*Annalen der Physik und Chemie*
CR	*Comptes Rendus Hebdomadaires des Séances de l'Académie des Sciences*
CUL	Cambridge University Library
DBS	*Dictionary of Scientific Biography*
HSPBS	*Historical Studies in the Physical and Biological Sciences*
HSPS	*Historical Studies in the Physical Sciences*
LR	*Journal de Physique et le Radium*
NA	*Nature*
PLC	*Leiden University Physical Laboratory Communications*
PM	*The London, Edinburgh, and Dublin Philosophical Magazine*
PR	*The Physical Review*
PRS	*Proceedings of the Royal Society of London*
PT	*Physics Today*
PTR	*Physikalisch-Technische Reichsanstalt*
PTRS	*Philosophical Transactions of the Royal Society of London*
PZ	*Physikalische Zeitschrift*
RMP	*Reviews of Modern Physics*
RN	*Akademie der Wetenschappen, Amsterdam Proceedings*
ZP	*Zeitschrift für Physik*

NOTES

Notes section 1.1

1-1 Glasser O 1934 The discovery of the Röntgen rays *Wilhelm Conrad Röentgen and the Early History of the Röntgen Rays* (Springfield, IL: Thomas) ch 1, with a chapter on personal reminiscences of W C Röntgen by Margaret Boveri p 3

1-2 Dam H J W 1896 The new marvel in photography *McClure's Magazine* 6 April, reprinted in Glasser: see note 1-1, pp 5–12

1-3 Both are preserved in the Deutsches Museum in Munich. The room in which the discovery was made, Room 119A of the Technische Hochschule, is nowadays a one-room museum, albeit without the original apparatus.

1-4 Dam: see note 1-2, p 8

1-5 O Lummer of the Physikalisch-Technische Reichsanstalt (the proffered directorship of which Röntgen declined in 1894) claimed that although Lenard's rays showed a visible fluorescent effect only for a distance of a few centimeters, they 'still showed electrical discharge effects at much greater distances'—something Röntgen may have been attempting to pin down. Lummer, in translation of lectures by Silvanus P Thompson on 'Visible and invisible light' before the Royal Institution, quoted by Glasser: see note 1-1, p 13. Many accounts credit the initial discovery to the chance observation of a fluorescent screen lying on a table some distance from the tube; see, e.g. Glasser: see note 1-1, p 3

1-6 Röntgen W C 1895 Über eine neue Art von Strahlen, Vorläufige Mittheilung *Sitzungsberichte der Würzburger Physikalischen-Medicinschen Gesellschaft* **137**; English translation in Magie W F 1965 *A Source Book in Physics* (Cambridge: Harvard University Press) pp 600–7. Only a photograph of the first page of the manuscript survived World War II; it is retained at the Deutsches Röntgen Museum in Remscheid-Lennep.

1-7 English translation by Barker G F 1899 *Röntgen Rays* (New York: Harper's Scientific Memoirs) reproduced in Shamos M H (ed) 1959 *Great Experiments in Physics: Firsthand Accounts from Galileo to Einstein* (New York: Dover) pp 201–2. The version of Magie (note 1-6) pp 600–10, omits several footnotes to Röntgen's original paper.

1-8 Shamos: see note 1-7, p 205

1-9 Shamos: see note 1-7, p 205

Notes section 1.2

1-10 Nye M J 1984 *The Question of the Atom: From the Karlsruhe Congress to the First Solvay Conference, 1860–1911* (Los Angeles, CA: Tomash) p xxv

1-11 Maxwell J C 1954 *A Treatise on Electricity and Magnetism* unabridged 3rd edn, vol 1 (New York: Dover) article 260, p 380

1-12 Planck M *Scientific autobiography*, cited by Segrè E 1980 *From X-rays to Quarks: Modern Physicists and their Discoveries* (Berkeley, CA: University of California Press) p 7

1-13 Heilbron J L 1986 *The Dilemmas of an Upright Man: Max Planck as Spokesman for German Science* (Berkeley, CA: University of California Press) p 49

1-14 Duhem P 1974 *The Aim and Structure of Physical Reality* translation by Wiener P P (New York: Atheneum) p 71
Heilbron J L 1977 Lectures in the history of atomic physics 1900–1922 *History of Twentieth Century Physics* ed C Weiner (New York: Academic)

1-15 Gesellschaft Deutscher Naturforscher und Ärzte

1-16 Stuewer R H 1974 Jean Baptiste Perrin *DSB* **10** 525

1-17 Perrin J B 1895 Nouvelles propriétés des rayons cathodiques *CR* **121** 1130 English translation in Magie: see note 1-6, pp 580–3

1-18 Magie: see note 1-6, p 582

1-19 Whittaker Sir E 1989 *A History of the Theories of Aether and Electricity: Volume I, The Classic Theories: Volume II, The Modern Theories 1900–1926* (New York: Dover) p 410

1-20 Zeeman P 1896 *RN* **5** 181; translation in 1897 *PM* **43** 226. In 1900 Zeeman accepted a post as professor at the University of Amsterdam where he continued his spectral studies for the balance of his career and where in time he became director of the physics department.

1-21 Lorentz H A 1892 *Arch. Neerl* **25** 363, reprinted in Zeeman P and Fokker A D (eds) 1935–1939 *H A Lorentz, Collected Papers* 9 vols, vol 2 (The Hague: Martinus Nijhoff) p 164

1-22 Lorentz H A 1895 *Versuch einer Theorie der Electrischen und Optischen Erscheinungen in Bewegten Körpern; Collected Papers* vol 5 (Leiden: Martinus Nijhoff) p 1. Lorentz wrote his formula, as an adjunct to Maxwell's equations, in compact vector notation, as we have also done. He used $\boldsymbol{H}$ for the magnetic 'force' or field, while we have substituted $\boldsymbol{B}$ in conformity with modern notation. In later applications of equation (1), we use $\boldsymbol{H}$ or $\boldsymbol{B}$ interchangeably, depending on the original notation. The latter-day formal distinction between $\boldsymbol{B}$ for flux density or magnetic induction, and $\boldsymbol{H}$ for field strength, is in any case mainly of importance in problems involving magnetized media.

1-23 In Zeeman's first measurements only a broadening of the spectral lines was observed, adequate to yield the all-important e/m value. Subsequent measurements in 1897 revealed the *splitting* of the spectral lines known as the normal Zeeman effect. The normal Zeeman effect can be roughly explained by classical electrodynamics (it was indeed predicted by Lorentz) and rigorously with the aid of Bohr's atom. Most substances, however, reveal an anomalous Zeeman effect—a much more complicated pattern only explainable by quantum mechanics and invoking the concept of electron spin.

1-24 Zeeman P 1896 *RN* **5** 242; translation in 1897 *PM* **43** 226; also in Magie: see note 1-6, pp 384–6. Zeeman and Lorentz shared the 1902 Nobel Prize for Physics 'in recognition of the extraordinary services they rendered by their researches into the influence of magnetism upon radiation phenomena'. A word about units. At the turn of the century, the equations of electrostatics were commonly written using the *electrostatic* system (esu), while equations relating to magnetic phenomena used the *electromagnetic* system (emu). Units of the same quantity are not of the same magnitude in the two systems. Thus, the electromagnetic unit of charge is approximately 3×10^{10} times as great as in the electrostatic system, where 3×10^{10} cm s^{-1} is the velocity of light in vaccum. Both the electrostatic and the electromagnetic systems use the centimeter, gram, and second as their fundamental units of length, mass, and time, respectively.

Notes section 1.3

1-25 Thomson J J 1936 *Recollections and Reflections* (London: Bell)

1-26 Thomson: see note 1-25, p 12

1-27 Thomson: see note 1-25, p 36. There were Triposes in the other fields of learning as well, e.g. Law Tripos, Classic Tripos, Natural Science Tripos.

1-28 Routh himself was Senior Wrangler in the year when Maxwell was Second Wrangler.

1-29 Fellow and Lecturer at Trinity, Glaisher accepted a tutorship at the College in 1883, which led to his professional undoing and redirected his life. As a tutor he was continously interrupted from his mathematical researches and, when the Sandleirian Professorship in Pure Mathematics came up for renewal, he was passed over. As his lectureship also ran out at about this time, he took up collecting china and pottery to fill the gap. During his last 30 years he gathered, at great expense, a vast and famous collection which is nowadays exhibited in the Fitzwilliam Museum. Thomson: see note 1-25, pp 45–6

1-30 Thomson: see note 1-25, p 76. He had this basic notion already at Manchester, where Balfour Stewart stressed the principle of the conservation of energy in his lectures. Thomson found the convertibility of different forms of energy, say kinetic into potential energy, hard to take, and argued instead that the 'transformation' of energy could be more correctly described as the transference of kinetic energy from one system to another, the physical effects of the transfer depending on the nature of the new system. Stewart was non-comittally sympathetic to his notion, which, however, Thomson found no time for developing while preparing for the Tripos.

1-31 Thomson J J 1888 *Applications of Dynamics to Physics and Chemistry* (London: Macmillan) p 15; cited by Topper D R 1980 To reason by means of images: J J Thomson and the mechanical picture of nature *Ann. Sci.* **37** 36

1-32 Thomson J J 1885; 1887 Some applications of dynamical principles to physical phenomena (parts I and II) *PTRS* **176** 307–42; *PTRS* **178A** 471–526; see note 1-31

1-33 Thomson: see note 1-25, p 78

1-34 Thomson: see note 1-25, p 93; 1881 On the electric and magnetic effects produced by the motion of electrified bodies *PM* **11** 229–49. The problem was further treated by Oliver Heavyside in 1889 and G F C Searle in 1897. See also Okun L B 1989 The concept of Mass *PT* **42** 31–6

1-35 For an in-depth analysis of these researches, see Whittaker: note 1-19, pp 306–10

1-36 Thomson: see note 1-25, p 94

1-37 Adams was Senior Wrangler in 1843, and remained at Cambridge throughout his life. He was still an undergraduate when he demonstrated (between 1843 and 1845) that the observed perturbation in the orbital motion of Uranus had to be due to the presence of an eighth, undiscovered planet. George Airy, the Astronomer Royal, ignored his prediction. Only when Urbain Le Verrier in France announced a similar result in 1846 did Airy initiate a search at Cambridge Observatory, though the actual discovery of Neptune, based on Le Verrier's figures, was made in Berlin, also in 1846. A bitter controversy over credit for the prediction ensued, and Adam's priority was in due course recognized. Nevertheless, Adams declined both a knighthood and the post of Astronomer Royal after Airy's retirement in 1881.

1-38 Thomson: see note 1-25, p 94; 1883 *A Treatise on the Motion of Vortex Rings* (London: Macmillan). This *Treatise* was Thomson's first book.

1-39 Thomson's affection for Maxwell, or rather his electromagnetism, stemmed from his undergraduate years at Owens College, when he avidly read Maxwell's early papers; his enthusiasm was reinforced in his final year at Owens by his attendence of Arthur Schuster's lectures there on Maxwell's newly published *Treatise*. He was one of three students attending the course!
Schuster A 1911 *The Progress of Physics during 33 Years (1875–1908)* (Cambridge: Cambridge University Press) p 6
Topper D R 1971 Commitment to mechanism: J J Thomson, the early years *AHES* 7 397–8

1-40 Thomson: see note 1-25, p 94. A succinct discussion of Thomson's mathematical works is found in Heilbron J L 1976 Thomson, Joseph John *DSB* **13** 362–72

1-41 Topper: see note 1–31, p 40

Notes section 1.4

1-42 Thomson: see note 1-25, p 95. Thomson notes that Maxwell's own lectures at the laboratory were sparsely attended toward the end. The only lecture by Maxwell he himself heard was Maxwell's 1879 Rede Lecture in the Senate House. That lecture was delivered over the newfangled telephone, whose working 'was demonstrated by an experiment in which a tune was played in the Geological Museum and heard in the Senate House' p 96.

1-43 1873 A voice from Cambridge *NA* **8** 21

1-44 Thomson: see note 1-25, pp 101–2

1-45 Thomson: see note 1-25, p 96

1-46 Lord Rayleigh 1943 *The Life of Sir J J Thomson, OM* (Cambridge: Cambridge University Press) p 15. For an interesting discussion of Maxwell and the Cavendish Laboratory, see Schuster: note 1-39, pp 23–31

1-47 Rayleigh's wife, Evely Balfour, was the sister of the first Earl of Balfour, Arthur James Balfour, prominent British statesman and Prime Minister during 1902–1906.

1-48 Glazebrook and Shaw were among the first to give college lectures in a university building at Cambridge. Both were college lecturers, receiving most of their renumeration from their respective colleges. Until that time, there had been a sharp distinction between a university professor and a college teacher. With the rapid growth of the natural sciences, it became impractical for each college to maintain its own staff and laboratories in all the new subjects, and the plan of giving university recognition to college lecturers gradually evolved. Glazebrook R 1926 The Cavendish Laboratory: 1876–1900 *NA* **118** Supplement p 54

1-49 The first was an inconclusive search for consequences of Maxwell's theory that changes in electric forces in a dielectric produced magnetic forces. The second, on the suggestion of Rayleigh, was of effects in the operation of induction coils by the electrostatic capacity of the primary and secondary of the induction coil. Thomson: see note 1-25, p 97; 1881 On some electromagnetic experiments with open circuits *PM* **12** 49–60

1-50 Thomson: see note 1-25, p 97; 1883 On the determination of the number of electrostatic units in the electromagnetic unit of electricity *PRS* **35** 346–7

1-51 Schuster began his work on an experimental test of Ohm's law for rapidly alternating currents while a visitor of Wilhelm Weber at Göttingen; subsequently he continued it at Manchester. These experiments appeared to indicate that a circuit had different resistances in different directions, at variance with Ohm's law. His announcement of this finding at a meeting of the British Association at Belfast in 1874 caught Maxwell's attention. Before long the anomaly was experimentally discredited at the Cavendish under Maxwell, Schuster's results being an artifact of uneven magnetization in the magnet and coil that generated the alternating current. Feffer S J 1989 Arthur Schuster, J J Thomson, and the discovery of the electron *HSPBS* **20** 35–36

1-52 Lord Rayleigh: see note 1-46, p 18. Most of the apparatus is on display in section 3 of the museum in the modern Cavendish Laboratory in West Cambridge.

1-53 Thomson: see note 1-25, p 98

1-54 Feffer: see note 1-51, p 39, citing Sviedry R 1970 The rise of physical science at Victorian Cambridge *HSPBS* **2** 143

See also, Wise M N and Smith C 1986 Measurement, work and industry in Lord Kelvin's Britain *HSPBS* **17** 147–73, in which the 'factory mentality' or industrial vision of the Victorian age is attributed to Kelvin's 'priority of measurement over mechanical models', p 148

1-55 Glazebrook to Thomson, December 25 1884 (Lord Rayleigh: see note 1-46, p 20)

1-56 Schuster to Thomson, not dated (Lord Rayleigh: see note 1-46, p 21)

1-57 Thomson to Schuster, December 26 1884 (Lord Rayleigh: see note 1-46, p 21)

1-58 Reynolds to Thomson, December 26 1884 (Lord Rayleigh: see note 1-46, p 21)

Allen J 1970 The life and work of Osborne Reynolds *Osborne Reynolds and Engineering Science Today* (Manchester: Manchester University Press) p 5

1-59 Thomson: see note 1-25, p 117

Notes section 2.1

2-1 Whittaker: see note 1-19, vol 1, p 35. I am indebted to Whittaker for much of the present material.

2-2 Bede 1964 *A History of the English Church and People* translation by Sherley-Price L (Baltimore, MD: Penguin) p 38; cited by Whittaker: see note 1-19, p 35

2-3 Gilbert W 1600 *De Magnete, Magneticique Corporibus, et de Magno Magnete Tellure; Physiologia Nova, Plurimis & Argumentic, & Experimentis Demonstratia* (London); 1900 English translation by Thompson S P (London: Chiswick). See also English version Price D J (ed) 1958 (New York: Basic)

2-4 The term *electricity* was introduced by Browne Sir T 1646 *Pseudodoxia Epidemica* (London) p 79

2-5 Dibner B 1957 *Early Electrical Machines* (Norwalk, CT: Burndy Library) p 14

2-6 Cabeo N 1629 *Philosophia Magnetica in qua Magnetis Natura Penitus Explicatur* (Cologne: Ferrara)
Hauksbee F 1705 Several experiments on the attrition of bodies in vacuo *PTRS* **24** 2165–74

2-7 Whittaker: see note 1-19, 37

2-8 Whittaker: see note 1-19

2-9 'sGravesande W J 1720 *Physices Elementa Mathematica Experimentis Confirmata* (Leiden)

2-10 A story has it that Newton delayed Gray's entry into the Charterhouse, an institution devoted to the support of impoverished gentlemen, out of hatred for John Flamsteed, the Astronomer Royal and a friend of Gray; however, it it seems to be without foundation. Heilbron J L 1972 Stephen Gray *DSB* **5** 516

2-11 Gray S 1731–1732 A letter... containing several experiments concerning electricity *PTRS* **37** 18

2-12 Desaguliers J T 1739–1740 Some thoughts and conjectures concerning the cause of electricity *PTRS* **41** 175-85; Some thoughts and experiments concerning electricity *PTRS* **41** 186, 193; Experiments made before the Royal Society, February 2 1737–38 *PTRS* **41** 193–208

2-13 Du Fay C-F 1733–1734 Letter to the Duke of Richmond and Lenox concerning electricity, December 27 1733 A letter from Mons. Du Fay... concerning Electricity *PTRS* **38** 258–66; extract in Magie: see note 1-6, p 399

2-14 Nollet J A 1743–1748 *Leçons de Physique Expérimentale* 6 vols (Paris)

2-15 Ewald Georg von Kleist of Camin in Pomerania independently invented the Leiden jar, a device for storing electricity generated by a Hauksbee-type machine. The jar was of glass, partially filled with water and contained a brass wire projecting through its cork stopper. An experimenter produced static electricity by friction, and used the wire to store it inside the jar. It was improved by William Watson, who coated its inside and outside with metal foil. The Hauksbee machine was an improvement on a device producing frictional electricity first developed by Otto von Guericke. It consisted of a glass globe which was spun rapidly on its axis while rubbed, and equipped with a valve allowing air to be extracted or introduced.

2-16 Nollet J A 1746 Observations sur quelques nouveaux phénomènes d'électricité *Mémoires de l'Académie Royale des Sciences* (Paris) pp 1–23; English translation of extract in Magie: see note 1-6, pp 403–6

2-17 Daintith J, Mitchel S and Tootill E 1986 *A Biographical Encyclopedia of Scientists* (New York: Facts on File) p 598

2-18 Watson W 1750 Some further inquiries into the nature and properties of electricity *PTRS* **45** 104

2-19 Watson, Sir W 1747 *PTRS* **44** 718. Watson also attempted, one year later, to measure the speed of electricity over a 6.4 km circuit. Though failing to yield a finite value, he sensibly concluded that the speed is too high to be measured. Daintith, Mitchel and Tootill: see note 2-17, p 827. Renewed attempts were made to measure the speed soon after Volta announced his pile in 1800. Giovanni Aldini, a nephew of Galvani, stretched a wire across the bay at Calais, with the water completing the closed circuit. A Voltaic pile produced the electric current, and a dissected animal registered the electric shock. His conclusion could, of necessity, be no more quantitative: the current traveled with 'astounding rapidity'. Dibner B 1962 *Ørsted and the Discovery of Electromagnetism* (New York: Blaisdell) p 10

2-20 The glass tube was actually presented to the Library Company of Philadelphia, a company founded by Franklin which collected scientific apparatus in addition to books and journals. Whittaker: see note 1-19, p 46
Heilbron J L 1976 Franklin's physics *PT* **29** 32

2-21 Letter dated June 1 1747 and published in W Watson: see note 2-18, p 98; extract in Magie: see note 1-6, 401. Whittaker quotes similar but not identical wording in a letter to Collinson July 11 1747; Whittaker: see note 1-19, pp 46-47

2-22 Magie: see note 1-6, p 402

2-23 Franklin B 1751 *Experiments and Observations on Electricity, made in Philadelphia in America* (London: Cave)

2-24 Æpinus F U T 1759 *Tentamen Theoriae Electricitatis et Magnetismi* (Saint Petersburg); English translation by Connor P J 1979 *Æpinus's Essay on the Theory of Electricity and Magnetism* (Princeton, NJ: Princeton University Press), introductory monograph and notes by R W Home. For a review of Connor's transaction by Bruce R Wheaton, see 1980 *AJP* **48** 502–3

2-25 Newton Sir I 1739–1742 *Philosophiae Naturalis Principia Mathematica* 3 vols, ed T Le Seur and F Jacquier (Geneva); translation by Motte A 1966, revised and annotated by Cajori F (Berkeley, CA: University of California Press)

2-26 Priestly J 1767 *The History and Present State of Electricity, with Original Experiments* (London) p 732

2-27 Robinson J 1882 *A System of Mechanical Philosophy* 4 vols, ed D Brewster, vol 4 (Edinburgh: Murray) pp 73–4

2-28 Concern over what to do about Cavendish's unpublished papers was raised by members of the Royal Society within weeks of Cavendish's death, including Sir Joseph Banks, president of the Society, and Lord George Cavendish, principal heir to Cavendish. They remained problematic for years to come, partly due to their great state of disorder, although they were examined off and on by, among others, William Snow Harris and Lord Kelvin. Maxwell expended prodigious energy on sorting, analyzing, editing, and publishing them. Russell McC 1988 Henry Cavendish on the theory of heat *Isis* **79** 37–8
J Clerk Maxwell (ed) 1879 *The Unpublished Electrical Writings of the Honorable Henry Cavendish* (Cambridge: Cambridge University Press)

2-29 Memoirs presented to the French Academy 1788 *Mémoires de l'Académie Royale des Sciences*; English translation of extracts in Magie: see note 1-6, pp 408–20

2-30 Magie: see note 1-6, p 411

2-31 Magie: see note 1-6, p 413

2-32 Symmer R 1759 Of two distinct powers in electricity *PTRS* **51** 371–89

2-33 Whittaker: see note 1-19, vol 1, p 58

Notes section 2.2

2-34 Whittaker: see note 1-19, vol 1, p 32

2-35 Sarton G 1947 Magnetism: the discovery of the compass *Science and Learning in the Fourteenth Century (Introduction to the History of Science, III)* (Baltimore, MD: Williams and Wilkins) pp 714–6

2-36 A fact known to the Chinese. Needham J 1969 *The Grand Titration: Science and Society in East and West* (London: Allen and Unwin)

2-37 *Epistola Petri Peregrini de Maricourt ad Sygerum de Foucaucourt, De Magnete*

2-38 Translated excerpts by E Fleury Mottelay in Magie: see note 1-6, pp 387–8

2-39 *Principia* book 3, prop 6, corr 5, cited by Whittaker: see note 1-19, p 55

2-40 Michell J 1750 *A Treatise of Artificial Magnets; in which is shown an Easy and Expeditious Method of Making them Superior to the Best Natural Ones* (London) p 17, cited by Whittaker: see note 1-19, p 56

2-41 Magie: see note 1-6, p 417

2-42 Volta A 1800 On the electricity excited by the mere contact of conducting substances of different kinds *PTRS* **90** 403–31; extract in Magie: see note 1-6, pp 427–31

2-43 Napoléon proposed to his minister that a prize be established and awarded each year to one who contributed experiments or discoveries in electricity or magnetism comparable to those of Franklin and Volta.

2-44 Nicholson W 1800 Account of the new electrical or galvanic apparatus of Sig. Alex. Volta, and experiments performed with the same *Journal of Natural Philosophy, Chemistry and the Arts* **4** 179; extracts in Magie: see note 1-6, pp 432–6. See also Nicholson W, Carlisle A and Cruickshank W 1800 Experiments on galvanic electricity *PM* **7** 337–50. Nicholson founded and published his journal, of which 41 volumes were published, and for which he is best remembered. The journal was eventually ingested by Alexander Tilloch's competing *Philosophical Magazine* which flourishes to this day

2-45 Nicholson, Carlisle and Cruickshank: see note 2-44, p 337

2-46 Wollaston W H 1801 *PTRS* **91** 427

2-47 Martin T 1961 *The Royal Institution* (London: The Royal Institution) p 19. Although, as noted, France and England were at war, the French Institute awarded him the medal and prize instituted by Bonaparte. Coleridge, the poet and a friend of the versatile Davy, is reputed to have said that if he had not been the first chemist he would have been the first poet of his age, p 20.

2-48 Davy H 1807 The Bakerian Lecture, on some chemical agencies of electricity *PTRS* **97** 1–56 (read November 20 1806)

2-49 Hall A R 1954 *The Scientific Revolution 1500–1800: the Formation of the Modern Scientific Attitude* (New York: Longmanns Green) pp 361–2

2-50 Davy's cousin Edmund, his assistant in these experiments, has described him dancing in ecstacy in the laboratory on the appearance of minute globules of bright metal bursting through the crust of potash. Martin: see note 2-47, p 16

2-51 In 1801 Ritter discovered ultraviolet rays through their darkening of silver chloride. Ritter J W 1801 Von den Herres Ritter und Rickmann *APC* **7** 527–8; 1803 Versuch aber das Sonnenlichs *APC* **12** 409–15

2-52 Quoted in K Meyer's 1920 essay H C Ørsteds Arbejdsliv i det danske Samfund (H C Ørsted's varied activities in the Danish community) vol 3, p xvii of *H C Ørsted, Naturvidenskabelige Skrifter* 3 vols, ed K Meyer (Copenhagen: Royal Danish Academy of Sciences and Letters) from Ostwald W 1904 *Vorträge u Reden* (Leipzig) p 368

2-53 Stauffer R C 1957 Speculation and experiment in the background of Ørsted's discovery of electromagnetism *Isis* **48** 35, from Schelling F W J 1856–1861 *Sämmtliche Werche* vol 3 ed K F A Schelling (Stuttgart and Augsburg: Cotta) p 320
2-54 Stauffer: see note 2-53, p 41, quoting letter by Ritter to Ørsted in Harding M C (ed) 1920 *Correspondance de H C Ørsted avec divers savants* vol 2 (Copenhagen: Aschehoug & Co.) pp 35–6
2-55 Dibner: see note 2-19, quoting Mottelay P F 1922 *Biographical History of Electricity and Magnetism* (London: Charles Griffith) p 376
2-56 Stauffer: see note 2-53, citing Young T 1807 *A Course of Lectures on Natural Philosophy* vol 1 (London: Johnson) p 694
2-57 Stauffer: see note 2-53, p 43, citing J B Biot at the end of his article on magnetism in Brewster D 1819 *Edinburgh Encyclopedia* vol 13, p 277
2-58 Gautherot N 1801 Mémorie sur le Galvanisme, lu à la classe des sciences de l'Institut, le 26 ventóse an 9 *ACP* **39** 203–19
2-59 Ørsted: see note 2-52, vol 2, p 356
2-60 Ørsted H C 1812 *Ansicht der chemischen Naturgesetze durch die neueren Entdeckungen gewonnen* (Berlin: In der Realschulbuchhandlung)
2-61 Ørsted H C 1813 *Recherches sur l'Identité des Forces Chimiques et Électriques* (Paris)
2-62 Ørsted: see note 2-52, vol 2 pp 223–5, i.e. 1821 Betrachtungen ueber den Electromagnetismus *Journal für Chemie und Physik* **32** 199–231. French version: 1821 Considerations sur l'électro-magnétisme *Journal de Physique* **93** 161–80. Ørsted wrote both the German and French versions of this article and, on examining them both, Robert C Stauffer has provided a fresh translation, from which the extract is quoted. Stauffer: see note 2-53, p 45
2-63 *Experimenta Circa Effectum Conflictus Electrici in Acum Magneticam*; translated as *Experiments on the Effect of a Current of Electricity on the Magnetic Needle* in Dibner: see note 2-19, pp 71–6. The statement circulated that Ørsted's discovery was wholly accidental originated with Gilbert, editor of *Annalen der Physik* who was unable to fully follow Ørsted's original Latin text.
2-64 Christopher Hansteen to Michael Faraday, December 30 1857; quoted in Ørsted: see note 2-52, vol 2, pp 356–8
2-65 Dibner: see note 2-19, p 75
2-66 Dibner: see note 2-19, p 75
2-67 Ørsted H C 1820 *Journal für Chemie und Physik* **29** 364–69
2-68 Biot J B and Savart F 1820 Note sur le Magnétisme de la pile de Volta *ACP* **15** 222–3; translated extract in Magie: see note 1-6, pp 441–2
2-69 Biot and Savart: see note 2-68
2-70 Arago D-F J 1820 *ACP* **15** 93
2-71 Ampère A-M 1820 Mémoire sur l'action mutuelle de deux courants électriques *ACP* **15** 59; translated extract in Magie: see note 1-6, 447–60. This was the first of five papers on the mutual action of conductors carrying electric currents.
2-72 Ampère A-M 1825 *Mémoire de l'Académie* **6** 175
2-73 Maxwell: see note 1-11, vol 2, article 528, p 175

Note section 3.1

3-1 Cohen I B 1972 Benjamin Franklin *DSB* **5** 129–39
3-2 Cohen: see note 3-1

3-3 Heilbron: see note 2-20, p 36
3-4 Hauksbee F 1719 *Physico-Mechanical Experiments on Various Subjects* 2nd edn (London: Senex and Taylor) p 9, quoted by Guerlac H 1972 Francis Hauksbee *DSB* **6** 169
3-5 Guerlac: see note 3-4, p 170
3-6 Hauksbee: see note 2-6, p 2167
3-7 Hauksbee: see note 2-6, p 2170
3-8 Hauksbee: see note 2-6, p 2172
3-9 Hauksbee: see note 2-6, p 2173
3-10 Watson: see note 2-18
3-11 Watson: see note 2-18
3-12 Whittaker: see note 1-19, vol 1, p 349, quoting Nollet J A 1749 *Recherches sur les Causes Particulières des Phénomènes Électriques* (Paris) p 44
3-13 Chalmers T W 1952 Conduction of electricity through gases *Historic Researches: Chapters in the History of Physical and Chemical Discovery* (New York: Scribner) ch 10. Another excellent introduction to the subject is found in Brown S C 1966 *Introduction to Electrical Discharges in Gases* (New York: Wiley)
3-14 Andrade E N da C 1959 The history of the vacuum pump *Vacuum* **9** 45 See also T E Madey and W C Brown (eds) 1984 *History of Vacuum Science and Technology; a Special Volume Commemorating the 30th Anniversary of the American Vacuum Society, 1953–1983* (New York: American Institute of Physics). The term 'pump' for such devices, instead of, say, rarefied gas compressor, is due to their early modification of existing water pumps. Hablanian M H Comments on the history of vacuum pumps *History of Vacuum Science and Technology; a Special Volume Commemorating the 30th Anniversary of the American Vacuum Society, 1953–1983* p 19
3-15 Chalmers: see note 3-13, p 189
3-16 Chalmers: see note 3-13, p 189

Notes section 3.2

3-17 Pierre Louis Dulong, the discoverer of nitrogen chloride, was seriously injured by nitrogen chloride, and Faraday, too, in due course, was slightly hurt by the explosive substance
3-18 Martin: see note 2-47, p 22
3-19 Williams L P 1971 Michael Faraday *DSB* **4** 527–40
James F A J L 1992 The tales of Benjamin Abbott: a source for the early life of Michael Faraday *The British Journal for the History of Science* **25** 229–40
3-20 The portable laboratory or chest, vulgarized in the form of a chemical amusement chest, became quite popular in the first decades of the 19th century. Gee B 1989 Amusement chests and portable laboratories: practical alternatives to the regular laboratory *The Development of the Laboratory: Essays on the Place of Experiment in Industrial Civilization* ed F A J L James (New York: American Institute of Physics) ch 3
3-21 Martin: see note 2-47, p 24. Lady Davy was unable to rid herself of the notion that Faraday was merely a servant of her husband.
3-22 Faraday refused all sorts of offers, including a government pension, twice the offer of presidency of the Royal Society, and, of all things, a knighthood.
3-23 Faraday M 1839–1855 On some new electro-magnetical motions, and on the theory of magnetism *Quarterly Journal of Science*. This paper

also contains the seed of Faraday's 'line of magnetic force' around a current-carrying conductor—a concept entirely new and different from Ampère's rigorous mathematical description of electromagnetism. Faraday collected his papers in four volumes: 1839–1855 *Experimental Researches in Electricity* 3 vols (London) and 1859 *Experimental Researches in Chemistry and Physics* (London: R and J E Taylor). He also kept meticulous notes of his daily observations, which have been preserved in 1932–1936 *Faraday's Diary, Being the Various Philosophical Notes of Experimental Investigations Made by Michael Faraday* 7 vols and index, ed T Martin (London: Bell)

3-24 This neglect may have been the reason why Davy thought him unqualified for Fellowship in the Royal Society.

3-25 Extract in Magie: see note 1-6, p 479. The page in Faraday's diary recording this memorable experiment is reproduced by Martin: see note 2-47, p 33. The iron ring, and well-nigh all of Faraday's apparatus, are preserved in Faraday's still extant laboratory in the basement of the Royal Institution

3-26 Despite his conviction of some great unity in nature, Faraday advised against presentation of Joule's first definitive paper on the mechanical equivalent of heat at the annual meeting of the British Association at Cork, Ireland, in 1843. Dahl P F 1972 *Ludvig Colding and the Conservation of Energy Principle: Experimental and Philosophical Contributions* (New York: Johnson) p xxiv

3-27 Magie: see note 1-6, pp 495–8

3-28 Stoney G J 1881 On the physical units of nature *PM* **11** 381
von Helmholtz H 1928 On the modern development of Faraday's conception of electricity *Faraday Lectures, 1869–1928* (London) p 145

3-29 Pais A 1986 *Inward Bound: On Matter and Forces in the Physical World* (Oxford: Clarendon and New York: Oxford University Press) pp 71–2

3-30 Williams: see note 3-19, p 535
Faraday M 1851 On the physical lines of magnetic force *Proceedings of the Royal Institution*; extract in Magie: see note 1-6, pp 506–11. A typical letter from Faraday to his friend Richard Phillips in November of 1831 shows how closely the electrotonic state, electromagnetic induction, and electrochemistry were interrelated in Faraday's thinking. Faraday to Phillips, November 29 1831, cited by Crawford E 1985 Learning from experience *Faraday Rediscovered: Essays on the Life and Work of Michael Faraday 1791–1867* ed D Gooding and F A J L James (New York: Stockton) ch 11. Richard Phillips, an editor of the *Philosophical Magazine*, persuaded Faraday to take up the subject of electromagnetism, by reviewing and making sense out of recent work in this field.

3-31 Magie: see note 1-6, p 500

3-32 Chalmers: see note 3-13, p 193. Sir William Snow Harris's electrical contributions were mainly in designing lightning conductors for naval vessels, both British and Russian. Harris received the Copley Medal in 1835 for his paper on the 'Laws of Electricity by High Tension', and was knighted in 1847.

3-33 Chalmers: see note 3-13, p 194. Teyler's Museum, Haarlem, was organized entirely by van Marum.

3-34 Wheatstone C 1833 Remarks on one of Mr Talbot's proposed philosophical experiments *PM* **3** 204–5. Wheatstone, Professor of Experimental Philosophy at King's College, London, is most readily identified with the bridged instrument that bears his name; the Wheatstone bridge was actually invented by Samuel Hunter Christie in 1833.

3-35 Chalmers: see note 3-13
Faraday *Experimental Researches*: see note 3-23, vol 1 pp 490–6

Notes section 3.3

3-36 Burau W 1975 Plücker, Julius *DSB* **11** 44–7

3-37 Williams: see note 3-19, p 413. Plücker's letters to Faraday are found among the letters to Faraday in the archives of the Institution of Electrical Engineers; those from Faraday to Plücker are in the possession of the National Research Council of Canada.
Williams L P 1965 *Michael Faraday: a Biography* (New York: da Capo) p 460

3-38 Faraday to Plücker, July 27 1857; cited in Williams: see note 3-37, p 502

3-39 Kangro H 1972 Geissler, Johann Heinrich Wilhelm *DSB* **5** 340–1

3-40 Kangro: see note 3-39, p 340

3-41 Shiers G 1971 The induction coil *Scientific American* **224** 80–7

3-42 Finn B S 1975 Rühmkorff, Heinrich Daniel *DSB* **11** 603–4. One problem remained with Rühmkorff's coil: it was prone to contact wear from sparking between the breaker contacts. The solution, a condenser shunted across the contacts, was provided by the French physicist Armand Fizeau, better known for his measurement of the speed of light with a system of mirrors. Several years later, Fizeau's colleague Jean Bernard Léon Foucault, whose ubiquitous pendulum is a constant reminder of the rotation of the Earth, introduced another improvement in the mercury interrupter operated by an automatic hammer break. Shiers: see note 3-41, p 83

3-43 Andrade: see note 3-14, p 45

3-44 Plücker J 1858 Ueber die Einwirkung des Magnets auf die elektrischen Entladungen in verdünnten Gasen *APC* **103** 88–106; reprinted in Pockels F 1892 *Julius Plückers Gesammelte Physikalische Abhandlungen* (Leipzig: Teubner) p 475; transl. Guthrie F 1858 On the action of the magnet upon the electrical discharge in rarefied gases *PM* **16** 119–35

3-45 Plücker transl. Guthrie: see note 3-44, p 119

3-46 Plücker transl. Guthrie: see note 3-44, p 126

3-47 Plücker transl. Guthrie: see note 3-44, p 126

3-48 Plücker transl. Guthrie: see note 3-44, p 130

3-49 Chalmers: see note 3-13, p 198

3-50 Plücker 1858 Observations on the electrical discharge through rarefied gases *PM* **16** 409

3-51 Plücker: see note 3-50, p 410

3-52 Plücker: see note 3-50, p 410

3-53 Drennan O J 1972 Hittorf, Johann Wilhelm *DSB* **6** 438–40

3-54 Hittorf J W 1869 Ueber die Elektricitätsleitung der Gase *APC* **136** 1-3; extract in Magie: see note 1-6, pp 562–3; 1869 Ueber die Elektricitätleitung der Gase *APC* **136** 197–234

3-55 Magie: see note 1-6, p 563

3-56 Plücker transl. Guthrie: see note 3-44, p 131

3-57 Anderson D L 1972 Goldstein, Eugen *DSB* **5** 458–9

3-58 Goldstein E 1876 Vorläufige Mittheilungen über elektrische Entladungen in verdünnten Gasen *Monatsbericht der Könglich Akademie der Wissenschaften zu Berlin* pp 279–95

3-59 Goldstein: see note 3-58, p 286

3-60 Goldstein: see note 3-58, p 285

3-61 Keller A 1983 *The Infancy of Atomic Physics: Hercules in his Cradle* (Oxford: Clarendon) p 40

Notes section 4.1

4-1 Thomson: see note 1-25, p 384. De La Rue W and Müller H W 1878 Experimental researches on the electric discharge with the chloride of silver battery; Part I—The discharge at ordinary atmospheric pressure *PTRS* **169** 55–121 (submitted August 23 1877); Part II—The discharge in exhausted tubes *PTRS* **169** 155–242 (submitted April 10 1878); Part III—Tube potentials; potential at a constant distance and various pressures; nature and phenomena of the electric arc *PTRS* **171** 65–116 (submitted August 7 1879). De La Rue succeeded Henry Bence Jones as Secretary of the Royal Institution.

4-2 Quoting Thomson: see note 1-25, p 385
Spottiswoode W and Moulton J F 1879 On the sensitive state of electrical discharges through rarefied gases *PTRS* **170** 165–229; 1880 On the sensitive state of vacuum discharges *PTRS* **171** 561–652

4-3 Grove graduated from Oxford in 1835, after which he became a barrister. However, ill health caused him to switch to physics, specializing in electrochemistry. In 1839, he produced an improved version of the Voltaic battery which became quite popular as the *Grove cell*. He also devised the first fuel cell, and later invented the first electrical filament lamp. Among other things, he was an early espouser of the conservation of energy principle. In 1840, he was elected to the Royal Society, and the following year was appointed professor of physics at the Royal Institution. Soon thereafter, he returned to the legal profession, rising to a high judgeship, while still maintaining his scientific interests as well.

4-4 Grove W R 1852 On the electro-chemical polarity of gases *PTRS* **142** p 100–1. The striated discharge was subsequently observed in Paris by Rühmkorff, and again by Masson, Du Moncel, Quet, and other Continental electricians, and by the Rev Dr Robinson in Ireland.

4-5 Grove: see note 4-4, p 96

4-6 Morse E W 1972 Gassiot, John Peter *DSB* **5** 292–3

4-7 Gassiot J P 1858 On the stratification and dark bands in electrical discharges as observed in the Torricellian vacua *PTRS* **148** 1-16, p 13. This paper had the distinction of being the Royal Society's Bakerian Lecture for 1858, read on March 4 of that year.

4-8 Gassiot: see note 4-7, p 15

4-9 Faraday: see note 3-23, vol 7, p 412, paragraph 1

4-10 Gassiot J P 1859 On the stratification in electrical discharges observed in Torricellian and other vacua—second communication *PTRS* **149** 137

4-11 Gassiot J P 1860 On the electrical discharge *in vacuo* with an extended series of the Voltaic battery *PRS* **10** 36–7

4-12 Gassiot: see note 4-10

4-13 Gassiot J P 1860 On the interruption of the Voltaic discharge *in vacuo* by magnetic force *PRS* **10** 269–74

4-14 Gassiot J P 1861 *British Association Report* section 2, pp 38–9

4-15 Grove: see note 4-4, p 100

4-16 Keller: see note 3-61, p 40

4-17 Varley C F 1871 Some experiments on the discharge of electricity through rarefied media and the atmosphere *PRS* **19** 236

4-18 Varley: see note 4-17, p 239

4-19 Varley: see note 4-17, pp 236–43
4-20 Varley: see note 4-17, pp 239–43. The italics are Varley's.
4-21 Varley: see note 4-17, p 240
4-22 Andrade: see note 3-14, p 46
4-23 Sprengel was born near Hanover in 1834. He eventually became an expert on safety explosives.

Notes section 4.2

4-24 Fournier d'Albe E E 1923 *The Life of Sir William Crookes OM FRS* (London: Fisher Unwin)
Brock W H 1971 Crookes, William *DSB* **3** 474–82
4-25 Brock: see note 4-24, p 474
4-26 For the protracted scientific correspondence between Stokes and Crookes, see 1907 *Memoir and Scientific Correspondence of the Late Sir George Gabriel Stokes* selected and arranged by Joseph Larmor, vol 2 (Cambridge: Cambridge University Press) pp 362–94. The correspondence between the two, over the period 1856–1901, occupies 132 pages. Sometimes there were two letters per day.
4-27 Strutt R J 1924 Fourth Baron Rayleigh *Life of John William Strutt* (London: University of Wisconsin Press) p 265, quoted by Thomson: see note 1-25, p 380. See also Oppenheimer J 1986 Physics and psychic research in Victorian and Edwardian England *PT* 62–70; Gorden Stein 1993 *The Sorcerer of Kings: The case of Daniel Dunglas Home and William Crookes* (Buffalo, NY: Prometheus)
4-28 Thomson: see note 1-25, p 383
4-29 Brock: see note 4-24, p 476
4-30 Maxwell: see note 1-11. Radiation pressure is discussed under 'Energy and Stress of Radiation' vol 2, articles 792, 793, pp 440–1
4-31 Crookes W 1874–1875 On the attraction and repulsion resulting from radiation *PRS* **23** 373–8. The bulk of Crookes's apparatus, specimens, and manuscripts are preserved at the Science Museum in London.
Greenway F G 1962 A Victorian scientist: the experimental researches of Sir William Crookes *Royal Institution Proceedings* **39** 198
4-32 Quoted by Albe: see note 4-24, p 255
4-33 Reynolds O 1876 On the forces caused by the communication of heat between a surface and a gas *PTRS* **165**
Gibson A H 1946 *Osborne Reynolds and his Work in Hydraulics and Hydrodynamics* (London: Longmans) p 28
4-34 Stokes to Crookes: see note 4-26, p 373
4-35 Thomson: see note 1-25, pp 373–4
Schuster A 1876 On the nature of the force producing the motion of a body exposed to rays of heat and light *PRS* **24** 391–2; 1877 *PTRS* **166** 715–24; 1876 *PM* **2** 313–4. A quantitative theoretical explanation of the radiometer behavior proved stubbornly elusive, even for Maxwell.
Brush S G and C W F Everitt 1969 Maxwell, Osborne Reynolds, and the radiometer *HSPS* **1** 105–25. The pressure of radiation was first detected during 1900–1901 by P N Lebedev (1866–1911) in Russia and by E F Nichols (1869–1924) and G F Hull (1870–) at Dartmouth College in America. If the gas pressure in the radiometer is of the order of 0.02 to 0.03 mm Hg, the net force experienced by the vane is directed towards the source of the incoming radiation; if the pressure is reduced to what

we nowadays consider a high vacuum, the thrust is oppositely directed and much smaller.
Worthing A G and D Halliday 1948 *Heat* (New York: Wiley) and (London: Chapman and Hall) pp 423–4

4-36 Crookes W The Bakerian Lecture 1879 On the illumination of lines of molecular pressure, and the trajectory of molecules *PTRS* **170** part I 135–64, p 135; extract in Magie: see note 1-6, pp 564–76. The unusual title may have been modeled on Faraday's title of his announcement of the rotary polarization of light in a magnetic field: 'The illumination of lines of magnetic force and the magnetisation of light'; Albe: see note 4-24, p 279

4-37 Magie: see note 1-6, pp 575–6

4-38 Magie: see note 1-6, p 576

4-39 Alice Bird to Crookes, April 1879, cited by Albe: see note 4-24, pp 283–4

4-40 Crookes W 1879 Radiant matter *Chemical News* **40** 91–3, 104–7, 127–31, 130–1; 1879 *NA* **20** 419–23, 436–40, 439–40; 1879 *American Journal of Science* **18** 241–62

4-41 Magie: see note 1-6

4-42 Magie: see note 1-6, p 565. The U-shaped tube was suggested by Stokes in a letter in 1878: 'What would take place in a small U-shaped tube, so that a good region around the positive should be well out of the line of fire from the negative, when the exhaustion was of this very high order? Would the positive branch be quite dark? Perhaps you have tubes which enable you to answer this question at once'. Stokes to Crookes: see note 4-26, p 413

4-43 Stokes to Crookes: see note 4-26, pp 568–70

4-44 Crookes: see note 4-40, p 253

4-45 Magie: see note 1-6, 571

4-46 Stokes to Crookes: see note 4-26, p 420

4-47 Crookes 1879 On the illumination of lines of molecular pressure, and the trajectory of molecules *PM* **7** 60

4-48 In 1880 Crookes moved from Mornington Road into a new and very large house in Kensington Park Gardens. It is said to have been the first house in England to be lighted by electricity, though Lord Salisbury's Hatfield House was also one of the first private houses in England to be equipped with a telephone and electric lights–both installations improvised by the owner whose hobby lay in science.

4-49 Tait P 1880 Note on the velocity of gaseous particles at the negative pole of a vacuum tube *Proc. Royal Society of Edinburgh* **10** 430

4-50 Goldstein E 1880 On the electric discharge in rarefied gases—part II. On luminous phenomena in gases caused by electricity *PM* **10** p 244; originally in 'Ueber die Entladung der Elektricität in verdünnten Gasen *Monatsberichte der Königlischen Akademie der Wissenschaften zu Berlin* (January 1880) pp 82–124

4-51 1800 Ueber die Entladung der Elektricität in verdünnten Gasen *Monatsberichte der Königlischen Akademie der Wissenschaften zu Berlin* p 241

Notes section 5.1

5-1 McCormach R 1972 Heinrich Rudolf Hertz *DSB* **6** 340–50

5-2 McCormach: see note 5-1, p 341

5-3 Hertz's work on Maxwell's electrodynamics had its start at Kiel. Earlier Helmholtz had, in fact, suggested that he undertake an experimental investigation of the critical assumptions underlying Maxwell's theory: in particular, of the relationship between electromagnetic forces and the dielectric polarization of insulators. Hertz declined the invitation at the time, feeling the experimental difficulties were too formidable. In the event, the lack of a physics laboratory at Kiel led him to publish several theoretical papers there, including one on Maxwell's electrodynamics. However, at Karlsruhe the physical cabinet was amply equipped for him to tackle a problem related to the one suggested by Helmholtz (see note 5-15, below) and by the end of 1888 he had performed his celebrated experiment confirming Maxwell's prediction of electromagnetic waves in the æther.

5-4 Hertz H 1883 Versuche übėr die Glimmentladung *APC* **19** 782–816

5-5 Hertz: see note 5-4, p 790

5-6 Hertz: see note 5-4, p 805

5-7 Hertz: see note 5-4, p 816

5-8 Hertz: see note 5-4, p 814. Buchwald has dealt with Hertz's conclusion as a case of the historiography of experiment, and shows the pitfalls of viewing his experiment in hindsight, without taking into account 'retrospective guidance provided by restraint'. Hertz assumed *a priori*, Buchwald maintains, that cathode rays were a new phenomenon, and Hertz showed that their behavior differed significantly from known electric and magnetic effects. Whatever their real nature, as seen within the context of his own scientific circumstances and traditions, the rays 'were not electric currents'. Buchwald J Z 1995 Why Hertz was right about cathode rays *Scientific Practice: Theories and Stories of Doing Physics* ed J Z Buchwald (Chicago: University of Chicago Press) pp 151–69

5-9 Helmholtz H von 1881 On the modern development of Faraday's conception of electricity (Faraday Lecture, delivered in the Royal Institution on April 5 1881, *Chemical Society (London) Journal* **39** 277–304; reprinted in Nye: see note 1-10, p 266. Neither was the German wave hypothesis accepted by Gustav Wiedemann or Emil Wiechert.

5-10 Schuster, stating that he was 'assured' on this point by Helmholtz. Schuster: see note 1-39, p 55

5-11 Cited by Keller: see note 3-61, p 47

5-12 Goldstein E 1886 Über eine noch nicht untersuchte Strahlungsform an der Kathode inducirter Entladungen *Sitzungsberichte der Königlichen Akademie der Wissenschaften zu Berlin* pp 693, 694, 697; extract in Magie: see note 1-6, pp 576–8

5-13 Goldstein: see note 5-12, p 696

5-14 Goldstein: see note 5-12, pp 698–9

5-15 The prize competition was for the experimental verification of the relation between electromagnetic action and the polarization of a dielectric. At Karlsruhe, Hertz decided to forego this experiment in favor of another he felt was important in deciding between Maxwell's concept of the electromagnetic field and the action-at-a-distance theories of Franz Neumann and Wilhelm Weber, namely, whether empty space behaves like other dielectrics. He proposed to establish this by measuring the speed of electromagnetic waves in air—a medium well approximating free space—which, according to Maxwell, should equal the speed of light. Mulligan J F 1989 Heinrich Hertz and the development of physics *PT* **42** 52

5-16 Hertz H 1888 Ueber Strahlen elektrischer Kraft, originally published in *Sitzungsberichte der Berliner Akademie der Wissenschaften*; 1889 *APC* **36** 769.

The citation is found in the translation (by D E Jones) in Magie: see note 1-6, p 561

5-17 Magie: see note 1-6, p 554–61

5-18 McCormach: see note 5-1, pp 342–3

5-19 Hertz H 1892 Ueber den Durchgang der Kathodenstrahlen durch dünne Metallschichten *APC* **45** 28–32

5-20 Hertz: see note 5-19, p 29

5-21 Hertz: see note 5-19, p 29

5-22 Hertz: see note 5-19, p 28

5-23 Hermann A 1973 Lenard, Philipp *DSB* **8** 180–3

5-24 Lenard P 1943 Erinnerungen eines Naturforschers, der Keiserreich, Judenherrschaft, und Hitler erlebt hat, unpublished manuscript, cited by Wheaton B R 1978 Philipp Lenard and the photoelectric effect, 1889–1911 *HSPS* **9** 304

5-25 Lenard P 1967 On cathode rays *Nobel Lectures including Presentation Speeches and Laureates' Biographies; Physics 1901–1921* (Amsterdam: Elsevier) p 107

5-26 Lenard: see note 5-25, pp 107–8

Notes section 5.2

5-27 Lenard: see note 5-25, p 108

5-28 Lenard P 1894 Über Kathodenstrahlen in Gasen von atmosphärischem Druck und im äussersten Vakuum *Sitzungsberichte der Berliner Akademie der Wissenschaften; APC* **51** 225–67; reprinted in Lenard P 1944 *Wissenschaftliche Abhandlugen, band 3, Kathodenstrahlen, Elektronen, Wirkungen ultravioletten Lichtes* (Leipzig: Hirzel)

5-29 Lenard P 1894 *APC* **51** pp 229–30; see note 5-25, pp 109–10

5-30 Lenard: see note 5-25, p 110

5-31 Lenard: see note 5-28; see note 5-25, p 113

5-32 Lenard: see note 5-25, p 113

5-33 Lenard: see note 5-28, p 248; see note 5-25, p 114
In retrospect, Lenard denies that he 'had concluded beforehand that cathode rays were 'waves in the aether' (see note 5-25), though in his conclusion to his paper of 1894 he implies as much: see note 5-28, pp 266–7

5-34 Hertz's collected scientific works occupy three volumes, edited by Philipp Lenard.
1895 *Schriften vermischten Inhalts* (Leipzig: Johann Ambrosius Barth) translated by D E Jones and G A Schott 1896 as *Miscellaneous Papers* (London: McMillan). For a contemporary discussion of the volume, see Fitzgerald G F 1896 Hertz's miscellaneous papers *NA* **55** 6–9
1892 *Untersuchungen über die Ausbretung der elektrischen Kraft* (Leipzig: Johann Ambrosius Barth) translated by D E Jones 1893 as *Electric Waves* (London: Macmillan); 1962 paperback reprint of the 1893 English edn (New York: Dover); 2nd German edition published in 1894
1894 *Die Prinzipien der Mechanik, in neuen Zusammenhange dargestellt* (Leipzig: Johann Ambrosius Barth) translated by D E Jones and J T Walley 1899 as *The Principles of Mechanics, Presented in a New Form* (London); 1956 reprint of 1899 English edn (New York: Dover)

5-35 Lenard P 1933 Heinrich Hertz *Great Men of Science* translated by H S Hatfield (New York: Macmillan) pp 358–71

5-36 Lenard to Max Wolf, January 27 1894, in Anon 1957 Hertz und Lenard *Physikalische Blätter* **13** 567–9. In his letter to Wolf, Lenard reminisces about Hertz, his relationship with his assistants, and his painful last days.

5-37 Wheaton: see note 5-24

5-38 Hertz H 1887 Ueber einen Einfluss des ultravioletten Lichtes auf die elektrische Entladung *APC* **31** 983–1000

5-39 See, for example, Hallwachs W 1888 Ueber den Einfluss des Lichtes auf elektrostatisch geladene Körper *APC* **33** 301–12; translated in Magie: see note 1-6, pp 578–9

5-40 Wheaton: see note 5-24, p 302

5-41 Lenard P and Wolf M 1889 Zerstäuben der Körper durch das ultraviolette Licht *APC* **37** 443–56

5-42 Lenard: see note 5-25, p 122

5-43 Wheaton: see note 5-24, p 303

5-44 Lenard P 1899 Erzeugung von Kathodenstrahlen durch ultraviolettes Licht *Sitzungsberichte der Kaiserl. Akademie der Wissenschaften zu Wien*; reprinted 1900 *AP* **2** 359–75; also reprinted in *Wissenschaftliche Abhandlungen*: see note 5-28

5-45 Lenard: see note 5-25, p 115
In fact, Röntgen did not stress the role of platinum, but the importance of higher vacuum and higher voltages in maximizing the penetrating power of the rays. Etter L E 1946 Some historical data relating to the discovery of the Roentgen rays *American Journal of Roentgenology and Radium Therapy* **56** 229

5-46 Lenard to Röntgen, May 7 1894. English translation in Etter: see note 5-45, p 222. This is one of several letters between Lenard and Röntgen in this period, preserved at the Physical Institute at Würzburg; Etter: see note 5-45, p 221.
See also Zehnder L 1935 *W C Röntgen. Briefe an L Zehnder* (Zürich: Racher)

5-47 Lenard to Röntgen: see note 5-46. Röntgen received the Lenard tube in May 1894.

5-48 Lenard: see note 5-25, p 115
O Glasser, Röntgen's biographer, claims Röntgen used a Hittorf–Crookes tube when the discovery was made. Glasser: see note 1-1, p 3
In his 'Ueber eine neue Art von Strahlen' Magie: see note 1-6, p 600, Röntgen writes 'If the discharge of a fairly large induction coil be made to pass through a Hittorf vacuum tube, or through a Lenard tube, a Crookes tube, or other similar apparatus... there is observed at each discharge a bright illumination of a paper screen covered with barium platino-cyanide, placed in the vicinity of the induction coil'.
According to Etter, the very sequence in which the names are listed is given as evidence by the physics faculty at Würzburg of the order in which Röntgen first observed x-rays. Etter: see note 5-45, p 230

5-49 Etter: see note 5-45, p 224
Half of the Vienna Aacademy's Baumgartner Prize for 1896 went to Lenard, the other half to Röntgen. The Royal Society awarded Rumford Medals to both the same year and the Paris Academy's La Caze Prize of 1897 was awarded to Lenard for physics and to Röntgen for physiology. Wheaton: see note 5-24, p 307

5-50 Freund F 1946 Lenard's share in the discovery of x-rays *British Journal of Radiology* p 19

5-51 Lenard: see note 5-25, p 115
Etter: see note 5-45, p 222, claims that Lenard is reported to have said, at about the time of his Nobel Prize, 'that anyone who was wide awake

and using a Lenard tube could have discovered the x-rays'.

5-52 Schuster: see note 1-39, pp 77–8

5-53 Magie: see note 1-6, p 604

5-54 Magie: see note 1-6, p 605

Notes section 6.1

6-1 Kargon R H 1975 Schuster, Arthur *DSB* **12** 237–9
Simpson G C 1935 Sir Arthur Schuster 1851–1934 *Obituary Notices of Fellows of the Royal Society, 1932–1935* (London: Harrison) pp 409–23
Crowther J G 1970 Arthur Schuster 1851–1934 *Scientific Types* (Chester Springs, PA: Dufour) pp 333–58

6-2 Schuster: see note 1-39, pp 20–1

6-3 Simpson: see note 6-1, p 410

6-4 Schuster A 1872 On the spectrum of nitrogen *PRS* **20** 484–7
At the time of Schuster's analysis, a controversy embroiled the field of spectroscopy over whether chemical elements had uniquely characteristic spectra. Adolph Wüllner claimed to observe four spectral lines in hydrogen and two in nitrogen, something Anders Jonas Ångström was unable to confirm. Ångström dismissed the lines as belonging to impurities or other effects. Feffer: see note 1-51, p 34. Schuster went along with Ångström, maintaining that Wüllner's spectrum in nitrogen arose from impurities. A second paper by Schuster reached the same conclusion for the multiple spectrum of hydrogen.

6-5 Thomson: see note 1-25, p 12

6-6 Schuster: see note 1-39, p 22

6-7 Schuster: see note 1-39, p 17

6-8 Lockyer's greatest claim to fame was identifying a spectral line observed in the 1868 solar eclipse with an unknown element which he named helium (after *helios*, the Greek for Sun). The existence of the element was only confirmed in 1895 when William Ramsay isolated it from atmospheric gases. Lockyer was a great populizer of science, founded the Science Museum in London, was one of the founders of the journal *Nature* and its editor for 50 years, and virtually created the discipline of astroarcheology singlehandedly. He was, on the whole, less successful as a practicing scientist, as evidenced by his two major, albeit bogus, research efforts. The first, his chemical dissociation theory, is discussed below. The second, his meteoric hypothesis, had stellar spectra, novae, comets, aurora, and lightning all produced by colliding meteorites!
Meadows A J 1972 *Science and Controversy* (Cambridge, MA: MIT Press) reviewed by Hawkins G S 1978 *American Journal of Physics* **41** 1377–8

6-9 Simpson: see note 6-1, p 411. Schuster took part in a number of subsequent eclipse expeditions: in Colorado (1878), Egypt (1882), and the West Indies (1886).

6-10 Feffer: see note 1-51, p 36
Davies P J and Marsh J D 1953 Ohm's law and the Schuster effect *Institution of Electrical Engineers Proceedings* A **132** 525–32

6-11 Clerk-Maxwell to Schuster, May 3 1876, cited by Schuster A 1910 The Clerk-Maxwell period *A History of the Cavendish Laboratory 1871–1910* (London: Longmans Green) ch 2, pp 24–5

6-12 Schuster: see note 1-39, pp 27–8

6-13 Thomson: see note 1-25, p 111. The Rayleigh apparatus may still be seen in the Museum of the modern Cavendish Laboratory in Cambridge.

6-14 Lodge's principal contributions lay in the area of electromagnetic waves. His initial discovery, that electromagnetic waves could be generated electrically and transmitted along conducting wires, was somewhat overshadowed by Hertz's 'wireless' waves. Eventually, however, Lodge's contributions would prove pivotal in the infancy of radio broadcasting. Though prolific in many other areas of physics, Lodge, like Crookes, was also deeply involved in psychical research, especially telepathy. The first experiments in this area were conducted in 1884 on girls employed in a draper's shop in Liverpool; Lodge tested the girls in his laboratory at Liverpool University. Thomson: see note 1-25, p 154

6-15 Feffer: see note 1-51, p 37
Schuster had submitted testimonials from continental physicists, including, naturally, Kirchhoff, but none from Cambridge scholars. Roscoe sensed that this might backfire and instructed him to solicit testimonials from high wranglers at Cambridge, whose endorsement 'would count for more than a thousand Kirchhoffs'. Heilbron: see note 1-14, p 43. Schuster was indeed able to secure the backing of wranglers from St John's College—happily, since his chief opponent, J J Thomson, was a Trinity man! Heilbron: see note 1-14, p 43

6-16 Kargon: see note 6-1, p 238
Schuster A 1881 On harmonic ratios in the spectra of gases *PRS* **31** 337–47

6-17 Lockyer J N 1878 Atoms and molecules spectroscopically considered *Studies in Spectrum Analysis* 2nd edn, ed J N Lockyer (London: Kegan Paul) ch 4 pp 113–44; reprinted in Nye: see note 1-10, p 188

6-18 Nye: see note 1-10, p 190

6-19 Schuster A 1884 Experiments on the discharge of electricity through gases. Sketch of a theory (The Bakerian Lecture) *PRS* **37** 317–39

6-20 Schuster: see note 6-19, p 323
Hittorf and Schuster were by no means the only ones dwelling on these ideas. The increasingly popular analogy of gaseous conduction to electrolysis was expounded, particularly by Wilhelm Giese of Berlin in 1882, to account for the conductivity of the hot gases of flames (something known for over two centuries). Giese, too, whose work Schuster was unaware of at the time, argues that the conductivity is associated with the *ions* of dissociated molecules. Giese W 1889 Grundzuge einer einheitlichen Theorie der Electricitätsleitung *APC* **37** 576–7, and earlier references therein. Another enthusiast for ionic dissociation was J J Thomson, as we shall see.

6-21 Schuster: see note 6-19, p 324

6-22 Schuster: see note 6-19, p 329

6-23 Schuster: see note 6-19, p 330

6-24 Schuster: see note 6-19, p 331

6-25 Schuster: see note 6-19, p 331

6-26 Schuster A 1890 The discharge of electricity through gases (The Bakerian Lecture) *PRS* **47** 526–59

6-27 Schuster: see note 1-39, pp 66–7

Notes section 6.2

6-28 Thomson: see note 1-25, p 117; see note 6-11, Survey of the last twenty-five years, pp 82–3

6-29 Eve A S 1939 *Rutherford: Being the Life and Letters of the Rt Hon Lord Rutherford OM* (New York: Cambridge University Press) p 14

6-30 Thomson: see note 1-25, p 118

6-31 Thomson J J 1887 On the dissociation of some gases by electric discharge *PRS* **42** 343–45
Feffer: see note 1-51, p 43

6-32 Thomson J J and Newall H F 1887 On the rate at which electricity leaks through liquids which are bad conductors of electricity *PRS* **42** 410–29
Newall: see note 6-11, 1885–1894, p 152

6-33 Thomson to Threlfall, August 7, September 4 1887; cited by Feffer: see note 1-51, pp 43–4

6-34 Thomson: see note 1-25, p 122

6-35 Rayleigh: see note 1-46, p 42

6-36 Thomson to Threlfall, August 7 1887: see note 1-46, p 29

6-37 Thomson to Threlfall, December 11 1887: see note 1-46, p 29

6-38 On January 2, 1890. Rayleigh: see note 1-46, 34

6-39 Feffer: see note 1-51, pp 44–5

6-40 Thomson: see note 1-25, p 94; 1891 On the illustration of the properties of the electric field by means of tubes of electrostatic induction *PM* **31** 149–71

6-41 Feffer: see note 1-51, pp 44–5

6-42 Thomson 1891 On the discharge of electricity through exhausted tubes without electrodes *PM* **32** 321–36, 445–64
Newall: see note 6-11, p 154. Two of the electrodeless discharge tubes are displayed in the Museum of the modern Cavendish Laboratory in west Cambridge.

6-43 For a prolonged dispute between Thomson and Nikola Tesla about the electrodeless discharge, see Martin T C 1992 *The Inventions, Researches and Writings of Nikola Tesla* 2nd edn (New York: Barnes and Noble) pp 396–406

6-44 Helmholtz: see note 5-9

6-45 Schuster 1887 Experiments on the discharge of electricity through gases *Electrician* **19** 353–5. Schuster points to similar experiments the same year leading to similar conclusions in Svante Arrhenius's 'On the dissociation of substances in aqueous solution', reprinted in Nye: see note 1-10, pp 285–309. From his observation that the electrical conductivity of an electrolytic solution is proportional to the excess number of particles over the normal solution, Arrhenius introduced his own concept of 'ionization'; all molecules of electrolytes in water dissociate into charged ions and exert equal action as conductors, p 284. Among others holding similar views on the ionic theory of the gaseous discharge, in Schuster's opinion, were Johann Elster and Hans Geitel, who, at about the same time, were investigating the electrical conductivity of gases in the vicinity of incandescent electrodes, and Emil Warburg, who studied the fall of potential at the cathode.

6-46 Schuster: see note 6-26, p 557

6-47 Schuster: see note 6-26, p 545

6-48 Schuster: see note 6-26, p 547

6-49 Schuster: see note 1-39, p 64
Feffer: see note 1-51, p 49

6-50 Schuster: see note 6-26, p 558

6-51 Schuster: see note 6-26, p 559

6-52 Schuster: see note 1-39, p 67
The year before, 1907, Schuster, feeling the strain of his academic burdens and wishing for more time for personal research and international science, had resigned his Langworthy chair at Owens College, then

Manchester University. On his initiative, Ernest Rutherford succeeded him at Manchester. Schuster's stay as 'Reader' at the University of Calcutta in 1908 was the culmination of a fascination with India dating from a stop there during his return from the eclipse expedition to Siam in 1875. Simpson: see note 6-1
Schuster A 1906 *The Physical Laboratories of the University of Manchester, A Record of 25 Years' Work* (Manchester: University Press)

6-53 Thomson J J 1894 On the velocity of the cathode rays *PM* **38** 365

6-54 Feffer: see note 1-51, p 51

6-55 Thomson to Schuster, January 20 1895, Thomson J J 1896 Letter to the editor *PM* **40** 151
Rayleigh: see note 1-46, pp 39–40

6-56 Rayleigh: see note 1-46, pp 36–7
Thomson J J 1893 *Notes on Recent Researches in Electricity and Magnetism* (Oxford: Clarendon)

6-57 Rayleigh: see note 1-46, pp 39–40
It was thanks to the joint effort of Oliver Lodge, Oliver Heavyside, and George Francis FitzGerald in interpreting, correcting, and extending the *Treatise* that what we know as Maxwell's electromagnetic theory became widely known and intellectually accessible. For instance, Maxwell's famous vector equations are nowhere to be found in his *Treatise*; it was Heavyside who elegantly condensed Maxwell's cumbersome analysis into the four equations. Hunt B J 1991 *The Maxwellians* [e.g. Lodge, Heavyside, and FitzGerald] (Ithaca, NY: Cornell University Press)

6-58 Stanton A 1890 The discharge of electricity from glowing metals *PRS* **47** 559–61

6-59 Röntgen: see note 1-6
Schuster: see note 1-39, p 72. The photograph, bearing an image of his wife's left hand, was taken by Röntgen on December 22, and was the first such x-ray photo of a part of the human body.

6-60 Schuster: see note 1-39, p 76
Lord Rayleigh, the son, happened to be with Schuster on one occasion, observing a demonstration of x-ray apparatus. He recalls Schuster's remarks: 'I owe a heavy grudge to Röntgen's discovery. It came out just at the time when I was working at the magnetic deflection of cathode rays. I laid my work aside and devoted all the resources of the laboratory to making and exhausting x-ray tubes, and taking photographs for the hospital at Manchester'. Rayleigh Lord 1936 Some reminiscences of scientific workers of the past generation, and their surroundings *The Proceedings of the Physical Society* **48** 245

6-61 Schuster: see note 1-39, p 76

Notes section 7.1

7-1 Perrot A 1861 *ACP* **61** 161

7-2 Thomson J J 1893 The electrolysis of steam *PRS* **53** 90–110

7-3 Rayleigh: see note 1-46, p 38

7-4 Rayleigh: see note 1-46, p 40. The tea ceremony, at 4 PM sharp, is rigorously adhered to at the Cavendish to this day, in the new Cavendish Laboratory in west Cambridge. One table in the modern cafeteria of the Bragg Building is set aside for each department. Coffee is served in the morning.

7-5 Thomson: see note 6-53, p 360

7-6 Thomson J J 1891 On the rate of propagation of the luminous discharge of electricity through a rarefied gas *PRS* **44** 84–100
7-7 Rayleigh: see note 1-46, p 38. Rayleigh was first introduced to Thomson by his father, as soon to be his pupil at Cambridge, at the British Association meeting in Oxford in 1894. Rayleigh: see note 1-46, p 39
7-8 Thomson: see note 6-53, p 364
7-9 Stuewer: see note 1-16, p 524
7-10 Nye M J 1972 *Molecular Reality: a Perspective on the Scientific Work of Jean Perrin* (London: Macdonald and New York: Elsevier) p 41
7-11 Nye: see note 1-10, pp xxv–xxvi
7-12 Nye: see note 7-10, p 60
7-13 Perrin 1948 Les progrès de la Science, preface to Perrin's 1929 *Les Éléments de la Physique* (Paris: Albin-Michel) in *La Science et L' Espérancé* (Paris: Presses Universitaires de France) p 65; quoted by Nye: see note 7-10, p 65
7-14 Perrin: see note 1-17
7-15 Magie: see note 1-6, p 580
7-16 Thomson J J 1896–1897 *NA* **55** 453; 1897 Cathode rays *PM* **44** 293–316, 294–6
7-17 Thomson: see note 7-16, Cathode rays, pp 295–6
Chalmers: see note 3-13, p 204

Notes section 7.2

7-18 Munro H 1896 *The Story of Electricity* (London: Newnes) p 175
7-19 Wilson D 1983 *Rutherford: Simple Genius* (Cambridge, MA: MIT Press) p 110
7-20 Rutherford to Mary Newton, January 25 1896, cited by Eve: see note 6-29, p 26
7-21 Thomson states that the lecture was delivered on January 28, whereas the archives of Thomson's papers and correspondence held in the Library, Trinity College, Cambridge, show that it was given on Monday, January 27. Thomson's communication to the Society was entitled 'Longitudinal electric waves, and Röntgen's rays'; Thomson's personal copy contains various marginal entries. TCL E.2, vol 2
7-22 E.g. Dam: see note 1-2
7-23 Thomson: see note 1-25, pp 402–3
7-24 Rayleigh: see note 1-46, pp 65–6
7-25 Thomson J J 1896 On the discharge of electricity produced by the Röntgen rays, and the effects produced by these rays on dielectrics through which they pass *PRS* **59** 274–6. Received by the Royal Society on February 7. Thomson's personal copy of this paper is preserved in the archives of Thomson's papers and correspondence held in the Library, Trinity College, Cambridge. TCL, E.2, vol 2
7-26 The paper acknowledges the contributions of McClelland and 'Mr. E. Everitt'—in point of fact Mr Everett.
7-27 Elster and Geitel formed a remarkable partnership unique in the history of physics, it would seem, and we will have more to say about them, and about a particular investigation of theirs, in section 13.1.
7-28 Thomson: see note 7-25, pp 275–6
7-29 Badash L 1975 Rutherford, Ernest *DSB* **12** 25–36
7-30 The 1851 Exhibition Scholarship had been established by Queen Victoria's Prince Consort to dedicate funds remaining from the exhibition to

scholarships for subjects of the British Empire—a scholarship still in force. In point of fact, the Commissioners had chosen MacLaurin, a chemist, for the 1895 nomination, with Rutherford a close second. In the event, MacLaurin decided to accept a post in the Civil Service, and the award went to Rutherford.

7-31 William Henry Bragg was a student (and tennis partner) of J J Thomson at Cambridge, where he graduated in 1884. After a year's research under Thomson, his mentor suggested he apply for a chair in mathematics and physics recently vacated by Horace Lamb at the Australian University of Adelaide; Bragg gained the post without difficulty. At Adelaide he led a leisurely life for 18 years, mostly teaching and playing tennis. What few papers he did publish were rather lackluster. However, an event in 1904 shook him out of his scientific doldrums. That year he was to give the presidential address to Section A of the Australian Association for the Advancement of Science. To enliven his talk, he decided to brush up on frontier topics in physics; namely, the newly discovered electron and the phenomenon of radioactivity. The latter topic caught his interest at once, and he soon immersed himself in studies of α-radiation with his assistant Richard Daniel Kleeman. They found that α-particles of a given energy have a unique range. With his son William Lawrence Bragg, he later extended these studies.

7-32 Thomson to Rutherford, September 24 1895, CUL, Add 7653, T9

7-33 Rutherford to Mary Newton, October 3 1895, cited by Eve: see note 6-29, p 15

7-34 George Paget Thomson was born and educated at Cambridge, where he taught from 1914 to 1922 when he assumed the chair of physics at Aberdeen University. In 1930, he accepted the chair at Imperial College, London, where he remained until 1952 when he returned to Cambridge as Master of Corpus Christi College. His 1937 Nobel Prize in physics was based on his experiment in which he passed a beam of electrons through a gold foil onto a photographic plate. The plate revealed a bright central spot surrounded by concentric circles with alternating darker and lighter rings—that is, a diffraction pattern. The experiment showed that each electron is associated with a wave whose wavelength is in good agreement with a formula derived by Louis de Broglie. Thomson shared the Nobel Prize with Clinton J Davisson who, with Lester H Germer, had independently made the same discovery the same year by a slightly different route—by the interference of electrons reflected from a nickel crystal.

William Henry Bragg and his son William Lawrence Bragg shared the 1915 Nobel Prize in physics for work in the field in which the elder Bragg's reputation mainly rests: x-ray crystallography. Their work followed hard on the heels of Max von Laue's discovery of x-ray diffraction by crystals in 1912, which showed beyond doubt that x-rays are waves (and which earned him the Nobel Prize in 1914). The Braggs used crystals as gratings in the first true x-ray spectrometer and made an extensive study of spectral lines and crystal structures.

7-35 von Engel A 1957 John Sealy Edward Townsend *Royal Society Biographical Memoirs of Fellows* **3** 257–72

7-36 Rutherford to Mary Newton, December 8 1895, cited by Eve: see note 6-29, p 20

7-37 Rayleigh: see note 1-46, p 63

Notes section 7.3

7-38 Thomson J J 1896 The Röntgen rays *NA* **53** 391–2

7-39 Wilson's original cloud chamber, built by him in 1911, is among the apparatus exhibited in the Museum of the Cavendish Laboratory in Cambridge. Apart from some equipment built to demonstrate the method of making visible the paths of ionizing particles in traversing a gas, this was his original and only chamber, though it was altered and refined somewhat over the years.

7-40 Wilson C T R 1959 Reminiscences of my early years *Royal Society Notes and Records* **14** 163–73
Crowther: see note 6-1, p 25–55. Wilson, 90 years old, died within weeks of recording his reminiscences.

7-41 Wilson had offered himself as a candidate for the then vacant Clerk Maxwell Scholarship, proposing to investigate subjects related to theories of solutions and osmotic pressure. Wilson to Thomson, November 8 1893, CUL, Add 7654, W36

7-42 Crowther: see note 6-1, pp 34–5

7-43 Wilson C T R 1954 Ben Nevis sixty years ago *Weather* **9** 309–11

7-44 Wilson C T R 1965 On the cloud method of making visible ions and the tracks of ionizing particles (Nobel Lecture, December 12, 1927) *Nobel Lectures, including Presentation Speeches and Laureates' Biographies; Physics 1922–1941* (Amsterdam: Elsevier) p 194

7-45 Crowther: see note 6-1, p 35

7-46 Wilson: see note 7-44, p 196
The actual working of Wilson's initial apparatus (figure 7.2) was as follows. The gas to be expanded was contained in the glass vessel *A*. Vessel *A*, in turn, was placed inside a glass bottle *B* partially filled with water so as to trap the gas in the inner vessel. The air above the water in the bottle was connected with an evacuated vessel *F* by tubes *D* and *G* to which were fitted valves *E* and *K*. Valve *K* was normally closed; however, when it was quickly opened the air at the top of bottle *B* rushed into the evacuated vessel *F* and the water in *B* rose until it filled the top of the bottle. In doing so, it closed the valve *E* thus stopping further expansion of the gas in *A*. By adjusting the initial volume of gas in *A* and water in *B* the relative expansion of the gas in *A* could be precisely controlled. Wilson: see note 7-40, pp 171–2

7-47 Crowther: see note 6-1, p 38

7-48 Thomson: see note 1-25, pp 419–20

7-49 Thomson: see note 1-31; Thomson J J 1893 On the effect of electrification and chemical action on a steam jet *PM* **36** 313–27

7-50 Thomson: see note 1-25, p 417

7-51 Rayleigh: see note 1-46, p 101. Kelvin himself was, in fact, involved in the background for Wilson's cloud chamber. His study of Aitken's results led him to conclude that the effectiveness of a dust particle in acting as a nucleus of condensation depended on its size; below a critical size a particle would be ineffective as a nucleus.

7-52 Thomson J J and McClelland J A 1896 On the leakage of electricity through dielectrics traversed by Röntgen rays *Cambridge Philosophical Society Proceedings* **9** 126–40
Thomson's personal copy of this paper is preserved in the archives of Thomson's papers and correspondence held in the Library, Trinity College, Cambridge. It contains various marginal entries by Thomson, including several sketches of apparatus. TCL, E.2, Pamphlets, vol 2

7-53 Whittaker: see note 1-19, vol 1, p 360

7-54 Falconer I 1989 J J Thomson and 'Cavendish physics' *The Development of the Laboratory: Essays on the Place of Experiment in Industrial Civilization* ed F A J L James (New York: American Institute of Physics) p 109

7-55 Rayleigh: see note 1-46, p 47

7-56 Rayleigh: see note 1-46, p 47

7-57 Rayleigh: see note 1-46, p 48

7-58 de Bruyne N 1984 A personal view of the Cavendish 1923–30 *Cambridge Physics in the Thirties* ed J Hendry (Bristol: Hilger) section 2.3
For accounts of the Laboratory in earlier years, see Glazebrook R 1926 The Cavendish Laboratory: 1876–1900 *NA* **118** (Supplement) 52–8 and Crowther J G 1926 Research work in the Cavendish Laboratory in 1900–1918 *NA* **118** (Supplement) 58–60

7-59 'The works of the Lord are great; sought out by all those that have pleasure therein'.

Notes section 8.1

8-1 Cited by Rayleigh: see note 1-46, p 54

8-2 Rayleigh: see note 1-46, p 54

8-3 The editors 1896 Recent work with Röntgen rays *NA* **53** 613–6

8-4 A thin plate of iron coated with a sensitized film, used for obtaining positive photographic images without a negative stage; also called *tintype*.

8-5 Recent work: see note 8-3, p 613

8-6 Recent work: see note 8-3, p 613

8-7 Recent work: see note 8-3, p 613. Their value for the refractive index was indeed an overestimate. Thus, about the same time Gouy showed that the index of refraction could not be greater than 1.000 005. The fact that the index is close to unity explains the lack of measurable refraction.

8-8 Recent work: see note 8-3, p 613

8-9 Recent work: see note 8-3, p 613

8-10 Recent work: see note 8-3, p 614

8-11 Glasser: see note 1-1, p 23

8-12 Recent work: see note 8-3, pp 614–5

8-13 Recent work: see note 8-3, p 615–6

8-14 Recent work: see note 8-3, p 616

8-15 Thomson J J 1896 The Röntgen rays *NA* **54** 302–6

8-16 Thomson: see note 8-15, p 302

8-17 Thomson: see note 8-15, p 303

8-18 Thomson: see note 8-15, pp 303–4

8-19 Thomson: see note 8-15, p 304

8-20 Thomson: see note 8-15, p 304

8-21 Thomson J J 1896 Presidential Address *Reports of the British Association for the Advancement of Science* pp 699–706

8-22 Vilhelm Friman Koren Bjerknes (1862–1951) began his career as assistant to Heinrich Hertz. In 1895, he became professor of applied mechanics and mathematical physics at Stockholm; he returned to Norway in 1907 and in 1912 was made professor of geophysics at the University of Leipzig. His important contributions to meteorology and weather forecasting include his models of atmospheric and oceanic motions. He founded the Bergen Geophysical Institute during World War I. Together with his son Jacob Bjerknes, Tor Bergeron, and Carl-Gustav Rossby, he developed the 'Bergen theory' of polar fronts—an essential feature of modern weather forecasting.

8-23 James Edward Keeler (1857–1900) was professor of astronomy in the Western University of Pennsylvania. He alternated between the Allegheny Observatory and Lick Observatory, becoming director of Lick in 1898. He began the Lick program on Doppler measurements of the radial velocity of stars, and used the Doppler shift to determine the complex rotation of Saturn's rings. He also undertook a photographic and spectroscopic survey of nebulae.

8-24 Maximilian Franz Joseph Cornelius Wolf (1863–1932) was appointed to the chair of astronomy and astrophysics at Heidelberg in 1901. He is best known for his discovery of over 500 new asteroids by a clever photographic technique, and for his study of dark nebular regions in the Milky Way.

8-25 1896 Physics at the British Association *NA* **54** 565–7

8-26 Thomson: see note 8-21, p 699

8-27 Thomson: see note 8-21, p 700. Section L of the British Association, Educational Science, was organized in 1901.

8-28 Thomson: see note 8-21, p 701

8-29 Thomson: see note 8-21, p 703

8-30 Thomson: see note 8-21, p 704

8-31 Thomson: see note 8-21, p 703
The National Physical Laboratory was established in 1900, not at Kew but at Bushy Park, under management of the Royal Society. Its first director was R Glazebrook, who at the Cavendish had become an expert on the problem of obtaining accurate electrical standards. Moseley R 1978 The origins and early years of the National Physical Laboratory: a chapter in the pre-history of British science policy *Minerva* **16** 222–50

8-32 Lenard to Thomson, November 20 1896, CUL, Add 7654, L29

8-33 Physics at the British Association: see note 8-25, p 565

8-34 Physics at the British Association: see note 8-25, p 565

8-35 Physics at the British Association: see note 8-25, p 565. The Hall effect was discovered by the American physicist Edwin Herbert Hall in 1879. The effect is the development of a transverse electric field in a current-carrying conductor placed in a magnetic field, when the conductor is positioned so that the direction of the magnetic field is perpendicular to the direction of current flow. The electric field is a result of the force that the magnetic field exerts on the moving positive or negative particles that constitute the electric current. The sign of the resultant potential voltage determines whether positive or negative charges are carrying the current. In metals, the Hall voltages are generally negative, indicating that the electric current is composed of moving negative charges. The Hall voltage is positive for a few metals such as beryllium, zinc, and cadmium, indicating that these metals conduct electric currents by the movement of positively charged carriers called holes.

8-36 Thomson: see note 8-21, pp 704–5

8-37 Thomson J J and Rutherford E 1896 On the passage of electricity through gases exposed to Röntgen rays *PM* **42** 392–407

8-38 Wilson: see note 7-19, pp 117–8

8-39 Rutherford to Mary Newton, June 6 1896, cited by Eve: see note 6-29, p 36

8-40 Rutherford to his mother, Martha Rutherford, July 1896, cited by Eve: see note 6-29, p 37

8-41 Rutherford to Mary Newton, August 12 1896, cited by Eve: see note 6-

29, p 38. In the same letter, Rutherford proudly informs Mary that his paper on his magnetic detector has been accepted for publication by the *Philosophical Transactions of the Royal Society*.

8-42 Rutherford to Mary Newton, August 27 1896, cited by Eve: see note 6-29, p 38

8-43 Thomson and Rutherford: see note 8-37, pp 393–4

8-44 Wilson: see note 7-19, p 114

8-45 Thompson S P 1897 Cathode rays and some analogous rays *PTRS* A **190** 471–90. Silvanus P Thompson (1851-1916) twice narrowly missed out in landmark discoveries in physics, somewhat like Lenard. He almost discovered electromagnetic waves in the 1870s, and almost discovered radioactivity in 1896, as we shall see shortly.

8-46 Eve: see note 6-29, p 44

8-47 Feather N 1940 *Lord Rutherford* (Glasgow: Blackie) reprinted 1973

8-48 Thomson: see note 1-25, p 168

8-49 Thomson: see note 1-25, pp 165–6

8-50 Thomson: see note 1-25, p 171

8-51 A notebook by Thomson contains lecture notes on the electric spark, and on the potential gradient of a current passing through air exposed to Röntgen rays, among other things. The same notebook includes a journal by Mrs Rose Thomson of their voyage to America on board the *Campania* on this occasion. CUL, Add 7654, NB 39 A

8-52 Thomson: see note 1-25 p 177

8-53 Thomson: see note 1-25, pp 178–9. Henry Van Dyke (1852–1933) was a prominent American clergyman and writer. He was Professor of English Literature at Princeton from 1899 to 1923, served under Woodrow Wilson as American Minister to the Netherlands during 1913–1916, and received the Legion of Honour for his service as naval chaplain in World War I.

8-54 Thomson: see note 1-25, pp 189–93

8-55 Thomson: see note 1-25, p 189

8-56 Thomson: see note 1-25, p 180

Notes section 8.2

8-57 Romer A 1970 Becquerel, [Antoine-] Henri *DSB* **1** 558

8-58 Pais: see note 3-29, p 43, citing Becquerel H 1903 Recherches sur une propriété nouvelle de la matiére. Activité radiante spontanée ou radioactivité de la matiére *Mémoires de l'Académie des Sciences, Paris* **46**

8-59 Romer: see note 8-57, p 558. The museum was created in June 1793 on the site of a herbal garden set up in 1635 by Louis XIII for the use of medical and apothecary students.

8-60 Romer: see note 8-57, p 558

8-61 Perrin J 1896 *CR* **122**. Perrin received his doctorate in 1897 for his work on cathode rays and x-rays. The École Normale Supérieure was, like the École Polytechnique, established in 1794. While the latter became an engineering school under the Ministry of Defence, the École Normale was intended as a model for teacher-training colleges.

8-62 Nye: see note 7-10, p 68

8-63 Poincaré H 1896 *Revue Générale des Sciences* **7** 43 (the italics are Poincaré's)

8-64 Badash L 1965 Radioactivity before the Curies *AJP* **33** 130

8-65 Becquerel H 1896 Sur les radiations émises par phosphorescence *CR* **122** 420, reproduced in *Great Experiments in Physics*: see note 1-7, pp 212–5, as well as in Magie: see note 1-6, pp 610–3

8-66 Becquerel H 1896 Sur les radiation invisibles émises par les corps phosphorescents *CR* **122** 501
8-67 Becquerel: see note 8-65, p 501
8-68 Crookes W 1909–1910 Antoine Henry Becquerel, 1852–1908 Obituary Notices of Fellows deceased *Royal Society Proceedings* A **83** 22
8-69 Crookes: see note 8-68, p 22
8-70 Becquerel H 1896 Émission de radiations nouvelles par l'uranium métallique *CR* **122** 1086–8
8-71 Badash: see note 8-64, p 128. One who was thrown off by this unfortunate claim was Silvanus P Thompson. In the course of repeating Röntgen's newly announced experiments in January 1896, he placed a variety of luminous materials on a sheet of aluminum under which was placed a photographic plate. Badash: see note 8-64, p 129
The arrangement was 'left for several days upon the sill of a window facing south to receive so much sunlight as penetrates in February into a back street in the heart of London'. Thompson S P 1896 *PM* **42** 103
On developing the plate, he found that only uranium nitrate and uranium ammonium fluoride produced photographic images, something he informed Sir George Stokes of without delay. Stokes urged him to publish his findings, but within a week drew his attention to Becquerel having anticipated him across the Channel. In the event, Thompson put off his own publication, particularly in view of Becquerel's significant (if spurious) proof of the refraction and polarization of the rays. At any rate, Thompson *did* discover radioactivity, but failed to recognize that it had nothing to do with luminescence. Badash: see note 8-64, p 130
8-72 Becquerel: see note 8-70, p 1086
8-73 Becquerel: see note 8-70, p 1086
8-74 Becquerel J 1946 *Conférences Prononcées à l'Occasion du Cinquantième Anniversaire de la Découverte de la Radioactivité* (Muséum d'Histoire Naturelle), cited by Pais: see note 3-29, p 47
See also Jauncey G E M 1946 The early years of radioactivity *AJP* **14** 226–41
8-75 Poincaré H 1897 *Revue Scientifique* **7** 72, cited by Pais: see note 3-29, p 49
8-76 Badash: see note 7-29, p 26
8-77 Feather: see note 8-47, pp 45–8
8-78 Feather: see note 8-47, p 46
8-79 Eve: see note 6-29, p 55
8-80 Rutherford E 1899 Uranium radiation and the electrical conduction produced by it *PM* **47** 109–63
8-81 Feather: see note 8-47, pp 50–1
8-82 Rutherford: see note 8-80, p 116
8-83 Feather: see note 8-47, p 53

Notes section 9.1

9-1 Röntgen in Magie: see note 1-6, p 607
9-2 Rayleigh: see note 1-46, p 66
9-3 Bullen K E 1976 Wiechert, Emil *DSB* **14** 327–8
9-4 Wiechert E 1896 Die Theorie der Elektrodynamik und die Röntgen'sche Entdeckung *Schriften der physikalisch-ökonomischen Gesellschaft zu Köningsberg, Abhandlungen* **37** 1; 1896 *Sitzungsberichte* **37** 29; 1896 Über die Grundlagen der Elektrodynamik *APC* **59** 321
9-5 Schuster: see note 1-39, p 76

9-6 With regard to the interference of x-rays, even diffraction gratings proved inadequate for the task, due to the extremely short wavelengths involved. In 1912 Max von Laue, at Munich, suggested using crystals instead. The actual experiment, using zinc sulfide, was performed by Walter Friedrich and Paul Knipping; it worked very well, proving that x-rays are, indeed, electromagnetic vibrations of very short, definite, wavelengths, and earned Laue the Nobel Prize in 1914. In 1913, the Braggs used crystal diffraction to determine the actual wavelengths, which varied from $\frac{1}{50}$ to 1/50 000 times the wavelength of visible light. For this the Braggs, too, shared the Nobel Prize—in 1915. See also note 7-34.

9-7 Rayleigh: see note 1-46, p 48

9-8 Rayleigh: see note 1-46, p 49

9-9 Rayleigh: see note 1-46, p 51

9-10 Rayleigh: see note 1-46, p 51

9-11 Rayleigh: see note 1-46, p 50

9-12 Rayleigh: see note 1-46, pp 50–1

9-13 Cahan D 1985 The institutional revolution in German physics, 1865–1914 *HSPS* **15** 19

9-14 Cahan: see note 9-13, p 38

9-15 Cahan: see note 9-13, p 53

9-16 Pyenson L 1979 Physics in the shadow of mathematics: the Göttingen electron-theory seminar of 1905 *Archive for History of Exact Sciences* **21** 55–89

9-17 Wiechert E 1897 Ueber das Wesen der Elektricität *Schriften der physikalisch-ökonomischen Gesellschaft zu Köningsberg, Sitzungsberichte* **38** 3

Notes section 9.2

9-18 Campbell J T 1973 Kaufmann, Walter *DSB* **7** 263–5

9-19 In 1902 Seitz showed that when the pressure is sufficiently low, the kinetic energy is indeed given by $\frac{1}{2}mv^2 = eV$. Seitz 1902 *AP* **8** 233

9-20 Kaufmann W 1897 Die magnetische Ablenkbarkeit der Kathodenstrahlen und ihre Ablängigkeit vom Entladungspotential *APC* **61** 552. The deflection of the rays in the z-direction of a right-handed coordinate system, in which the rays travel in the x-direction and a uniform magnetic field points in the y-direction, is given in Gaussian (electromagnetic) units, by Kaufmann, as

$$m\frac{\mathrm{d}^2 z}{\mathrm{d}t^2} = He\left(\frac{\mathrm{d}x}{\mathrm{d}t}\right).$$

If x_0 denotes the path length of the rays, the solution for z is

$$z = H_0\frac{x_0^2}{2}\sqrt{\frac{e}{2mV}}.$$

Hence, if z is proportional to $1\sqrt{v}$ as found by Kaufmann, e/m is a constant.

9-21 Kaufmann: see note 9-20, p 552; translated in Pais: see note 3-29, p 84

9-22 Kaufmann: see note 9-20, p 552

Kaufmann's e/m ratio was approximately 1.86×10^7 emu g^{-1}; Thomson J J 1906 *Conduction of Electricity through Gases* 2nd edn (Cambridge:

Cambridge University Press) p 128. This book, first published in 1903, became a standard reference work for many years. G P Thomson helped his father in producing the third edition in two volumes, one in 1928 and one in 1932.

9-23 Kaufmann: see note 9-20

9-24 Jaffé G 1952 Recollections of three great laboratories *Journal of Chemical Education* **29** 236

9-25 Weinberg S 1990 *The Discovery of Subatomic Particles* (New York: Freeman) p 71

9-26 Jungnickel C and McCormmach R 1990 New foundations for theoretical physics at the turn of the twentieth century *Intellectual Mastery of Nature: Theoretical Physics from Ohm to Einstein; Vol 2, the Now Mighty Theoretical Physics 1870–1925* paperback edn (Chicago: University of Chicago Press) ch 24 p 219

9-27 Rayleigh: see note 1-46, p 82

9-28 Thomson J J 1897 On the cathode rays *Cambridge Philosophical Society Proceedings* **9** 244

9-29 In a footnote to his April *APC* manuscript (see note 9-20), Kaufmann claims he only learned of Thomson's conclusion regarding the lack of influence of the chemical nature of the residual gas shortly before he finished his own experiments.

9-30 Thomson G P 1964 *J J Thomson and the Cavendish Laboratory in his Day* (London: Nelson) p 45

9-31 Owen G E 1955 The discovery of the electron *Annals of Science* **11** 177. Thomson's Princeton lectures were revised for publication, with some additions, as noted, in August 1897, and published as 1898 *The Discharge of Electricity through Gases* (Westminster: Constable) and (Cambridge, MA: Scribner) 'dedicated to the members of the class who attended the author's lectures at Princeton, whose sympathy and kindness he can never forget'. A brief preface notes that the author has 'added some results which have been published between the delivery and the printing of the lectures... in the hope of making the work more useful'.

9-32 Thomson: see note 1-25, p 339

9-33 Thomson J J 1897 Cathode rays *Electrician* **39** 104–22; reprinted in Watson E C 1947 Jubilee of the electron *AJP* **15** 458–64

Thomson's personal copy of this important paper is preserved in the archives of Thomson's papers and correspondence held in the Library, Trinity College, Cambridge. It contains various marginal entries by Thomson, some editorial corrections also in his hand, and one correction of substance. On page 17 the published value for m/e, 1.6×10^{-7}, has been 'corrected' to 1.5×10^{-7}. The significance of this reevaluation is noted in section 9.3. TCL, E.2, vol 2

9-34 Birkeland, struck by the resemblance between cathode ray phenomena and the aurora borealis, had suggested the year before, 1896, that the aurora and magnetic storms are the result of charged rays emitted by the Sun and trapped in the Earth's magnetic field near the poles. In 1897, Birkeland embarked on his first northern light expedition to Finnmark in northern Norway to make a reconnaissance for establishing a mountaintop observatory. The same year he devised the original form of his experimental model of the earth and a cathode-ray-emitting sun: a metal sphere covered with a phosphorescent coating along with a cathode in an evacuated chamber. As the air became increasingly rarefied, electrons emitted from the 'solar' filament were sucked into the 'earth's' magnetic field, producing 'auroral' rings around the polar areas. Together with an assistant, he built ever larger models in the cellars of the old University buildings in Oslo (then Kristiania) culminating in a 100 l 'world room'—the world's largest discharge tube, driven by a giant Swiss high-tension, direct-current generator capable of 300 mA at 15 000 V. Friedman R M 1995 Civilization and National Honour: The rise of Norwegian geophysical and cosmic ray science *Making Sense of Space: the History of Norwegian Space Activities* ed J P Collett (Oslo: Scandinavian University Press) ch 1

9-35 Thomson in Watson: see note 9-33, p 460

9-36 Thomson in Watson: see note 9-33, p 460

9-37 Thomson in Watson: see note 9-33, pp 460–1

9-38 The tube is on display as item 93 in the Cavendish Museum in the Cavendish Laboratory, Cambridge. Note that figure 9.2 is a highly stylized rendering of the tube; in the actual tube the angle between the path of the undeflected rays and the centerline of the Faraday cup extension is not nearly as great as depicted. For more on the museum, see note 9-55, below.

9-39 Thomson in Watson: see note 9-33, p 461

9-40 Goldstein: see note 4-50, pp 257–8

9-41 Thomson: see note 1-25, p 334

9-42 Andrade: see note 3-14, p 45

The Kaufmann pump was, however, soon supplanted by Gaede's rotary mercury pump. Andrade: see note 3-14

9-43 Thomson: see note 1-25, p 335

9-44 Thomson in Watson: see note 9-33, p 462

9-45 Thomson in Watson: see note 9-33, pp 463–4. In the table of Lenard's results, p 463, note a misprint of 0.760 mm for 760 mm in the case of 'air'.

9-46 An error in Q arises from the aforesaid conductivity of the gas, caused by the passage of the cathode rays. The resultant leakage of charge from the Faraday cylinder causes the electrometer to underestimate the accumulated charge. This was compensated for by connecting the inner cylinder to the largest possible capacity, and limiting the duration of a given exposure to the beam.

9-47 See, e.g., Harnwell C P and Livingood J J 1933 *Experimental Atomic Physics* (New York: McGraw-Hill) p 118. Some uncertainty in the radius occurs due to the aforesaid spreading out of the deflected rays (really a spread in particle velocities) which causes an uncertainty in d, resulting in an error of about 20% in the value for r.

9-48 Thomson in Watson: see note 9-33, p 464

9-49 Thomson: see note 1-25, p 341

9-50 Rayleigh: see note 1-46, p 84. Strutt's writings include excellent biographies of his father and of J J Thomson—the latter a constant source of reference in the present monograph.

9-51 Rayleigh: see note 1-46, p 94. 'Unfortunately', adds Rayleigh, 'I cannot fix exactly when this occurred. It was probably in the summer term of 1897, and it is rather difficult to understand why he did not find plenty of more experienced listeners'.

Notes section 9.3

9-52 Thomson: see note 7-16. Curiously, this famous paper is totally devoid of formal references to earlier work, including Lenard's results (albeit with a tabular error) that figure so prominently in Thomson's argument for the smallness of his corpuscles.

9-53 Pencilled corrections to Thomson's ink-written manuscript of this famous paper, preserved in the archives of Cambridge University Library, show that he originally intended to include a figure of the tube in question (i.e. that shown in our figure 9.3) but then omitted it, opting to refer to it as a modified version of his subsequent, famous tube with crossed electromagnetic fields (our figure 9.7) but with a Faraday cup in place of the deflection plates D and E. CUL, Add 7654, PD13

9-54 Thomson: see note 7-16

9-55 The museum is divided into 15 sections, as follows: Maxwell apparatus; early apparatus; standards; electrical discharge in gases: the discovery of the electron; positive rays and isotopes; cloud chambers; α-particles and early nuclear physics; the Cockroft–Walton machine; electron microscopy; Sweepnik (high-voltage physics); surface physics; amorphous materials; radio astonomy; x-ray crystallography; low-temperature physics. The apparatus is described in two booklets, both by I J Falconer, dated 1980, and available at the Cavendish Museum: *Apparatus from the Cavendish Museum* and *The Cavendish Laboratory: an Outline Guide to the Museum*.

9-56 Since the time for the rays to traverse the distance l is l/v, the velocity in the direction of E is

$$v_E = \frac{Ee}{m}\frac{l}{v}$$

and

$$\theta = \frac{Ee}{m}\frac{l}{v^2}.$$

Moreover, if ϕ is the angle through which the rays are deflected upon leaving the magnetic field,

$$\phi = \frac{He}{m}\frac{l}{v}.$$

From the last two equations,

$$v = \frac{E}{H}\frac{\phi}{\theta}.$$

If now H is adjusted so that $\phi = \theta$,

$$v = E/H$$

as in (9.10).

9-57 Thomson: see note 7-16, p 310

9-58 Thomson: see note 7-16, p 307

9-59 TCL, E.2, vol 2

9-60 CUL, Add 7654, PD13, or p 309 in the printed version: see note 7-16

9-61 Thomson: see note 7-16, p 310

Notes section 10.1

10-1 Thomson: see note 7-16, p 312
10-2 Thomson: see note 1-25, pp 338–9
10-3 Chalmers: see note 3-13, p 211
10-4 Rutherford E 1938 The development of the theory of atomic structure *Background to Modern Science* eds J Needham and W Pagel (Cambridge: Cambridge University Press) pp 64–5, cited by Trenn T J 1976 Townsend, John Sealy Edward *DSB* **13** pp 445–6
10-5 The well known Stokes' law for the force on a sphere moving through a viscous medium made its first appearance in 1854, as a proposition to be proved in a Smith Prize examination. Barr E S 1969 Stokes, Sir George Gabriel *AJP* **37** 6
10-6 Rayleigh: see note 1-46, p 102
10-7 Townsend J S E 1897 On electricity in gases and the formation of clouds in charged gases *Camb. Phil. Soc. Proc.* **9** 244–58
10-8 Townsend: see note 10-7, p 257
10-9 Morrow R A 1969 On the discovery of the electron *Journal of Chemical Education* **46** 587
10-10 Wilson C T R 1897 Condensation of water vapour in the presence of dust-free air and other gases *PTRS* A **188** 265–307
10-11 Equally curious, Chalmers also, in his *Historic Researches* (note 3-13) neglects to mention Townsend in all of this.
10-12 Thomson J J 1898 On the charge of electricity carried by the ions produced by Röntgen rays *PM* **46** 534–5
10-13 Thomson: see note 10-12, pp 535–6
10-14 Thomson: see note 10-12, p 536
10-15 Thomson: see note 10-12, p 536
10-16 Thomson: see note 10-12, p 536
10-17 Thomson: see note 10-12, pp 537–8
10-18 Thomson: see note 10-12, p 544
10-19 Thomson: see note 10-12, pp 544–5

Notes section 10.2

10-20 Rutherford E 1898 The discharge of electrification by ultra-violet light *Cambridge Philosophical Society Proceedings* **9** 401
10-21 Rutherford: see note 10-20, cited by Wilson: see note 7-19, p 121
10-22 Magie: see note 1-6, p 579
10-23 Elster J and Geitel H 1890 *APC* **41** 166
10-24 Rayleigh: see note 1-46, p 108–9
10-25 Rayleigh: see note 1-46, p 109
10-26 Thomson J J 1899 On the masses of the ions in gases at low pressure *PM* **48** 548–9
10-27 The reader is reminded that a cycloid is generated by a circle rolling on a straight line, with a point on the circle describing the cycloid. The distance between the point and the straight line cannot exceed the diameter of the generating circle.
10-28 Thomson: see note 10-26, p 554
10-29 Rayleigh: see note 1-46, p 111
10-30 Thomson: see note 10-26, p 556–7
10-31 Thomson: see note 10-26, p 563
Thomson to Rutherford, July 23 1899, in Eve: see note 6-29, p 68

In the experiments of Townsend alluded to here, a direct comparison was made between e for gaseous ions and the charge E carried by a monovalent ion, such as the hydrogen ion, in electrolysis. From measured coefficients of diffusion of ions into gases, and their velocities under a potential gradient, he deduced Ne, where N is the number of molecules in unit volume of gas, and found it to be equal to NE, where E is the aforesaid hydrogen ionic charge. Thus $e = E$, or the charges of ions in the liquid and the gaseous state are equal. These results were communicated to the Royal Society in April 1899. Townsend J S E 1899 Diffusion of ions into gases *PTRS* **193** 129; 1899 *PRS* **65** 192

10-32 Thomson: see note 10-26, p 563

10-33 Thomson: see note 10-26, p 565

10-34 Kragh H 1989 Concept and controversy: Jean Becquerel and the positive electron *Centaurus* **32** 210

As early as 1874 the Irish physicist George Johnstone Stoney (1826–1911), uncle of FitzGerald, expressed his faith in Maxwell's 'molecules of electricity' before a meeting of the British Association at Belfast, as follows: 'For each chemical bond which is ruptured within an electrolyte a certain quantity of electricity traverses the electrolyte, which is the same in all cases. This definite quantity of electricity *I* shall call *E1*. If we make this our unit quantity of electricity, we shall probably have made a very important step in our study of molecular phenomena'. Stoney G J 1874 *British Association Belfast Report*

In 1881, Stoney estimated his value for the 'definite quantity of electricity' to be approximately 3×10^{-11} esu—about two orders of magnitude low, yet not bad for the time. Stoney G J 1881 On the physical units of nature *PM* **11** 381

He returned to the subject in an address before the Royal Dublin Society in 1891, in which he first used the term 'electron'. Stoney G J 1891 On the cause of double lines and of equidistant satellites in the spectra of gases *Sci. Trans. R. Dublin Soc.* **4** 563; 1894 On the 'electron' or atom of electricity *PM* **38** 418

10-35 1899 Physics at the British Association *NA* **60** 586

10-36 E.g. see note 10-26

The handwritten manuscript of this famous paper is preserved in the archives of Thomson's papers held at the University Library, Cambridge. The ink-written manuscript contains various pencilled corrections in Thomson's hand. Moreover, in the archival listing, the term 'temperatures' is erroneously substituted for 'pressures' in the title of the paper. CUL, Add 7654, PD 17

Portions of the paper also appear in one of Thomson's notebooks for 1899; namely, notebook No 45, listed under the same title as the paper in P Spitzer's handlist of Thomson's notebooks, CUL, Add 7654, NB 45

10-37 Thomson: see note 10-26, p 565

10-38 Poynting J H 1899 Opening address of Prof. J. H. Poynting *NA* **60** p 472

10-39 Lodge: see note 10-35, p 587

Hunt: see note 6-57, that discusses the interaction between Oliver Lodge, George FitzGerald, and Oliver Heavyside in promoting Maxwell's electrodynamics.

10-40 Henry Edward Armstrong studied at the Royal College of Chemistry and received his doctorate in 1870 under A Kolbe at Leipzig. After various teaching posts, he served as professor at the Central Technical College in London (later Imperial College of Science and Technology).

His research included aromatic substitution, crystallography, stereochemistry, terpenes, and enzymes. He was a major contributor in British technical education, and a forceful advocate for environmental concern and the reduction of wasteful practices in manufacturing industry and energy generation.

10-41 Rayleigh: see note 1-46, p 114

10-42 Lodge: see note 10-35, p 587

10-43 Howarth O J R 1931 *The British Association for the Advancement of Science: a Retrospect 1831–1931* 2nd edn (London: British Association)

10-44 Thomson: see note 1-25, p 341

Notes section 11.1

11-1 Keesom W H 1926 Prof Dr H Kamerlingh Onnes. His life-work, the founding of the Cryogenic Laboratory *PLS* **57** 3

11-2 Keesom: see note 11-1

11-3 Keesom: see note 11-1, p 4

11-4 Most of the papers of Kamerlingh Onnes and his collaborators were first published in the *Proceedings of the Royal Academy* in Amsterdam.
An excellent overview of the earlier researches in the Leiden laboratory is found in two commemorative volumes. The first one, with an introduction by J Bosscha, was published in 1904, on the occasion of Kamerlingh Onnes' silver doctorate; the second one, with an introduction by H A Lorentz, on the 40th anniversary of his inaugural lecture at Leiden in 1922. 1904 *Het Natuurkundig Laboratorium der Rijksuniversiteit te Leiden in de jaren 1882–1904* (Leiden: Ijdo); 1922 *Het Natuurkundig Laboratorium der Rijksuniversiteit te Leiden in de jaren 1904–1922* (Leiden: Ijdo)

11-5 Cohen E 1927 Kamerlingh Onnes memorial lecture *Chemical Society Journal* **1** 1199

11-6 Keesom: see note 11-1, p 11

11-7 Gorter C J and Taconis K W 1964 The Kamerlingh Onnes Laboratory *Cryogenics* **4** 346

Notes section 11.2

11-8 Freely translated from the Latin in Watson E C 1954 Reproductions of prints, drawings, and paintings of interest in the history of physics: the discovery of the Zeeman effect *AJP* **22** 635. The artist was Harm K Onnes, Kamerlingh Onnes' nephew.

11-9 Henry Augustus Rowland (1848–1901) was perhaps the most influential American experimental physicist in the last quarter of the 19th century, despite the fame of Albert Michelson. He graduated in civil engineering from Rensselaer Technical Institute in 1870, after which, for want of a scientific position to his liking, he spent some time doing researches on magnetic materials in his mother's home. This work soon put him in contact with Maxwell who was quite receptive to his ideas. In 1875, he chanced across Daniel C Gilman, president of the newly founded Johns Hopkins University, who hired the young engineer (on the strength of his correspondence with Maxwell) to organize a physics department at the University. First, however, Rowland spent a year at Helmholtz's laboratory, performing experiments tending to confirm that the mere motion of electric charge could generate magnetic effects analogous to those of a current in a wire.

At Johns Hopkins, Rowland worked on more accurate values for the basic physical constants, in particular the mechanical equivalent of heat and the standard ohm. By the late 1870s, he had assembled what was perhaps the most extensive collection of apparatus then extant. Rowland is best remembered for the concave grating which bears his name, and for the ruling engine which allowed him to produce diffraction gratings with 20 000 grooves in^{-1}, eliminating lenses and mirrors in spectrometers and giving greater resolving power and dispersion. Miller J D 1976 Rowland's physics *PT* **29** 39–45

11-10 Zeeman P 1896 Over den invloed eener magnetisatie op den aard van het door een stof uitgezonden licht *Koninklijke Akademie van Wetenschappen te Amsterdam, Verslag* **5** 181–3

Kamerlingh Onnes H 1921 Zeeman's Ontdekking van het naar hem genoemde Effect *Physica* **1** 241–50

11-11 1896 *Nature* **55** 192

11-12 Zeeman P 1897 On the influence of magnetism on the nature of the light emitted by a substance *PM* **43** 232

11-13 Spencer J B 1976 Zeeman, Pieter *DSB* **14** 597–9

11-14 Faraday: see note 3-23, *Diary*, paragraph 7504

11-15 Maxwell J C 1954 *The Scientific Papers of J C Maxwell* vol 2, ed W P Niven (New York: Dover) p 225

11-16 Jones H B 1870 *Life and Letters of Faraday* 2 vols, vol 2 (London: Longmans) p 449

11-17 Faraday: see note 3-23, *Diary*, paragraph 465

11-18 McCormmack R 1973 Lorentz, Hendrik Antoon *DSB* **8** 487–500

11-19 McCormmack: see note 11-18

11-20 de Haas-Lorentz G L (ed) 1957 *H A Lorentz. Impressions of his Life and Work* (Amsterdam: North-Holland)

11-21 Lorentz H A 1892 La théorie électromagnétique de Maxwell et son application aux corps mouvants *NA* **25** 363; reprinted in Zeeman P and Fokker A D (eds) 1935–1939 *H A Lorentz, Collected Papers* 9 vols, vol 2 (The Hague: Martinus Nijhoff) pp 164–343. Vols 1–8 of the *Collected Papers* contain technical writings in Dutch, English, or French. Vol 9 contains Lorentz's non-mathematical writings, mostly in Dutch, including his 1878 inaugural lecture (also in English translation); it also contains an extensive biography, including papers *not* reprinted in the *Collected Papers*.

11-22 Lorentz H A 1895 Versuch einer Theorie der electrischen und optischen Erscheinungen in bewegted Körpern *Collected Papers* (note 11-21), vol 5, 1–155

11-23 McCormmack: see note 11-18, p 495. See also note 1-22

11-24 Pais: see note 3-29, p 76

A quotation by Albert Einstein is also appropriate in this connection. In the slender volume on Lorentz edited by Lorentz's daughter, he wrote as follows: 'No longer... do physicists of the younger generation fully realize, as a rule, the determinant part which H. A. Lorentz played in the formation of the basic principles of theoretical physics. The reason for this curious fact is that they have absorbed Lorentz' fundamental ideas so completely that they are hardly able to realize to the full the boldness of these ideas, and the simplification which they brought into the foundations of the science of physics'. Einstein A in de Haas-Lorentz: see note 11-20, pp 5–9

11-25 Zeeman: see note 11-12, p 235

11-26 Zeeman: see note 11-12, p 235
11-27 Zeeman P 1896 Over den invloed eener magnetisatie op den aard van het door een stof uitgezonden licht *Koninklijke Akademie van Wetenschappen te Amsterdam, Verslag* **5** 242–8
11-28 Zeeman P 1896 Ueber einen Einfluss der Magnetisirung auf die Natur des von einer Substanz emittirten Lichtes *Verhandlungen der Physikalischen Gesellschaft zu Berlin* **7** 128; translated version in Magie: see note 1-6, pp 384–6
11-29 Zeeman P 1897 The effect of magnetization on the nature of light emitted by a substance *NA* **55** 347
11-30 Zeeman: see note 11-12, p 226

Notes section 11.3

11-31 Lodge O 1922 The history of Zeeman's discovery, and its reception in England *NA* **109** 66–9
11-32 Lodge O 1897 The influence of a magnetic field on radiation frequency *PRS* **60** 513
11-33 Lodge: see note 11-32, p 513
11-34 Larmor J 1897 The influence of a magnetic field on radiation frequency *PRS* **60** 514–5
11-35 Woodruff A E 1973 Larmor, Joseph *DSB* **8** 39
11-36 Woodruff: see note 11-35, p 39
11-37 Romer A 1948 Zeeman's discovery of the electron *AJP* **16** 217
11-38 Larmor J 1894 A dynamical theory of the electric and luminiferous medium *PTRS* A **185** 719–822; 1895 A dynamical theory..., Part II. Theory of electrons *PTRS* **186** 695–743
11-39 Larmor J 1929 *Mathematical and Physical Papers* 2 vols, vol 2 (Cambridge: Cambridge University Press) p 140
11-40 Lodge: see note 11-31, p 67
11-41 Lodge O 1897 The latest discovery in physics *Electrician* **38** 568–70
11-42 Zeeman: see note 11-12
11-43 Lodge: see note 11-41, p 569
11-44 Lodge: see note 11-31, pp 67–8
11-45 Lodge O 1897 Further note on the influence of a magnetic field on radiation frequency *PRS* **61** 414; 1897 *NA* **56** 238
11-46 Lodge: see note 11-31, p 68
11-47 Lodge: see note 11-31, pp 68–9
11-48 Larmor: see note 11-38, p 720
11-49 Tait P 1875 On a possible influence of magnetism on the absorption of light, and some correlated subjects *Proceedings of the Royal Society of Edinburgh* **9** 118
11-50 Tait: see note 11-49, p 118
11-51 Crowther: see note 6-1, p 344
Stewart was deeply into metaphysics, and, with P G Tait, published in 1875 *The Unseen Universe,* which attempted to establish a physical basis for immortality. Such was the public response that they published a sequel *Paradoxical Philosophy*—not nearly as successful. Thomson: see note 1-25, p 22
11-52 Schuster: see note 1-39, p 22
11-53 Schuster: see note 1-39, p 23

11-54 Balmer J J 1885 Die Spectrallinien der Wasserstoffs *Verhandelung der Naturforschenden Gesellschaft in Basel* **7** 894, reported to the Basel *Gesellschaft* June 25 1884.
Stoney G J 1871 *PM* **41** 291. Stoney suggested that studying the line spectra was best accomplished by mapping them on a scale of wave-numbers (reciprocals of wavelengths).

11-55 Stoney G J 1891 On the cause of double lines and of equidistant satellites of the spectra of gases *Trans. R. Dublin Soc.* **4** 563–608

11-56 Fievez C 1885 De l'influence du magnétisme sur les caractères des raies spectrales *Bulletin de l'Académe des Sciences de Belgique* **9** 381; 1886 Essai sur l'origine des raies de Fraunhofer, en rapport avec la constitution du soleil *Bulletin de l'Académe des Sciences de Belgique* **12** 30
The Belgian metallurgist E van Aubel of Ghent drew attention to these papers in a letter to Kamerlingh Onnes, who read it at the January 1897 meeting of the Amsterdam Academy; Zeeman: see note 11-12, p 237

11-57 Zeeman: see note 11-12, p 238
For further discussion of Fievez's 'Qualified discovery', see Brookes Spencer J 1970 On the varieties of nineteenth-century magneto-optical discovery *Isis* **61** 44–5

11-58 Lodge: see note 11-41, p 569

11-59 Miller: see note 11-9, p 44

11-60 Miller: see note 11-9, p 45

11-61 Zeeman P 1897 Doublets and triplets in the spectrum produced by external magnetic forces *PM* **44** 55–60

11-62 Zeeman: see note 11-61, p 59

11-63 Zeeman: see note 11-61, p 58

11-64 Zeeman: see note 11-61, p 60

11-65 Zeeman P 1897 Doublets and triplets in the spectrum produced by external magnetic forces—II *PM* **44** 256

11-66 1897 The radiation of light in the magnetic field *NA* **56** 420

11-67 see note 11-66

11-68 Thomas Preston was born in Kilmore, County Armagh, Ireland, in 1860. He studied physics at the University of Dublin, where he became Professor of Natural Philosophy in 1891, and remained until his untimely death in 1900. His contributions lay in the areas of heat, magnetism, and spectroscopy.

11-69 Preston T 1897 The Zeeman effect photographed *NA* **57** 173

11-70 Preston T 1898 Radiation phenomena in the magnetic field *PM* **55** 325–39

11-71 Preston: see note 11-70, p 329

11-72 Preston: see note 11-70, p 329

11-73 Preston: see note 11-70, pp 330–1

11-74 Preston: see note 11-70, p 338

11-75 Cornu A 1897 Sur l'observation et l'interprétation cinématique des phénomènes découverts par M. le Dr. Zeeman *CR* **125** 555–61
Herivell J W 1971 Cornu, Marie Alfred *DSB* **3** 419–20

11-76 Cornu A 1898 Sur quelques résultats nouveaux relatifs au phénomène découverts par M. le Dr. Zeeman *CR* **126** 181–6
Wrote Lorentz himself in 1921, on the occasion of the 25th anniversary of the Zeeman effect 'Unfortunately... theory could not keep pace with experiment and the joy aroused by [Zeeman's] first success was but short-lived. In 1898 Cornu discovered—it was hardly credible at first!—that the line D_1 is decomposed into a quartet...' Lorentz H A 1921 The theoretical significance of the Zeeman effect *Physica* **1** 228; Engl. transl. in 1934 *H A Lorentz, Collected Papers* vol 7 (The Hague: Martinus Nijhoff) p 87

11-77 Michelson A A 1898 Radiation in a magnetic field *PM* **55** 348–56

11-78 Zeeman P 1967 Light radiation in a magnetic field, Nobel Lecture, December 11 1902 *Nobel Lectures, including Presentation Speeches and Laureates' Biographies: Physics 1901–1921* (Amsterdam: Elsevier) pp 33–40
Lorentz H A 1967 The theory of electrons and the propagation of light *Nobel Lectures, including Presentation Speeches and Laureates' Biographies: Physics 1901–1921* (Amsterdam: Elsevier) pp 14–29

11-79 Preston T 1899 General law of the phenomena of magnetic perturbation of spectral lines *NA* **59** 248, cited by Mehra J and Rechenberg H 1982 *The Historical Development of Quantum Theory, Vol 1, the Quantum Theory of Planck, Einstein, Bohr and Sommerfeld: its Foundations and the Rise of its Difficulties 1900–1925* (New York: Springer) part 1, p 448; 1970 Magnetic perturbation of the spectral lines, Friday Evening Discourse, Royal Institution, May 12, 1899 *The Royal Institution Library of Sciences, Physical Sciences* vol 5, eds W L Bragg and G Porter (Barring: Elsevier) p 274

11-80 Spencer: see note 11-13, p 598
See also the commemorative volume 1921 *Verhandelingen van Dr P Zeeman over Magneto-Optische Verschijnselen* (Leiden: Ijdo) introduction signed by H A Lorentz, H Kamerlingh Onnes, I M Graftdijk, J J Hallo and H R Woltjer
For Zeeman on Michelson's echelon spectroscope, see Zeeman P 1901 Some observations on the resolving power of the Michelson echelon microscope *Versl. Kon. Akad. v. Wet. Amsterdam* **10** 136–42
For Zeeman on Voigt versus Lorentz, see Zeeman P 1899 Some observations concerning an asymmetrical change of the spectral lines of iron radiating in a magnetic field *Verslag van de gewone Vergadering der wis-en natuurkundige Afdeeling Koninklijke Akademie van Wetenschappen te Amsterdam* **8** 122–8

11-81 Watson: see note 11-8, p 634

11-82 Pais: see note 3-29, p 77

11-83 Lorentz H A 1934 Optical phenomena connected with the charge and mass of the ions I, II *Collected Papers* vol 3 (The Hague: Martinus Nijhoff) p 17, 30

11-84 Thomson: see note 10-26, p 567

11-85 Thomson: see note 10-26, p 567

11-86 Zeeman: see note 11-78

Notes section 12.1

12-1 Lenard: see note 5-44
Thomson: see note 10-26

12-2 Wheaton: see note 5-24, p 311. Note: Thomson J J 1900 Über die Masse der Träger der negativen Electisierung in Gasen von niederen Drucken *PZ* **1** 20–2 is erroneously dated 1889 in Wheaton: see note 5-24. Perhaps Thomson published his results in a German journal to safeguard his position in light of Lenard's ongoing program, of which Thomson undoubtedly was quite aware.

12-3 Lenard P 1900 Erzeugung von Kathodenstrahlen . . . *AP* **2** 359–75
In the reprint of this paper in *Wissenschaftliche Abhandlungen* Lenard added a long footnoted tirade wherein he castigates Thomson for his 'contrary' or unethical conduct on receiving a copy of the paper from Lenard himself. Thomson, Lenard claims, belittled his work in favor

of Thomson's own 'extemporized' experiment which postdated Lenard's publication in print, and appeared without even extending due credit to Lenard's work. Worse, continues Lenard, was the humiliating treatment he suffered at the hands of his countrymen at the Bad Neuheim meeting of the German Physical Society after the Great War in 1920, where Lenard fruitlessly attempted to uphold wronged German science in the face of 'misguided scientific attitudes' abroad. Lenard: see note 5-28, pp 237–8
See also Lenard, note 5-24, p 105
Beyerchen A D 1977 *Scientists under Hitler: Politics and the Physics Community in the Third Reich* (New Haven, CT: Yale University Press) p 81

12-4 Lenard P 1920 *Ueber Kathodenstrahlen* (Berlin: Vereinig Wissenschaftl. Verleger)
Wheaton: see note 5-24, p 303
The annotated version is reprinted in *Wissenschaftliche Abhandlungen* with an addendum drawing further attention to the originality of Lenard's work on cathode rays *vis-à-vis* that of his contemporaries at home and abroad, and to more information in the same vein in his unpublished 'Erinnerungen'. Lenard: see note 5-28, p 197; see note 5-24

12-5 Hertz: see note 5-38, p 983; translated in Margaritondo G 1988 100 years of photoemission *PT* **41** 67

12-6 Hertz to Gustav F Hertz, July 7 1887 in Hertz J and Süsskind C (ed) 1977 *Heinrich Hertz: Erinnerungen, Briefe, Tagebücher* 2nd revised edition (San Francisco: San Francisco Press) p 226; cited by Jungnickel and McCormmach: see note 9-26, p 80

12-7 Nahrwold R 1887 Ueber Luftelectricität *APC* **31** 469, 473; cited by Wheaton: see note 5-24, p 302

12-8 Lenard: see note 5-41, p 444; cited by Wheaton: see note 5-24, p 302

12-9 For a derivation of R from the x-and y-coordinates of β, d_1, and d_2, see the earlier discussion on p 165 and Harnwell and Livingood: see note 9-47

12-10 Values for e/m are tabulated in the fourth column of Lenard: see note 12-3, table III, p 368.

12-11 Lenard: see note 12-3, pp 370–1

12-12 Lenard: see note 12-3, p 374

12-13 In considering Lenard, his former student at Heidelberg, for a theoretical position at Heidelberg in 1896, Georg Quincke required of Lenard no more evidence of mathematical ability than his editorial role in connection with Hertz's *Mechanics* (volume 3 of Hertz's *Gesammelte Werke*). Jungnickel and McCormmach: see note 9-26, p 42

12-14 Lenard P 1898 *Ueber die electrostatischen Eigenschaften der Kathodenstrahlen APC* **64** 279, translated in Wheaton: see note 5-24, p 309. Lenard termed the electrical entities 'elementary quanta' after Helmholtz in his Faraday Lecture (p 80), not to be confused with Planck's subsequent 'quantum of action'.

12-15 Wheaton: see note 5-24

12-16 Lenard P 1902 Ueber die lichtelektrische Wirkung *AP* **8** 149–98, with additional table 1, figures 1 and 2

12-17 Lenard: see note 12-16, p 154

12-18 Lenard: see note 12-16, p 166

12-19 Lenard: see note 12-16, pp 169–70

12-20 Landenburg E 1907 Über Anfangsgeschwindigkeit und Mengde der photoelektrischen Elektronen in ihrem Zusammenhange mit der

Wellenlänge des auslösenden Lichtes *Verhandlungen der Deutschen Physikalischen Gesellschaft* **9** 504–14

12-21 Lenard: see note 12-16, p 168
Wheaton: see note 5-24, pp 317–8

12-22 Einstein A 1905 Über einen die Erzeugung und Verwandlung des Lichtes betreffenden heuristischen Gesichtspunkt *AP* **17** 132–48, submitted to *Annalen* on March 18 1905. (This work earned Einstein the Nobel Prize in physics for 1921.) The second paper, 'Über die von der molekularkinetischen Theorie der Wärme gerforderte Bewegung von in ruhenden Flüssigkeiten suspendierten Teilchen', submitted on May 11 1905, contains the theory of Brownian motion. (This work earned him the doctorate from the University of Zürich.) The third, 'Zur Elektrodynamik bewegeter Körper', reached *Annalen* on June 30; it contains Einstein's special theory of relativity. It is interesting to note that only one of eight sections of the first paper, known today almost entirely for its treatment of the photoelectric effect, is devoted to that effect; the rest of the paper deals with the black-body radiation law, the entropy of radiation, and photoluminescence.

12-23 In Einstein's paper, the quantum of energy is obscured as $(R/N)\beta\nu$ where $R/N = k$, the Boltzmann constant, R is the molar gas constant, and N Avogadro's number); thus $\beta = h/k$. Einstein's photoelectric equation, which we may write as $\frac{1}{2}mv_m^2 = h\nu - \omega_0 = eV_0$, where v_m is the maximum electron velocity, ω_0 the work function of the emitting material, and V_0 the retarding potential, explains Lenard's finding that v_m is independent of the light intensity. The retarding potential, Einstein deduced, agreed 'in order of magnitude' with Lenard's result. 'As far as I can see our ideas are not contrary to Mr. Lenard's observations on the photoelectric effect', he continued, stressing particularly that his light-quantum hypothesis explained Lenard's finding that the number of photoelectrons emitted was proportional to the intensity of the incident radiation. Einstein: see note 12-22, p 147
Lenard: see note 12-16, pp 150, 166–8. Einstein's equation was first verified by Robert Millikan in 1915, when he showed that the slope of a curve plotting eV_0 against frequency gives Planck's constant h.

12-24 Wheaton: see note 5-24, p 317

12-25 Lenard P 1903 Über die Absorption von Kathodenstrahlen verschiedener Geschwindigkeit *AP* **12** 714–44

12-26 Wheaton: see note 5-24, pp 318–22
Rudolf Landenburg, younger brother of Erich Landenburg, became one of the first physicists to accept Einstein's view of the constitution of radiation as it applied to the photoelectric effect, in Landenburg R 1909 Die neueren Forschungen über die durch Licht und Röntgenstrahlen hervorgerufene Emission negativer Elektronen *Jahrbuch der Radioaktivität und Elektronik* **6** 425–84
Shenstone A G 1973 Ladenburg, Rudolf Walther *DSB* **7** 522–56

Notes section 13.1

13-1 Malley M 1971 The discovery of the beta-particle *AJP* **39** 1454

13-2 Rutherford: see note 8-80, pp 162–3

13-3 Badash L 1965 Radioactivity before the Curies *AJP* **33** 131

13-4 Curie M 1898 On a new radioactive substance contained in pitchblende *CR* **127** 175

13-5 A largely forgotten contemporary claim for a new radioactive element was W Crookes's 'monium', also called 'Victorium' in honor of the Diamond Jubilee, that proved to be a rare-earth mixture. Crookes W 1899 Photographic researches on phosphorescent spectra: on Victorium, a new element associated with Yttrium *Chemical News* **ixxx 31** 49
DeKosky R K 1972–1973 Spectroscopy and the elements in the late nineteenth century: the work of Sir William Crookes *The British Journal for the History of Science* **6** 418

13-6 Gerlach W 1971 Elster, Johann Philipp Ludwig Julius *DSB* **4** 354–7; Geitel, F. K. Hans *DSB* **5** 341–2

13-7 Gerlach: see note 13-6, p 355

13-8 Elster J and Geitel H 1889 Elektrisierung der Gase durch glühende Körper *APC* **37** 315

13-9 Trenn T J 1971 Giesel, Friedrich Oskar *DSB* **5** 394–5

13-10 Hahn O 1966 *A Scientific Autobiography* (New York: Scribner) p 271. A radiation-induced carcinoma on his left index finger is blamed for his death in 1927. Aside from Giesel, the only suppliers of radium were the Curies in Paris and the chemical firm of E de Haën of List near Hanover.

13-11 Elster J and Geitel H 1899 Ueber den Einfluss eines magnetischen Feldes auf die durch die Becquerelstrahlen bewirkte Leitfähigkeit der Luft *Verhandlungen Deutsche Physikalische Gesellschaft* **1** 138

13-12 Meyer S 1949 Zur geschichte der entdeckung der Natur der Becquerel Strahlen *Die Naturwissenschaften* **36** 129

13-13 Meyer S and von Schweidler E 1899 Über des Verhalten von Radium und Polonium im magnetischen Felde *Anzeiger der Kaiserlichen Akademie der Wissenschaften zu Wien* **36** 309–10, 323–4 (received November 3 and 9 1899); *PZ* **1** 90–1, 113–4 (publ. November 25 and December 2 1899)

13-14 Badash L 1966 An Elster and Geitel failure: magnetic deflection of beta-rays *Centaurus* **11** 238

13-15 Meyer: see note 13-12, p 130
Malley: see note 13-1, p 1456

13-16 Meyer, see note 13-12, p 131

13-17 Giesel F 1899 Ueber die Ablenbarkeit der Becquerelstrahlen im magnetischen Felde *APC* **69** 834-6 (submitted November 2 1899)
Meyer: see note 13-12, p 129

13-18 Giesel to Meyer, October 22 1899, Meyer: see note 13-12, p 129

13-19 Giesel to Meyer, October 26 1899, Meyer: see note 13-12, p 130

13-20 Malley: see note 13-1, p 1460
Giesel F 1906 *Berlin, Deutsche Chemische Gesellschaft* **39** 780, 1014

Notes section 13.2

13-21 Meyer: see note 13-12

13-22 For his photographic results, see Giesel: see note 13-17, pp 835-6

13-23 Meyer and von Schweidler: see note 13-13

13-24 We are reminded of the importance of the strength of available electromagnets in another, related context: confirmation of the Zeeman effect, then a much hotter topic of research (chapter 11).

Notes section 13.3

13-25 Romer: see note 8-57, p 559

13-26 Becquerel H 1899 Note sur quelques propriétés du rayonnement de l'uranium et les corps radio-actifs *CR* **128** 771–7
Heilbron J L 1968 The scattering of α and β particles and Rutherford's atom *Archive for History of Exact Sciences* **4** 251

13-27 Jauncey: see note 8-74, p 229

13-28 Malley: see note 13-1, p 457

13-29 Becquerel H 1899 Sur le rayonnement des corps radioactifs *CR* **129** 1205–7

13-30 Curie M 1899 *Revue Generales des Sciences Pures et Appliqués et Bulletin de l'Association Française pour l'Advancement des Sciences* **10** 41

13-31 Becquerel H 1899 Influence d'un champ magnétique sur le rayonnements des corps radio-actifs *CR* **129** 996–1001; see note 13-29. That the rays from uranium consisted of deviable and non-deviable rays was also reported by P Curie on January 8 1900

13-32 Becquerel H: see note 13-29
Heilbron: see note 13-26, p 252

13-33 Rutherford to Elster and Geitel, June 12 1899, cited by Malley: see note 13-1, p 1460

13-34 Thomson to Rutherford, December 21 1899, CUL, Add 7653, T13

13-35 Rutherford to Thomson, January 9 1900, cited by Rayleigh: see note 1-46, pp 132–3

13-36 Rayleigh: see note 1-46, p 132

13-37 Rayleigh: see note 1-46, p 133. Becquerel H 1970 Sur la radio-activité de la Matière, Friday Evening Discourse, Royal Institution, March 7 1902 *The Royal Institution Library of Science: Physical Sciences* vol 5, ed W L Bragg and G Porter (Amsterdam: Elsevier) 508–24

13-38 Rutherford to his mother, January 5 1902, cited by Eve: see note 6-29, pp 80–1

13-39 Thomson: see note 1-25, p 413

13-40 Thomson: see note 1-25, pp 413–4

13-41 The term 'excited radioactivity' was Rutherford's, in keeping with his simple pleasure of assigning the phenomenon the same initials as his own, Keller: see note 3-61, p 96; 'induced radioactivity' was coined by Pierre Curie. Since in time it was realized that the affected surfaces did not actually become radioactive themselves, Rutherford also introduced the term 'active deposit', as a substitute for his 'excited radioactivity'.

13-42 Zeleny to Rutherford, March 25 1900, cited by Feather: see note 8-47, p 64

13-43 Callendar to Rutherford, January 19 1900, cited by Feather: see note 8-47, p 71

13-44 FitzGerald to Rutherford, May 5 1900, cited by Feather: see note 8-47, p 73

13-45 Strutt to Rutherford (date uncertain, but spring 1900), cited by Feather: see note 8-47, p 74

13-46 Rutherford to his mother, January 5 1902: see note 13-38. Early notes and calculations on radioactivity still extant, in Rutherford's hand and written at McGill, include the following packets among the Rutherford archives at Cambridge: 'Set of summaries of nine papers by the Curies and associates, ca. 1902, CUL, Add 7653, PA166; References and abstracts of papers on radioactivity, 32 leaves. 1903, CUL, Add 7653, PA168

13-47 Shaw N 1937 *McGill News* December, cited by Eve: see note 6-29, p 88

13-48 Curie P 1900 Action du champ magnétique sur les rayons de Becquerel. Rayons déviés et rayons non déviés *CR* **130** 73

13-49 Curie P and Curie M 1900 Sur la charge électrique des rayons déviables du radium *CR* **130** 647

13-50 Thomson to Rutherford, February 15 1901, CUL, Add 7653, T14

13-51 Dorn E 1900 *Abhandlungen der Naturforschenden Gesellschaft in Halle* **22** 47 (signed March 11 1900); *PZ* **1** 337

13-52 Becquerel H 1900 Contribution a l'étude du rayonnement du radium *CR* **130** 206–11 (received January 29 1900); Déviation du rayonnement du radium dans un champ électrique' *CR* **130** 809–15 (received March 26 1900); *Rapports présentés au Congrés International de Physique à Paris* vol 3, (Paris: Gauthier-Villars) p 47

13-53 Thomson J J 1906 *Conduction of electricity through gases* 2nd edn (Cambridge: Cambridge University Press) pp 139–44

13-54 A more elegant method than invoking Becquerel's electrostatic method for obtaining a second relation between v and e/m has been suggested by Thomson. Place the radium on a photographic plate, encapsulated in such a manner that the rays are emitted at right angles to the plate. Let a uniform field act parallel to the plate, and mount a metal plate above and parallel to the photographic plate. The plate can be charged to a high potential, producing an electric field parallel to the direction of projection of the rays and at right angles to the magnetic field. Photographs taken with the metal plate charged and uncharged will then give two sets of expressions involving v and e/m, which may be solved for the ratio e/m alone. Thomson: see note 13-53, p 142–4

13-55 H A Boorse and L Motz (eds) 1966 Mass changes with velocity *The World of the Atom* (New York: Basic Books) chap 34, p 503

13-56 Kaufmann W 1901 Die magnetische und elektrische Ablenkbarkeit der Becquerelstrahlen und die scheinbare Masse der Elektronen *Nachrichten von der Gesellschaft der Wissenschaften zu Göttingen* Math.–Phys. Kl. **2** 143–55; transl. Magnetic and electric deflectability of the Becquerel rays and the apparent mass of the electron: see note 13-55, p 506–12

13-57 Curie M 1900 Action du champ magnétique sur les rayons de Becquerel. Rayons déviés *CR* **130** 73–6 (submitted January 8 1900)

13-58 Curie M 1900 Sur la pénétration des rayons de Becquerel non déviables par le champ magnétique *CR* **130** 76–9 (submitted January 8 1900)

13-59 Heilbron: see note 13-26, p 253

13-60 In either instrument, electrometer or electroscope, two surfaces, physically connected but electrically insulated, are charged up with electricity of the same polarity, so that they repel each other. Normal air acts as an insulator, but as charge leaks away slowly the repulsion between the two surfaces diminishes; the rate of leakage is measured quantitatively by observing the movement of the charged surfaces with a microscope or telescope. The familiar gold-leaf electroscope has, in one variation, a vertical stationary metal piece and a single movable strip of gold leaf fastened near the top. The lead-in is insulated from the metal housing with an amber or sulfur bushing. The most common electrometer was the Dolezalek quadrant electrometer, consisting of a cylindrical brass 'pillbox' divided into four equal and insulated quadrants, opposite quadrants being connected together. Typically, a torsion-suspended needle is kept at a high potential with respect to ground, and the charge to be measured is applied to one pair of quadrants while the other pair is grounded. Strong J 1946 *Procedures in Experimental Physics* (New York: Prentice Hall) ch 6

Eve relates that Rutherford asked him, about 1904, to make a small gold-leaf electroscope that would remain charged for 2 to 3 days, a task which he botched. Annoyed, Rutherford told him to get Jost, the laboratory mechanic, to make one. Jost 'made a beauty to look at, but a bad one to go. Its leaf collapsed in twelve hours'. Puzzled, Eve arose one night, unable to sleep, and concocted one from 'a tobacco tin, an amber mouthpiece of a tobacco pipe, and some Dutch metal foil; charged it with sealing wax and went to sleep'. It worked like a charm, and solved their unexpected problem; everything inside the laboratory building was contaminated and coated with active deposits. 'Rutherford said "Good Boy!" though I was eight or nine years his senior in age'. Eve: see note 6-29, p 101

13-61 Rutherford to Thomson, December 26 1902, CUL, Add 7654, R67

13-62 Rutherford E 1903 The magnetic and electric deviation of the easily absorbed rays from radium *PM* **5** 178

13-63 Eve: see note 6-29, p 86, citing Soddy F 1933 *Old McGill* **36** 19

13-64 Terroux F R 1938 *The Rutherford Collection of Apparatus at McGill University* printed for the Royal Society of Canada, from 1938 *Transactions of the Royal Society of Canada* 3rd series, section III, vol 32, CUL, Add 7653, PA315

13-65 Rutherford E 1906 The mass and velocity of the α-particles expelled from radium and actinium *PM* **12** 348–71
This apparatus is also preserved at McGill, Terroux: see note 13-64

13-66 Rutherford E and Geiger H 1908 The charge and nature of the α-particle *Proceedings of the Royal Society of London* A **81** 162–73

13-67 Villard P 1900 Sur la réflexion et la réfraction des rayons cathodiques et des rayons déviables du radium *CR* **130** 1010–12; Sur le rayonnement du radium *CR* **130** 1178–9

13-68 Wilson: see note 7-19, p 127

13-69 Rutherford: see note 8-80, p 116

13-70 Rutherford: see note 8-80, p 122

13-71 Rutherford: see note 8-80, p 123

13-72 Jauncey 1904: see note 8-74, p 234
Rutherford E 1904 *Radio-activity* (Cambridge: Cambridge University Press)
According to Eve, 'It says a good deal for both men that [the Rutherford–Soddy] friendship survived this strain' of their near-simultaneous books on radioactivity; Eve: see note 6-29, p 99

13-73 Rutherford E and Andrade E N da C 1914 The wave-length of the soft γ rays from Radium B *PM* **27** 854–68

Notes section 14.1

14-1 Nye M J 1980 N-rays: an episode in the history and psychology of science *HSPS* **11** 125–56
Lagemann R T 1977 New light on old rays: N rays *AJP* **45** 281–4
Weart S 1978 A little more light on N rays *AJP* **46** 306
Klotz I M 1980 The N-ray affair *Scientific American* **242** 168–75
Firth I 1969 N-rays—ghost of scandal past *New Scientist* **44** 642–3

14-2 Gough J B 1970 Blondlot, René-Prosper *DSB* **2** 202–3

14-3 Gain A 1934 L'enseignement supérieur à Nancy de 1789–1896 *L'Université de Nancy (1572-1934)* (Nancy) pp 25-42
The loss of Strasbourg to the Germans in the war of 1870 resulted in the transfer to Nancy of Strasbourg faculty members who chose to remain

French. The Nancy faculty and curriculum expanded even more with the decentralization of the national university system in the 1880s—physical sciences particularly so under the direction of Albin Haller and Blondlot's colleague Ernest Bichat. Bichat was also professor of physics at Nancy; this unusual arrangement of two professors of physics in a French university came about when Bichat became dean of science in 1882. Lagemann: see note 14-1, p 284, note 15

14-4 Nye: see note 14-1, pp 129–30
Blondlot R-P 1881 *Recherches Expérimentales sur la Capacité de Polarisation Voltaique* (Paris); 1891 Determination de la vitesse de propagation des ondes électromagnétiques *CR* **113** 628–31; 1893 Détermination de la vitesse de propagation d'une perturbation électrique le long d'un fil de cuivre *CR* **117** 543–6
Poincaré H 1904 Rapport sur le prix Leconte attribué a M. Blondlot *Revue Scientifique* **2** 843–4
Thomson J J 1910 Electric waves *Encyclopedia Britannica* 11th edn (Cambridge: Cambridge University Press) p 208

14-5 Nye: see note 14-1, p 126

14-6 This was a widely discussed explanation for the recurrent failure to polarize x-rays at the time; one argued, for instance, by J J Thomson. Nye: see note 14-1
Thomson J J 1898 Theory on the connection between cathode and Röentgen rays *PM* **45** 172–83. Not until 1906 was the partial polarization of x-rays achieved by Charles Barkla (1877–1944); the question of the transverse wave interpretation of x-rays was not fully settled in the affirmative until 1912 when Walter Friedrich and Paul Knipping, on the suggestion of Max von Laue, performed the first x-ray refraction experiments on crystals.

14-7 Nye: see note 14-1, pp 127–8
Blondlot R-P 1903 Sur la polarisation des rayons x *CR* **136** 284–6, reprinted in 1905 *'N' Rays, a Collection of Papers Communicated to the Academy of Sciences...* transl. J Garcin (London: Longmans Green). Blondlot's own depiction of his apparatus (reproduced as figure 14.1 here) lacks pertinent details, but they are provided in a good reconstruction of the experimental arrangement in Nye: see note 14-1, p 127

14-8 Nye: see note 14-1, p 131
Blondlot carried out systematic measurements of N-ray wavelengths with an iron-enclosed Auer lamp equipped with an aluminum window and a quartz lens. Heinrich (Leopold) Rubens, the guru of infrared radiation research, argued vehemently against the possibility of radiation of such long wavelength traversing an aluminum window fully 0.1 mm thick and, with Otto Lummer (1860) and Ernst Hagen (1851–1923), was unable to confirm Blondlot's results. Nor, according to Georges Sagnac, should quartz pass the rays. Blondlot countered that they were not, strictly, heat rays, but radiation outside the spectral range explored by Rubens and his associates. Indeed, later, revised measurements by Blondlot found the rays not to lie in the infrared region, after all, but in the far ultraviolet. Nye: see note 14-1, p 139

14-9 Thomson: see note 1-25, p 396

14-10 Lagemann: see note 14-1, p 281

14-11 Firth: see note 14-1, p 642

14-12 Bordier H 1905 *Les Rayons N et les Rayons* N_1*: les Actualités Medicales* (Paris: Baillère) noted by Lagemann: see note 14-1, p 284, note 7

14-13 Nye: see note 14-1, p 145
14-14 D'Arsonval 1904 Remarques à propos des communications de M. A. Charpentier et des revendications de priorité auxquelles elles ont donné lieu *CR* **138** 884–5
14-15 Discussion of the Groupe d'Etudes des Phénomènes Psychiques, Institut Général Psychologique 1904 *Bulletin* **4** 149–63, p 154; cited by Nye: see note 14-1, p 134
14-16 Nye: see note 14-1, pp 144–5
14-17 Becquerel J 1934 *Notice sur les Travaux Scientifiques de M Jean Becquerel* (Paris: Hermann) pp 89–90, cited by Nye: see note 14-1, p 152. Becquerel's father, Henri Becquerel, served on the committee for the Academy's prestigious Prix Le Conte awarded to Blondlot in 1904, including a cash award of 50 000 francs. Becquerel prepared the original report on Blondlot's work, including N-rays, but Henri Poincaré rewrote the report, mentioning Blondlot's 'new ray' only guardedly at the very end of a three-page account of his researches.
14-18 Whetham to Rutherford, February 13 1904; cited by Eve: see note 6-29, p 102. Whetham, a Trinity man who worked under J J at the Cavendish from 1889, made his initial mark in researches on the velocity of ions in electrolysis in the early 1890s, and became Fellow of the Royal Society in 1901.
14-19 Seabrook W B 1941 *Doctor Wood: Modern Wizard of the Laboratory* (New York: Harcourt Brace) p 237. Seabrook's semi-popular biography must be read with caution. Thus, his claim (p 239) that exposure of the N-ray affair 'led to Blondlot's madness and death' is highly dubious.
14-20 Lindsay R B 1970 Wood, Robert Williams *DSB* **13** 497–9
14-21 Seabrook: see note 14-19, p 234
14-22 Wood, cited by Seabrook: see note 14-19, p 236
14-23 Wood, cited by Seabrook: see note 14-19, p 238
14-24 Wood R W 1904 The n-rays *NA* **70** 530; 1904 La question de l'existence des rayons N *Revue Scientifique* **2** 536–8; also translated as 1904 Die N-Strahlen *PZ* **5** 789–91. On reading Wood's letter, the French chemist Joseph Achille LeBel exclaimed: 'What a spectacle for French science when one of its distinguished *savants* measures the position of spectrum lines, while the prism reposes in the pocket of his American colleague!' Seabrook: see note 14-19, p 239
14-25 Among the few influential defenders of Blondlot abroad had been Pierre Weiss at the Zürich Polytechnik; however, he was subsequently, with Walther Ritz, unable to duplicate Blondlot's work. Nye: see note 14-1, p 141

Supposed observers of N-rays outside France included the Englishman J S Hooker and F E Hackett, a student at the Royal University of Ireland; in addition, the London instrument maker Leslie Miller sold apparatus for observing the rays. Lagemann: see note 14-1, p 283
14-26 Cited by Klotz: see note 14-1, p 175
14-27 The editors proposed that two small, sealed wood boxes be submitted to Blondlot, both of equal weight but one containing a piece of tempered steel (supposedly an N-ray source) and one a piece of lead. Using either a spark gap or a phosphorescent screen, Blondlot would be asked to identify the box emitting N-rays.
14-28 Cited in Klotz: see note 14-1, p 175
14-29 Lagemann: see note 14-1, p 284

Weart: see note 14-1

Blondlot's retirement at age 61 was somewhat younger than the customary age of retirement, apparently at the insistence of a committee of inquiry; Lagemann: see note 14-1

Notes section 14.2

14-30 For an overview of these researches, the reader may wish to consult Dahl P F 1992 *Superconductivity: its Historical Roots and Development from Mercury to the Ceramic Oxides* (New York: American Institute of Physics) pp 13–22

14-31 Woodruff A E 1970 Weber, Wilhelm Eduard *DSB* **13** 203–9

14-32 Weber W 1875 Ueber die Bewegungen der Elektricität in Körpen von molecularer Constitution *APC* **156** 1–16

14-33 Wiederkehr K H 1967 *Wilhelm Eduard Weber. Erforscher der Wellenbewegung und der Elektrizität 1804–1891 (Grosse Naturforscher 32)* (Stuttgart: Wissenschaftliche Verlagsgesellschaft) p 106, cited by Jungnickel and McCormmach: see note 9-26, p 76

14-34 Giese W 1889 Grundzüge einer einheitlichen Theorie der Electricitätsleitung *APC* **37** 576–7

14-35 Schuster A 1884 Experiments on the discharge of electricity through gases. Sketch of a theory *PRS* **37** 318. Schuster's interaction with Weber at Göttingen late in the latter's career had launched him on his investigation of a possible breakdown of Ohm's law.

14-36 Goldberg S 1970 Riecke, Eduard *DSB* **11** 445–7

14-37 In one experiment, an aluminum cylinder was inserted between two copper cylinders in a circuit, and a current was passed for long enough that the amount of copper deposited by electrolysis would have amounted to over one kilogram. In fact, however, the weight of each of the three cylinders showed no measurable change, rendering it unlikely that metallic conduction is accompanied by the transport of metallic ions. Riecke E 1901 Ist die metallische Leitung verbunden mit einem Transport von Metallionen? *PZ* **2** 639, cited by Wittaker: see note 1-19, p 420.

14-38 Riecke E 1898 Zur Theorie des Galvanismus und der Wärme *APC* **66** 353–89, 545–81

14-39 Bohr N 1911 *Studier over Metallernes Elektrontheori* (Copenhagen: Thaning & Appel); reprinted with English translation in Rosenfeld L and Nielsen J R (eds) 1971 *Niels Bohr Collected Works* vol 1 (Amsterdam). Bohr's doctoral dissertation on the electron theory of metals played a minor role in these developments, because he failed in his attempts to have it translated into English from the Danish.

14-40 Goldberg S 1971 Drude, Paul Karl Ludwig *DSB* **6** 189–93

14-41 Drude P 1900 Zur Elektronentheorie. I *AP* **1** 566–613; 1900 Zur Elektronentheorie. II *AP* **3** 369–402; 1902 Zur Elektronentheorie. III *AP* **7** 687–92

14-42 Drude: see note 14-41, p 566

14-43 Dahl: see note 14-30, p 25

14-44 Drude: see note 14-41, p 579

14-45 Lorentz H A 1905 The motion of electrons in metallic bodies. I, II and III *RN* **7** 438–53, 590–3, 684–91

14-46 Lorentz: see note 14-45, pp 438–9

14-47 The Hall effect (note 8-35) was sometimes positive for certain metals, implying that the moving carriers of charge are positive. Lorentz himself had discussed the positive Hall effect in his papers of 1905 without getting very far with it. Lorentz H A 1938 Ergebnisse und Probleme der

Eletronentheorie *Collected Papers* vol 8 (The Hague: Martinus Nijhoff) 76–124

Paul Ehrenfest always emphasized in his lectures at Leiden that the occurrence of a positive Hall effect was 'one of the major mysteries'. Casimir H B G 1977 Development of solid-state physics *History of Twentieth Century Physics, Proceedings of the International School of Physics 'Enrico Fermi'* course 57, ed C Weiner (New York: Academic) pp 162–3

14-48 Helge Kragh, in his excellent treatment of the positive electron, notes several defenders of the concept. Though mindful of the apparent absence of true free positive electrons, Oliver Lodge, in his review of atomic theory in 1906, pointed out that 'the bulk of the atom may consist of a multitude of positive and negative electrons, interleaved, as it were, and holding themselves together in a cluster by their mutual attractions'. Kragh H 1989 Concept and controversy: Jean Becquerel and the positive electron *Centaurus* **32** 211

In 1901, James Jeans proposed an atomic model based on positive and negative electrons (though he called them 'ions'); Kragh, p 211. We, too, have noted Lenard's 'dynamids' of 1903—atoms arranged in electrical dipoles containing equal charges of opposite sign (p 220; Kragh, p 211). Electrical pairs or doublets of this kind were very popular; J J Thomson, too, had his version, unfortunately, as we shall see before long.

Notes section 14.3

14-49 Lorentz H A 1906 Positive and negative electrons *Proc. Am. Phil. Soc.* **45** 130; cited by Kragh: see note 14-48, p 212

14-50 Kragh: see note 10-34, p 215

14-51 Spencer J B 1970 On the varieties of nineteenth-century magneto-optical discovery *Isis* **61** 34–51

14-52 Kragh: see note 10-34, p 214

Jenkins F A and White H E 1976 *Fundamentals of Optics* 4th edn (New York: McGraw-Hill) ch 32

14-53 Becquerel J 1906 Sur les variations des bandes d'absorption d'un cristal dans un champ magnétique *CR* **142** 874–6; Sur la corrélation entre les variations des bandes d'absorption des cristaux dans un champ magnétique et la polarisation rotatoire magnétique *CR* **142** 1144–6

14-54 Lilienfeld J E 1907 Über neuartige Erscheinungen in der positiven Lichtsäule der Glimmentladung *Verhandlungen der deutschen physikalischen Gesellschaft* **9** 125–35. This was actually a follow-up report on a preliminary communication at the meeting of the German Physical Society the previous November.

14-55 Bestelmeyer A and Marsh S 1907 Über das gemeinsame Auftreden von Strahlen positiver und negativer Elektrizität in verdünnten Gasen *Verhandlungen der deutschen physikalischen Gesellschaft* **9** 758–63

Goldstein E 1907 Über das Auftreten roten Phosphorezenslichtes an Geisslerschen Röhren 1907 *Verhandlungen der deutschen physikalischen Gesellschaft* **9** 598–605

14-56 Lilienfeld J E 1910 Die Elektrizitätleitung im extremen Vakuum *AP* **32** 673–738

Lilienfeld is a little known physicist, concerning whom one would like to know more. Born in Lemberg, Poland, in 1882, he obtained his doctorate at the University of Berlin in 1905; under whom, it is not clear. He was a *Privatdozent* at Leipzig from 1910 to 1916, and Professor *Extradordinarius* from 1916 to 1926. In 1927, he emigrated to the USA, where he was Director of the Ergon Research Laboratories in Malden, MA, for a period. In the mid-1930s, he settled in the Virgin Islands with his American-born wife, the former Beatrice Ginsburg. From that point on, his activities as a physicist are rather obscure. It is known that in the late 1920s he filed three patents for a prototype of what is now known as the field-effect transistor. In the words of John Bardeen 'Lilienfeld's patents predated the work of Brattain, Shockley, and me by almost 20 years'. He also contributed to the development of electron field emission, and appears to have been occupied with various inventions and patent claims until his death in 1963. Sweet W 1988 American Physical Society establishes major prize in memory of Lilienfeld *PT* **41** 87–9 (Bardeen's quotation is from this source)

14-57 Becquerel J 1907 Sur les variations d'absorption des cristaux de parisite et de tysonite dans un champ magnétique, à la température de l'air liquide *CR* **145** 413–6

14-58 Becquerel J 1907 Recherches sur les phénomènes magnéto-optiques dans les cristaux *Le Radium* **4** 49–57 (German translation 1907 *PZ* **8** 632–56)

14-59 Becquerel J 1908 On the nature of charges of positive electricity and on the existence of positive electrons *The Electrician* **61** 525; translated from Sur la nature des charges d'électricité et sur l'existence des électrons positifs *CR* **146** 1308–11

14-60 Becquerel: see note 14-59

14-61 Becquerel J cited by Stranges A N 1982 *Electrons and Valence: Development of the Theory 1900–1925* (College Station: Texas A&M University Press) p 78

14-62 Wood R W 1908 On the existence of positive electrons in the sodium atom *PM* **15** 279

14-63 Casimir H B G 1983 *Haphazard Reality: Half a Century of Science* (New York: Harper & Row) p 332. The term 'helium day' came from the fact that, because of its scarcity, liquid helium was only available on select days, usually just once a week.

14-64 Onnes to Voigt, January 9 1908, SON, item 5694

14-65 The hapless physicist, H Bremmer, 'later distinguished himself in mathematical physics', Casimir: see note 14-63, p 333

14-66 Becquerel J and Kamerlingh Onnes H 1908 The absorption spectra of the compounds of the rare earths at the temperatures obtainable with liquid hydrogen, and their change by the magnetic field *RN* **10** 598; *PLC* **103** 3–16

14-67 Becquerel and Kamerlingh Onnes: see note 14-66, p 599

14-68 Onnes to Voigt, November 5 1908, SON, item 5698

14-69 Dufour A 1908 Sur quelques exemples des raies présentant le phénomène de Zeeman anormal dans le sens des lignes de force magnétiques *CR* **146** 634–5

14-70 Becquerel to Lorentz, July 11 1908, AHQP, Lorentz correspondence, cited by Kragh: see note 10-34, p 216

14-71 Dufour A 1909 Sur les phénomènes de Zeeman normaux et anormaux *Le Radium* **6** 44–5; 1909 Modifications normales et normales, sous l'influence d'un champ magnétique, de certaines bandes des spectres d'émission de molécules de divers corps à l'état gaseux *Journal de Physique* **9** 237–64

14-72 Dufour A 1909 Sur l'existence d'électrons positifs dans les tubes à vide *CR* **148** 481–4
For additional references to the ongoing dispute between Dufour and Becquerel, see Kragh: see note 10-34, pp 219, 233

14-73 Righi A 1909 *Strahlende Materie und magnetische Strahlen* (Leipzig: Johan Ambrosius Barth)
Bestelmeyer A 1908 Positive Elektronen? *PZ* **9** 541–2
Moulin A 1909 Sur la déviation des rayons positifs *Le Radium* **6** 4–5; Sur la déviation des rayons canaux *Le Radium* **6** 78–9

14-74 Wood R W and Hackett F E 1909 The resonance and magnetic rotation spectra of sodium vapor photographed with the concave grating *Astrophysical Journal* **30** 339–72 cited by Kragh: see note 10-34, p 217

14-75 Voigt W 1909 Remarks on the Leyden observations of the Zeeman-effect at low temperatures *RN* **11** 365

14-76 Voigt: see note 14-75, p 365
See also Onnes to Voigt, June 23 1908, SON, item 5697
Onnes to Voigt, July 15 1908, SON, item 5696

14-77 Becquerel to Lorentz, February 18 1909 and July 4 1909, AHQP, Lorentz correspondence, cited by Kragh: see note 10-34, p 233

14-78 Thomson J J 1909 Inaugural address to the British Association meeting at Winnipeg, 1909 *NA* **81** 252

14-79 Rutherford E 1909 Opening address to Section A, Mathematics and Physics, of the British Association meeting at Winnipeg, 1909 *NA* **81** 262

14-80 Van den Handel J 1956 Low temperature magnetism *Kaltephysik II (Handbuch der Physik 15)* ed S Flügge (Berlin: Springer) pp 15–6

14-81 Becquerel J 1922 Absorption de la lumière et phénomènes, magnéto-optiques dans les composés de terres rares aux très basses températures *Het Natuurkundig Laboratorium der Rijksuniversiteit te Leiden in de Jaren 1904–1922, Gedenkboek aangeboden aan H Kamerlingh Onnes, Directeur van het Laboratorium, bij Gelegenheid van zijn Veertigwarig Professoraat op 11 November 1922* (Leiden: Ijdo) pp 350–1
The same year Oliver Lodge, among a handful of prominent scientists (Einstein, for one) still bothered by the apparent lack of a true positive counterpart to the electron, speculated that positive electrons do indeed exist, but only within the proton. Lodge O 1922 Speculation concerning the positive electron *NA* **110** 696–7
Kragh: see note 10-34, p 228

Notes section 15.1

15-1 Thomson: see note 13-53

15-2 Kangro H 1972 Wien, Wilhelm Carl Werner Otto Fritz Franz *DSB* **5** 337–42

15-3 Wien W 1967 On the laws of thermal radiation *Nobel Lectures, including Presentation Speeches and Laureates' Biographies, Physics 1901–1921* (Amsterdam: Elsevier) pp 275–86
It says much for the PTR that Wien's Nobel-Prize-winning research was performed at an institute dedicated to technical problems of industry. The term 'displacement law' (*Verschiebungsgesetz*) was coined by Otto Lummer and Ernst Pringsheim in 1899; Kangro: see note 15-2, p 341

15-4 In the reprint of his first cathode ray paper of 1898, Lenard refers in a footnote to Wien's indebtness for the latter use of Lenard's apparatus in his own cathode ray investigations, as also testified to in Wien's obituary of 1930. Lenard P 1898 Ueber die electrostatischen Eigenschaften der Katodenstrahlen *APC* **64** 279–89, reprinted in Lenard: see note 5-28, *Wissenschaftliche Abhandlungen*, p 74

Wien W 1930 *Aus dem Leben und Wirken eines Physikers* (Leipzig: Barth) pp 139–89

15-5 Wien W 1898 Untersuchungen über die elektrische Entladung in verdünnten Gasen *APC* **65** 440–52; extract in Magie: see note 1-6, pp 598–600; 1898 Die elektrostatische und magnetische Ablenkung der Kanalstrahlen *Berlin Physikalische Gesellschaft Verhandlungen* pp 10–2

15-6 Magie: see note 1-6, p 598. It is perhaps worthy of note that, like one of Lenard's tubes (or, at least, one of his design) in the hands of Röntgen earlier, once again a colleague of Lenard made good use of another tube of his.

15-7 The deflections are discussed in note 9-56, above, and in greater detail in the next section. For a derivation of the expressions, the reader may wish to refer to chapter 2 of Wehr M R, Richards J A and Adair T W 1984 *Physics of the Atom* 4th edn (New York: Addison–Wesley)

15-8 Thomson: see note 10-26, p 557

15-9 Rayleigh: see note 1-46, p 167

15-10 Thomson J J 1907 On rays of positive electricity *PM* **13** 562

15-11 Rayleigh: see note 1-46, p 167

15-12 Thomson: see note 13-53, p 149

15-13 Thomson: see note 13-53, p 149

15-14 Heilbron: see note 1-14, 45
Falconer I 1988 J. J. Thomson's work on positive rays, 1906–1914 *HSPBS* **18** 271

15-15 Stranges: see note 14-61, p 34

15-16 Thomson J J 1906 On the number of corpuscles in an atom *PM* **11** 769–81
The three lines of reasoning were based, respectively, on the dispersion of light in gases (using Rutherford's and Ketteler's data) the scattering of x-rays (using Barkla's data) and the absorption of β-rays (using Becquerel's and Rutherford's data). Thomson's personal notebook for 1906 is also available, and shows the evolution of his thoughts on the subject. Notebook No 54, also listed as 'On the number of corpuscles in the atom, 1906' in P Spitzer's handlist of Thomson's notebooks, CUL, Add 7654, NB61

15-17 Lord Kelvin 1902 Æpinus atomized *PM* **3** 257–83

15-18 Thomson J J 1903 The magnetic properties of systems of corpuscles describing circular orbits *PM* **6** 673–93; 1904 On the structure of the atom: an investigation of the stability and periods of oscillation of a number of corpuscles arranged at equal intervals around the circumference of a circle; with application of the results to the theory of atomic structure *PM* **7** 237–65; 1907 The arrangement of corpuscles in the atom *The Corpuscular Theory of Matter* (New York: Scribner) pp 103–67, excerpted in Boorse and Motz: see note 13-55, pp 616–24 (Mayer's magnets)

15-19 Heilbron: see note 1-14, p 57
Falconer: see note 15-14, p 271

15-20 Stranges: see note 14-61, p 56

15-21 Thomson: see note 15-18, *The Corpuscular Theory of Matter* p 23; cited by Falconer: see note 15-14, pp 271–2

Notes section 15.2

15-22 Thomson: see note 15-10
Thomson to Rutherford, December 18 1906, CUL, Add 7653, T28

For an authentic, retrospective account of these researches, see Thomson J J 1921 *Rays of Positive Electricity and their Application to Chemical Analyses* 2nd edn (London: Longmans Green)
For a contemporary review of the first edition of this book (published in 1913) see 1914 *NA* **92** 549–50

15-23 Thomson came across Ebenezer Everett in 1886, when Everett was working in the Chemical Laboratory under G D Liveing. As Thomson's technician, he built apparatus and got things ready for experiments. He had poor appreciation of science, and of the aim of the experiments. 'He considered it his duty to do what the Professor wanted, not to understand why'. Crowther J G 1974 *The Cavendish Laboratory 1874–1974* (New York: Science History Publications) p 293
For some reason, Everett's name recurs repeatedly spelled Everitt in many of the accounts and papers of the time–often by Thomson himself. It cannot be due to some careless copyeditor down the line; we find 'Everitt' in the acknowledgements in the original, handwritten manuscript of Thomson's famous paper, 'Cathode rays', in the *Philosophical Magazine* for 1897, CUL, Add 7654, PD13

15-24 Notebook 57, listed as 'Canal rays (positive rays) 1906' in P Spitzer's handlist of Thomson's notebooks, CUL, Add 7654, NB57. The first entries in this notebook, apparently the first covering explicitly the positive ray experiments, are dated October 30 1906, and the last ones late December 1906.

15-25 Thomson: see note 15-10, p 562

15-26 For more on the Cavendish Museum, see note 9-55 above.

15-27 The Gaede rotary mercury pump was expensive, though one was certainly in use at the Cavendish by mid-1910; Thomson: see note 9-30, p 175
See also interview with G P Thomson by J L Heilbron, June 20 1963; AHQP, tape 72, p 6 of transcript.

15-28 Thomson: see note 15-10, p 564

15-29 The first tissue paper tracing, in the notebook for October–December 1906, is dated December 1 1906, and the last dated December 17. Notebook 57: see note 15-24; CUL, Add 7654, NB57. The scraps pick up again in the first notebook for 1907, dated between January 3 and February 8. Notebook 59, listed as 'Experimental notebook. (Gas discharges using calcium cathode) 1907' in Spitzer's handlist, CUL, Add 7654, NB59
They continue in the next notebook; the first of the third and last batch of tissues (now for 'low-pressure' traces) is dated February 13 1907 and the last scrap was pinned in place on March 12. Notebook 60, continuation of notebook 59, CUL, Add 7654, NB60

15-30 The physics principles behind the method are lucidly explained Trigg G L 1975 *Landmark Experiments in Twentieth Century Physics* (New York: Crane Russak) p 15

15-31 Note that the practicality of the scheme depended on the fact that only the vertical deflection of the wire at the position of the screen was needed. In 1909 Thomson devised a simple method for determining the field integral along the ray path, by measuring the magnetic induction through a triangular coil with its base at the end of the narrow tube *F* and its apex at the screen; the induction was measured with a Grassot fluxmeter. Thomson J J 1909 Note on a method of measuring the effective magnetic field in the magnetic deflexion of the canalstrahlen *PM* **18** 844–5

15-32 CUL, Add 7654, NB60. The measurements occupy but four notebook pages, covering March 18–19 1907

15-33 Crookes (Honorary Secretary of the Royal Institution) to Thomson, March 13 1905, CUL, Add 7654, C47

15-34 Griffiths E 1941 George William Clarkson Kaye *Royal Society Obituary Notices of Fellows* **3** 881–95

15-35 Thomson to Rutherford, October 25 1908, CUL, Add 7653, T31

15-36 We now know that hydrogen from water vapor, driven into the glass walls during tube operation, is desorbed from the surface of the glass and can only be removed by prolonged 'outbaking' (heating in a vacuum)—a procedure not possible with all the sealing wax joints in the discharge tubes in Thomson's day and, in any case, a vacuum technique developed somewhat later.

15-37 Thomson: see note 15-10, p 571

15-38 CUL, Add 7654, NB55

15-39 Falconer: see note 15-14, p 279
Thomson: see note 9-30, pp 130–2

15-40 Thomson J J 1908 Positive rays *PM* **16** 657–91

15-41 Falconer: see note 15-14, p 285

15-42 Thomson: see note 15-40, p 691

15-43 Thomson J J 1909 Positive electricity *PM* **18** 821–44
The final draft of this paper, including the two-page 'Note' on an effective method of field measurement: see note 15-31, is preserved in the archives of Thomson's papers held at the University Library, Cambridge. The typescript contains several major pencilled additions, as well as numerous corrections, all in Thomson's hand. CUL, Add 7654, PD42

15-44 In the discussion which followed Thomson's opening remarks, 'many points were raised dealing with side-issues, but the main question [is there a definite unit of positive electricity, and what is its size?] did not receive much fresh elucidation', Thomson J J 1909 Mathematics and physics at the British Association *NA* **81** 471
Among the foreign visitors attending the Winnipeg meeting was E Goldstein from Berlin, who had been awarded the Hughes Medal by the Royal Society the year before. Goldstein himself read a paper to the Association, but it was on the phosphorescent spectra of certain organic compounds when cooled to the temperature of liquid air, not canal rays. Goldstein to Thomson, June 21 1910, CUL, Add 7654, G25

15-45 Thomson: see note 15-43, p 835

15-46 Thomson: see note 15-43, p 830

15-47 Thomson: see note 15-43, p 831

15-48 Thomson: see note 15-43, p 835

15-49 Thomson: see note 15-43, p 835

15-50 Thomson J J 1910 Rays of positive electricity *PM* **19** 433

15-51 Thomson: see note 15-50, p 424
The Wehrsen induction machine is dealt with in Thomson's laboratory notebook covering the period October 27–November 12. Notebook No 61, listed as 'Measurements of retrograde positive rays. 1909' in P Spitzer's handlist of Thomson's notebooks, CUL, Add 7654, NB61

15-52 Thomson to Rutherford, December 20 1909, CUL, Add 7653, T33

15-53 Thomson: see note 15-52

15-54 Thomson: see note 15-50, p 433

15-55 Thomson: see note 15-51, Notebook 61, CUL, Add 7654, NB61

15-56 Falconer: see note 15-14, p 289

Notes section 15.3

15-57 Thomson to Rutherford, December 20 1909, CUL, Add 7653, T33
15-58 Thomson G P 1961 Francis William Aston *Great Chemists* ed E Farber (New York: Interscience) p 1455
15-59 Thomson: see note 15-58
Brook W H 1970 Aston, Francis William *DSB* **1** 320–2
von Hevesy G 1945–1948 Francis William Aston *Royal Society Obituary Notices of Fellows* **5** 635–50
Curiously J J Thomson says very little about Aston as a person in his *Recollections* even though Aston figured so importantly in the later stages of Thomson's positive ray program.
15-60 Aston to Thomson, June 13 1907, CUL, Add 7654, A37
15-61 Thomson J J 1910 Rays of positive electricity *PM* **20** 752–67
15-62 The first measurements with Aston on board are recorded, partly in Aston's hand, in the notebook covering the summer experiments of 1910, with the first entry marked 'June 1910'. The last data entered were recorded on August 8. Notebook No 62, listed as 'F W Aston—Positive ray experiments 1910' in P Spitzer's handlist of Thomson's notebooks, CUL, Add 7654, NB62
15-63 One of Aston's large bulbs is among the positive ray apparatus exhibited in the museum of the Cavendish Laboratory. Even with this improvement, and liquid air–charcoal cooling, we find recurring reference in the notebook to the effect 'hardening of the tube during running'. CUL, Add 7654, NB62
15-64 Hint of parabolas appear in the data for late 1909, e.g. data entries for November 12 1909 in Notebook 61, as also reflected in the published paper of March 1910. CUL, Add 7654, NB61; Thomson: see note 15-50, p 425
15-65 Thomson J J 1911 Rays of positive electricity *PM* **21** 225–49
15-66 Dewar J 1906 Studies on charcoal and liquid air, Friday Evening Discourse June 8 *The Royal Institution Library of Science, Physical Sciences* vol 6, pp 210–31
15-67 Thomson to Rutherford, April 8 1904, CUL, Add 7653, T24
15-68 Notebook No 62: see note 15-62, CUL, Add 7654, NB62
15-69 Occasional reference to the technique is made in Thomson's notebook covering the positive ray experiments during January and early February 1907, but not in the succeeding notebook. Notebook No 59, listed as 'Experimental notebook (gas discharges using calcium cathode)' in P Spitzer's handlist of Thomson's notebooks, CUL, Add 7654, NB59
Falconer: see note 15-14, p 293
15-70 Rayleigh: see note 1-46, p 170
15-71 Thomson: see note 15-65, pp 225–6
15-72 Thomson: see note 15-65, pp 227–8
15-73 Oliphant M L 1972 *Rutherford: Recollections of Cambridge Days* (Amsterdam: Elsevier) p 45
15-74 Rayleigh: see note 1-46, p 164
15-75 Schuster: see note 6-11
15-76 Rayleigh: see note 1-46
15-77 Thomson J J 1912 Further experiments on positive rays *PM* **24** 209–53
15-78 Thomson: see note 15-77, pp 210–1
15-79 Thomson J J 1913 Rays of positive electricity Bakerian Lecture *PRS* **89** 3–4

The handwritten manuscript of this paper is preserved in the archives of Thomson's papers held at the University Library, Cambridge. The ink-written manuscript contains various pencilled corrections in Thomson's hand. CUL, Add 7654, PD54.1

15-80 Falconer: see note 15-14, p 303
The last time he dwelt on the concept was in the third paper (1909) in the *Philosophical Magazine* series: see note 15-43

15-81 Thomson: see note 15-61, p 758

15-82 Costa A B 1971 Dewar, James *DSB* **4** 80. In 1908 he used the carbon-absorption technique in making the first direct measurement of the rate of production of helium from radium.

15-83 Thomson: see note 15-79, p 12

15-84 Thomson: see note 15-79, p 12

15-85 Thomson: see note 15-79, p 12

15-86 Trenn T J 1975 Soddy, Frederick *DSB* **12** 504–9

15-87 Soddy F 1913 The radio-elements and the periodic law *Chemical News* **107** 97–9

15-88 Schuster A 1913 Letter to the editor *NA* **91** 30–1

15-89 Soddy F 1913 Letter to the editor *NA* **91** 57–8

15-90 1913 Chemistry at the British Association meeting *NA* **92** 331–2
The term 'isotopes' was introduced by Soddy in *Nature* for December 4 1913: Soddy F 1913 Intra-atomic charge *NA* **92** 400
The term had been suggested to him by his friend Dr Margaret Todd of Edinburgh. Howorth M 1958 *The Life Story of Frederick Soddy* (London: New World), cited by Boorse and Motz: see note 13-55, p 780. Kasimir Fajans's term 'pleiads', proposed earlier in March, did not stick.

15-91 See note 15-90, p 331. Thus, Soddy competed with Thomson for attention in parallel sessions of this meeting of the British Association on September 16 1913. Thomson spoke on 'X_3 and the evolution of helium'. According to the résumé of that session in *Nature*, 'A number of chemists came to hear this fresh invasion of their territory by Sir J. J. Thomson', p 308
'It is remarkable', adds Trigg, 'that Soddy seems to have taken as little note of Thomson's work at this time as Thomson did of his'; Trigg: see note 15-30, p 25. The ion X_3, of mass 3 and first seen in 1911, preoccupied Thomson to the extent that he left neon pretty much to Aston. Thomson concluded that it was an H_3^+ molecule.

15-92 E.g. Cranston J A 1964 The group displacement law *Chemical Society Proceedings* pp 104–7

15-93 An argument against the line of mass 22 being due to NeH_2 was that it was frequently accompanied by a faint line corresponding to a mass-to-charge ratio of 11, the value it would have with a double charge. Such a charge state was common for atoms, but rare for molecules. Carbon dioxide was open to the same objection.

15-94 Thomson to von Hevesy, June 2 1913, Hevesy collection, Niels Bohr Library, American Institute of Physics, College Park, MD
See also interview with Hevesy by Kuhn T S, Segrè E G and Heilbron J L, May 25 1962; OH 215, Niels Bohr Library, American Institute of Physics, College Park, MD

15-95 Thomson: see note 9-30, p 137
When the war ended, Aston and Lindemann wrote a joint paper on the separation of isotopes. Aston F W and Lindemann F A 1919 The possibility of separating isotopes *PM* **37** 523

15-96 The principle of Aston's mass spectrograph, as the name implies, is based on an optical analogy. Just as white light is dispersed into an optical spectrum by a prism, so a beam of heterogeneous positive ions can be dispersed into an electrical spectrum by an electric field. That is, the bending as a function of wavelength in the optical case is analogous to the bending as a function of velocity in the electrical case. Aston's instrument did not, therefore, use the parabolic method. Instead, it brought to a single focus ions of different velocities but of the same charge-to-mass ratio, the focal point for different e/m-ratios being distributed linearly along a photographic plate. For this purpose, the electric and magnetic fields, instead of being superposed as in the parabola method, were arranged in series, first the electric and then the magnetic field. The magnetic field was perpendicular, not parallel, to the electric field; hence, the magnetic deflection was parallel to (and, by design, in the opposite direction to) that of the electric field. The ability to focus rays of uniform mass *irrespective* of their velocity, onto a photographic plate, was a great advantage over Thomson's apparatus, which yielded parabolas depending on the velocity of the rays. Aston F W 1920 Positive rays and isotopes *NA* **105** 617–9

For a description of Aston's mass spectrograph, see also Harnwell and Livingood: see note 9-47, pp 138–40

15-97 Rayleigh: see note 1-46, p 174

Notes section 16.1

16-1 Curiously, the most accurate value for e in 1900 was that estimated on purely theoretical grounds by Max Planck that year. In the course of deriving his celebrated expression for the energy density of black-body radiation, he introduced and evaluated two new universal constants h and k, nowadays known as Planck's and Boltzmann's constants, respectively. Planck interpreted k as being the number which, multiplied by Avogadro's constant N, gives the gas constant per gram molecule R, or $R = Nk$. The gas constant was well known at the time, as was the Faraday constant F, or the amount of charge needed to plate one gram molecule of monatomic ions, given by $F = Ne$, where e is, as usual, the electronic charge. Avogadro's constant N was less well known. However, knowing k, the first relation gave Planck the value for N. Knowing N (Planck's value was 6.175×10^{23} molecules mol^{-1}), he then obtained e from F and the second relation. Planck's value for e 4.69×10^{-10} esu, differs from the currently accepted value (4.8×10^{-10} esu) by only 2.3%. Planck M 1901 Ueber die Elementarquanta der Materie und der Elektricität *AP* **4** 564–6

See also note 12-23

16-2 Du Bridge L A and Epstein P S 1959 Robert Andrews Millikan *Biographical Memoirs, National Academy of Sciences* **33** 241–82

Epstein P S 1948 Robert Andrews Millikan as physicist and teacher *Reviews of Modern Physics* **20** 10–25 (a volume honoring Millikan's eightieth birthday)

Kevles D J 1974 Millikan, Robert Andrews *DSB* **9** 395–400

16-3 Millikan R A 1950 *Autobiography* (New York: Prentice-Hall) p 14

16-4 Millikan: see note 16-3, p 58–60

16-5 Millikan R A and Winchester G 1907 Upon the discharge of electrons from ordinary metals under the influence of ultra-violet light *Physical Review*

24 116–8; 1907 The influence of temperature upon photo-electric effects in a very high vacuum, and the order of photo-electric sensitiveness of the metals *PM* **14** 188–210

16-6 Millikan: see note 16-3, p 69
The unfortunate 'Nernst episode' concerned Millikan's aforesaid paper on his work on the Clausius-Mossotti formula at Göttingen. Millikan's manuscript, as mailed to Nernst, was expanded on by the addition of certain theoretical considerations of Millikan's, but these additions were summarily stripped by Nernst from the version published in *Annalen der Physik und Chemie*. Nernst returned the theoretical part to Millikan, advising him to subject it to further experimental tests and publish it on his own volition. Meanwhile, Paul Drude, the editor of *Annalen*, developed Millikan's idea into a theory of his own, publishing it and leaving Millikan more or less in the lurch. Epstein: see note 16-2, p 11
The 'failures at Chicago', referring to the negative photoelectric results, would in time be shown to stem from the degenerate state of metal electrons. Epstein: see note 16-2, p 17
Millikan returned to the photoelectric effect after 1912. By then, quite aware of Einstein's interpretation of the effect, he undertook to verify Einstein's photoelectric equation, which he and his students duly accomplished by 1916 in a series of elegant experiments. See note 12-23, above
Epstein: see note 16-2, pp 17–9

16-7 Millikan: see note 16-3, p 68

16-8 Millikan: see note 16-3, p 72

16-9 Harold Albert Wilson, born in York, England, in 1874, was the only son of a North Eastern Railway clerk who later became a district manager. (His one sister, Lilian, married Owen Williams Richardson, the Nobel-Prize-winning physicist whose work was closely related to Wilson's.) Wilson displayed an early interest in science and technology. He received his BSc degree in physics and chemistry at Victoria University College at Leeds, where he began his work on the electrical conductivity of flames that would occupy him off and on for more than 30 years. During his years at the Cavendish, from 1897 to 1904, he continued his work on the mobility of ions in flames, studied the electric discharge in rarefied gases, and the Hall effect, and, following J J Thomson, carried out his noteworthy attempt to determine the charge of the electron. From Cambridge, he went to King's College, London, as professor and head of the department of physics. In 1909, he accepted a professorship at McGill University, Canada, and in 1912, when the Rice Institute (now Rice University) opened in Houston, TX, he settled there for the rest of his academic career, except for one year at Glasgow. After 1932, nuclear physics was his primary focus. Szymborski K 1970 Wilson, Harold Albert *DSB* **18** 992–3

16-10 Wilson H A 1903 Determination of the charge on the ions produced in air by Röntgen rays *PM* **5** 429–40
See also Thomson J J 1903 On the charge of electricity carried by a gaseous ion *PM* **5** 346–55

16-11 Louis Begeman (1865–1958), born in Evansville, IN, in 1865, received his BS and MS degrees from the University of Michigan, and his PhD in physics at Chicago in 1910. Except for a period as professor of physics and chemistry at Parson's College in Fairfield, IA, during 1895–1910, his career was spent on the faculty of Iowa State Teacher's College. He was the author of two undergraduate texts on physics.

16-12 Millikan R A and Begeman L 1908 On the charge carried by the negative ion of an ionized gas *PR* **26** 197–8

16-13 Millikan R A 1963 *The Electron: its Isolation and Measurement and the Determination of some of its Properties* facsimile of original 1917 edition (Chicago: University of Chicago Press) p 55

16-14 Trenn T J 1972 Geiger, Hans (Johannes) Wilhelm *DSB* **5** 330–3. Hans Geiger was born in Neustad, Germany, in 1882, the son of a professor of philology at the University of Erlangen. He obtained his doctorate under Eilhard Wiedemann at Erlangen in 1906. At Manchester, while investigating with Rutherford the charge and nature of the α-particle, he noticed that a beam of α-particles had a tendency to spread. Teaming with Ernest Marsden, and using a scintillation detector, they undertook a study of the scattering of the particles that ultimately led Rutherford to propose his nuclear atom of 1911, as we relate shortly. While still at Manchester, Geiger and John Nuttal established their well-known relation between the decay constant of a radioactive nucleus and the range of the emitted α-particles. In 1912, Geiger became director of the newly established laboratory for radium research at the Physikalisch-Technischen Reichsanstalt in Berlin, where he remained until 1925 when he accepted the chair of physics at the University of Kiel. It was there that he developed in 1928, with his student Walther Müller, the counter for which he is best known. From Kiel, he went to Tübingen where he began research on cosmic rays. His final chair was at the Technische Hochschule in Berlin. There he remained until his death in 1945 from rheumatoid arthritis—a condition suffered from the time of his front-line duty as an artillery officer in World War I.

16-15 Rutherford E and Geiger H 1908 An electrical method of counting the number of α-particles from radio-active substances *PRS* **81** 141–61
Original notes and calculations in connection with this piece of investigation are among the Rutherford archives at Cambridge: 'Charge on alpha-particle', readings, graphs, and a sketch; 32 leaves, January 11 1908; with Geiger H, CUL, Add 7653, PA 181
The original, handwritten manuscript, with numerous corrections in Rutherford's hand, is also extant, CUL, Add 7653, PA 11. Thus, the word 'electrical' is absent from the title in the first draft of the manuscript. Absent from the draft manuscript: a final section titled *Summary of Results*.

16-16 Rutherford and Geiger: see note 13-66

16-17 Rutherford and Geiger: see note 13-66, p 168

16-18 Millikan: see note 16-13, p 55

16-19 Rutherford and Geiger: see note 13-66, p 170

16-20 Rutherford and Geiger: see note 13-66, pp 170–1

16-21 Rutherford and Geiger: see note 13-66, p 171; see also Planck: note 16-1

16-22 Eve: see note 6-29, p 176

16-23 Millikan: see note 16-3, p 73

16-24 Millikan: see note 16-3, p 73

16-25 Millikan: see note 16-3, p 74

16-26 Millikan to his wife, the former Greta Irvin Blanchard, August 25 1909, cited by Holton G 1978 Subelectrons, presuppositions, and the Millikan–Ehrenhaft dispute *HSPS* **9** 187–8

16-27 Rutherford E 1909 The British Association at Winnipeg. Section A. Mathematics and Physics. Opening Address by Prof. E Rutherford *NA* **81** 258

16-28 Rutherford: see note 16-27, p 261

16-29 Rutherford: see note 16-27, p 261
16-30 Rutherford: see note 16-27, p 261
16-31 Rutherford: see note 16-27, pp 261–2
16-32 Millikan: see note 16-3, p 75
One of Larmor's students, E Cunningham, published a result on the topic the following year, cited by Holton: see note 16-26, p 190, note 92. The applicability of Stokes' law in these experiments would preoccupy Millikan off and on over the ensuing several years, as we shall see shortly.
16-33 Millikan R A 1909 A new modification of the cloud method of measuring the elementary electrical charge, and the most probable value of that charge *PR* **29** 560–1
16-34 Millikan R A 1910 A new modification... *PM* **19** 209–28
16-35 Millikan: see note 16-34, p 220
16-36 Millikan: see note 16-34, p 224
16-37 Millikan: see note 16-34, p 224
16-38 Millikan: see note 16-34, p 225
In the *Philosophical Magazine* article Millikan actually gives 4.67×10^{-10} for Begeman's value, up from 4.66×10^{-10} in the abstract read at Princeton. Begeman L 1910 The value of *e* by Wilson's method *PR* **30** 131; An experimental determination of the charge of an electron by the cloud method *PR* **31** 41. A chief source of error in the falling cloud method was that different parts of the cloud moved with different velocities, the reason being that the droplets of the cloud were not all singly, but multiply charged. The remedy was measuring the electronic charge separately from the singly, doubly, and triply charged parts of the cloud. Epstein: see note 16-2, p 15
16-39 Millikan: see note 16-34, pp 225–6
16-40 Millikan: see note 16-34, p 226
16-41 Millikan's objections were fourfold: he questioned the validity of Stokes' law for Ehrenhaft's particles; the two velocity measurements were not made on one and the same particle; he questioned Ehrenhaft's radius determination; there was the possibility of multiple charges carried by some of the particles.
16-42 G Holton, whose exhaustive review of Millikan's *Philosophical Magazine* results is a must for scholarly analysis, notes that they were vulnerable to criticism. The measurements were difficult, Millikan relied to a considerable extent on personal judgment of individual observations, and this paper was really his first major publication. Holton: see note 16-26, pp 192–3

Notes section 16.2

16-43 Ehrenhaft F 1909 Eine Methode zur Bestimmung des elektrischen Elementarquantums, I *PZ* **10** 308–10, received April 10 1909
16-44 Holton G 1970 Ehrenhaft, Felix *DSB* **17** (Supplement 2) 256–8
Much of Ehrenhaft's private correspondence, lectures, manuscripts and clippings is held in the archives of the Niels Bohr Library, American Institute of Physics, College Park, MD. Included is the start of an unpublished autobiography furnished by his son, Johann L Ehrenhaft; AR59, Box 1
16-45 Jungnickel and McCormmach: see note 9-26, pp 184–5
16-46 Klein M J 1970 *Paul Ehrenfest. Volume 1: the Making of a Theoretical Physicist* (Amsterdam: Elsevier) p 49

16-47 Colloidal particles range in size from some tens of atomic radii to somewhat greater than the wavelength of light. Kerker M 1974 Movement of small particles by light *American Scientist* **62** 92–8

16-48 The reality of atoms was hotly debated at the scientific congress in St Louis in 1904; Millikan: see note 16-3, p 84

The first Solvay Congress, as late as 1911, was, in many respects, devoted to an impasse in the face of the atomic–molecular hypothesis; Nye: see note 7-10, p 154

16-49 Ehrenhaft F 1909 Eine Methode zur Messung der elektrischen Ladung kleiner Teilchen zur Bestimmung der elektrischen Elementarquantums *Anzeiger der Kaiserlichen Akademie der Wissenschaften, Mathematisch-Naturwissenschaftliche Klasse* **7** 72, dated March 4 1909

Holton: see note 16-26, p 186

16-50 Ehrenhaft F 1909 Eine Methode zur Bestimmung des elektrischen Elementarquantums, I *Sitzungsberichte der Mathematisch-Naturwissenschaftlichen Klasse der Kaiserlichen Akademie der Wissenschaften* **118** 321–30, dated March 18 1909: see note 16-43

16-51 Ehrenhaft: see note 16-50, pp 328–9, p 310

16-52 Ehrenhaft: see note 16-50, p 328

16-53 Ehrenhaft F 1910 Über die kleinsten messbaren Elektrizitätsmengen. Zweite vorläfige Mitteilung der Methode zur Bestimmung des elektrischen Elementarquantums *Anzeiger der Kaiserlichen Akademie den Wissenschaften, Mathematisch–Naturwissenschaftliche Klasse (Vienna)* No 10, pp 118–9

16-54 His supporters included mainly students and co-workers at the University of Vienna, among them Alfred Lechner, D Konstantinowsky and Karl Przibram. We will have more to say about Przibram presently.

16-55 Ehrenhaft F 1910 Über die Messung von Elektrizitätsmengen, die Ladung des einwertigen Wasserstoffions oder Elektrons zu unterschreiten scheinen. Zweite vorläufige Mitteilung seiner Methode zur Bestimmung des elektrischen Elementarquantums *Anzeiger Akad. Wiss. (Vienna)* No 13, pp 215–9; 1910 Über die Messung von Elektrizitätsmengen, die kleiner zu sein scheinen als die Ladung des einwertigen Wasserstoffions oder Elektrons und von dessen Vielfachen abweichen *Sitzungberichte Mathematisch–Naturwissenschaftlichen Klasse der Kaiserlichen Akademie der Wissenschaften (Vienna)* **119** 815–66

16-56 Ehrenhaft F 1910 Über eine neue Methode zur Messung von Elektrizitätsmengen an Einzelteilchen, deren Ladungen die Ladung des Elektrons erheblich unterschreiten und auch von dessen Vielfachen abzuweichen scheinen *PZ* **11** 619–30

16-57 Ehrenhaft: see note 16-55, *Anzeiger*, p 218. This time a simplistic application of Stokes' law was not the culprit. Using newly introduced corrections to the law for small particles by Larmor's student E Cunningham (stimulated by Millikan's Winnipeg report, see note 16-32, above) would have led to even smaller charges. Nor did Ehrenhaft doubt that the density of the particles from his metal arc was representative of the density of the basic electrode materials

See also Holton, note 16-26, p 199

16-58 Ehrenhaft: see note 16-55, *Anzeiger*, pp 217–8

16-59 Ehrenhaft: see note 16-55, *Sitzungberichte*, p 866; see note 16-56, p 630

16-60 Holton: see note 16-26, pp 199–200

The sensitivity to the selection of data is also treated by Bär R 1922 Der Streit um das Elektron *Die Naturwissenschaften* **10**, 322–7, 340–50

Notes section 16.3

16-61 Millikan: see note 16-3, p 75

16-62 Karl Przibram, *Privatdozent* at Vienna and a younger colleague of Ehrenhaft, had repeated Millikan's experiments with water drops (as opposed to Ehrenhaft's metallic particles) with much the same results as Ehrenhaft's; namely, that downward deviations from a mean value for *e* exceeded experimental errors. However, Przibram remained cautious in his conclusions. At the Solvay Congress in 1911, Friedrich Hasenöhrl, ordinary professor of theoretical physics at Vienna, reported that Przibram was distancing himself from Ehrenhaft's conclusion regarding *e*, something Przibram seems to have confirmed in letters to Millikan during late 1912. Holton: see note 16-26, pp 198, 200
The letters are also referred to by Millikan R A 1916 The existence of a subelectron? *PR* **8** 604
Hasenöhrl F 1912 Discussion du Rapport de M Perrin *La Théorie du Rayonnement et les Quanta, Rapports et Discussions de la Réunion Tenue à Bruxelles, du 30 Octobre au 3 Novembre 1911 sous let Auspices de M E Solvay* ed P Langevin and M de Broglie (Paris: Gauthier-Villars) p 252, hereafter referred to as Solvay 1911.
In the same Solvay discussion session, Albert Einstein drew attention to the fact that Edmund Weiss, *Assistent* in physics at the German University of Prague, had recently shown that one of Ehrenhaft's key assumptions, the validity of Stokes' law for his metal particles, was not justified; Einstein A 1912 The problem of specific heat *Solvay 1911* p 251
See also Einstein to Heinrich Zangger (1874–1957), April 7 1911, 1993 *The Collected Papers of Albert Einstein, Vol 5, the Swiss Years, Correspondence 1902–1914* (Princeton, NJ: Princeton University Press) pp 289–90

16-63 Fletcher H 1982 My work with Millikan on the oil-drop experiment *PT* **35** 43–7
Fletcher's account is found in a manuscript autobiography left in the care of his friend Mark B Gardner of Spanish Fork, UT, with instructions to publish it only posthumously; thereby Fletcher sought to make it clear that he had no personal interest at stake in its publication, nor wished to cast doubt on Millikan's reputation. The account was published in *Physics Today* with the consent of Fletcher's family. Much the same ground is covered in an oral history interview with Fletcher conducted by Vern Knudsen (protégé of Fletcher) and W James King on May 15 1964; audiotape and transcript, OH143, Niels Bohr Library, American Institute of Physics, College Park, MD

16-64 Fletcher: see note 16-63, p 44

16-65 Fletcher: see note 16-63, p 46
Slightly different versions of the remark are attributed to Steinmetz in the transcript of Fletcher's oral history interview, OH143: see note 16-63, p 24, as well as in a set of autobiographical notes in the archives of the Niels Bohr Library; Fletcher, MB164, item E

16-66 Fletcher: see note 16-63, p 46

16-67 An article, 'Substance of address', was published in *The Daily Maroon* of the University of Chicago, May 25 1910, under the heading 'Millikan makes great scientific discovery. Associate Professor in Physics Department succeeds in isolating individual ion. Holds it under observation. Proves truth of kinetic theory of matter—result of four years of research'. Cited by Holton: see note 16-26, pp 200–1

16-68 Millikan R A 1910 The isolation of an ion and a precision measurement of its charge *PR* **31** 92
16-69 Fletcher: see note 16-63, p 47
16-70 Millikan R A 1910 The isolation of an ion, a precision measurement of its charge, and the correction of Stokes' law *Science* **32** 436–48
The paper was republished in December 1910 as Das Isolieren eines Ion, eine genause Messung der daran gebundenen Elektrizitätsmenge und die Korrektion des Stokesschen Gesetzes *PZ* **11** 1097–109, and in abridged form as Obtention d'un ion isolé, mesure précise de sa charge; correction à loi de Stokes *Le Radium* **7** 345–50
The paper must have met the approval of the faculty committee at the University of Chicago as well; soon after its appearance, Millikan found himself a full professor. Millikan: see note 16-3, p 89
16-71 Millikan: see note 16-70, p 436
16-72 Millikan R A 1911 The isolation of an ion... *PR* **32** 352
The figure is also reproduced in *The Electron*: see note 16-13, p 65. Millikan refused to preserve the original oil-drop equipment for posterity, arguing that 'once the experiment has been performed and the conclusions have been reached and published so that anyone can repeat and check them, the apparatus with which the work was done is just *so much junk*. I don't believe in glorifying or worshipping junk'. DuMond: see note 16-13, p xiv
16-73 Millikan: see note 16-70, p 440
16-74 In his autobiography, Millikan illustrates his preoccupation with keeping watch on the droplets by the following amusing incident. 'One night Mrs. Millikan and I had invited guests to dinner. When six o'clock came I was only half through with the needed data on a particular drop. So I had Mrs. Millikan appraised by phone that "I had watched an ion for an hour and a half and had to finish the job", but asked her to please go ahead with dinner without me. The guests complimented me on my domesticity because what they said Mrs. Millikan had told them was that Mr. Millikan had "washed and ironed for an hour and a half and had to finish the job"'. Millikan: see note 16-3, p 83
16-75 Millikan: see note 16-70, p 438
16-76 Some selectivity among the data was again resorted to, as in 1909 and as was Millikan's research policy (in contrast to that of Ehrenhaft). In the final averaging, a few drops were omitted because they 'yielded values of e_1 from two to four per cent too low to fall upon a smooth $e_1 v_1$ curve...'. They were attributed to 'two drops stuck together'. Millikan: see note 16-70, p 446. Some observations corresponding to very fast or very slow drops were omitted as a matter of principle; while not affecting the final value of e the attendant experimental errors were judged unacceptable.
16-77 Fletcher: see note 16-63, transcript of interview with Knudsen and King, p 34
16-78 Millikan: see note 16-70, p 447
16-79 Millikan: see note 16-72
16-80 Millikan: see note 16-72, p 384
16-81 Ehrenhaft: see note 16-56
16-82 Holton: see note 16-26, p 186
16-83 Millikan: see note 16-72, pp 393–4
16-84 Millikan: see note 16-72, p 395
16-85 Svedberg T 1913 *Jahrbuch der Radioaktivität und Elektronik* **10** 513
16-86 'We see from this formula [wrote Einstein in a paper for the benefit of chemists with less background in mathematical physics] that the

path described by a molecule on the average is not proportional to the time, but proportional to the square root of the time. This follows because the paths described during two consecutive unit time-intervals are not always to be added, but just as frequently have to be subtracted'. Einstein A 1909 The elementary theory of the Brownian motion *Zeitschrift für Elektrochemie* **14** 235–9; reprinted in Boorse and Motz: see note 13-55, p 594

16-87 Fletcher H 1911 Einige Beiträge zur Theorie der Brownischen Bewegung mit experimentellen Anwendungen *PZ* **12** 202–8
A brief summary of this work was given by Millikan at the annual meeting of the American Association for the Advancement of Science, a joint meeting with the American Physical Society, at Minneapolis, December 28–30 1910; Millikan R A 1911 The isolation of ions *Science* **33** 256–7

16-88 Fletcher H 1911 A verification of the theory of Brownian movements and a direct determination of the value of Ne for gaseous ionization *PR* **33** 81–110

16-89 Millikan: see note 16-3, p 84
With regard to Begeman, Millikan's first junior partner in the electron experiments, his most recent contribution, as may be recalled, had been a more reliable value of e by the 'regular Wilson' water drop method. As Fletcher tells it, Begeman 'was disgruntled with Millikan', thinking that he, Fletcher, 'had stolen the show'. Interview with Fletcher: see note 16-63, p 22

16-90 Ehrenhaft F 1910 Über eine neue Methode zur Messung von Elektrizitätsmengen, die kleiner zu sein scheinen als die Ladung des einwertigen Wasserstoffions oder Elektrons und von dessen Vielfachen abweichen *PZ* **11** 940–52
Przibram K 1910 Über die Ladungen im Phosphornebel *PZ* **11** 630–2

16-91 Arnold Sommerfeld opened the discussion by questioning the sphericity of the particles employed, and Siedentopf and Kalähne expressed further concern about the particle geometry. Bestelmeyer asked about possible convection currents in the observation cell from inhomogeneous electric fields, and Edgar Meyer dwelt on the possibility of charge exchange and neutralization. Max Born commented on Brownian motion in an ionized gas, and Walter Kaufmann picked up on the topic as well. The discussion concluded with Max Planck's concern about the applicability of Stokes' formula for such small particles—i.e. the degree of fluidity achieved by Przibram in his experiment. Ehrenhaft: see note 16-90, pp 949–52. In countering this distinguished audience, we might add, Ehrenhaft bore himself well.

16-92 Ehrenhaft: see note 16-90, p 949

16-93 Ehrenhaft F 1914 Über die Quanten der Elektrizität. Der Nachweis von Elektrizitätsmengen, welche das Elektron unterschreiten, sowie ein Beitrag zur Brownischen Bewegung in Gasen *AP* **44** 699

16-94 Bär: see note 16-60, p 326

16-95 Kerker: see note 16-47, p 95

16-96 Millikan: see note 16-3, p 89. His stay in Berlin afforded an emotional reunion with Planck, Rubens, and particularly Nernst, old friends from his postdoctoral stint during 1895–1896. Nernst had by now moved up from Göttingen to the directorship of the Laboratory of Physical Chemistry in Berlin.

16-97 Millikan: see note 16-3, p 96

16-98 1912 The British Association at Dundee *NA* **90** 6, 19–27. The account of preparations for the meeting at Dundee does not sound very different from similar woes of organizing committees nowdays; thus: 'Every nook and corner of the town is filled almost to overflowing, and members who arrive without having made their arrangements beforehand will have little chance of finding even the simplest houseroom. Private hospitality has provided for between 700 and 800 guests, and every hotel in the town and in the near neighbourhood was filled up many days ago'; p 6. That year the opening address for Section A was delivered by H L Callendar.

16-99 Millikan R A 1913 On the elementary electrical charge and the Avogadro constant *PR* **2** 109–43

16-100 Millikan: see note 16-99, pp 140–1

16-101 'So far as I am aware', continued Millikan, 'there is at present no determination of *e* or *N* by any other method which does not involve an uncertainty of at least 15 times as great as that represented in the above measurements'. Millikan: see note 16-99, p 141. The 'other methods' referred to included the radioactive method of Regener (though Geiger and Rutherford could, he admitted, do better before long) the Brownian movement method (Perrin, Fletcher, and Svedberg) and the radiation method of Planck and of W W Coblentz (1873–1962).

16-102 Interview with Fletcher: see note 16-63, p 26

The paper was completed on August 31 1915, after Fletcher had returned to Brigham Young University in Provo. Fletcher H 1915 Über die Frage der Elektrizitätsladungen, welche die der Elektronen unterschreiten *PZ* **16** 316–8

Fletcher's brief paper is followed by Smoluchowski M von 1915 Notiz über die Berechnung der Brownischen Molekularbewegung bei der Ehrenhaft–Millikanschen Versuchsanordnung *PZ* **16** 318–21

Millikan himself wrote a much longer critique of Ehrenhaft's subelectron in 1916; Millikan: see note 16-62

16-103 Millikan R A 1917 A new determination of *e*, *N* and related constants *PM* **34** 1–30. The 1913 and 1917 values for *e* were, respectively, $(4.774 \pm 0.009) \times 10^{-10}$ and $(4.774 \pm 0.004) \times 10^{-10}$ esu. The 1917 value for *N* was 6.062×10^{23}.

16-104 Epstein: see note 16-2, p 15

16-105 Epstein: see note 16-2, p 15. For a different slant on Millikan's program, B J Sokol has drawn recent attention to the American poet Robert Frost *vis-à-vis* the oil drop experiments. Frost takes them to task in his penultimate collection of poems, *Steeple Bush* of 1947, in which he confronts various issues of the 'new physics' and raises epistemological questions that arise with experimental science. Sokol B J 1966 Poet in the atomic age: Robert Frost's 'That Millikan Mote' expanded *Annals of Science* **53** 399–411

16-106 Epstein: see note 16-2, p 16. The modern value, $e = (4.8025 \pm 0.0005) \times 10^{-10}$ esu, derives from an indirect x-ray spectroscopic determination at Uppsala.

16-107 Lorentz H A 1952 *The Theory of Electrons and its Application to Phenomena of Light and Radiant Heat* 2nd edn (New York: Dover) p 251

16-108 Holton: see note 16-44, p 257

16-109 Bär: see note 16-60, p 327

As an aside A H Compton is said, about this time, to have argued in favor of a *large* electron at a meeting of the Physical Society at Cambridge. Rutherford, hotly opposed to the notion, is reported to have declared, 'I will not have an electron as big as a balloon in my laboratory'. Eve: see note 6-29, p 285

16-110 Ehrenhaft F 1941 The microcoulomb experiment *Philosophy of Science* **8** 403–57

16-111 Holton: see note 16-26, p 222

16-112 Lampa to Mach, March 1 1910, Ernst Mach archives, Freiburg; cited by Holton: see note 16-26

16-113 Lampa to Mach: see note 16-112. Lampa reports Viktor Lang, Ehrenhaft's old mentor, telling him of Ehrenhaft's newly discovered fractional electronic charges. The work of Planck referred to is his book 1910 *Acht Vorlesungen über theoretische Physik* (Leipzig: Hirzel); 1915 *Eight Lectures on Theoretical Physics Delivered at Columbia University in 1909* transl. A P Wills (New York: Columbia University Press)

16-114 Ehrenhaft F 1926 Ernst Mach's Stellung im wissenschaftlichen Leben *Neue Freie Presse* Supplement, June 12, p 12; cited by Holton: see note 16-26, p 221. 'Ehrenhaft had indeed touched a key point', adds Holton. 'Whatever else the controversy was about, it was also about two ancient sets of thematically antithetical positions: the concepts of atomism and of the continuum as basic explanatory tools in electrical phenomena, and the use of methodological pragmatism versus an ideological phenomenology'; p 222

16-115 Peterson I 1992 Particles of history: chronicling the emergence of the Standard Model *Beam Line* **22** 1–8. The designation 'quark' is due to Murray Gell-Mann who, in 1964 with George Zweig, postulated a triplet of hypothetical particles to account for certain symmetries in particle physics. Tradition has it that Gell-Mann borrowed the name for his particles from a phrase in James Joyce's *Finnegan's Wake*. Gell-Mann corrects this attribution slightly, stating that he felt the need for a somewhat playful tag for the particles, and fell into using the term 'quork'. Only later did he come across the alternative spelling for his hypothetical triplet in Joyce's novel; p 3. Zweig dubbed his particles 'aces', a term that did not stick. Note that the standard model of the strong, electromagnetic, and weak forces is actually a well founded theory with great predictive power—one of the salient achievements of elementary particle physics.

16-116 Ehrenhaft: see note 16-110, p 433

16-117 Millikan: see note 16-34, p 220

16-118 Dirac P A M Ehrenhaft, the subelectron and the quark: see *History of Twentieth Century Physics*, note 14-47, p 293. While the reality of electrons (and atoms) was a firmly entrenched concept in twentieth century science by ca. 1915, the reality of fractional charges in the guise of quarks would have a checkered history from the introduction of the concept in 1964. An excellent account of the search for free quarks is found in Riordan M 1987 *The Hunting of the Quark: a True Story of Modern Physics* (New York: Simon and Schuster). Soon after Gell-Mann launched his strictly hypothetical concept, the Italian physicist Giacomo Morpurgo of the University of Genoa, undertook a search for 'real heavy quarks'. Albeit a theorist, he had the gumption to initiate the search utilizing an apparatus designed along the principles of Millikan's levitation concept: granules of graphite were levitated in a magnetic field, and charges carried by them measured by their deflection in a superimposed electric field. Later the Genoa team utilized small iron cylinders spun at high velocity. In the event, the search lasted fully a decade, without so much as a hint of a quark—that

is, fewer than one quark in 3×10^{21} nucleons. Nor was a team at Moscow State University any more successful. The explanation, as Gell-Mann had argued from the outset, was simple: 'Quarks are not real', pp 115–6.
The pros and cons of their reality were debated vociferously at the so-called Rochester Conference of 1966, actually held at Berkeley's Lawrence Radiation Laboratory. One who argued vehemently in favor of free quarks before a largely unconvinced audience at Berkeley was the Oxford theorist Richard Dalitz; 'Gell-Mann left abruptly during the first hour of the talk', p 119
Despite the mood swinging more and more in a direction akin to the positivism of Ernst Mach and followers in bygone years, the experimental search was resumed in the 1970s by William Fairbank of Stanford University. Fairbank and collaborators would claim to have observed fractional charges on balls of superconducting niobium floating in a magnetic field; however, nobody else could confirm their findings.
Lubkin G B 1977 Stanford group shows apparent evidence for quarks *PT* **30** 17–20
See also Fairbank W M and Franklin A 1982 Did Millikan observe fractional charges on oil drops? *AJP* **50** 394–7. Today the search for free quarks is all but abandoned in the face of strong theoretical evidence that quarks must inevitably remain confined within the nucleons—the so-called bag model of quantum chromodynamics.

Notes section 17.1

17-1 Heilbron: see note 1-14, p 58

17-2 Heilbron: see note 1-14
Boorse and Motz: see note 13-55, p 616

17-3 Kelvin: see note 15-17
Kelvin's term 'electrion' was an amalgam of Faraday's *ion* and Stoney's *electron*; he coined it in 1897 Contact electricity and electrolysis according to Father Boscovich *NA* **56** 84
Static atoms composed of equal parts of negative and positive point charges interacting according to the law of inverse squares were inherently unstable, according to the old theorem of the Reverend Samuel Earnshaw of Cambridge, and thus called for forces other than Coulomb's.
Earnshaw S 1842 On the nature of the molecular forces which regulate the constitution of the luminiferous ether *Trans. Camb. Phil. Soc.* **7** 97–114
See also Scott W T 1959 Who was Earnshaw? *AJP* **27** 418–9
Heilbron: see note 1-14, p 52

17-4 Soddy F 1906 The recent controversy on radium *NA* **74** 516
The controversy pitted Oliver Lodge and R J Strutt, among others, against Kelvin, and was finally laid to rest by Rutherford in a letter to *Nature* on October 25 1905. It is well summarized by Eve: see note 6-29, 140–2, who quotes Rutherford to the effect that 'when a single experimental fact is established which does not conform with the disintegration theory it will be time to abandon it'. It is also summarized by Soddy in *Nature*.

17-5 Perrin J 1901 Les hypothèses moléculares' *Revue Scientifique* **15** 460; cited by Nye: see note 7-10, p 84

17-6 Nye: see note 7-10. Nye cites Rutherford as later commenting that he was unaware of Perrin's hypothesis when he first published his own atomic version.

17-7 Pingree D 1974 Nagaoka, Hantaro *DSB* **9** 606–7

17-8 Nagaoka to Tanakadate, 1888; cited by Koyzumi K 1975 The emergence of Japan's first physicists: 1868–1900 *HSPS* **6** 87

17-9 Nagaoka to Rutherford, February 22 1911, CUL, Add 7653, N1

17-10 Nagaoka H 1904 Kinematics of a system of particles illustrating the line and band spectrum and the phenomena of radioactivity *PM* **7** 445–55; 1904 *NA* **69** 392–3; 1904 *Proceedings of the Tokyo Mathematico-Physical Society* **2** 92–107, 129–31. Nagaoka had been inspired, in part, by Poincaré's remark that the study of atomic structure would be facilitated by proper attention to the spectral properties of an atom. Nagaoka H 1950 Gemshikaku tankyú no omoide *Nagaku Asahi* **10** 23; cited by Koyzumi: see note 17-8, p 90
Poincaré himself referred to Nagaoka's model as 'a very interesting attempt but not yet wholly satisfactory', and thought that 'this attempt should be renewed', Poincaré H 1958 *The Value of Science* transl. G B Halsted (New York: Dover) p 109

17-11 Schott G A 1904 On the kinematics of a system of particles illustrating the line and band spectra *PM* **8** 384–7
For the subsequent exchange between Nagaoka and Schott, see 1904 *NA* **70** 124–5, 176
Schott, a student of J J Thomson and former wrangler, became a sometime professor of applied mathematics at the University of Wales. Nagaoka's atom was brought to Rutherford's attention by the elder Bragg, who wrote him that '[Lewis] Campbell tells me that Nagaoka once tried to deduce a big positive centre in his atom in order to account for the optical effects. He thinks Nagaoka, but it was a Jap anyway. Time about 5 or 6 years ago, when Schott and others were on the subject'. W H Bragg to Rutherford, March 11 1911, CUL, Add 7653, B385
Wilson: see note 7-19 gives March 7 for the date of Bragg's postcard

17-12 Heilbron: see note 1-14, p 53

17-13 Thomson J J 1904 *Electricity and Matter* (New Haven, CT: Yale University Press)

17-14 Thomson: see note 15-18, presented initially as a Friday Evening Discourse at the Royal Institution on Friday, March 10 1905. It is perhaps unnecessary to note that Thomson knew only too well that a 'Solar' atom is inherently unstable; Rutherford's subsequent atom is possible only because Newtonian mechanics do not apply in the case of atoms, as Niels Bohr came to accept in 1913.

17-15 Thomson: see note 1-25, pp 181–6

17-16 Thomson to Lodge, April 11 1904, cited by Rayleigh: see note 1-46, pp 140–1

17-17 Heilbron: see note 1-14, p 53

17-18 Thomson to Rutherford, February 18 1904, CUL, Add 7653, T23

17-19 Thomson: see note 15-18

17-20 Mayer A M 1878 On the morphological laws of the configurations formed by magnets floating vertically and subjected to the attraction of a superposed magnet... *American Journal of Science* **116** 247–56; 1877 Floating magnets *NA* **17** 487–8; 1878 *NA* **18** 258–60
Thomson had appealed to Mayer's magnets much earlier, in connection with the vertex atom: see note 1-38, p 107, and more recently in 'Cathode rays' of 1897: see note 7-16, pp 313–4. Kelvin had come across and drawn attention to the floating magnet experiments even earlier, in 1878, the year of Mayer's publications. Mayer's experiments were refined by R W Wood with iron spheres floating on mercury; Wood R W 1898 On the equilibrium figures formed by floating magnets *PM* **46** 162–4

17-21 Of course, Thomson (and Mayer) did not imply that atomic electrons are necessarily constrained to concentric *rings*; Mayer's planar scheme was simply a convenient heuristic substitute for the presumed electronic shells of real atoms.

17-22 Rayleigh: see note 1-46, p 140. In fact, Mayer had devised his floating magnet experiments partly for teaching purposes, their being simple, entertaining, and inexpensive.

17-23 Wilson to Thomson, December 18 1913, CUL, Add 7654, W43

17-24 Wilson to Thomson: see note 17-23. Wilson is referring to the then commonly accepted theory of chemical valence, as due to transfer of a negative electron between atoms. Stranges: see note 14-61, p 39
In Thomson's scheme, the role of core and valence electrons was interchanged, from a modern point of view. Atoms of a given period in Mendeléev's table were characterized by the same number of *external* electrons, and differed only in the number of electrons in their inner rings. Thus, chemical properties were determined primarily by the innermost electrons. Heilbron: see note 1-14, p 55

17-25 Thomson: see note 15-16, p 771
This important paper has been ranked by A Pais as 'the first paper of substance on the physics of atomic structure'. Pais: see note 3-29, p 187

17-26 Heilbron's expression, Heilbron J L 1977 J J Thomson and the Bohr atom *PT* **30** 23. Decreasing n had another consequence: there were not enough electrons to account for the number of observed spectral lines, on the assumption that they originated in the oscillations of negative electrons with three degrees of freedom. Thomson, nevertheless, as was his habit, avoided details of atomic spectra all his career.

17-27 Lenard: see note 12-25

17-28 Rayleigh recalls that both Rutherford and Thomson were at first loath to admit that β-rays were electrons precisely because of their great penetrating power. Rayleigh: see note 1-46, p 133
Max Born, in his *Atomic Physics*, implies that Lenard should perhaps be credited with the first suggestion of the modern model of the atom. Born M 1962 *Atomic Physics* 7th edn (London: Blackie) p 64
Eve: see note 6-29, p 197. Perhaps we have here one more example of narrowly missed discoveries by the hapless physicist!

17-29 Pais: see note 3-29, p 149. The purported exponential fall-off with distance can be viewed as the 'sudden death' mechanism akin to bullets penetrating a rigid mesh or screen. Those that strike the mesh go no further; the rest continue on their way with unchanged velocity.

17-30 Schuster: see note 6-11, ch 8, p 236

17-31 That is, on the basis of Thomson's model, the net deflection was the average deflection per collision not multiplied by the number of collisions, but multiplied by the square root of the number of collisions. Thomson J J 1910 The scattering of rapidly moving electrified particles *Proc. Camb. Phil. Soc.* **15** 465–71
For a comprehensive treatment of the subject, consult Heilbron: see note 13-26

17-32 Friedman F L and Sartori L 1965 *Origins of Quantum Physics: the Classical Atom* (New York: Addison-Wesley) p 88
James Arnold Crowther, a Yorkshireman, was born in 1883 in Sheffield, where he received his early education at the Royal Grammar School. He entered St John's College, Cambridge on a Mackinnon Scholarship from the Royal Society, and held a Fellowship at St John's between 1908 and 1912. He was appointed demonstrator and lecturer in physics at the

Cavendish Laboratory in 1912, a post he held until 1924. During 1921–1924 he was, as well, lecturer in medical radiology at the University of Cambridge. He occupied the chair of physics at the young and expanding University of Reading from 1924 until 1946, when he retired as professor emeritus. He authored several influential textbooks, was president of the British Institute of Radiology, founding Fellow and honorary secretary of the Institute of Physics, and vice-chairman of the Parliamentary Science Committee late in his career. He died in 1950. His only obituary of note is one by Whiddington R 1950 *NA* Prof. J. A. Crowther **165** 750–1

17-33 Crowther J A 1910 On the scattering of homogenous β-rays and the number of electrons in the atom *PRS* **84** 226–47

17-34 Feather: see note 8-47, p 117

Notes section 17.2

17-35 The Darwins provide an interesting example of a scientific dynasty, if not quite in the league of the Becquerels. Himself grandson of Erasmus Darwin (1731–1802), the physician and botanist, Charles R Darwin had seven surviving children, among them Sir George H Darwin, Plumian Professor of Astronomy and Experimental Philosophy at Cambridge and father of Charles G Darwin; the botanist (and curator of his father's work) Sir Francis Darwin of Cambridge; and Sir Horace Darwin, founder of the Cambridge Scientific Instrument Co. Charles G Darwin took honors in the mathematical tripos of 1910. After some work on the absorption of α-rays under Rutherford at Manchester, he turned to x-ray diffraction, joining H G J Moseley in certain experiments, and then produced a series of lasting papers forming the foundation for all subsequent interpretation of x-ray crystal diffraction phenomena. He was, successively, Fellow of Christ College, Cambridge; Tait Professor of Natural Philosophy at Edinburgh; Master of Christ College; and Director of the National Physical Laboratory. Lindsay R B 1971 Darwin, Charles Galton *DSB* **3** 563–5

17-36 Darwin to Eve, February 4 1938, CUL, Add 7653, D32

17-37 Rutherford E 1906 Some properties of the α rays from radium *PM* **11** 166–76; 1906 Retardation of the α particle from radium in passing through matter *PM* **12** 134–46

17-38 Rutherford: see note 17-37, Retardation of the α particle... p 145

17-39 Heilbron: see note 13-26, p 260

17-40 Schuster to Rutherford, July 7 1906, CUL, Add 7653, S31

17-41 1906 *The Physical Laboratories of the University of Manchester: a Record of 25 Years' Work, Prepared in Commemoration of the 25th Anniversary of the Election of Dr Arthur Schuster FRS, to a Professorship in the Owens College, by his Old Students and Assistants* (Manchester: Manchester University Press). Schuster had succeeded Balfour Stewart to the chair at Manchester in 1887. At the time he was preoccupied with cathode ray studies which he pursued until 1896; in so doing, he nearly anticipated J J Thomson in the discovery of the electron. His studies came to an abrupt halt when he received a personal copy of Röntgen's famous paper on x-rays, mailed by Röntgen himself on New Year's Day, 1896. As related elsewhere, Schuster repeated Röntgen's experiment, with the result that the laboratory was inundated with requests from the medical community for assistance in instituting x-ray practice. With the death two years later of his personal assistant, Arthur T Stanton, he withdrew from active experimental work, turning increasingly to purely mathematical and theoretical investigations, along with his manifold duties in university administration.

17-42 Rutherford to Schuster, September 26 1906; included in Birks J B (ed) 1963 *Rutherford at Manchester* (New York: Benjamin) pp 47–8

17-43 Thomson to Rutherford, December 18 1906, CUL, Add 7653, T28

17-44 Cited by Eve: see note 6-29, p 156. In addition to the Manchester offer, and the feeler from King's College, Rutherford had been approached twice for the Secretaryship of the Smithsonian Institution in Washington, and for a chair at Yale.

17-45 Thomson: see note 1-25, p 8

17-46 Rutherford to Hahn, July 14 1908; cited by Eve: see note 6-29, p 180

17-47 Rutherford and Geiger: see note 13-66, p 172
The apparatus is preserved in the museum of the Cavendish Laboratory in West Cambridge: see note 9-55
The original, handwritten manuscript, with corrections in Rutherford's hand, is in principle also extant in the Rutherford archives at Cambridge, CUL, Add 7653, PA12. In fact, the draft manuscript was missing from the file at the time of my examination; it could not be ascertained whether it was truly missing, or displaced or on loan elsewhere within the CUL archival system
The actual measurements were performed on June 10–11 1908, according to surviving notes, PA181, note 16-15, chapter 16

17-48 Rutherford E and Royds T 1909 The nature of the α particle from radioactive substances *PM* **17** 284
The fragile apparatus is still intact in the museum of the Cavendish Laboratory in West Cambridge. The handwritten manuscript, with corrections in Rutherford's hand, is also among the Rutherford archives at Cambridge, CUL, Add 7653, PA8

17-49 Rutherford to Bumstead, July 11 1908; cited by Heilbron: see note 13-26, p 262

17-50 Bragg to Rutherford, October 1 1908, CUL, Add7 653, B376
Bragg's experiments during 1904–1906 had uncovered the 'characteristic range' for α-particles. Unlike β-particles, α-particles were, on account of their great energy of motion, not readily deflected from their path, but travelled in nearly a straight line, ionizing the molecules in their path. They lost their ionizing power suddenly while still relatively energetic, after traversing a definite distance in air. In 1907, Bragg, characteristically, shifted his interest abruptly to γ- and x-rays; the following year, he was appointed Cavendish Professor of Physics at the University of Leeds, England. Bragg would remain an intimate friend of Rutherford, and the two provided mutual support and understanding of their respective research programs in radioactivity. Bragg's son, later Sir Lawrence Bragg, would succeed Rutherford in his chairs at both Manchester and Cambridge.

17-51 Geiger H 1908 On the scattering of the α-particles by matter *PRS* **81** 174–7

17-52 Fleming C A 1971 Ernest Marsden *Biographical Memoirs of Fellows of the Royal Society, 1971* vol 17 (London: The Royal Society) p 464. Marsden was just turning 19 in 1909, the year he joined Geiger. He was the second of four sons and a daughter of Thomas Marsden, a household draper and hardware dealer, born at Rishton, Lancashire, in 1889. He won an Entrance Scholarship to Queen Elizabeth's Grammar

School in Blackburn, where he came under the influence of F Allcroft, a physics teacher, and F H Peachell, who taught mathematics. From the Grammar School he won a Lancashire County Council Scholarship to the University of Manchester, which he entered in 1906. During his first year at Manchester, he attended lectures in physics, mathematics, and electrical engineering, before deciding to concentrate on the course of Honours Physics. After completing his graduate work at Manchester, he became lecturer in physics at East London College, before returning to Manchester as John Harling Fellow and to assist Geiger in testing Rutherford's new theory of α-particle scattering.

17-53 Eve: see note 6-29, p 181

17-54 Rutherford to Hahn, November 29 1908; cited by Eve: see note 6-29, p 183

17-55 Eve: see note 6-29. Curiously, the physics prize in 1908 went to Lippmann for inventions in color photography, instead of to Max Planck who was a strong runner-up that year. Planck himself, in fact, had, on the express invitation of the Swedish Academy, nominated Rutherford for the 1908 prize for physics.

17-56 Geiger: see note 17-51, cited by Heilbron: see note 13-26, p 263

17-57 Marsden E 1963 Rutherford at Manchester *Rutherford at Manchester*: see note 17-42, p 8

17-58 Marsden: see note 17-57
Robinson, in his account, has Rutherford saying that 'I agreed with Geiger that young Marsden, whom he had been training in radioactive methods, ought to begin a research. Why not let him see if any α-particles can be scattered through a large angle'. Robinson H R 1963 Rutherford: life and work to the year 1919, with personal reminiscences of the Manchester period *Rutherford at Manchester*: see note 17-42, p 68

17-59 Badash: see note 7-29, p 31
Marsden: see note 17-57

17-60 Marsden: see note 17-42, pp 8–9

17-61 Geiger H and Marsden E 1909 On a diffuse reflection of the α-particles *PRS* **82** 500

17-62 Cited, without a source given, by Andrade E A da C 1964 *Rutherford and the Nature of the Atom* (New York: Anchor) p 111

17-63 Heilbron: see note 13-26, p 265

17-64 Rutherford E 1912 History of the alpha particles from radioactive substances *Lectures Delivered at the Celebration of the Twentieth Anniversary of the Foundation of Clark University, September 7–11, 1909* (Worcester, MA: Clark University) pp 83–95, cited by Heilbron: see note 13-26

17-65 Heilbron: see note 13-26, p 281, on Geiger H 1910 The scattering of α-particles *PRS* A **83** 492–504

17-66 Using gold foil, Geiger had determined that the angle Φ, beyond which half the α-particles were scattered, was approximately 1°. According to the multiple-scattering theory, the corresponding mean square angle of deflection: see note 17-31, is given by $\theta_{rms}^2 = 1.4°^2$. If θ_0 is the mean deflection caused by a single atom, θ_{rms} is related to θ_0 (again according to the multiple-scattering theory) by $\theta_{rms} = \sqrt{N}\theta_0$, where N is the total number of atoms encountered. Geiger's foil was about 1500 atoms thick, so that $\theta_0 \sim 0.03°$. With such a small value for θ_0, the chance of an α-particle undergoing large-angle deflections was minute. Long before the particles had experienced the required number of collisions (9×10^6 atoms, on average, to attain a total deflection of 90°, or a path length of

0.3 cm), they would have lost their energy and come to rest. Friedman and Sartori: see note 17-32, pp 89–90

17-67 Geiger: see note 17-65

Notes section 17.3

17-68 Rutherford to Boltwood, December 14 1910; cited by Badash L 1969 *Rutherford and Boltwood, Letters on Radioactivity* (New Haven, CT: Yale University Press) p 235

17-69 Geiger H Recollections of my years in Manchester from 1906–1917, CUL, Add 7653, PA312/9

17-70 Theory of structure of atoms. Notes and calculations: 35 leaves, undated, ca. 1910–1911, CUL, Add 7653, PA194. (Only 15 leaves of PA194 were present when I studied the file during June 19–22 1996.) Large scattering of alpha particles. Notes and calculations: 17 leaves, undated, ca. 1910–1911, CUL, Add 7653, PA198. (Only five leaves present in June, 1996). Theory of deflection of alpha particles through angles large compared with small scattering. Calculations: 36 leaves, including one graph paper, undated, ca. 1910–1911, CUL, Add 7653, PA198A

17-71 Bragg to Rutherford, December 21 1910, CUL, Add 7653, B380A

17-72 Rutherford to Bragg, February 8 1911; cited by Eve: see note 6-29, p 194–5
In this letter, by 'large scattering', Rutherford is really referring to *single* scattering, and by 'small scattering' multiple scattering. Heilbron: see note 13-26, p 293
In this connection, C G Darwin explains that 'at first Rutherford talked about [large angle scattering] as *simple scattering* and then he realized that this was ambiguous, and I can still recall the satisfaction in his voice when he hit on the name *single scattering* since nobody could mistake what that meant'. Darwin C G 1963 Moseley and the atomic numbers of the elements *Rutherford at Manchester*: see note 17-42, p 18

17-73 Rutherford to Bragg, February 9 1911; cited by Eve: see note 6-29, p 195
The reference is to a plot of I/I_0 (normalized ionization current) against foil thickness t on p 242 in Crowther: see note 17-33; the double inflection in figure 5 of that paper was accounted for by the exponential decrease $I/I_0 = 1 - e^{-k/t}$, where k is a constant.

17-74 Rutherford to Boltwood, February 1 1911; cited by Badash: see note 17-68, p 242

17-75 Darwin to Eve, February 4 1938, CUL, Add 7653, D32

17-76 Darwin: see note 17-72, p 19

17-77 Calculations by C G Darwin, relating to PA194. Notes and calculations: 14 leaves, undated, 'probably Darwin', CUL, Add 7653, PA196. The calculations cover the α-particle scattering for inverse square, as well as inverse cube, forces, including corrections when the target atom is also set in motion. Darwin discusses his calculations in some length in his letter to A S Eve many years later. Darwin to Eve, February 4 1938, CUL, Add 7653, D32

17-78 Rutherford E 1911 The scattering of the α and β rays and the structure of the atom *Manchester Literary and Philosophical Society Proceedings* pp xviii–xx. March 7 was, by chance, also the day Rutherford learned from Bragg of Nagaoka's Saturnian atom: see note 17-11.

17-79 Keller: see note 3-61, p 143

17-80 Draft manuscripts of published papers, CUL, Add 7653, PA16. The first two pages are typewritten, double spaced; the rest are handwritten.

The reference to Geiger's contribution in the opening paragraph of the published paper was added by Rutherford late in the manuscript stage. Despite Rutherford's copy editing, the manuscript as submitted to *Philosophical Magazine* is still riddled with typos.

17-81 Nagaoka to Rutherford: see note 17-9

17-82 In his letter of February 22, Nagaoka added that 'Righi in Bologna was much interested with my model of Saturnian atom published in 1904'; Nagaoka to Rutherford: see note 17-9

17-83 Cited by Wilson: see note 7-19, p 301

17-84 Tago E 1964 On Nagaoka's Saturnian atomic model *Japanese Studies in the History of Science* **3** 47

17-85 Rutherford E 1911 The scattering of α and β particles by matter and the structure of the atom *PM* **21** 669

17-86 Rutherford's formula for the number of particles N scattered into the solid angle ϕ, equation (5), may be written (using conventional notation) as follows:

$$\frac{dN}{d\phi} = \frac{N_0 nt Z_1^2 Z_2^2 e^4}{64\pi^2 \varepsilon_0^2 m_1^2 v_0^4 \sin^4(\phi/2)}$$

where N_0 is the number of incident particles, n the number of scattering centers per unit volume, t the foil thickness, $Z_1 = 2$ the atomic number of the incident particles, Z_2 the atomic number of the foil material, m_1 the mass of the incident particles, v_0 the speed of the incident particles, e the electronic charge, and ε_0 the permittivity of free space.

17-87 Rutherford: see note 17-85, p 683

17-88 'As to what you say about the nucleus being negative [wrote Darwin to Eve in 1938], I do not think you are right. I have no recollection of it myself and am sure that I should have *at once* pointed out to him that repulsion would work just as well as attraction and could not be discriminated'. Darwin to Eve, February 4 1938, CUL, Add 7653, D32. Darwin refers to Eve: see note 6-29, p 199

17-89 Rutherford: see note 17-85, p 670

17-90 Andrade E N da C 1956 The birth of the nuclear atom *Scientific American* **195** 93–104

Andrade obtained his doctorate at Heidelberg in 1911, and spent one year, 1913–1914, in Rutherford's laboratory. He has much of interest to say about the stark contrast in working under Lenard at Heidelberg and under Rutherford at Manchester. Andrade E N da C 1963 Rutherford at Manchester, 1913–14 *Rutherford at Manchester*: see note 17-42, pp 27–9

17-91 Thomson's lectures are found in Thomson J J 1913 *Engineering* **95** 232, 266, 300, 328, 356, 397

17-92 Thomson J J 1921 Structure of the atom *La Structure de la Matière, Rapports et Discussions du Conseil de Physique tenu à Bruxelles du 27 au 31 Octobre 1913* (Paris: Gauthier-Villars). World War I delayed publication of the reports of the Congress.

See also Mehra J 1975 *The Solvay Conferences on Physics: Aspects of the Development of Physics since 1911* (Dordrecht: Reidel) pp 76, 78

17-93 Thomson: see note 17-92

17-94 E.g., Soddy to Rutherford, April 26 1911: 'I… shall await your paper with much interest', CUL, Add 7653, S149

Hahn to Rutherford, May 8 1911: 'I am looking forward to your new theory', CUL, Add 7653, H52

17-95 Rutherford to Boltwood, April 11 1911; cited by Badash: see note 17-68, p 246

17-96 Rutherford to Boltwood, March 18 1912; cited by Badash: see note 17-68, p 265

17-97 Rosenfeld L 1970 Bohr, Niels Henrik David *DSB* **2** 239–54

17-98 Bohr N 1911 *Studier over Metallernes Elektrontheori* (Copenhagen: Thaning & Appel)

17-99 Bohr to his mother, October 4 1911, cited by Rosenfeld L and Rüdinger E 1967 The decisive years 1911–1918 *Niels Bohr: his Life and Work as Seen by his Friends and Colleagues* ed S Rozental (Amsterdam: North Holland) and (New York: Wiley) p 41

17-100 Bohr to his fiancée, Margrethe Nørlund, cited by Rosenfeld and Rüdinger: see note 17-99, p 40

17-101 Nor did J J appreciate being criticized, e.g. for certain passages in his *Conduction of Electricity through Gases*, by the brash visitor whose broken English he could scarcely follow. Before long Thomson got into the habit of making detours to avoid the impetuous young Dane. The only physicist with whom Bohr managed to discuss his ideas on atomic physics during his stay at Cambridge was Samuel B McLaren, a lecturer at Birminham nine years his senior, who was killed in World War I. Earlier in 1911, McLaren had concluded that any attempt to explain radiation phenomena on purely classical grounds without invoking Planck's quantum led to contradictions—a view increasingly Bohr's as well, dating from his thesis research in Copenhagen. For Bohr on McLaren's views, see Bohr to Rutherford, October 17 1913, CUL, Add 7653, B108

17-102 Bohr N 1963 Reminiscences of the founder of nuclear science and of some developments based on his work *Rutherford at Manchester*: see note 17-42, p 115

17-103 Rosenfeld L 1963 Bohr in Manchester *On the Constitution of Atoms and Molecules* papers of 1913 reprinted from the *Philosophical Magazine* with an introduction by L Rosenfeld (Copenhagen: Munksgaard) and (New York: Benjamin)

17-104 Bohr to his brother, Harald, June 12 1912; Bohr N 1972 *Collected Works* vol 1, ed L Rosenfeld, J Rud Nielsen, and E Rüdinger (Amsterdam: North Holland) p 555

17-105 'When I came up [to Manchester]', recalled Bohr, 'I can say that Rutherford was very nice and arranged that I could take part in courses [on radioactive techniques] by Geiger. But it was so that just a few weeks later on I said to Rutherford that it would not work to go on making experiments and that I would better like to concentrate on the theoretical things'. Bohr, interview with Thomas Kuhn, January 1 1962; AHQP, transcript of tape No 34b, p 11

17-106 Bohr to his brother, June 19 1912; Bohr: see note 17-104, vol 1, p 559

17-107 Rosenfeld: see note 17-103, pp xv–xvi

17-108 Bohr: see note 17-102, pp 116–7

17-109 As an interesting aside, when the displacement law was publicly announced a few months later by Soddy in Glasgow and Kasimir Fajans (1887–1975) in Karlsruhe, the authors failed to recognize its close tie to Rutherford's model, and even touted the law as evidence *against* the model!

17-110 Rosenfeld and Rüdinger: see note 17-99, p 50. As noted earlier: see note 17-101, McLaren had already stressed the importance of invoking Planck's quantum for explaining radiation phenomena. Bohr only learned later that the Austrian physicist (and historian of physics) Arthur Erich Haas (1884–1941) had attempted in 1910, on the basis of Thomson's atomic model, to fix dimensions and periods of electronic motion by means of Planck's relation between the energy and frequency of the harmonic oscillator. Furthermore, John William Nicholson, then at Cambridge, had in 1912 utilized quantized angular momenta in his search for the origin of certain spectral lines in stellar nebulæ and the Solar corona. Haas' paper, incidentally, came about in a curious way. After receiving his doctorate in physics at Vienna in 1906, he turned to the history of physics, submitting a paper on the history of the energy principle to the philosophy faculty at the University. The paper puzzled faculty members, who urged him to prepare an additional work on pure physics—what became the aforesaid quantum-theoretical study. Alas, his ideas in this area were termed a 'carnival joke' by the Vienna physicists.

17-111 Bohr: see note 17-104, *Collected Works* vol 2, pp 136–58. The memorandum is also reproduced in Rosenfeld: see note 17-103, pp xxi–xxviii
See also Heilbron: see note 17-26, for discussion of the memorandum. Curiously, the stability condition (or '*special* hypothesis') given in the memorandum, $E = k\nu$, linking the kinetic energy E of the orbiting electron with the rotation frequency ν, does not explicitly involve the quantum of action h, but an unspecified constant K of the same order of magnitude. The actual form of the relationship between K and h is thought to have been spelled out in a page missing (as indicated by the pagination) from the document. The constant K was subsequently shown by Bohr to be given by $h/2$. Rosenfeld: see note 17-103, pp xxix–xxxi

17-112 Rutherford to Bohr, November 11 1912; Bohr: see note 17-104, *Collected Works* vol 2, p 578

17-113 Bohr to Rutherford, March 6 1913, CUL, Add 7653, B101

17-114 Rutherford to Bohr, March 20 1913, CUL, Add 7653, B102

17-115 Bohr N 1913 On the constitution of atoms and molecules, part I *PM* **26** 1–25; On the constitution..., part II, systems containing only a single nucleus *PM* **26** 476–502; On the constitution..., part III, systems containing several nuclei *PM* **26** 857–75

17-116 1913 Physics at the British Association *NA* **92** 304–7

17-117 See note 17-116, p 306

17-118 Hevesy to Bohr, August 6 1913; Bohr: see note 17-104, *Collected Works*, vol 2, pp 531–2
Rosenfeld: see note 17-103, p li

17-119 Sommerfeld to Bohr, September 4 1913; Bohr: see note 17-104, p 603
Rosenfeld: see note 17-103, p lii

17-120 Hevesy to Rutherford, October 14 1913; cited by Eve: see note 6-29, p 224

17-121 Thomson J J 1913 On the structure of the atom *PM* **26** 792–9

17-122 See note 17-116, p 305
Bohr himself characterized Thomson's apparent agreement with his own results at Birmingham as follows: '... as to the theory of the structure of the atom of Sir J. J. Thomson, I did not realize in Birmingham how similar many of his results are to those I had obtained. As it can be deduced from considerations of dimensions, this agreement, however, has no foundation in the special atom-model used by Thomson, but will follow from any theory which considers electrons and nuclei and makes use of Planck's relation $\varepsilon = h\nu$'. Bohr to Rutherford, October 17 1913, CUL, Add 7653, B108

17-123 Thomson: see note 1-25, p 425

17-124 Interview with G P Thomson by J L Heilbron, June 20 1963; AHQP, transcript of tape 72b, p 6

17-125 Rutherford to Geiger, March 18 1914, CUL, Add 7653, G50

Notes section 18.1

18-1 Charles G Barkla's career was a curious one, on several accounts. Born in Widnes, Lancashire, in 1877, he initially studied mathematics and physics at University College, Liverpool, then experimental physics under Oliver Lodge, receiving the BSc with First-Class Honours in 1898, and the MSc the next year. He then entered Trinity College, Cambridge, on an 1851 Exhibition Scholarship, where he began researches at the Cavendish Laboratory. After 18 months at Trinity, with the Cambridge BA in hand, he switched to King's College in order to sing in its famous chapel choir! He received his doctorate under Lodge at Liverpool in 1904, and in 1909 succeeded H A Wilson as Wheatstone Professor of Physics at King's College, London. In 1913 he became professor of natural philosophy at the University of Edinburgh, where he remained until his death in 1944. Despite receiving the Nobel Prize for physics in 1917 for his discovery that each element emits a characteristic spectrum of x-rays, at about that time he rapidly began isolating himself from the professional community of physicists, committing himself largely to the pursuit of a will-o'-the-wisp he called the 'J-phenomenon'. Forman P 1970 Barkla, Charles Glover *DSB* **1** 456–9

See also Stephenson R J 1967 The scientific career of Charles Glover Barkla *AJP* **35** 140–52

18-2 Henry Gwyn Jeffreys Moseley (Harry to his friends) was born in 1887 in Weymouth, Dorsetshire, the son of Henry Nottidge Moseley (1844–1891), the founder of a strong school of zoology at Oxford. His paternal grandfather, Canon Henry Moseley (1801–1872) the first professor of natural philosophy at King's College, London, was an eminent authority on naval architecture; his maternal grandfather, John Gwyn Jeffreys (1809–1885), was the dean of England's conchologists. Young Harry, with his two sisters, was raised in Chilworth, near Guildford, in Surrey, where they received their elementary education. At the age of nine, Harry was sent to Summer Fields, near Oxford, where he was groomed for a scholarship to Eton. Understandably, Harry took an early interest in natural history. His last letters from Gallipoli are filled with observations of the natural habitat around his encampment. Heilbron J L 1974 Moseley, Henry Gwyn Jeffreys *DSB* **9** 542–5

18-3 Darwin: see note 17-72, p 21. Rutherford was at first luke warm, because of lack of x-ray experience at Manchester compared with, say, Bragg's laboratory at the University of Leeds.

18-4 Moseley, recalls Darwin, 'was without exception the hardest worker I had ever known'. Darwin: see note 17-72, p 27

18-5 Darwin: see note 17-72, p 23

18-6 Bragg, and Laue before him, had found that the reflecting crystal behaved like a family of semitransparent, parallel reflecting planes, producing interference maxima whenever the wavelength λ, the angle θ between the incident beam and normal to the planes, and the spacing d between planes, are related by $n\lambda = 2d\cos\theta$, where n is a positive integer.

18-7 Moseley H G J 1913 The high-frequency spectra of the elements *PM* **26** 1031

Moseley's follow-up paper under the same title appeared in 1914 *PM* **27** 703–13
They are excerpted by Boorse and Motz: see note 13-55, vol 2, pp 874–83

18-8 Notes and calculations by Moseley on the Canadian–Australian Royal Mail Line, 1914, CUL, Add 7653, PA375/4
Moseley to Rutherford, 1914, CUL, Add 7653, M258

18-9 Darwin to Rutherford, February 24 1917, October 26 1917, CUL, Add 7653, D28, D29

18-10 Marsden to Rutherford, November 28 1915, CUL, Add 7653, M65

18-11 Bohr: see note 17-102, pp 137–8

18-12 Thomson: see note 1-25, pp 228–9

18-13 Among the signers of the Manifesto *Aufruf an die Kulturweld*, were Röntgen, Planck, Nernst, Wien, Lenard, Ostwald, Emil Fischer, and the mathematician Felix Klein. However, Einstein, ill at ease in imperial Germany in any case, refused to sign the document, and even considered organizing an opposing manifesto, but then discarded the idea.

18-14 Planck to Wien, November 8 1914; cited by Jungnickel and McCormmach: see note 9-26, p 349

18-15 Voigt W 1914 Ansprache gelegentlich der Zusammenkunft der Lehrer der Georgia-Augusta am 31 Oktober 1914 Jungnickel and McCormmach: see note 9-26

18-16 Einstein to Lorentz, December 18 1917; cited by Jungnickel and McCormmach: see note 9-26, p 350

18-17 Forman P and Hermann A 1975 Sommerfeld, Arnold Johannes Wilhelm *DSB* **12** 525–32

18-18 Sommerfeld to Bohr: see note 17-119
The French physicist L Brillouin remembers entering Sommerfeld's office right after Sommerfeld had seen the issue of the *Philosophical Magazine* with Bohr's article in it. 'There is a most important paper here by N. Bohr', said Sommerfeld; 'it will mark a date in theoretical physics'. Cited by Rosenfeld and Rüdinger: see note 17-99, p 56

18-19 Bohr to Sommerfeld, March 19 1916; cited by Bohr: see note 17-104, *Collected Works*, vol 2, p 603

18-20 The Manchester notebooks relevant in the present context are the following. Production of H-atoms and other high-speed atoms, March 8–September 4 1917, CUL, Add 7653, NB21. Range of high-speed atoms in air and other gases, September 8 1917–February 27 1918, NB22. Theory of scattering of recoil atoms in hydrogen, March 3 1918; Magnetic deflection of hydrogen atoms from paraffin, June 20 1918; Redetermination of magnetic deflection of H rays, August 2–December 3 1918, all in NB23. Calculations on the passage of α-particles through hydrogen, June 13 1918, NB24. Experiments on H-particles, with Marsden and Kay, January 3–March 8 1919, NB25. The first Cambridge notebook is Experiments on alpha particles in hydrogen, November 14 1919–July 25 1920, preceded by a short note on the question of gases in silver, Manchester, August 14 1918, NB26. NB27–NB36 are all on 'artificial disintegration of light elements'.

18-21 Marsden E and Lantsberry W C 1915 The passage of α-particles through hydrogen, II *PM* **30** 243
Marsden finished the paper shortly after arriving at Wellington, barely recovered from a 'recurring swelling nose between his eyes... all the way until Tasmania'. Marsden to Rutherford, March 21 1915, CUL, Add 7653, M56

18-22 Marsden E 1963 Rutherford at Manchester *Rutherford at Manchester*: see note 17-42, p 12

18-23 The nuclear disintegration chamber on view in the museum of the Cavendish Laboratory: see note 9-55, is not the original apparatus of 1919, but a refined version built to investigate the effect further by Rutherford and Chadwick in 1921. Rutherford E and Chadwick J 1921 The artificial disintegration of light elements *PM* **42** 809
The apparatus is also described by Falconer I J 1980 *Apparatus from the Cavendish Museum* pp 10–1

18-24 Experiments on H-particles, January 3–March 18 1919, CUL, Add 7653, NB25

18-25 The reaction $\alpha + N^{14} \rightarrow p + O^{17}$, where p stands for proton, was only confirmed somewhat later by P M S Blackett in Cambridge.

18-26 'The results [of Aston's work on isotopes] thus show that the elements may be considered as being composed of... hydrogen nuclei, or protons as Sir Ernest Rutherford would have us call them, and we thus return to Prout's conception of the constitution of matter, modified only by the recent discoveries and ideas of modern physics'; 1920 Physics at the British Association [at Cardiff] *NA* **106** 357
See also CUL, Add 7653, PA322. Soddy proposed the name 'hydrion' to stress the particle's identity with the hydrogen ion.

18-27 Rutherford E 1919 Collision of a particles with light atoms. IV. An anomalous effect in nitrogen *PM* **37** 586–7. For a seminal work charting the way for the study of nuclear structure, this, part IV, must represent the extreme in modesty among the keystone scientific papers of the twentieth century. The original draft of this paper (and its three companion papers) mostly typewritten and thoroughly edited in Rutherford's hand, is found among the Rutherford archives, CUL, Add 7653, PA25

18-28 Kragh's term: see note 10-34, p 226

18-29 One difficulty came from applying Heisenberg's uncertainty principle to an electron in a nucleus. In terms of the De Broglie wavelength, the wave pattern for the electron was too distended for it to be contained within a nucleus. There were also problems with the intrinsic spin of a proton–electron model of the nucleus

18-30 Lecture delivered June 3 1920. Rutherford E 1920 Nuclear constitution of atoms *PRS* **97** p 396

18-31 Hanson N R 1963 *The Concept of the Positron: a Philosophical Analysis* (Cambridge: Cambridge University Press) p 158

18-32 Hendry J 1970 Chadwick, James *DSB* **17** (Supplement 2) 143–8

18-33 Wrote Geiger to Rutherford from the western front in 1915: 'I hear from Chadwick occasionally and of course we do for him what can be done under the present circumstances. But that is very little'. Geiger to Rutherford, March 26 1915, CUL, Add 7653, G52. In fact, Chadwick and Ellis even managed to do a little research at Ruhleben, assisted chiefly by Heinrich Rubens, Walther Nernst, and Emil Warburg.

18-34 Chadwick J 1964 Some personal notes on the search for the neutron *Proceedings of the 10th International Congress of the History of Science (Ithaca, NY, 1962)* (Paris: Hermann) p 161; cited by Weiner C 1972 1932—moving into the new physics *PT* **25** 334

18-35 Chadwick's neutron chamber is exhibited in the museum of the modern Cavendish Laboratory in West Cambridge: see note 9-55.

18-36 The elastic scattering of a photon by an electron that is effectively free; named after its discoverer, Arthur Holly Compton

18-37 Chadwick J 1932 Possible existence of a neutron *NA* **129** 312, submitted

on February 17 1932
Chadwick gave his first public account of his discovery to the Kapitza Club in Cambridge a few days before his letter appeared in *Nature*. His full paper, submitted on May 10, is 1932 The existence of a neutron *PRS* **136** 692–708

18-38 Chadwick to Bohr, February 24 1932; AHQP, in Chadwick, interview with Charles Weiner, April 15–16 1969, p 73

18-39 Translation from the German by Barbara Gamow in Gamow G 1966 *Thirty Years that Shook Physics; the Story of Quantum Theory* (New York: Doubleday) p 213. The reference to Wolfgang Pauli came about as follows. Several years before Chadwick's discovery, Pauli coined the name 'neutron' for his hypothetical neutral and *massless* particle necessary for explaining the apparent violation of energy conservation in radioactive β-decay (among other things). The name gave rise to heated discussions, but was never sanctified by appearing in print. Consequently, when Chadwick's heavy neutron was announced, Pauli's name was in jeopardy. Enrico Fermi suggested instead calling Pauli's particle the *neutrino* or 'little neutron' in Italian.

Notes section 18.2

18-40 The zealously guarded ritual of admission at Trinity is described by Thomson: see note 1-25, pp 241–2. In the other Colleges in Cambridge, the Master is elected by the Fellows

18-41 'I heard [wrote Geiger] that you have taken up already your new position and Chadwick writes to me that "the Cavendish people are doing more in a day now than they used to do in a week". I am sure they all enjoy your guidance and your power to push things ahead'. Geiger to Rutherford, November 19 1919, CUL, Add 7653, G54

18-42 It was not their first rift. They had a falling out ten years earlier over, among other things, how the Cavendish should be run. On that occasion, they were drawn together again by the need to defend their artificial disintegration results against an attack by Hans Pettersson and Gerhard Kirsch of Stefan Meyer's Radium Institute in Vienna. Both parties were to some extent proven right and wrong in this acrimonious dispute. Stuewer R H 1986 Rutherford's satellite model of the nucleus *HSPBS* **16** pp 331–4. The satellite model in this case was a short-lived but productive interpretation of the Cambridge disintegration experiments ca. 1919–1922 in terms of a nucleus consisting of a central core surrounded by loosely bound hydrogen nuclei ('H satellites') relatively easily knocked out under α-particle bombardment. The model was decisively blown asunder, not by the quarrelsome Viennese upstarts, but by George Gamow's quantum-mechanical tunneling in 1928.

18-43 Wrote Geiger to Rutherford at the end of hostilities: 'I take the first opportunity to write you a few lines just to say that all what has happened these last four years has had no influence on my personal feelings to you and I hope, Dear Prof. Rutherford, that you still take a little interest in your old pupil who keeps his years in Manchester always in pleasant memory'. Geiger to Rutherford, May 18 1919, CUL, Add 7653, G53

18-44 Geiger joined Heisenberg and Wien on a position paper, or memorandum, reflecting the opinion of the majority of German physicists on theoretical physics in the face of Nazi ideology; it was prepared on the express request of the Reich Education Ministry, signed by almost all of Germany's most notable physicists, and circulated in 1936. Beyerchen A D 1977 *Scientists under Hitler: Politics and the Physics Community in the Third Reich* (New Haven, CT: Yale University Press) pp 148–50

18-45 Einstein to Janos Plesch, February 3 1944. Plesch correspondence, held by his family; cited by Clark R W 1971 *Einstein, the Life and Times* (New York: World Publishing) p 568. Plesch was a wealthy Hungarian who built up a fashionable medical practice in Berlin. Einstein was his client and friend in Berlin.

SELECT BIBLIOGRAPHY

The following bibliography lists many, but not all, books, journal articles, and other sources cited in the notes. Thus, bibliographical entries in the *Dictionary of Scientific Biography* are not included. Abbreviations are those used in the notes.

For additional works pertinent to the subject at hand, or to the disovery of fundamental particles generally (particularly more recent developments), the reader may also consult Corby Hovis R and Kragh H 1991 Resource letter HEPP-1: history of elementary-particle physics *AJP* **59** 779–807.

Æpinus F U T 1759 *Tentamen Theoriæ Electricitatis et Magnetismi* (St Petersburg: Imperial Academy of St Petersburg)

—— 1979 *Æpinus's Essay on the Theory of Electricity and Magnetism* introductory monograph and notes R W Home, transl. P J Connor (Princeton, NJ: Princeton University Press)

Albe, E E Fournier d' 1923 *The Life of Sir William Crookes OM, FRS* (London: Fisher Unwin)

Allen J 1970 The life and work of Osborne Reynolds *Osborne Reynolds and Engineering Science Today* (Manchester: Manchester University Press)

Ampère A-M 1820 Mémoire sur l'action mutuelle de deux courants électriques *ACP* **15** 59

Andrade E M da C 1956 The birth of the nuclear atom *Scientific American* **195** 93–104

—— 1959 The history of the vacuum pump *Vacuum* **9** 41–7

—— 1964 *Rutherford and the Nature of the Atom* (New York: Anchor)

Aston F W 1920 Positive rays and isotopes *NA* **105** 617–9

Aston F W and Lindemann F A 1919 The possibility of separating isotopes *PM* **37** 523

Badash L 1965 Radioactivity before the Curies *AJP* **33** 128–35

—— 1966 An Elster and Geitel failure: magnetic deflection of beta rays *Centaurus* **11** 236–40

—— 1969 *Rutherford and Boltwood, Letters on Radioactivity* (New Haven, CT: Yale University Press)

Balmer J J 1885 Die Spectrallinien der Wasserstoffs *Verhandelung der Naturforschenden Gesellschaft in Basel* **7** 894

Bär R 1922 Der Streit um das Elektron *Die Naturwissenschaften* **10** 340–50

Becquerel H 1896 Sur les radiations émises par phosphorescence *CR* **122** 420

—— 1899 Note sur quelques propriétés du rayonnement de l'uranium et les corps radio-actifs *CR* **128** 771–7

—— 1899 Influence d'un champ magnétique sur le rayonnements des corps radio-actifs *CR* **129** 996–1001

—— 1900 Contribution à l'étude du rayonnement du radium *CR* **130** 206–11
—— 1900 Déviation du rayonnement du radium dans un champ électrique *CR* **130** 809–15
—— 1970 Sur la radio-activité de la Matière *The Royal Institution Library of Science: Physical Sciences* vol 5, ed W L Bragg and G Porter (Amsterdam: Elsevier) pp 508–24
Becquerel J 1906 Sur les variations des bandes d'absorption d'un cristal dans un champ magnétique *CR* **142** 874–6
—— 1906 Sur la corrélation entre les variations des bandes d'absorption des cristaux dans un champ magnétique et la polarisation rotatoire magnétique *CR* **142** 1144–6
—— 1907 Sur les variations d'absorption des cristaux de parasite et de tysonite dans un champ magnétique, à la température de l'air liquide *CR* **145** 413–6
—— 1907 Recherches sur les phénomènes magnéto-optiques dans les cristaux *LR* **4** 49–57
—— 1908 Sur la nature des charges d'électricité et sur l'existence des électrons positifs *CR* **146** 1308–11
—— 1922 Absorption de la lumière et phénomènes, magnéto-optiques dans les composés de terres rares aux très basses températures *Het Natuurkundig Laboratorium der Rijksuniversiteit te Leiden in de Jaren 1904–1922, Gedenkboek Aangeboden aan H Kamerlingh Onnes, Directeur van het Laboratorium bij Gelegenheid van zijn Veertigwarig Professoraat op 11 November 1922* (Leiden: Ijdo) pp 319–61
Becquerel J and Kamerlingh Onnes H 1908 The absorption spectra of the compounds of the rare earths at the temperatures obtainable with liquid hydrogen, and their change by the magnetic field *RN* **10** 592–603
Bestelmeyer A 1908 Positive electronen? *PZ* **9** 541–2
Bestelmeyer A and Marsh S 1907 Über das gemeinsame Auftreden von Strahlen positiver und negativer Elektrizität in verdünnten Gasen *Verhandlungen der Deutschen Physikalischen Gesellschaft* **9** 758–63
Beyerchen A D 1977 *Scientists under Hitler: Politics and the Physics Community in the Third Reich* (New Haven, CT: Yale University Press)
Biot J-B and Savart F 1820 Note sur le magnétisme de la pile de Volta *ACP* **15** 222–3
Birks J B (ed) 1963 *Rutherford at Manchester* (New York: Benjamin)
Blondlot R-P 1903 Sur la polarisation des rayons X *CR* **136** 284–6
—— 1905 *N Rays, a Collection of Papers Communicated to the Academy of Sciences* Transl. J Garcin (London: Longmans Green)
Bohr N 1911 *Studier over Metallernes Elektrontheori* (Copenhagen: Thaning & Appel)
—— 1913 On the constitution of atoms and molecules, part I *PM* **26** 1–25
—— On the constitution ..., part II. Systems containing only a single nucleus *PM* **26** 476–502
—— On the constitution ..., part III. Systems containing several nuclei *PM* **26** 857–75
—— 1963 *On the Constitution of Atoms and Molecules* papers of 1913 reprinted from the *Philosophical Magazine* introduced L Rosenfeld (Copenhagen: Munksgaard) and (New York: Benjamin)
—— 1972 *Collected Works* ed L Rosenfeld, J Rud Nielsen, and E Rüdinger (Amsterdam: North Holland)
Boorse H A and Motz L (eds) 1966 *The World of the Atom* (New York: Basic Books)
Bordier H 1905 *Les Rayons N et les Rayons N1: Les Actualités Medicales* (Paris: Baillère)

Brown S C 1966 *Introduction to Electrical Discharges in Gases* (New York: Wiley)
Bush S G and Everitt C W F 1969 Maxwell, Osborne Reynolds, and the radiometer *HSPS* **1** 105–25
Cabeo N 1629 *Philosophia Magnetica in qua Magnetis Natura Penitus Explicatur* (Ferrara)
Cahan D 1985 The institutional revolution in German physics, 1865–1914 *HSPS* **15** 1–65
Casimir H B G 1983 *Haphazard Reality: Half a Century of Science* (New York: Harper and Row)
Chalmers T W 1952 *Historic Researches: Chapters in the History of Physical and Chemical Discovery* (New York: Scribner)
Chadwick J 1932 Possible existence of a neutron *NA* **129** 312
—— 1932 The existence of a neutron *PRS* **136** 692–708
Clark R W 1971 *Einstein, the Life and Times* (New York: World Publishing)
Cohen E 1927 Kamerlingh Onnes memorial lecture *Chemical Society Journal* **1** 1193–209
Cornu M A 1897 Sur l'observation et l'interprétation cinématique des phénomènes découverts par M. le Dr. Zeeman *CR* **125** 555–61
—— 1898 Sur quelques résultats noveaux relatifs au phénomène découverts par M. le Dr. Zeeman *CR* **126** 181–6
Cranston J A 1964 The group displacement law *Chem. Soc. Proc.* 104–7
Crookes W 1874–1875 On the attraction and repulsion resulting from radiation *PRS* **23** 373–8
—— 1879 On the illumination of lines of molecular pressure, and the trajectory of molecules *Phil. Trans.* **170** 135–64
—— 1879 Radiant matter *Chemical News* **40** 91–3
—— 1879 Radiant matter *Chemical News* **40** 104–7
—— 1879 Radiant matter *Chemical News* **40** 127–31
—— 1879 Radiant matter *NA* **20** 419–23
—— 1879 Radiant matter *NA* **20** 436–40
Crowther J A 1910 On the scattering of homogeneous β-rays and the number of electrons in the atom *PRS* **84** 226–47
Crowther J G 1974 *The Cavendish Laboratory 1874–1974* (New York: Science History)
Curie M 1898 On a new radioactive substance contained in pitchblende *CR* **127** 175–8
—— 1900 Action du champ magnétique sur les rayons de Becquerel. Rayons déviés *CR* **130** 73–6
—— 1900 Sur la Pénétration des rayons de Becquerel non déviables par le champ magnétiques *CR* **130** 76–9
Dahl P F 1972 *Ludvig Colding and the Conservation of Energy Principle: Experimental and Philosophical Contributions* (New York: Johnson)
—— 1992 *Superconductivity: its Historical Roots and Development from Mercury to the Ceramic Oxides* (New York: American Institute of Physics)
Daintith J, Mitchel S, and Tootill E 1986 *A Biographical Encyclopedia of Scientists* (New York: Facts on File)
Davies P J and Marsch J D 1953 Ohm's law and the Schuster effect *Institution of Electrical Engineers Proc.* **132** 525–32
Davy H 1807 The Bakerian Lecture, on some chemical agencies of electricity *PTRS* **97** 1–56
De Kosky R K 1972–1973 Spectroscopy and the elements in the late nineteenth century: the work of Sir William Crookes *The British Journal for the History of Science* **6** 400–23
Desaguliers J T 1739–1740 Some thoughts and conjectures concerning the cause

of electricity *PTRS* **41** 175–85
—— 1739–1740 Some thoughts and experiments concerning electricity *PTRS* **41** 186
—— 1739–1740 Some thoughts and experiments concerning electricity *PTRS* **41** 193
—— 1739–1740 Experiments made before the Royal Society, February 2, 1737–38 *PTRS* **41** 193–9
—— 1739–1740 Experiments made before the Royal Society, February 2, 1737–38 *PTRS* **41** 200–8
Dewar J 1970 Studies on charcoal and liquid air *The Royal Institution Library of Science, Physical Sciences* vol 6 (Amsterdam: Elsevier)
Dibner B 1957 *Early Electrical Machines* (Norwalk, CT: Burndy Library)
—— 1962 *Oersted and the Discovery of Electromagnetism* (New York: Blaisdell)
Dirac P A M 1977 Ehrenhaft, the subelectron and the quark *History of Twentieth Century Physics. Proceedings of the International School of Physics 'Enrico Fermi'* course 57, ed C Weiner (New York: Academic) pp 290–3
Drude P 1900 Zur Elektronentheorie. I *AP* **1** 566–613
—— 1900 Zur Elektronentheorie. II *APC* **3** 369–402
—— 1902 Zur Elektronentheorie. III *APC* **7** 687–92
Du Bridge L A and Epstein P S 1959 Robert Andrews Millikan *Biographical Memoirs, National Academy of Sciences* **33** 241–82
Du Fay C-F 1733–1734 A letter from Mons. Du Fay . . . concerning electricity *PTRS* **38** 258–66
Dufour A 1908 Sur quelques exemples des raies présentant le phénomène de Zeeman anormal dans le sens des lignes de force magnétiques *CR* **146** 634–5
—— 1909 Sur les phénomènes de Zeeman normaux et anormaux *LR* **6** 44–5
—— 1909 Modifications normales et anormales, sous l'influence d'un champ magnétique, de certaines bandes des spectres d'émission de molécules de divers corps à l'état gaseux *Journal de Physique* **9** 237–64
—— 1909 Sur l'existence d'électrons positifs dans les tubes à vide *CR* **148** 481–4
Earnshaw S 1842 On the nature of the molecular forces which regulate the constitution of the luminiferous ether *Trans. Camb. Phil. Soc.* **7** 97–114
Ehrenhaft F 1909 Eine Methode zur Bestimmung des elektrischen Elementarquantums. I *PZ* **10** 308–10
—— 1909 Eine Methode zur Messung der elektrischen Ladung kleiner Teilchen zur Bestimmung der elektrischen Elementarquantums *Anzeiger der Kaiserlichen Akademie der Wissenschaften Mathematisch-Naturwissenschaftlichen Klasse* **7** 72
—— 1909 Eine Methode zur Bestimmung des elektrischen Elementarquantums. I *Sitzungsberichte der Mathematisch-Naturwissenschaftlichen Klasse der Kaiserlichen Akademie der Wissenschaften* **118** 321–30
—— 1910 Über die kleinsten messbaren Elektrizitätsmengen. Zweite vorläufige Mitteilung der Methode zur Bestimmung des elektrischen Elementarquantum *Anzeiger der Kaiserlichen Akademie der Wissenschaften, Mathematisch-Naturwissenschaftlichen Klasse* **10** 118–9
—— 1910 Über die Messung von Elektrizitätsmengen, die Ladung des einwertigen Wasserstoffions oder Elektrons zu unterschreiten scheinen. Zweite vorläufige Mitteilung seiner Methode zur Bestimmung des electrischen Elementarquantums *Anzeiger der Kaiserlichen Akademie des Wissenschaften* **13** 215–9
—— 1910 Über die Messung von Elektrizitätsmengen, die kleiner zu sein scheinen als die Ladung des einwertigen Wasserstoffions oder Elektrons und von dessen Vielfachen abweichen *Sitzungsberichte der Mathematisch-Naturwissenschaftlichen Klasse der Kaiserlichen Akademie der Wissenschaften* **119**

815–66
—— 1910 Über eine neue Methode zur Messung von Elektrizitätsmengen an Einzelteilchen, deren Ladungen die Ladung des Elektrons erheblich unterschreiten und auch von dessen Vielfachen abzuweichen scheinen *PZ* **11** 619–30
—— 1910 Über eine neue Methode zur Messung von Elektrizitätsmengen, die kleiner zu scheinene als die Ladung des einwertigen Wasserstoffions oder Elektrons und von dessen Vielfachen abweichen *PZ* **11** 940–52
—— 1914 Über die quanten der Elektrizität. Der Nachweis von Elektrizitätsmengen, welche das Elektron unterschreiten, sowie ein Beitrag zur Brownischen Bewegung in Gasen *AP* **44** 657–700
—— 1941 The microcoulomb experiment *Philosophy of Science* **8** 403–57
Einstein A 1905 Über einen die Erzeugung und Verwandlung des Lichtes betreffenden heuristischen Gesichtpunkt *AP* **17** 132–48
—— 1909 The elementary theory of the Brownian motion *Zeitschrift für Elektrochemie* **14** 235–9
Elster J and Geitel H 1889 Elektrisierung der Gase durch glühende Körper *APC* **37** 315
—— 1899 Ueber den Einfluss eines magnetischen Feldes auf die durch die Becquerelstrahlen bewirkte Leitfähigkeit der Luft *Verhandlungen Deutschen Physikalischen Gesellschaft* **1** 136–8
Epstein P S 1948 Robert Andrews Millikan as physicist and teacher *Reviews of Modern Physics* **20** 10–25
Etter L E 1946 Some historical data relating to the discovery of the Roentgen rays *American Journal of Roentgenology and Radium Therapy* **56** 220–31
Eve A S 1939 *Rutherford: Being the Life and Letters of the Rt Hon Lord Rutherford OM* (Cambridge: Cambridge University Press)
Fairbank W M and Franklin A 1982 Did Millikan observe fractional charges on oil drops? *AJP* **50** 394–7
Falconer I 1988 J. J. Thomson's work on positive rays, 1906–1914 *HSPBS* **18** 265–310
—— 1989 J. J. Thomson and 'Cavendish Physics' *The Development of the Laboratory: Essays on the Place of Experiment in Industrial Civilization* ed F A J L James (New York: American Institute of Physics) pp 104–17
Faraday M 1839–1855 *Experimental Researches in Electricity* 3 vols (London: Taylor)
—— 1851 On the physical lines of magnetic force *Proc. of the Royal Institution*
—— 1859 *Experimental Researches in Chemistry and Physics* (London: Taylor and Francis)
Farber E (ed) 1961 *Great Chemists* (New York: Interscience)
Feffer S J 1989 Arthur Schuster, J. J. Thomson, and the discovery of the electron *HSPBS* **20** 33–61
Fievez Ch 1885 De l'influence du magnétisme sur les caractéres des raies spectrales *Bulletin de l'Académe des Sciences de Belgique* **9** 381
—— 1886 Essai sur l'origine des raies de Fraunhofer, en rapport avec la constitution du soleil *Bulletin de l'Académe des Sciences de Belgique* **12** 30
Firth I 1969 N-rays—ghost of scandal past *New Scientist* **44** 642–3
Fleming C A 1971 Ernest Marsden *Biographical Memoirs of Fellows of the Royal Society* **17** 463–96
Fletcher H 1911 Einige Beiträge zur Theorie der Brownischen Bewegung mit experimentellen Anwendungen *PZ* **12** 202–8
—— 1911 A verification of the theory of Brownian Movements and a direct determination of the value of Ne for gaseous ionization *PR* **33** 81–110
—— 1915 Über die Frage der Elektrizitätsladungen, welche die der Elektronen

unterschreiten *PZ* **16** 316–8

—— 1982 My work with Millikan on the oil-drop experiments *PT* **35** 43–7

Franklin B 1751 *Experiments and Observations on Electricity, Made in Philadelphia in America* (London: Cave)

Freund F 1946 Lenard's share in the discovery of X-rays *British Journal of Radiology* **19**

Friedman F L and Sartori L 1965 *Origins of Quantum Physics: the Classical Atom* (New York: Addison-Wesley)

Gamow G 1966 *Thirty Years that Shook Physics; the Story of Quantum Theory* (New York: Doubleday)

Gassiot J P 1859 On the stratification and dark bands in electrical discharges as observed in the Torricellian vacua *PTRS* **148** 1–16

—— 1859 On the stratification... second communication *PTRS* **149** 137–60

—— 1860 On the electrical discharge *in vacuo* with an extended series of the Voltaic battery *PRS* **10** 36–7

—— 1860 On the interruption of the Voltaic discharge *in vacuo* by magnetic force. *PRS* **10** 269–74

—— 1861 *British Association Report* section 2, pp 38–9

Geiger H 1910 The scattering of α particles *PRS* **83** 492–504

Geiger H and Marsden E 1909 On a diffuse reflection of the α-particles *PRS* **82** 495–500

Gibson A H 1946 *Osborne Reynolds and his Work in Hydraulics and Hydrodynamics* (London: Longmans Green)

Giese W 1889 Grundzuge einer einheitlichen Theorie der Electricitätsleitung *APC* **37** 576–7

Giesel F 1899 Ueber die Ablenbarkeit der Becquerelstrahlen im magnetischen Felde *APC* **69** 834–6

Gilbert W 1600 *De Magnete Magneticique Corporibus et de Magno Magnete Tellure; Physiologia Nova Plurimis & Argumentic, & Experimentis Demonstratia* (London) (1958 Engl. transl. D J Price (New York: Basic Books))

Gillispie C C (ed) 1970 *Dictionary of Scientific Biography* 16 vols (New York: Scribner)

Glasser O 1934 *Wilhelm Conrad Röntgen and the Early History of the Röntgen Rays* (Springfield, IL: Thomas)

Glazebrook R 1926 The Cavendish Laboratory: 1876–1900 *NA* **118** (Supplement) 52–8

Goldstein E 1876 Vorläufige Mittheilungen über elektrische Entladungen in verdünnten Gasen *Monatsbericht der Könglich Akademie der Wissenschaften zu Berlin* pp 279–95

—— 1880 Ueber die Entlandung der Elektricität in verdünnten Gasen *Monatsbericht der Könglich Akademie der Wissenschaften zu Berlin* pp 82–124

—— 1880 On the electric discharge in rarefied gases—part II. On luminous phenomena in gases caused by electricity *PM* **10** 234–44

—— 1886 Über eine noch nicht unterschte Strahlungsform an der Kathode inducirter Entladungen *Sitzungsberichte der Königlichen Akademie der Wissenschaften zu Berlin* pp 691–9

Gooding D and James F A J L (eds) 1985 *Faraday Rediscovered*: *Essays on the Life and Work of Michael Faraday 1791–1867* (New York: Stockton)

Gorter C J and Taconis K W 1964 The Kamerlingh Onnes Laboratory *Cryogenics* **4** 345–53

Gravesande W J van s' 1720 *Physices Elementa Mathematica Experimentis confirmata sive Introductid ad Philosophiam Newtonianam* 2 vols (Leiden)

Gray S 1731–1732 A letter ... containing several experiments concerning

electricity *PTRS* **37** 18–44
Greenway F G 1962 A Victorian scientist: the experimental researches of Sir William Crookes *Royal Institution Proc.* **39** 172–98
Grove W R 1852 On the electro-chemical polarity of gases *PTRS* **142** 87–101
Hall A R 1954 *The Scientific Revolution 1500–1800: the Formation of the Modern Scientific Attitude* (New York: Longmans Green)
Hallwachs W 1888 Ueber den Einfluss des Lichtes auf electrostatisch geladene Körper *APC* **33** 301–12
Hanson N R 1963 *The Concept of the Positron* (Cambridge: Cambridge University Press)
Harnwell C P and Livingood J J 1933 *Experimental Atomic Physics* (New York: McGraw-Hill)
Hauksbee F 1705 Several experiments on the attrition of bodies in vacuo *PTRS* **24** 2165–74
Heilbron J L 1968 The scattering of α and β particles and Rutherford's atom *AHES* **4** 247–307
—— 1976 Franklin's physics *PT* **29** 32–7
—— 1977 Lectures in the history of atomic physics 1900–1922 *History of Twentieth Century Physics* ed C Weiner (New York: Academic) 40–108
—— 1977 J. J. Thomson and the Bohr atom *PT* **30** 23–30
—— 1986 *The Dilemmas of an Upright Man: Max Planck as Spokesman for German Science* (Berkeley, LA: University of California Press)
Heilbron J L and Wheaton B R with the assistance of May J G, Rider R, and Robinson D 1981 *Literature on the History of Physics in the 20th Century* (Berkeley, LA: Office of History of Science and Technology, University of California, Berkeley)
Helmholtz H von 1881 On the modern development of Faraday's conception of electricity *Chemical Society (London) Journal* **39** 277–304
—— 1928 *Faraday Lectures, 1869–1928* (London)
Hendry J (ed) 1984 *Cambridge Physics in the Thirties* (Bristol: Hilger)
Hertz H 1887 Ueber einen Einfluss des ultravioletten Lichtes auf die electrische Entaladung *APC* **31** 983–1000
—— 1889 Ueber Strahlen elektrischer Kraft *APC* **36** 769
—— 1892 Ueber den Durchgang der Kathodenstrahlen durch dünne Metallschichten *APC* **45** 28–32
—— 1893 *Unterschungen über die Ausbetung der elektrischen Kraft* ed P Lenard (Transl. D E Jones *Electric Waves*) (London: MacMillan)
—— 1894 *Die Prinzipien der Mechanik, in neuen Zusammenhange dargestellt* ed P Lenard (Leipzig: Johan Ambrosius Barth). (1899 Transl. D E Jones and J T Walley *The Principles of Mechanics, Presented in a New Form* (London))
—— 1895 *Schriften vermischten Inhalts* ed P Lenard (Leipzig: Johan Ambrosius Barth) (1896 Transl. D E Jones and G A Schott *Miscellaneous Papers* (London))
1904 *Het Natuurkundig Laboratorium der Rijksuniversiteit te Leiden in de Jaren 1882–1904* (Leiden: Ijdo)
1922 *Het Natuurkundig Laboratorium der Rijksuniversiteit te Leiden in de jaren 1904–1922* (Leiden: Ijdo)
1910 *A History of the Cavendish Laboratory 1871–1910* (London: Longmans Green)
Hitterof J W 1869 Ueber die elektricitätsleitung der Gase *APC* **136** 1–3
—— 1869 Ueber die elektricitätsleitung der Gase *APC* **136** 197–234
Holton G 1978 Subelectrons, presuppositions, and the Millikan–Ehrenhaft dispute *HSPS* **9** 161–224
Howarth O J R 1931 *The British Association for the Advancement of Science: a Retrospective 1831–1931* 2nd edn (London: The British Association)
Howorth M 1958 *The Life Story of Frederick Soddy* (London: New World)

Hunt B J 1991 *The Maxwellians* (New York: Cornell University Press)
Jaffé G 1952 Recollections of three great laboratories *Journal of Chemical Education* **29** 230–8
James F A J L (ed) 1989 *The Development of the Laboratory: Essays on the Place of Experiment in Industrial Civilization* (New York: American Institute of Physics)
—— 1992 The tales of Benjamin Abbott: a source for the early life of Michael Faraday *The British Journal for the History of Science* **25** 229–40
Jauncey G E M 1946 The early years of radioactivity *AJP* **14** 226–41
Jones H B 1870 *Life and Letters of Faraday* 2 vols (London: Longmans)
Jungnickel C and McCormmach R 1986 *Intellectual Mastery of Nature: Theoretical Physics from Ohm to Einstein; Vol 2, the Now Mighty Theoretical Physics 1870–1925* (Chicago: University of Chicago Press)
Kamerlingh Onnes H 1921 Zeeman's Ontdekking van het naar hem genoemde Effect *Physica* **1** 241–50
Kaufmann W 1897 Die magnetische Ablenkbarkeit der Kathodenstrahlen und ihre Abl̈angigkeit vom Entladungspotential *APC* **61** 544–52
—— 1901 Die magnetische und electrische Ablenkbarkeit der Becquerelstrahlen und die scheinbare Masse der Electronen *Nachrichten von der Gesellschaft der Wissenschaften zu Göttingen, Mathematisch-Physikalische Klasse* **2** 143–55
Keesom W H 1926 Prof. Dr. H. Kamerlingh Onnes. His life-work, the founding of the cryogenic laboratory *PLS* **57** 3–21
Keller A 1983 *The Infancy of Atomic Physics: Hercules in his Cradle* (Oxford: Clarendon)
Kelvin Lord 1902 Æpinus atomized *PM* **3** 257–83
Klein M J 1970 *Paul Ehrenfest Vol I: the Making of a Theoretical Physicist* (Amsterdam: Elsevier)
Klotz I M 1980 The N-ray affair *Scientific American* **242** 168–75
Koyzumi K 1975 The emergence of Japan's first physicist: 1868–1900 *HSPS* **6** 3–108
Kragh H 1989 Concept and controversy: Jean Becquerel and the positive electron *Centaurus* **32** 203–40
Langemann R T 1977 New light on old rays: N rays *AJP* **45** 281–4
Larmor J 1894 A dynamical theory of the electric and luminiferous medium *PTRS* **185** 719–822
—— 1895 A dynamical theory..., part II. Theory of electrons *PTRS* **186** 695–743
—— 1897 The influence of a magnetic field on radiation frequency *PRS* **60** 514–5
Lenard P 1898 Ueber die electrostatischen Eigenschaften der Kathodenstrahlen *APC* **64** 279–89
—— 1902 Ueber die lichtelektrische Wirkung *AP* **8** 149–98
—— 1903 Über die Absorption von Kathodenstrahlen verschiedener Geschwindigkeit *AP* **12** 714–44
—— 1920 *Ueber Kathodenstrahlen* (Berlin: Vereinig Wissenschaftl.)
—— 1933 *Great Men of Science; a History of Scientific Progress* transl. H Stafford Hatfield (New York: Macmillan)
—— 1944 *Wissenschaftliche Abhandlugen* 3 vols (Leipzig: Hirzel)
—— 1967 On cathode rays *Nobel Lectures, including Presentation Speeches and Laureates' Biographies; Physics, 1901–1921* (Amsterdam: Elsevier)
Lenard P and Wolf M 1889 Zerstäuben der Körper durch das ultraviolette Licht *APC* **37** 443–56
Lilienfeld J E 1907 Über neuartige Erscheinungen in der positiven Lichtsäule der Glimmentladung *Verhandlungen der Deutschen Physikalischen Gesellschaft* **9** 125–35
—— 1910 Die Elektrizitätleitung im extremen Vakuum *AP* **32** 673–738

Lockyer J N 1878 *Studies in Spectrum Analysis* (London: Kegan Paul)
Lodge O 1897 On the influence of a magnetic field on radiation frequency *PRS* **60** 513–4
—— 1897 Further note on the influence of a magnetic field on radiation frequency *PRS* **61** 413–5
—— 1897 The latest discovery in physics *Electrician* **38** 568–70
—— 1922 The history of Zeeman's discovery, and its reception in England *NA* **109** 66–9
Lorentz H A 1905 The motion of electrons in metallic bodies. I *RN* **7** 438–53
—— 1905 The motion of electrons in metallic bodies. II *RN* **7** 590–3
—— 1905 The motion of electrons in metallic bodies. III *RN* **7** 684–91
—— 1906 Positive and negative electrons *Proc. Am. Phil. Soc.* **45** 103–39
—— 1935–1939 *H A Lorentz, Collected Papers* 9 vols, eds P Zeeman and A D Fokker (The Hague: Martinus Nijhoff)
—— 1952 *The Theory of Electrons and its Application to Phenomena of Light and Radiant Heat* (New York: Dover)
—— 1967 The theory of electrons and the propagation of light *Nobel Lectures, including Presentation Speeches and Laureates' Biographies: Physics 1901–1921* (Amsterdam: Elsevier)
—— 1957 *Lorentz, H A Impressions of his Life and Work* ed G L de Haas-Lorentz (Amsterdam: North-Holland)
Lubkin G B 1977 Stanford group shows apparent evidence for quarks *PT* **30** 17–20
Madey T E and Brown W C (eds) 1984 *History of Vacuum Science and Technology; a Special Volume Commemorating the 30th Anniversary of the American Vacuum Society, 1953–1983* (New York: American Institute of Physics)
Magie W F 1965 *A Source Book in Physics* (Cambridge, MA: Harvard University Press)
Malley M 1971 The discovery of the beta particle *AJP* **39** 1454–61
Margaritondo G 1988 100 years of photoemission *PT* **41** 66–72
Marsden E and Lantsberry W C 1915 The passage of α-particles through hydrogen, II *PM* **30** 240–3
Martin T (ed) 1932–1936 *Faraday's Diary, being the Various Philosophical Notes of Experimental Investigations Made by Michael Faraday* 7 vols (London)
—— 1961 *The Royal Institution* (London: The Royal Institution)
Martin T C 1992 *The Inventions Researches and Writings of Nikola Tesla* (New York: Barnes and Noble)
Maxwell J C (ed) 1879 *The Electrical Researches of the Honorable Henry Cavendish* (Cambridge: Cambridge University Press)
—— 1954 *A Treatise on Electricity and Magnetism* 3rd edn (New York: Dover)
—— *The Scientific Papers of J C Maxwell* 2 vols ed W P Nevin (New York: Dover)
Mayer A M 1877 Floating magnets *NA* **17** 487–8
—— 1878 On the morphological laws of the configurations formed by magnets floating vertically and subjected to the attraction of a superposed magnet *American Journal of Science* **116** 247–56
McCormmach R 1988 Henry Cavendish on the theory of heat *Isis* **79** 37–67
Meadows A J 1972 *Science and Controversy* (Cambridge, MA: MIT Press)
Mehra J 1975 *The Solvay Conferences on Physics: Aspects of the Development of Physics since 1911* (Dordrecht: Reidel)
Meyer K (ed) 1920 *H C Ørsted, Naturvidenskabelige Skrifter* 3 vols (Copenhagen: Høst)
Meyer S 1949 Zur geschichte der entdeckung der Natur der Becquerel Strahlen *Die Naturwissenschaften* **36** 129–32

Meyer S and von Schweidler E 1899 Über des Verhalten von Radium und Polonium im magnetischen Felde *PZ* **1** 90–1

Michelson A A 1898 Radiation in a magnetic field *PM* **55** 348–56

Miller J D 1976 Rowland's physics *PT* **29** 39–45

Millikan R A 1909 A new modification of the cloud method of measuring the elementary electrical charge, and the most probable value of that charge *PR* **29** 560–1

—— 1910 The isolation of an ion and a precision measurement of its charge *PR* **31** 92

—— 1910 The isolation of an ion, a precision measurement of its charge, and the correction of Stokes's law *Science* **32** 436–48

—— 1913 On the elementary charge and the Avogadro constant *PR* **2** 109–43

—— 1916 The existence of a subelectron? *PR* **8** 595–625

—— 1917 A new determination of *e*, *N* and related constants *PM* **34** 1–30

—— 1950 *Autobiography* (New York: Prentice-Hall)

—— 1963 *The Electron: its Isolation and Measurement and the Determination of Some of its Properties* (Chicago: University of Chicago Press)

Millikan R A and Begeman L 1908 On the charge carried by the negative ion of an ionized gas *PR* **26** 197–8

Millikan R A and Winchester G 1907 Upon the discharge of electrons from ordinary metals under the influence of ultra-violet light *PR* **24** 116–8

—— 1907 The influence of temperature upon photo-electric effects in a very high vacuum, and the order of photo-electric sensitiveness of the metal *PM* **14** 188–210

Morrow R A 1969 On the discovery of the electron *Journal of Chemical Education* **46** 584–8

Moseley H G J 1913 The high-frequency spectra of the elements *PM* **26** 1024–34

—— 1914 The high-frequency spectra of the elements *PM* **27** 703–13

Moseley R 1978 The origins and early years of the National Physical Laboratory: a chapter in the pre-history of British science policy *Minerva* **16** 222–50

Moulin A 1909 Sur la déviation des rayons positifs *LR* **6** 4–5

—— 1909 Sur la déviation des rayons canaux *LR* **6** 78–9

Mulligan J F 1989 Henrich Hertz and the development of physics *PT* **42** 50–7

Munro H 1896 *The Story of Electricity* (London: Newnes)

Nagaoka H 1904 Kinematics of a system of particles illustrating the line and band spectrum and the phenomena of radioactivity *PM* **7** 445–55

Newton Sir I 1962 *Sir Isaac Newton's Principles of Natural Philosophy and His System of the World* 2 vols, ed F Cajori (Berkeley, CA: University of California Press)

Nicholson W 1800 Account of the new electrical or galvanic apparatus of Sig. Alex. Volta, and experiments performed with the same *Nicholson's Journal of Natural Philosophy, Chemistry and the Arts* **4** 179

Nicholson W, Carlisle A, and Cruickshank W 1800 Experiments on galvanic electricity *PM* **7** 337–50

Nollet J A 1746 Observations sur quelques nouveaux phénomènes d'électricité *Mémoires de l'Académie Royales des Sciences* (Paris)

Nye M J 1972 *Molecular Reality: a Perspective on the Scientific Work of Jean Perrin* (London: Macdonald) and (New York: Elsevier)

—— 1980 N-rays: an episode in the history and psychology of science *HSPS* **11** 125–56

—— 1984 *The Question of the Atom: from the Karlsruhe Congress to the First Solvay Conference, 1860–1911* (Los Angeles: Tomash)

Ørsted H A 1812 *Ansicht der chemischen Naturgesetze durch die neueren Entdeckungen gewonnen* (Berlin)

—— 1813 *Recherches sur l'Identité des Forces Chimiques et Électriques* (Paris)
—— 1821 Betrachtungen ueber den Electromagnetismus *Journal für Chemie und Physik* **32** 199–231
Okun L B 1989 The concept of mass *PT* **42** 31–6
Oliphant M L 1972 *Rutherford: Recollections of Cambridge Days* (Amsterdam: Elsevier)
Oppenheimer J 1986 Physics and psychic research in Victorian and Edwardian England *PT* **39** 62–70
Owen G E 1955 The discovery of the electron *Annals of Science* **11** 173–82
Pais A 1986 *Inward Bound: on Matter and Forces in the Physical World* (Oxford: Clarendon) and (New York: Oxford University Press)
Perrin J-B 1895 Nouvelles propriétés des rayons cathodiques *CR* **121** 1130
—— 1901 Les hypothèses moléculares *Revue Scientifique* **15** 449–61
—— 1929 *Les Éléments de la Physique* (Paris: Albin-Michel)
Peterson I 1992 Particles of history: chronicling the emergence of the Standard Model *Beam Line* **22** 1–8
Planck M 1901 Ueber die Elementarquanta der Materie und der Elektricität *AP* **4** 564–6
Plücker J 1858 Ueber die Einwirkung des Magnets auf die elektrischen Entlandungen in verdünnten Gasen *APC* **103** 88–106 (1858 Transl. F Guthrie On the action of the magnet upon the electrical discharge in rarefied gases *PM* **16** 119–35)
—— 1858 Observations on the electrical discharge through rarefied gases *PM* **16** 408–18
Poincaré J H 1958 *The Value of Science* transl. G B Halsted (New York: Dover)
Preston T 1897 The Zeeman effect photographed *NA* **57** 173
—— 1898 Radiation phenomena in the magnetic field *PM* **55** 325–39
—— 1899 General law of the phenomena of magnetic perturbation of spectral lines *NA* **59** 248
—— 1970 Magnetic perturbation of the spectral lines *The Royal Institution Library of Sciences, Physical Sciences* vol 5, eds Sir W L Bragg and G Porter (Barring: Elsevier) pp 264–76
Pyenson R L 1979 Physics in the shadow of mathematics: the Göttingen electron-theory seminar of 1905 *AHES* **21** 55–89
Rayleigh Lord 1936 Some reminiscences of scientific workers of the past generation, and their surroundings *Proc. Phys. Soc.* **48** 217–46
—— 1943 *The Life of Sir J J Thomson, OM* (Cambridge: Cambridge University Press)
Reynolds O 1876 On the forces caused by the communication of heat between a surface and a gas *PTRS* **166** 725–35
Riecke E 1898 Zur Theorie des Galvanismus und der Wärme *APC* **66** 353–89
Righi A 1909 *Strahlene Materie und Magnetische Strahlen* (Leipzig: Johan Ambrosius Barth)
Riordan M 1987 *The Hunting of the Quark: a True Story of Modern Physics* (New York: Simon and Schuster)
Ritter J W 1801 Von den Herres Ritter und Rickmann *APC* **7** 527–8
—— 1803 Versuch aber das Sonnenlichs *APC* **12** 409–15
Robinson J 1822 *A System of Mechanical Philosophy* ed D Brewster (Edinburgh)
Röntgen W C 1895 Ueber eine neue Art von Strahlen Erste Mitteilung *Sitzungsberichte der Würzburger Physikalischen-Medicinischen Gesellschaft* **137** (1899 Transl. G F Barker On a new kind of rays *Harper's Scientific Memoirs* (New York: Harper & Brothers))
Rozental S (ed) 1967 *Niels Bohr: his Life and Work as Seen by his Friends and*

Colleagues (Amsterdam: North-Holland) and (New York: Wiley)
Rue W de la and Müller H W 1878 Experimental researches on the electric discharge with the chloride of silver battery, part I. The discharge at ordinary atmospheric pressure *PTRS* **169** 55–121
—— 1878 Part II. The discharge in exhausted tubes *PTRS* **169** 155–242
—— 1880 Part III. Tube potentials; potentials at a constant distance and various pressures; nature and phenomena of the electric arc *PTRS* **177** 65–116
Rutherford E 1899 Uranium radiation and the electrical conduction produced by it *PM* **47** 109–63
—— 1903 The magnetic and electric deviation of the easily absorbed rays from radium *PM* **5** 177–87
—— 1906 Some properties of the α rays from radium *PM* **11** 166–76
—— 1906 Retardation of the α particle from radium in passing through matter *PM* **12** 134–46
—— 1906 The mass and velocity of the α-particles expelled from radium and actinium *PM* **12** 348–71
—— 1909 Opening address to Section A, Mathematics and Physics, of the British Association Meeting at Winnipeg, 1909 *NA* **81** 257–63
—— 1911 The scattering of the α and β rays and the structure of the atom *Manchester Literary and Philosophical Soc. Proc.* **55** 18–20
—— 1912 History of the alpha particles from radioactive substances *Lectures Delivered at the Celebration of the Twentieth Anniversary of the Foundation of Clark University, September 7–11, 1909* (Worchester, MA: Clark University)
—— 1919 Collision of α particles with light atoms. IV. An anomalous effect in nitrogen *PM* **37** 581–7
—— 1920 Nuclear constitution of atoms *PRS* **97** 374–400
—— 1938 The development of the theory of atomic structure *Background to Modern Science* eds J Needham and W Pagel (Cambridge: Cambridge University Press)
Rutherford E and Andrade E N da C 1914 The wave-length of the soft gamma rays from radium B *PM* **27** 854–68
Rutherford E and Chadwick J 1921 The artificial disintegration of light elements *PM* **42** 809
Rutherford E and Geiger H 1908 An electrical method of counting the number of α-particles from radio-active substances *PRS* **81** 141–61
—— 1908 The charge and nature of the α-particle *PRS* **81** 162–73
Rutherford E and Royds T D 1909 The nature of the α particle from radioactive substances *PM* **17** 281–6
Schott G A 1904 On the kinematics of a system of particles illustrating the line and band spectra *PM* **8** 384–7
Schuster A 1872 On the spectrum of nitrogen *PRS* **20** 484–7
—— 1884 Experiments on the discharge of electricity through gases. Sketch of a theory *PRS* **37** 317–39
—— 1887 Experiments on the discharge of electricity through gases *Electrician* **19** 353–5
—— 1890 The discharge of electricity through gases *PRS* **47** 526–59
—— 1906 *The Physical Laboratories of the University of Manchester. A Record of 25 Years' Work* (Manchester: Manchester University Press)
—— 1911 *The Progress of Physics during 33 Years (1875–1908)* (Cambridge: Cambridge University Press)
Scott W T 1959 Who was Earnshaw? *AJP* **27** 418–9
Seabrook W B 1941 *Doctor Wood: Modern Wizard of the Laboratory* (New York: Harcourt Brace)
Segrè E 1980 *From X-Rays to Quarks: Modern Physicists and Their Discoveries*

(Berkeley, CA: University of California Press)
Shamos M S (ed) 1959 *Great Experiments in Physics: First Accounts from Galileo to Einstein* (New York: Dover)
Shiers G 1971 The induction coil *Scientific American* **224** 80–7
Soddy F 1906 The recent controversy on radium *NA* **74** 516–8
—— 1913 The radio-elements and the periodic law *Chemical News* **107** 97–9
—— 1913 Intra-atomic charge *NA* **92** 399–400
Spencer J B 1970 On the varieties of nineteenth-century magneto-optical discovery *Isis* **61** 34–51
Spottiswoode W and Moulton J F 1879 On the sensitive state of electrical discharges through rarefied gases *PTRS* **170** 165–229
—— 1880 On the sensitive state of vacuum discharges *PTRS* **171** 561–652
Stanton A 1890 The discharge of electricity from glowing metals *PRS* **47** 559–61
Stauffer R C 1957 Speculation and experiment in the background of Oersted's discovery of electro-magnetism *Isis* **48** 33–50
Stein G 1993 *The Sourcerer of Kings: the Case of Daniel Dunglas Home and William Crookes* (Buffalo, NY: Promethius)
Stephenson R J 1967 The scientific career of Charles Glover Barkla *AJP* **35** 140–52
Stokes G G 1907 *Memoir and Scientific Correspondence of the Late Sir George Gabriel Stokes* selected and arranged J Larmor (Cambridge: Cambridge University Press)
Stoney G J 1881 On the physical units of nature *PM* **11** 381
—— 1891 On the cause of double lines and of equidistant satellites of the spectra of gases *Trans. R. Dublin Soc.* **4** 563–608
—— 1894 On the 'electron' or atom of electricity *PM* **38** 418
Stranges A N 1982 *Electrons and Valence: Development of the Theory 1900–1925* (College Station: Texas A&M University Press)
Strong J 1946 *Procedures in Experimental Physics* (New York: Prentice Hall)
Strutt R J, Fourth Baron Rayleigh 1924 *Life of John William Strutt* (London: University of Wisconsin Press)
Stuewer R H 1986 Rutherford's satellite model of the nucleus *HSPBS* **16** 321–52
Sviedry R 1970 The rise of physical science at Victorian Cambridge *HSPBS* **2** 127–51
Tait P 1880 Note on the velocity of gaseous particles at the negative pole of a vacuum tube *Proc. R. Soc. of Edinburgh* **10** 430–1
Tago E 1964 On Nagaoka's Saturnian atomic model *Japanese Studies in the History of Science* **3** 33
Thompson S P 1897 Cathode rays and some analogous rays *PTRS* **190** 471–90
Thomson G P 1964 *J J Thomson and the Cavendish Laboratory in His Day* (London: Nelson)
Thomson J J 1887 On the dissociation of some gases by electric discharge *PRS* **42** 343–5
—— 1891 On the rate of propagation of the luminous discharge of electricity through a rarefied gas *PRS* **44** 84–100
—— 1891 On the discharge of electricity through exhausted tubes without electrodes *PM* **32** 321–36
—— 1893 The electrolysis of steam *PRS* **53** 90–110
—— 1893 *Notes on Recent Researches in Electricity and Magnetism* (Oxford: Clarendon)
—— 1894 On the velocity of the cathode rays *PM* **38** 358–65
—— 1896 The Röntgen rays *NA* **54** 302–6
—— 1896 On the discharge of electricity produced by the Röntgen rays, and the effects produced by these rays on dielectrics through which they pass *PRS* **59** 274–6

—— 1896 Presidential address *Reports of the British Association of the Advancement of Science* 699–706
—— 1897 Cathode rays *Electrician* **39** 104–22
—— 1897 On the cathode rays *Proc. Camb. Phil. Soc.* **9** 243–4
—— 1897 Cathode rays *PM* **44** 293–316
—— 1898 Theory on the connection between cathode and Roentgen rays *PM* **45** 172–83
—— 1898 On the charge of electricity carried by the ions produced by Röntgen rays *PM* **46** 528–45
—— 1898 *The Discharge of Electricity Through Gases* (London: Constable) and (Cambridge, MA: Scribner)
—— 1899 On the masses of the ions in gases at low pressure *PM* **48** 547–67
—— 1903 On the charge of electricity carried by a gaseous ion *PM* **5** 346–55
—— 1903 The magnetic properties of systems of corpuscles describing circular orbits *PM* **6** 673–93
—— 1904 On the structure of the atom: an investigation of the stability and periods of oscillation of a number of corpuscles arranged at equal intervals around the circumference of a circle; with application of the results to the theory of atomic structure *PM* **7** 237–65
—— 1904 *Electricity and Matter* (New Haven, CT: Yale University Press)
—— 1906 *Conduction of Electricity Through Gases* 2nd edn (Cambridge: Cambridge University Press)
—— 1906 On the number of corpuscles in an atom *PM* **11** 769–81
—— 1907 On rays of positive electricity *PM* **13** 561–75
—— 1907 *The Corpuscular Theory of Matter* (New York: Scribner)
—— 1908 Positive rays *PM* **16** 657–91
—— 1909 Inaugural address to the British Association Meeting at Winnipeg, 1909 *NA* **81** 248–57
—— 1909 Positive electricity *PM* **18** 821–44
—— 1910 Rays of positive electricity *PM* **19** 424–35
—— 1910 Rays of positive electricity *PM* **20** 752–67
—— 1910 The scattering of rapidly moving electrified particles *Proc. Camb. Phil. Soc.* **15** 465–71
—— 1911 Rays of positive electricity *PM* **21** 225–49
—— 1912 Further experiments on positive rays *PM* **24** 209–53
—— 1913 Rays of positive electricity *PRS* **89** 1–20
—— 1921 *Rays of Positive Electricity and their Application to Chemical Analysis* 2nd edn (London: Longmans Green)
—— 1936 *Recollections and Reflections* (London: Bell)
Thomson J J and Rutherford E 1896 On the passage of electricity through gases exposed to Röntgen rays *PM* **42** 392–407
Thomson J J and McClelland J A 1896 On the leakage of electricity through dielectrics traversed by Röntgen rays *Proc. Camb. Phil. Soc.* **9** 126–40
Townsend J S E 1897 On electricity in gases and the formation of clouds in charged gases *Proc. Camb. Phil. Soc.* **9** 244–58
Trigg G L 1975 *Landmark Experiments in Twentieth Century Physics* (New York: Crane Russak)
Varley C F 1871 Some experiments on the discharge of electricity through rarefied media and the atmosphere *PRS* **19** 236–43
Voigt W 1909 Remarks on the Leyden observations of the Zeeman-effect at low temperatures *RN* **11** 360–5
Volta A 1800 On the electricity excited by the mere contact of conducting substances of different kinds *PTRS* **90** 403–31
Watson E C 1947 Jubilee of the electron *AJP* **15** 458–64

—— 1954 Reproductions of prints, drawings, and paintings of interest in the history of physics: the discovery of the Zeeman effect *AJP* **22** 633–5
Watson W 1750 Some further inquiries into the nature and properties of electricity *PTRS* **45** 93–120
Weart S 1978 A little more light on N rays *AJP* **45** 306
Weber W 1875 Ueber die Bewegungen der Elektricität in Körpen von molecularer Constitution *APC* **156** 1–16
Weinberg S 1990 *The Discovery of Subatomic Particles* (New York: Freeman)
Weiner C 1972 1932—moving into the new physics *PT* **25** 332–9
Weiner C (ed) 1977 *History of Twentieth Century Physics* (New York: Academic)
Wheaton B R 1978 Philipp Lenard and the photoelectric effect, 1889–1911 *HSPS* **9** 299–322
Wheaton B R and Heilbron J L 1982 *An Inventory of Published Letters to and from Physicists, 1900–1950* (Berkeley, CA: Office for History of Science and Technology, University of California, Berkeley)
Whittaker Sir E 1989 *A History of the Theories of Æether and Electricity; Vol I, the Classic Theories; Vol II, the Modern Theories 1900–1926* (New York: Dover)
Wiechert E 1896 Die Theorie der Elektrodynamik und die Röntgen'sche Entdeckung *Schriften der physikalisch-ökonomischen Gesellschaft zu Köningsberg, Abhandlungen* **37** 1–47
—— 1896 Ueber die Grundlagen der Elektrodynamik *APC* **59** 283–323
—— 1897 Ueber das Wesen der Elektricität *Schriften der physikalisch-ökonomischen Gesellschaft zu Köningsberg, Sitzungsberichte* **38** 3–12
Wiederkehr K H 1967 *Wilhelm Eduard Weber. Erforscher der Wellen Bewegung und der Elektrizität 1804–1891* (Stuttgart: Wissenschaftliche Verlagsgesellschaft)
Wien W 1898 Untersuchungen über die elektrische Entladung in verdünnten Gasen *APC* **65** 440–52
—— 1898 Die elektrostatische und magnetische Ablenkung der Kanalstrahlen *Berlin Physikalische Gesellschaft Verhandlungen* pp 10–2
—— 1930 *Aus dem Leben und Wirken eines Physikers* (Leipzig: Barth)
Wilson C T R 1897 Condensation of water vapour in the presence of dust-free air and other gases *Trans. R. Soc.* A **188** 265–307
—— 1954 Ben Nevis sixty years ago *Weather* **9** 309–11
—— 1959 Reminiscences of my early years *Royal Society Notes and Records* **14** 163–73
—— 1965 On the cloud method of making visible ions and the tracks of ionizing particles *Nobel Lectures including Presentation Speeches and Laureates' Biographies; Physics 1922–1941* (Amsterdam: Elsevier)
Wilson D 1983 *Rutherford: Simple Genius* (Cambridge, MA: MIT Press)
Wilson H A 1903 Determination of the charge on the ions produced in air by Röntgen rays *PM* **5** 429–40
Wise M N and Smith C 1986 Measurement, work and industry in Lord Kelvin's Britain *HSPBS* **17** 147–73
Wood R W 1898 On the equilibrium figures formed by floating magnets *PM* **46** 162–4
—— 1904 The n-rays *NA* **70** 530–1
—— 1904 La question de l'existence des rayons N *Revue Scientifique* **2** 536–8
—— 1908 On the existence of positive electrons in the sodium atom *PM* **15** 274–9
Wood R W and Hackett F E 1909 The resonance and magnetic rotation spectra of sodium vapor photographed with the concave grating *Astrophysical Journal* **30** 339–72
Zeeman P 1896 Over den invloed eener magnetisatie op den aard van het door een stof uitgezonden licht *Koninklijke Akademie van Wetenschappen te Amsterdam Verslag* **5** 181–3

—— 1896 Over den invloed eener magnetisatie op den aard van het door een stof uitgezonden licht *Koninklijke Akademie van Wetenschappen te Amsterdam Verslag* **5** 242–8
—— 1897 On the influence of magnetism on the nature of the light emitted by a substance *PM* **43** 226–39
—— 1897 Doublets and triplets in the spectrum produced by external magnetic forces *PM* **44** 55–60
—— 1897 Doublets and triplets in the spectrum produced by external magnetic forces *PM* **44** 255–259
—— 1967 Light radiation in a magnetic field *Nobel Lectures, including Presentation Speeches and Laureates' Biographies: Physics 1901–1921* (Amsterdam: Elsevier)

Name Index

Note: Years of birth and death (if known) are recorded after each name. Endnotes are indicated by a suffix 'n' and a note number.

Subject Index

Note: Endnotes are indicated by a suffix 'n' and a note number.

Milton Keynes UK
Ingram Content Group UK Ltd.
UKHW021927071024
449327UK00022B/1710